WILEY SERIES IN TELECOMMUNICATIONS AND SIGNAL PROCESSING

John G. Proakis, Editor
Northeastern University

Biomedical Signal Processing and Signal Modeling

Biomedical Signal Processing and Signal Modeling

Eugene N. Bruce

University of Kentucky

A Wiley-Interscience Publication

JOHN WILEY & SONS, INC.

New York • Chichester • Weinheim • Brisbane • Singapore • Toronto

Copyright © 2001 by John Wiley & Sons, Inc. All rights reserved.

Published simultaneously in Canada.

For ordering and customer service, call 1-800-CALL-WILEY.

Library of Congress Cataloging-in-Publication Data.

Bruce, Eugene N.
    Biomedical signal processing and signal modeling / Eugene N. Bruce
      p. cm. — (Wiley series in telecommunications and signal processing)
    Includes bibliographical references and index.
    ISBN 0-471-34540-7 (cloth : alk. paper)
    1. Signal processing. 2. Biomedical engineering. I. Title. II. Series.

    R857.S47 B78 2001
    610'.28—dc21

                                      00-039851

10 9 8 7 6 5 4 3 2

Contents

PREFACE

There are two general rationales for performing signal processing: (1) for the acquisition and processing of signals to extract a priori desired information; and (2) for interpreting the nature of a physical process based either on observation of a signal or on observation of how the process alters the characteristics of a signal. The goal of this book is to assist the reader to develop specific skills related to these two rationales. First, it is hoped that the reader will enhance skills for quantifying and (if necessary) compensating for the effects of measuring devices and noise on a signal being measured. Second, the reader should develop skills for identifying and separating desired and unwanted components of a signal. Third, the reader should enhance skills aimed at uncovering the nature of phenomena responsible for generating the signal based on the identification and interpretation of an appropriate model for the signal. Fourth, the reader should understand how to relate the properties of a physical system to the differences in characteristics between a signal used to excite (or probe) the system and the measured response of the system.

A core text in biomedical signal processing must present both the relationships among different theoretical measures of signals and an understanding of the information these measures provide regarding the sources of signals and the behaviors of their sources in response to natural or imposed perturbations. A useful didactic framework to achieve these goals is one in which signal processing is portrayed as the development and manipulation of a model of the signal. In this framework the "model" becomes a surrogate for the signal source and it is natural to ask to what degree this model is reasonable given one's knowledge of the signal source and the intended application. Thus this framework provides a smooth transition from theory to application. Furthermore, the topic of filtering, in which the model of the output signal is derived from the input signal model and the properties of the filter, is included naturally. This signal modeling perspective is the framework within which this book is developed.

Because biomedical engineering involves the application of engineering methods for the improvement of human health, the signals encountered by biomedical engineers are typically derived from biological processes. Often such signals are not well represented by the simple deterministic signals favored for illustrating the basic principles of signal processing. Real-world biomedical signals usually include large stochastic components; they also may be fractal or chaotic. Often we are too ignorant of the properties of their sources to know a priori what the character of the signal will be. Therefore the biomedical engineering student must first recognize the range of possible signal types and be able to determine the most appropriate type of analysis for the signal of interest. Unfortunately, this choice is not always clear. Of special importance is knowing not only the basic ways of processing each type of signal, but also recognizing indications that the selected analysis method may have been inappropriate. Even when the method is correct (or, more likely, when there is no indication that it is *not* correct), the best way to process the signal will depend on the user's objectives.

By presenting signal processing as the process of developing and manipulating a model of the signal, this book attacks the problems discussed above using an integrated framework. Three issues—(1) choosing a class of signal model, (2) selecting a specific form of the model, and (3) evaluating indicators of adequacy of the model—are emphasized in the book. The question of what information a given class of signal models can provide naturally lends itself to a format in which each chapter discusses one class of models and how to manipulate them. It is then straightforward to discuss the criteria for choosing a "better" model, although such a discussion will encompass the models from previous chapters as well. The question of which general class of model to utilize is more difficult to discuss because of its breadth, although this question is raised in each chapter in which a new class is introduced. Part of the motivation for the interactive exercise that starts each chapter is to demonstrate the limitations to the knowledge that the reader has hopefully accumulated to that point and motivate the introduction of a new way to look at signals. Nonetheless, this question of choosing an appropriate class of signal model for a specific application needs repeated emphasis by an instructor, who should raise the question in lectures and discussions.

This book has been developed completely from a biomedical engineering perspective, in the hope of conveying to the reader that measuring, manipulating, and interpreting signals is fundamental to every practitioner of biomedical engineering, and that the concepts being presented are fundamental tools needed by all biomedical engineers. With the breadth of the field of biomedical engineering and the wide availability of computer software for signal processing, the nuances and detailed mathematical variations of algorithms are less relevant to many practitioners than is the relationship between theoretical concepts and applications. This book strives to emphasize the latter issue by presenting many examples, sometimes at the expense of omitting certain details that are common in signal processing books—for example, the various algorithmic implementations of the Fast Fourier Transform (FFT). These details are readily available from other sources. Otherwise, I have attempted to present the material in a mathematically rigorous fashion.

Although the book is self-contained, the MATLAB files that implement many of the examples provide repetition of the material, enhanced visual feedback, more detailed explanations of the biomedical applications, and opportunities for exploration of related topics. These files are available from a web site of the publisher and from the author's anonymous ftp site (ondine.image.uky.edu, or 128.163.176.10). The optimal way to master the material in the textbook involves reading the text while sitting in front of a computer on which MATLAB is running. Hopefully the reader will call forth the MATLAB-based demonstrations and exercises as they are encountered in the text and experiment with them. Also the value of the examples of real-world biomedical signals that are available as .mat files is greatest if they are utilized at the time they are encountered in the text. The reader is encouraged to examine these signals when the text discusses them.

Unless stated otherwise, whenever the text uses a phrase like "we can show," it is assumed that the motivated reader will be capable of supplying the missing steps of a procedure. Exceptions are those situations in which the material is beyond the scope of the book, and these situations are noted.

Of the approximately twenty exercises at the end of each chapter, about half are intended primarily to reinforce the theoretical concepts through repetition. The remainder attempt to relate the theory to biomedical applications. My intent is to emphasize the process of translating qualitative descriptions of biomedical phenomena into quantitative descriptions amenable to the analyses developed in the book. Given this emphasis and the exigencies of the book format, the discussions of the applications often are simplistic. For those who wish for more detailed and insightful discussions of specific applications, I hope that the present material provides a suitable foundation upon which to build.

This book is written at a level for use in a first-semester graduate course in biomedical engineering or an advanced undergraduate course, with the expectation that interested students will take a subsequent course in biosystems analysis. Therefore the text focuses on signal processing with the intent to present material that is relevant to students in all subspecialties of biomedical engineering and to lay a foundation for a biosystems course. It assumes the reader has prior exposure to Fourier (and Laplace) transforms and has a basic understanding of human physiology. Ideally, a graduate from an ABET-accredited undergraduate engineering program would be familiar with the basic concepts in Chapters 3–5 and would have been introduced to the material of Chapters 6 and 7. In that case, the book could be covered in a three-hour, one-semester course. Alternatively, one might incorporate the computer exercises into a companion one-hour laboratory course. The text could be used for a one-semester, senior undergraduate course, but it is likely that the instructor would need to omit major sections.

This book also could serve as an introduction to biomedical engineering for engineers from other disciplines and as a reference book for practicing biomedical engineers. It presents biomedical applications of signal processing within a framework of signal modeling. Because both classical and modern signal processing methods are developed from this same framework, it is natural to answer the same two important questions about each method: (1) What assumptions does the method make about the signal? (2) What information can the method provide? These questions are especially important to the practitioner who needs assistance in choosing a method of analysis for his or her particular application. Physiologists and neuroscientists with a background in differential equations should find the text approachable. Physicists, mathematicians, or chemists who wish to apply their knowledge to biomedical problems may appreciate the breadth of biomedical examples.

Although one might be tempted to attribute the intellectual content of a book only to the creative efforts of its author, an author's teaching style and philosophy evolve through interactions with other teachers and students. I owe a debt to all of those who praised or critiqued my efforts, especially to the late Dr. Fred Grodins, whose clarity of explanation and rigor in applying theory to biomedical practice stimulated the interest of a young electrical engineer in the then new field of biomedical engineering. I also wish to recognize the insightful discussions of the role of engineering in physiology during my collaborations with Professor Curt von Euler of the Karolinska Institute, and Dr. Neil Cherniack, then at Case Western Reserve University. Although neither is a biomedical engineer by training, each is able to transcend mathematical formulas and apply theoretical concepts to real biomedical processes. Of course, one's thought processes are honed continually by graduate students. In particular, the challenge of quenching the intellectual thirsts of Mo Modarreszadeh, Mike Sammon, and Xiaobin Zhang has motivated my continual searching for new wellsprings of signal processing enlightenment!

Special thanks go to the consultants who guided my efforts to include examples of applications outside of my own fields of experience. I assume full responsibility for any errors in the presentation of these examples; the errors would have been much more numerous without their guidance. I wish to express special thanks to Tim Black and Tom Dolan, for their assistance in preparing the final typed manuscript and the artwork.

A book becomes reality only when sufficient resources are made available to the author. In this regard I am heavily indebted to the Whitaker Foundation for support through their Teaching Materials Program. This funding allowed me to reduce my academic responsibilities during the time of writing and to engage assistants who helped assemble the examples and exercises and verified that the exercises could be solved! Dr. Charles Knapp, Director of the Center for Biomedical Engineering at the University of Kentucky, has been very generous in providing relief from academic chores so that I could maintain the necessary writing schedule. More importantly, he has vigorously supported my vision of developing a textbook for this field. I also wish to thank Dr. Peter Katona, whose encouragement was the final push needed to embark on this project. Finally, without the understanding and continued support of my wife, Peggy, this book would still be a project for the future. Although she knew from experience the disruption it would cause, she also knew that some dreams have to be pursued.

EUGENE N. BRUCE
February 7, 2000

THE NATURE OF
BIOMEDICAL SIGNALS

1.1 THE REASONS FOR STUDYING BIOMEDICAL SIGNAL PROCESSING

It may seem obvious that signal processing concepts should be part of the core training in biomedical engineering because students in such an applied field must learn to "deal with" signals and systems. The motivation for studying this topic, however, is more profound and can be related to fundamental approaches to conceptualizing and solving biomedical problems. A fundamental construct for interpreting both quantitative and qualitative data in all of biomedical engineering is the conceptualization that measurable, real-world behavior results from interactions among sources of energy and modifiers (or dissipators) of energy. Since a signal is just a record of energy production by a process, this fundamental framework naturally gives rise to questions concerning: (i) the inferences that one can draw from a signal about the properties of the source of the energy, and; (ii) the relationships among simultaneously observed energy records (i.e., signals). These are exactly the questions that signal processing addresses.

The biomedical engineering student should understand that signal processing (and the closely related field of systems analysis) comprises more than just mathematical manipulations that are only useful to control systems engineers. Indeed, these topics provide a fundamental framework for conceptualizing and analyzing physical behavior in a rigorously organized manner, whether that behavior is the output of an electronic control system, the flow of blood through a defective aortic valve, or the reaction of an implant with surrounding tissue (to give a very few examples). Furthermore, while the computational side of signals/systems can produce precise analyses, a qualitative understanding of these subjects is just as important. For example, a student proposing to use wavelets to detect abnormalities in the electrocardiogram signal surely needs to understand the mathematics of wavelet transforms. On the other hand, a student with neurophysiological interests who wishes to study the effects of whole-body vibration on visual function needs to understand the qualitative concept of resonance (even if that student cannot remember the form of a second-order differential equation to quantify the phenomenon!). Similarly, a student who addresses neural control of heart rate needs to understand the concepts of memory or correlation and the causes of temporal variability in energy records, whether or not the student will utilize methods based on statistics of random processes to describe heart rate. *The primary objectives of this textbook are to instill in the student a qualitative understanding of signal processing as an essential component of knowledge for a biomedical engineer and to develop the student's skills in applying these concepts quantitatively to biomedical problems.*

A more direct approach to defining these objectives is to ask: "What skills should the student possess after studying this text?" Such skills fall into two broad categories: (1) skills related to the acquisition and processing of biomedical signals for extracting a priori desired information, and (2) skills related to interpretation of the nature of physical processes based either on observations of a signal or on observations of how a process alters the characteristics of a signal. More specifically, the objectives can be summarized by four phrases, which describe what the student should be able to do: (1) measure signals, (2) manipulate (i.e., filter) signals, (3) describe qualities of the source based on those of the signal, and (4) probe the source. Expanding these objectives, the student first should be able to quantify and (if necessary) compensate for the effects of measuring devices on the signal being measured. Second, the student should be able to identify and separate desired and unwanted components of a signal. Third, the student should be able to uncover the nature of phenomena responsible for generating the signal, as well as anticipate its future behavior, based on the identification and interpretation of an appropriate model for the signal. Fourth, the student should be able to relate the properties of a physical system to the differences in characteristics between a signal used to excite (or probe) the system and the measured response of the system.

The didactic framework adopted in this textbook to achieve the goals described above is one in which signal processing is presented as a process for the development and manipulation of a model of an observable variable (the signal). In this framework the "model" becomes a surrogate for the signal source, and one naturally asks to what degree this model is reasonable given one's knowledge of the signal source and one's application. Thus there is a natural connection to real-world problems. This approach is discussed in more detail later in this chapter.

Often in signal processing textbooks there is heavy emphasis on computation and algorithms. For students eager to expand their knowledge in signals and systems, this approach may become a boring repetition of familiar material; for students who see no need for this knowledge, the relevance of the material is not made apparent by such an approach. Yet both groups of students are missing the main message that physical (observable, real-world) behaviors can be described in an organized manner that permits both insights into the underlying physical processes and prediction of unobserved behavior. The "icing on the cake" is that these techniques are independent of the devices used to acquire signals. In other words, the methods are equally applicable to signals from heart rate monitors, signals from strain gauges in an Instron bone test device, spectra from chemical spectroscopy assays, data from a cell sorter, or essentially any other biomedical data. Biomedical engineers have the real-world problems, and signal processors have many excellent solutions. This text will attempt to unite the two by discussing many examples having a real-world connection and by emphasizing those aspects of the theory that are relevant for deciding whether the theory is truly applicable to a particular real-world problem.

1.2 WHAT IS A SIGNAL?

The electrical current induced in an antenna wire by the electromagnetic field transmitted from your favorite radio station, hourly readings of the atmospheric pressure, and the hum of a mosquito all are examples of signals (see Fig. 1.1). In each case the signal is the output of a sensing device (i.e., antenna, barometer, and ear) and the temporal variations in the signal convey information. To make sense of that information, however, may require some processing of the signal—for example, we learn to recognize the sound signal created by the pressure waves generated by the mosquito and associate its frequency with the insect and its itching bite. How the auditory system discriminates frequency is a complicated story that is not yet fully understood, but the underlying transduction of the pressure wave and manipulation of the resulting neural signals is an example of signal processing by a biological system.

   A *signal* is a single-valued representation of information as a function of an independent variable (e.g., time). The specific type of information being represented may have real or complex values. In the case of physical processes, the information

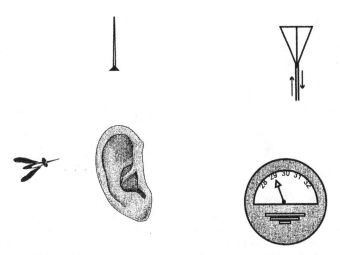

FIGURE 1.1 Examples of common signals: current induced in a radio antenna, humming of a mosquito, hourly readings from a barometer.

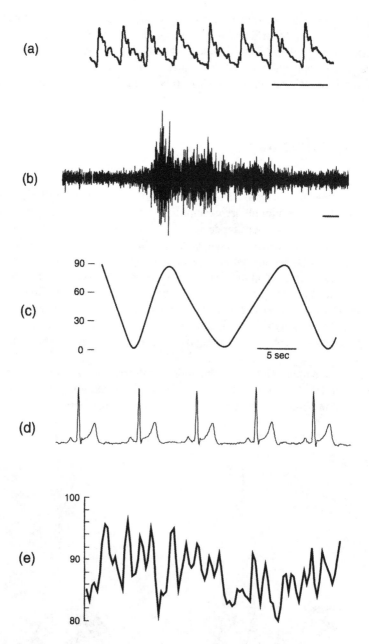

FIGURE 1.2 (a) Instantaneous mean blood flow velocity in the middle cerebral artery of a human subject obtained from the Doppler shifts of a reflected ultrasound beam. Time marker = 1 s; (b) Electromyogram (EMG) from two wires held firmly under the tongue by a mouthpiece. Subject contracted tongue, then relaxed. Time marker = 0.2 s; (c) Angle of rotation of knee obtained from an angle sensor (Data courtesy of James Abbas); (d) An electrocardiogram (ECG) recording (Data courtesy of Abhijit Patwardhan); (e) Instantaneous heart rate (beats/min, reciprocal of beat-to-beat interval) for 100 consecutive heart beats. Implicit independent variable is

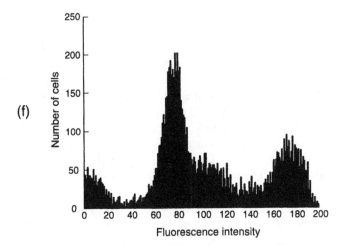

FIGURE 1.2 (*continued*) "heart beat number"; (f) *x*-axis: Intensity of fluorescence from cells excited by incident laser light beam. *y*-axis: Number of cells displaying a given intensity. From cells stained with a fluorescent dye that binds to phospholipids (Data courtesy of Margaret Bruce).

often is a measure of some form of energy produced by the process. Signals that arise from strictly mathematical processes and have no apparent physical equivalent are commonly considered to represent some form of mathematical energy. A signal may be a function of another variable besides time or even a function of two or more variables, but often we will consider time-dependent functions. Indeed, a great many signals in practice are real-valued, scalar functions of time. Nonetheless, the reader will be reminded throughout that many important biomedical signals are not time-dependent functions—for example, in image processing the intensity values are functions of x and y coordinates—and that the methods being presented apply equally to such signals. For practical purposes, often one can define a signal as the output from a measuring or sensing device (or the result of mathematical operations on such an output), although this definition is too restrictive for some biomedical applications. For example, I can create a function to represent the bases in a DNA molecule by assigning a unique number from 1 to 4 to each of the four bases. Then if I plot the position of each base along the DNA strand as the independent variable and the value of the function for that base as the dependent variable, I have created a signal that is not derived from a sensor or measuring device (nor is it a function of time). As we shall see, if I can represent the signal mathematically, then I can manipulate it using the concepts from this book.

1.3 SOME TYPICAL SOURCES OF BIOMEDICAL SIGNALS

The sources of biomedical signals are infinite in variety and we shall identify only a few as examples of the different types of these signals. A large class of biomedical signals comprises those which are electrical in nature. The electrocardiogram (ECG), which can be recorded by placing electrodes on the chest, is one such signal. It records the electrical potential at the electrode (or the potential difference between two electrodes) induced by the presence of time-varying electrical activity in cardiac muscle associated with the generation and propagation of action potentials. Each heartbeat produces a sequence of electrical waves (the P, Q, R, S, and T waves) and by examining the shape of the ECG waveforms (Fig. 1.2(d)) a physician can obtain considerable insight about whether the contractions of the heart are occurring normally. In practice, the physician usually compares the electrocardiogram signals from several sites on the body. Although visual inspection of ECG signals is useful clinically, analyses of electrocardiograms can be very sophisticated and involve considerable signal processing. For example, there is substantial interest in detecting small changes in the shape of the ECG waveform occurring after the large peak (the R-wave) during the time that the ventricles are recovering from their contraction. Several approaches being used involve spectral analysis of the ECG signal, a technique which will be discussed in Chapter 9. Another question of clinical relevance is whether the heart is beating regularly or irregularly. (Believe it or not, too much regularity of the heartbeat is considered unhealthy!) Some of the most sophisticated biomedical signal processing in the research literature is aimed at characterizing the type and degree of irregularity of the cardiac rhythm. A portion of the irregularity in heart rate seems to be spontaneous random changes caused partly by random events at the membranes of sinus node pacemaker cells. As we shall study in a later chapter, however, the variations in heart rate at times that are several tens of seconds apart are related. Recently developed signal analysis techniques for characterizing such temporal dependencies are among the most sophisticated and exciting new methods for biomedical signal processing. Chapter 10 will introduce these methods.

Another example of a bioelectrical signal is the electromyogram (EMG). EMG signals are recorded by placing electrodes in, on, or near a muscle and amplifying the electrical potential (or potential difference between two electrodes) that results from the generation and propagation of action potentials along muscle fibers. A multiple-unit EMG (MUEMG) records the potentials from many muscle fibers at once (Fig. 1.2(b)). The presence of MUEMG activity indicates that a muscle is actively contracting and many researchers have attempted to develop techniques for filtering of the MUEMG signal to produce another signal that is proportional to the force that the muscle is generating. This goal is important because it would allow one to estimate muscle force without having to connect a force transducer to the muscle—something that is often impossible to do on human subjects. (How, for example, could one attach a force transducer to the extraocular muscles?) MUEMG signals are utilized also in rehabilitation engineering for controlling a prosthetic device. The subject is trained to contract a muscle that he or she can still control voluntarily and the amplitude of the MUEMG is taken as a measure of the desired degree of activation of the prosthesis—for example, the gripping force to be produced by an artificial hand. But since the MUEMG is inherently noise-like, it must be filtered to extract a smoothed signal which varies with the amplitude of the MUEMG. Other examples of bioelectrical signals are the electroretinogram (ERG), electrogastrogram (EGG), and the electroencephalogram (EEG), which is a recording of brain electrical activity from electrodes on the scalp. Processing of EEG signals is an active research area because the EEG is used to determine if a person is asleep and to identify the sleep state. For example, the EEG contains predominantly oscillations in the 8–12 Hz range (alpha rhythm) when the person is awake. In contrast, predominant oscillations in the 1–4 Hz range (delta rhythm) indicate non-dreaming (also called slow-wave or non-REM) sleep.

In Chapters 6 and 8 we will study both analog filters and their digital counterparts that could be used to filter bioelectrical and other signals. We study digital filters because so much of biomedical signal processing is performed on sampled (i.e., digitized) signals. Furthermore, it will become apparent that digital filtering has an important advantage, in that there are many digital filters which have no analog counterpart.

The field of imaging provides many examples of both biomedical signals and biomedical signal processing. In magnetic resonance imaging (MRI) the basic signals are currents induced in a coil caused by the movement of molecular dipoles as the molecules resume a condition of random orientation after having been aligned by the imposed magnetic field. These current signals are complex and substantial. Signal processing is required just to properly detect and decode them, which is done in terms of the spatial locations of the dipoles that caused the currents and the rate of relaxation of the dipoles (which is related to the type of tissue in which they are located). Much of the associated signal processing is based on Fourier transforms, which will be studied in Chapters 5 and 7. Although this book will consider only one-dimensional Fourier transforms and MRI utilizes two-dimensional Fourier transforms, the basic concepts are the same. In addition, once an image is constructed it often is desirable to "process" it to enhance the visibility of certain features such as the edges of a tumor. Although there are many advanced methods of image processing involving techniques beyond the scope of this book, the basic techniques are based on concepts derived from digital filtering theory, which we shall encounter in Chapter 8. Indeed, even though other imaging modalities such as positron emission tomography (PET), ultrasound, and x-ray utilize different physical principles for acquisition of the image, the signal processing methods for enhancing images are similar.

When an ultrasound beam strikes a moving object, the frequency of the reflected beam differs from that of the incident beam in proportion to the velocity of the object (Doppler shift). Because high-frequency ultrasound signals can penetrate hard biological tissues (such as thin bones), this property provides a means of measuring velocities of inaccessible tissues such as blood cells (Fig. 1.2(a)). Although this measurement is not a direct assessment of the bulk flow of blood, it is used in humans to identify vessels in the brain in which flow is sluggish.

Sensors that transduce mechanical actions into electrical signals are common in biomedical engineering applications. In studies of whole-body biomechanics, accelerometers record the acceleration of limbs or joints (Fig. 1.2(c)) and force plates use strain gauge sensors to measure the reaction forces where the feet contact the ground. Although the outputs of such sensors are electrical signals, it is usual to calibrate the electrical signals in terms of the mechanical signal being measured and to think of the electrical signal as the mechanical signal. That is, one does not describe the accelerometer output as "the electrical equivalent of the acceleration signal" but as the acceleration signal itself. Since accelerometers are sensitive to noisy vibrations, it is usually necessary to remove the noise on (computer-sampled) accelerometer signals using digital filters, such as those described in Chapter 8.

Force transducers, displacement transducers, and pressure transducers are additional types of sensors which produce electrical signals that correspond to mechanical actions. Some of these transducers can be made quite small—for example, to fit inside an artery for measuring blood pressure. Some can be miniaturized for measurements, for example, from single excised muscle fibers or from cells grown in culture. In some of these situations the transducer modifies the measured behavior because its mechanical characteristics are not negligible compared to the system under study. In such cases it is necessary to correct the measured signal for this "loading" effect. Often this correction comprises designing a digital filter based on the frequency response of the transducer. The concept of frequency response will be examined in several chapters, beginning in Chapter 4, and the difficulties in designing these inverse filters will be discussed in Chapter 8.

As biomedical engineers become more involved in probing behavior at the cellular and molecular levels, sensors for events occurring at the microscopic level have assumed greater importance. Often these sensors directly or indirectly measure products of biochemical reactions. For example, the concentration of calcium ion in a cell can be measured indirectly by introducing into the cell a chemical (Flura2) which fluoresces with an intensity that is a function of the calcium concentration. One then focuses a microscope on the cell in question and directs the light at the eyepiece to a photocell through an optical system that is tuned to the frequency of the fluorescent light. The electrical output from the photocell is proportional to the calcium concentration. A similar approach using a voltage-sensitive dye permits one to measure intracellular potentials by optical means. Other types of optical biosensors measure the absorption (or reflectance) of an incident light beam containing one or a few frequencies of light. For example, the hydrogen ion concentration in a thin tissue can be determined by staining the tissue with a particular red dye and measuring the relative transmission of light through the tissue at two different frequencies. Some newer approaches to measuring the concentration of protein in a sample utilize a bioactive electrode that reacts with the desired protein. For certain reactions, the reflectance of an imposed laser light having a specific frequency is proportional to the concentration of the reactant. Again the signal of interest is derived by leading the reflected beam to a photocell or photosensitive solid-state device (i.e., CCD). Often in such sensor systems the analysis of the signal involves taking a ratio or some nonlinear processing such as logarithmic scaling. For ratios or nonlinear scaling the noise might be amplified disproportionately, so it becomes important to filter the signal from the photocell to reduce this effect. Furthermore, there may be contamination by background light which should be removed by filtering. The digital filters we will analyze in Chapter 8 could be used for these purposes. Finally, if there is motion of the environment, movement of such small sensors might introduce a time-varying artifact. In some cases one can reduce this latter type of artifact using adaptive filtering techniques, but these methods are beyond the scope of this book.

Up to this point the discussion has assumed implicitly that biomedical signals comprise a combination of a desired component which represents a true measure of a physical behavior and an undesired component categorized generally as noise. We need to recognize that certain biomedical signals which visually resemble noise actually represent true physical behavior. If, for example, one uses a fine-tipped glass microelectrode to measure the current flowing through a tiny (i.e., a few microns square) piece of a cell membrane, the fluctuations of the current with time appear to be rather random but this apparent randomness can be explained by applying certain simple rules which determine the conformation of proteins that form the channels for conducting ions through the membrane. The consequence of these rules is that the openings and closings of channels for ionic current flow exhibit a self-similarity across many scales of time. This self-similarity causes the appearance of rapid fluctuations in the signal, thus creating the illusion of unpredictableness (i.e., randomness) in a signal that is obeying simple rules. Such signals are said to be *fractal*. Chapter 10 will introduce the analysis of fractal signals.

1.4 CONTINUOUS-TIME AND DISCRETE-TIME

Many biomedical measurements such as arterial blood pressure or the torque at a joint are inherently defined at all instants of time. The signals resulting from these measurements are *continuous-time signals* and it is traditional to represent them as explicit functions of time in the form $x(t)$, $p(t)$, and so forth. On the other hand, we could sample the values of a continuous-time signal at integer multiples of some fundamental time increment and obtain a signal that consists of a sequence of numbers, each corresponding to one instant of time. Often we ignore the time increment and represent this *discrete-time signal* as a function of the integer multiple of the time increment, saying that the argument of the function can assume only integer values—that is, the first entry in the sequence corresponds to zero times the time increment and is called $x[0]$, the second entry corresponds to one times the time increment and corresponds to $x[1]$, and so on. Notice that in order to distinguish continuous-time and discrete-time signals, their arguments are enclosed in different types of brackets.

Example 1.1 An example of obtaining a discrete-time signal by sampling a continuous-time signal at a constant rate is shown in Fig. 1.3. Note the different ways of graphically representing continuous-time (CT) and discrete-time (DT) signals.

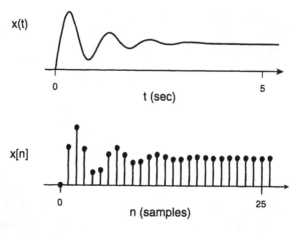

FIGURE 1.3 Examples of a continuous-time (CT) signal (top) and a discrete-time (DT) signal (bottom), the latter in this case being obtained by sampling the former at a uniform rate of 5 samples/s.

Discrete-time signals can arise from inherently discrete processes as well as from sampling of continuous-time signals. Within the biomedical realm one can consider the amount of blood ejected from the heart with each beat as a discrete-time variable, and its representation as a function of a time variable (which assumes only integer values and increments by one with each heartbeat) would constitute a discrete-time signal. If one were counting the number of cells in a culture dish at hourly intervals, then a plot of this count versus an integer index of the time would be a discrete-time signal. Figures 1.2(e) and 1.2(f) show "discrete-time" signals which, however, are displayed as if they were continuous-time by connecting consecutive points with interpolating straight lines. Often we aggregate data by counting events over a fixed time interval—for example, the number of breaths taken by a subject in each minute, the number of action potentials generated by a neuron in each second—and thus create a discrete-time signal from an inherently continuous-time signal (e.g., Example 1.2 and Fig. 1.4).

Example 1.2 Figure 1.4 demonstrates a common method for obtaining a DT signal from biomedical data. The top tracing is a representative extracellular recording from a neuron, showing its action potentials during a 10-second time interval. The number of action potentials which occur each second are counted and graphed in two different ways. In the middle tracing a CT signal is constructed by holding a constant value during each one-second period. This constant value equals the action potential count during that time period. In the bottom tracing a DT signal is constructed from these same data by plotting single values, representing the number of action potentials which occurred during the previous second, at one-second intervals.

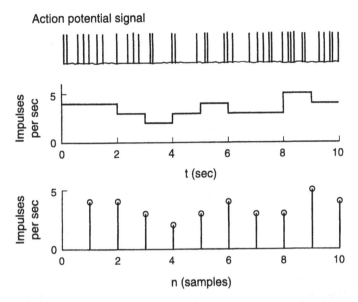

FIGURE 1.4 Constructing CT and DT representations of the firing rate (number of action potentials per second) of a neuronal spike train. See Example 1.2 for further explanation.

1.5 ASSESSING THE RELATIONSHIPS BETWEEN TWO SIGNALS

The above examples have the common theme of extracting information from observations (i.e., measurements) of spontaneous behavior. Another broad area of signal processing involves assessing the relationship between two signals. By discussing such a relationship, we are implying that there exists some interconnection between the sources of these signals. Figure 1.5 indicates the two general possibilities for this interconnection: (1) one signal directly activates the source of the second signal; (2) a third signal activates the separate sources of the initial two signals. To continue this discussion we need to formally define what the boxes in Fig. 1.5 represent. In the most general sense these boxes, which will be called *systems*, represent *any* mechanism through which one signal depends on (or is derived from) another signal, and we shall define a *system* as any physical device or set of rules which *transforms* one variable (i.e., signal) into another variable (or signal). A typical system has an *input* variable and an *output* variable, although some systems have an output with no explicit input. These latter systems are called *autonomous* systems, whereas those having an explicit input are called *non-autonomous* systems. It should be noted that a system may be a physical device (or collection of devices which interact with one another) or may simply be a set of rules which can be written down. Because this book is a signal processing text, we shall place considerable emphasis on the viewpoint that a system *transforms* one signal into another signal and we shall devote some effort to formal descriptions of various classes of trans-

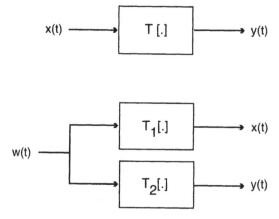

FIGURE 1.5 Two ways by which two signals, $x(t)$ and $y(t)$, could be related.

formations. Transformations will be symbolized as T[.], as shown in Fig. 1.5. Often we will utilize the attributes of linear transformations, which obey certain properties that are discussed in Chapter 2. An example of a physical device that can be considered a system is an electronic amplifier into which an electrical signal can be led, and which produces an output that is a scaled version of the input signal. A pressure transducer can be described as a system whose input and output are the pressure on the sensor and the electrical voltage at its output terminals, respectively. Analysis of the blood flow through an artery may be cast in the framework of a system in which the input is blood pressure, the system comprises a description of the mechanical properties of the arterial wall and the blood, and the output is the blood flow rate. Rule-based systems are common in the fields of artificial intelligence and fuzzy logic. In biomedical applications a simple model of information processing by the nervous system may be based on a set of rules since detailed neurophysiological information may be lacking. A system may have more than one input. Consider, for example, the system comprising the extraocular muscles and the eyeball. The inputs to this system are the neural signals that activate the muscles and the output is the position of the eyeball in its orbit. If one were to consider the control of both eyes simultaneously, then the previous multiple-input, single-output system would become a multiple-input, multiple-output system.

Example 1.3   The systems framework is very important in many applications of biomedical signal processing. In developing drug therapy regimens, for example, one needs to know how the target behavior will respond to a given amount of drug. Thus, when a drug such as sodium nitroprusside is given to maintain blood pressure, it is desirable to be able to predict the amount of drug needed to raise the blood pressure by the desired amount. Conceptually, one could inject several different levels of the drug, measure the blood pressure responses, and establish a look-up table for determining the drug dose. In some applications this approach may be viable, but a better alternative will be developed in Chapter 3. For many systems one can calculate the response of the output variable of a system (e.g., blood pressure) to *any* input signal (e.g., amount of drug injected into the blood) if one knows the response of the output to *certain specific* input signals. Even when this method is not directly applicable, it is often possible to approximate the behavior of a system over restricted ranges of its input and output signals.

Even in the case of observing spontaneous behavior (when there is no observed input signal) it is possible to apply signal processing methods by pretending that the signal being observed is the output of a system. The choice of a specific signal processing method will dictate what the fictitious "input" signal is assumed to be. This conceptual framework, which is developed later in this chapter, provides a common basis for comparing signal processing methods and explicitly highlights implicit assumptions underlying the use of each method.

Obviously the above illustrations are only a few of the many examples of biomedical signals and signal processing and many more will be introduced in the discussions and homework problems of this book. The student surely will encounter many others in his or her own work. The purpose of presenting these examples is in part to illustrate the breadth of biomedical engineering. Another purpose is to demonstrate that signal processing is a critical element of biomedical engineering activities, whether they address problems at the whole body, organ system, tissue, cell, or molecular level.

1.6 WHY SIGNALS ARE "PROCESSED"

Signal processing can be defined as the manipulation of a signal for the purpose of either extracting information from the signal, extracting information about the relationships of two (or more) signals, or producing an alternative representation of the signal. Most commonly the manipulation process is specified by (a set of) mathematical equations, although qualitative or "fuzzy" rules are equally valid. There are numerous specific motivations for signal processing but many fall into the following categories: (1) to remove unwanted signal components that are corrupting the signal of interest; (2) to extract information by rendering it in a more obvious or more useful form; and (3) to predict future values of the signal in order to anticipate the behavior of its source. The first motivation clearly comprises the process of filtering to remove noise and this motivation will be encountered in almost every chapter, as most methods of signal processing implicitly provide some basis for discriminating desired from undesired signal components. (An important issue is the basis on which the user decides what is desired signal and what is noise!) The general problem of discriminating noise from signal is discussed later in this chapter and the major focus of Chapter 8 will be the use of filters for noise removal.

The idea of applying signal processing to extract information is pervasive in biomedical applications. Often the objective is to discriminate abnormal from normal signals and on this basis diagnose the presence of disease. Just as the physician utilizes auditory discrimination to detect abnormal heart or lung sounds, biomedical engineers often separate a signal into a sum of basic signal types in order to detect the presence of abnormal signals that are suggestive of disease. Many of these approaches involve searching for unusual features in the Fourier transform of a signal. For example, in several cardiorespiratory diseases, such as congestive heart failure, blood pressure oscillates with a period of a few or several tens of seconds. To detect this oscillation one can examine the power spectrum (which is proportional to the squared magnitude of the Fourier transform) of blood pressure. On the other hand, we shall find that many biomedical signals do not adhere well (in a practical sense, not in a mathematical sense) to the basic premise of the Fourier transform: that the signal can be expressed as a sum of sinusoids. Consequently, an active and important area of research is to develop alternative methods for decomposing a signal into basic signal types that better represent the properties of biomedical signal sources.

The need to predict future values of a signal arises in two different situations. Most commonly this need occurs when one is interested in controlling a behavior—for example, regulating the blood glucose level by periodic injections of insulin. Because the result of any control action requires some finite time before its effect is evident, it is useful if one can predict the behavior of interest a short time in the future. In this way one can apply a control action based on the anticipated behavior at the time that the control action will be effective. The concepts of control theory underlying such applications are not discussed in this textbook, but in Chapters 3 and 9 we will address the question of how one processes a signal to predict its future values. The other biomedical situation in which the prediction of future values is important is the early detection of the onset of a disease process. Often in this case the problem is to predict the limits of future normal behavior so that small deviations from normalcy which might signal the onset of disease may be identified. This type of application has been a "Holy Grail" of biomedical signal processing since its inception. It is exemplified today, for example, by attempts to correlate indices of fractal behavior of heart rate with the presence or absence of disease. This area is an active research field whose theoretical advances are driven, in part, by the need for better methods for applications involving biomedical problems.

1.7   TYPES OF SIGNALS: DETERMINISTIC, STOCHASTIC, FRACTAL AND CHAOTIC

As one examines the examples of signals in Fig. 1.2, there is the impression of qualitative differences among them. Some appear to be smooth and one has the impression that it is possible to predict how these signals would behave beyond the period of measurement. In contrast, other signals are quite variable and give the impression that it would be difficult to predict their exact future behaviors. Such impressions are correct and reflect a fundamental problem for biomedical signal processing: There are different types (or classes) of signals. The class to which a signal belongs strongly influences the inferences one can draw from the signal regarding the source of the signal and determines, to a certain extent, the applicable signal processing methods.

We shall recognize four classes of signals: (1) deterministic, (2) stochastic, (3) fractal, and (4) chaotic. Examples of these types of signals are shown in Fig. 1.6. Deterministic signals are encountered commonly in textbook examples but less frequently in the real world. A *deterministic signal* is one whose values in the future can be predicted exactly with absolute confidence if enough information about its past is available. Often one only requires a small amount of past information. For example, once I know one cycle of a sine wave, I can predict its value at any time in the future. All of the common textbook signals such as impulses, steps, and exponential functions are deterministic. In fact, any signal which can be expressed exactly in closed mathematical form as a function of time is deterministic.

Stochastic signals are signals for which it is impossible to predict an exact future value even if one knows its entire past history. That is, there is some aspect of the signal that is random, and the name "random signal" is often used for these signals. This book will follow this practice but it should be noted that some random signals are completely unpredictable (i.e., uncorrelated), whereas others can be predicted with greater (but not absolute) confidence. For example, predictions of the *direction* of change of these latter signals may be right more often than not, even though predictions of their exact value will almost certainly be incorrect every time.

Random signals are abundant in physical processes. Noise generated by electronic components in instrumentation is a common type of random signal that is present in much biomedical data. Although contemporary electronic design mini-

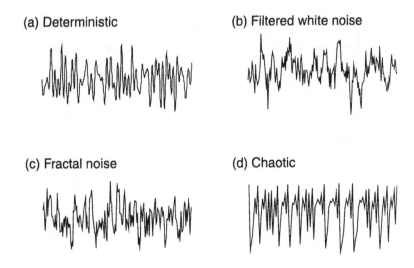

FIGURE 1.6 Examples of the four types of signals. Note the visual similarities. The reader should explore other examples of these signal types using the m-file sigtype.m. See text or m-file for description of how these signals were generated.

mizes this noise, it can be significant compared to signals from microminiature bio-medical sensors such as extracellular potentials recorded with glass microelec-trodes. Almost all bioelectrical measurements contain random noise caused by ran-dom fluctuations of ionic currents and by the stochastic nature of action potential generation. In addition, because most biomedical systems are complex, it is not pos-sible to appreciate all of the factors that influence a measurement. Often we classify those parts of a signal which are not understood as "noise." Because of our igno-rance, these signal components may appear to vary randomly relative to known mechanisms, thereby reinforcing the impression of stochastic behavior. Further-more, it is often valid to treat these signals as stochastic.

The presence of fractal signals in biomedical processes has become widely rec-ognized in the last decade. *Fractal signals* (Fig. 1.6(c)) have the interesting property that they look very similar at all levels of magnification, a property referred to as scale-invariance. For example, if I create a new signal from the one in Fig. 1.6(c) having one-fourth of the time resolution (i.e., by averaging the first four consecutive points, then the next four points, etc.), I cannot discriminate between that and the initial one-fourth of the original signal using many of the usual measures for quanti-fying a signal (Fig. 1.7). Visually they are not the same signal but they look very much alike. You might expect that random signals would exhibit this property also, but there are important quantitative differences between the scaling properties of fractal signals and of random signals. These differences will be discussed in Chap-ter 10. There is very good evidence that a part of the beat-to-beat heart rate signal (e.g., Fig. 1.2(e)) is fractal, as well as the signal representing current through a sin-gle ionic channel of a cell membrane. It is likely that many other biomedical signals are fractal. Furthermore, the concept of fractals can be applied to spatial variations, such as the branchings of blood vessels or airways or inhomogeneities in an elec-trode, as well as to temporal variations. More applications of spatial fractals in bio-medicine are likely to appear. If a signal is shown to be fractal, then the challenge is to understand how the structure and properties of its source could produce the scale

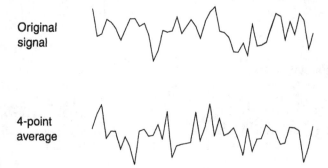

Original
signal

4-point
average

FIGURE 1.7 The similarity of fractal signals observed at different scales is seen by compar-ing (top) 50 points from the signal of Fig. 1.4(c) with (bottom) a 50-point signal constructed by averaging groups of four consecutive data points of the same signal.

invariance described above. Many of the simple physical processes that are often invoked as sources of biomedical signals cannot reproduce this property. Directly proving that a signal is fractal is difficult at this time, although there are fairly reproducible (but complex) techniques for determining indices of scale-invariance. One's confidence in a specific calculated value for any of these indices, however, may not be strong.

Chaotic signals are deterministic signals that cannot be predicted exactly in the future. The apparent contradiction in this definition is explainable on the basis of the "sensitive dependence on initial conditions" of chaotic signals. For some deterministic signals their trajectories in the future are so sensitive to their past values that it is impossible to specify those past values with sufficient accuracy that we can predict ahead in time with certainty. Thus, in theory, these signals are deterministic, but beyond a short time into the future the error of the prediction becomes very large. Because a chaotic signal is not fully predictable, visually its behavior has some characteristics of a random signal (but random signals are not chaotic and chaotic signals are not random!). As was the case with fractal signals, it is only in the last decade or so that researchers have recognized that biomedical systems could generate chaotic behavior that is reflected as chaotic signals. Again, the challenge is to understand the underlying structure and properties of a physical process which could produce this sensitive dependence on initial conditions and the unpredictable fluctuations in the signal.

Proving that a signal is chaotic is difficult at this time and developing new methods for such proof is an active research topic. One of the difficulties is the essentially universal presence of stochastic signal components, which seriously corrupt the analyses for chaotic signals. Another difficulty is that a process which is chaotic under some circumstances may not be chaotic under a different set of conditions. There is evidence, however, that some immunological and biochemical regulatory processes can exhibit chaotic behavior, and that EEG activity and breathing may have chaotic characteristics. Several neurophysiological systems, ranging from multicellular oscillators to single neurons, have been reported to exhibit chaos and the spread of disease in recurring epidemics has been assessed from this perspective also. Although evidence of chaotic behavior in real-world signals has been circumstantial, often it is possible to demonstrate the *potential* for chaotic behavior in a mathematical model of a biomedical process and then infer that the real-world process might also exhibit this behavior.

Example 1.4   Figure 1.6 displays signals of the four types discussed above which were generated using the file `sigtype.m`. The signals were constructed as follows: The deterministic signal is a sum of five sine waves having randomly selected values of amplitude and frequency. The stochastic signal is the output of a linear, lowpass filter whose input is uncorrelated white noise. The fractal Brownian noise signal was constructed using the spectral synthesis method with 300 frequency components and $H = 0.8$. The chaotic signal is the solution for the Henon map with parameters (1.4, 0.3) and randomly chosen initial conditions.

You may create other examples by running the file `sigtype.m`, which will generate other examples of these signal types. It is strongly recommended that you examine other examples of these types of signals because different realizations from these examples can be visually quite different. Use the MATLAB command `help sigtype` to retrieve information about the parameter values in `sigtype`.

1.8 SIGNAL MODELING AS A FRAMEWORK FOR SIGNAL PROCESSING

Signal processing was defined above as the manipulation of a signal for the purpose of either extracting information from the signal (or information about two or more signals) or producing an alternative representation of the signal. The parallels between this definition and the previous discussion of a system should be obvious. In signal processing the original signal is analogous to the input signal of a system, the rules specifying the manipulation process are analogous to the system itself, and the information (in the form of a derived signal) or the alternative representation of the original signal is analogous to the system output (Fig. 1.8). "Manipulation process" is simply another name for the transformation that specifies the input–output relationship of a system. The viewpoint of this book is that signal processing can be formulated as a process of applying a transformation to a signal. In this case, however, one does not necessarily describe the transformation in the same terms that would be used to describe an electronic amplifier, for example. A transformation is just a compact way to describe how to combine current and past (and perhaps future) values of the input signal in order to determine the current value of the output signal. The algorithms for signal processing often are not mathematically simple, but they can be expressed in a form that by analogy accomplishes this same end. An advantage of this framework is that signal processing algorithms can be grouped according to their similarities, and generalizations about the group can be derived. This structure will be called the *analysis* framework for signal processing. As an example, consider the construction of a Fourier series representation of a periodic signal. As shown in Fig. 1.9(a), the input to the "signal processing system" is the periodic signal, the system in this case is the set of equations for calculating the Fourier se-

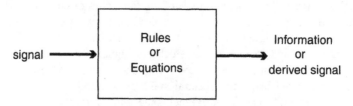

FIGURE 1.8 System model of signal processing.

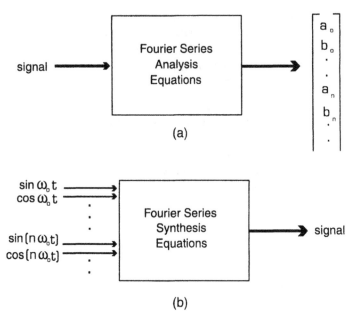

FIGURE 1.9 (a) Signal processing analysis model of Fourier series calculations; (b) Signal processing synthesis model of a periodic signal based on Fourier series summation of sinusoidal inputs.

ries coefficients, and the output is the ordered sequence of coefficients (which can be considered to be a discrete signal).

Another way to view this process is shown in Fig. 1.9(b). Here we represent not the analysis process but the model of the signal that is implied by the Fourier series analysis. This "signal modeling" structure will be called the *synthesis* framework for signal processing. This approach explicitly presumes that the observed periodic signal can be represented as a specific transformation of a large number of harmonically related sine waves. The transformation T[.] is exactly the usual equation for a Fourier series expansion with the appropriate values for the coefficients. It should be apparent that knowledge of the input signals and the transformation equation(s) is equivalent to knowledge of the output signal. Both the analysis and the synthesis frameworks will be important for developing signal processing methods and the reader is strongly urged to clarify in his own mind the distinctions between these two approaches. (This specific example will be developed in detail in Chapter 5.)

Example 1.5 The synthesis framework for Fourier series analysis can be carried a step further by utilizing a normalized time scale—that is, let the duration of one cycle be normalized to one and the frequencies of the input sine waves be integer multiples of unity. Then, in order to construct the output signal another parameter is needed to complete the description of the input signals, the fundamental frequency to be used. Now this model structure for Fourier series analysis is analogous to the ARMA model structure for random signals, which is studied in Chapter 9. In the latter case the input signals are sample functions of a white noise process and the additional parameter that is needed is the variance of the white noise.

A differential equation that expresses a relationship between a function, say $y(t)$, and its derivatives and a forcing function, say $x(t)$, and its derivatives may be considered to be a signal processing model in the synthesis form. That is, the differential equation relates how one "constructs" the current value of $y(t)$ by adding (or subtracting) current and past values of $x(t)$ and past values of $y(t)$. To visualize this result, think of the approximation

$$\frac{dy(t)}{dt} \approx \frac{y(t) - y(t - \Delta t)}{\Delta t}.$$

Consequently, any situation that can be described by a differential equation also can be considered as an example of signal processing.

1.9   WHAT IS NOISE?

To a considerable degree the direct answer to this question does not lie in the realm of signal processing. Rather, the ultimate basis for deciding what constitutes noise should derive from considerations about the experimental or clinical measurements and the known properties of the source of a signal. Ideally the biomedical engineer decides a priori the criteria for judging whether certain components of a signal represent the desired measurement or not. Then the signal processing method is chosen to enhance the desired signal and reduce any undesired signal components. That is not to say that the user should be able to visually identify the noise in the signal before any signal processing occurs; but he or she ought to be able to specify properties of the noise which will permit specification of signal processing parameters that will reduce it. In some cases this information may not be known a priori and it may be necessary to examine the results of the signal processing steps to assess whether the output signal exhibits some apparent separation into desired and noise components. The user is strongly cautioned, however, to try to understand this separation relative to the original signal through a process of "reversing" the signal processing steps, if at all possible. Any determination that a signal component is unwanted noise ideally should be a reflection of knowledge of the process under study. Signal processing should implement that knowledge rather than being the sole basis of its determination.

Example 1.6   In many cases it will be obvious that certain components of a signal are noise, and one will not even contemplate that this result is based on knowledge of the properties of the signal source. Consider, for example, a recording of

EMG activity from an intercostal muscle of the chest in Fig. 1.10. The ECG signal invariably is present within a chest muscle EMG recording, but we recognize from the shape of the waveform, from its regular timing that is asynchronous with muscle contraction, and from its amplitude, that the ECG is not part of the EMG. Note that those discrimination criteria are independent of the signal processing steps. (The reverse is not true, of course. The signal processing steps are likely to be designed to reduce the ECG signal components based on their waveshape and timing.)

It is important to reiterate two related points made earlier: First, just because a signal "looks like" noise, it may not be noise. An obvious possibility is that it has a fractal character or that it arises from a chaotic system. Second, by deciding what will be considered noise, the user may well have limited the choice of signal models that can reasonably be applied to the non-noise signal.

Example 1.7   Assume I have a measurement that appears to be a sine wave with a lot of added noise (Fig. 1.11(a)). With this model in mind I can devise a method of filtering that removes as much of the "noise" as I can. Probably the resultant filtered signal could reasonably be analyzed only as a sinusoid. On the other hand, I might assume that the original signal represents a filtered random signal that has a strong component at one frequency (Fig. 1.11(b)); therefore both the desired and noise components are random signals. Unfortunately, there are no easy guidelines to resolve dilemmas like this. One must incorporate other information about the process under study and even then it may be necessary to test several models of the signal

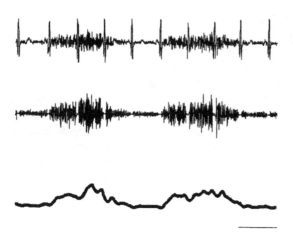

FIGURE 1.10   Example of a chest wall EMG during two breaths, showing contamination by the ECG signal (top). Signal processing can detect and remove the ECG signal (middle). After rectifying and lowpass filtering, one obtains a signal representing "amplitude" of the EMG (bottom). Short line (lower right) represents one second.

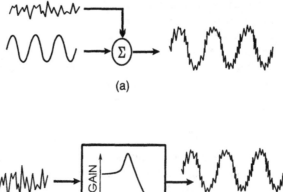

(a)

(b)

FIGURE 1.11 Two possible models of a noisy sinusoid-like signal: (a) as a pure sine wave plus added noise; (b) as white noise passed through a filter with a strong resonant behavior.

and try to judge which one is best. Once again, that type of assessment is an area of very active research.

Several examples of the need for noise removal by filtering were mentioned in the previous discussion of the sources of biomedical signals, although the causes of noise were not always discussed. Noise from electronic instrumentation is invariably present in biomedical signals, although the art of instrumentation design is such that this noise source may be negligible. Sometimes signals of interest are contaminated by signals of a similar type from another source. One example was given above: records of EMGs from muscles of the chest invariably contain an ECG signal which one wishes to minimize, usually by filtering. An extreme example of this type of noise contamination occurs during the recording of potentials from the scalp that are evoked by a brief sensory test stimulus such as a flash of light. Often the evoked potential is not even apparent in the recording because of the background EEG activity and a great deal of signal processing is necessary to permit visualization of it. Sometimes the motion of recording devices cannot be escaped and this motion adds a contaminating component to the signal being recorded. Other examples of unwanted signal components in biomedical applications are ubiquitous.

1.10 SUMMARY

This chapter has discussed the reasons for studying biomedical signal processing, presented some examples of biomedical signals, explained the different types of signals which appear in biomedical applications, established some basic definitions, and described a framework for the introduction and analysis of signal processing techniques. A signal is a representation of information as a function of an independent variable, which is often time. Signals may be either continuous-time or discrete-time. From the signals they acquire, biomedical engineers want either to extract particular information or to draw inferences about the properties of the sources of the signals. In other applications they need to determine the relationships between two signals. Signal processing is the manipulation of a signal for obtaining information from the signal, deriving an alternative representation of the signal, or ascertaining the relationships of two or more signals. Often the intent is to extract information by suppressing unwanted signal components, but the determination of what constitutes noise should be based on knowledge of the signal source, if possible. In some applications it is desired to predict future values of the signal.

There are four types of signals that might be encountered—(1) deterministic, (2) stochastic, (3) fractal, and (4) chaotic—and it is important to determine to which class a signal of interest should be assigned. Often, however, this task is very difficult because the methods for identification of fractal and chaotic signals are evolving.

Signal processing can be viewed in the same framework as systems analysis—the signal is the input to a "black box" which contains the rules for processing the signal, and the output is the desired information or derived signal representation. This framework would be considered the *analysis* model of signal processing. In addition, each signal processing method can be placed into a "signal modeling" or *synthesis* framework. In this framework the signal under study is the output and the inputs are the signal waveforms which, according to the rules of the signal processing method, are to be used to construct the original signal. The "black box" then contains the rules for this construction. We will utilize both frameworks in the ensuing chapters.

EXERCISES

1.1 Consider how each of the following situations can be put into a systems framework. Specify the input and output signals, and describe the contents of the system "box" (i.e., the transformation process):

a. A miniature, solid-state pH sensor is placed at the end of a catheter which is then inserted into an artery. Wires running through the catheter connect the sensor to an external amplifier to permit recording of the pH.

b. To determine sleep state a neurologist examines a 12-lead EEG recording to determine the predominant rhythms (called *alpha, beta, gamma,* and *delta*). She also looks at a chin EMG recording to see if continuous muscle activity is present and at EMG recordings from the extraocular muscles to detect rapid eye movements. From these observations, every 30 seconds she classifies the sleep state into one of six possible states which are referred to as: awake; stage I, II, II, or IV of nonREM; REM.

c. Nerve cells are grown in a culture dish. Every day the culture dish is exposed to an electromagnetic field of a particular frequency and intensity for four hours. Afterwards the lengths of any axonal processes are measured microscopically to see if the electromagnetic field promotes axonal growth.

1.2   As traffic passes by a research building, the floor vibrates. To minimize vibrations of a microscope sitting on it, a table in the building has each of its four legs sitting in a tub of sand. The viscous damping provided by the sand should reduce the vibration of the table top relative to that of the floor. To assess the degree of damping to be expected, a researcher wants to relate the displacement of the table top to that of the floor. Formulate this problem in a systems framework, identify the input and output signals, and discuss the system which relates these signals.

1.3   Draw a block diagram for both the analysis and synthesis signal processing models for each of the following:

a. Rectal temperature of a subject is measured continuously for four days and these data are approximated using four cycles of a sine wave having a period of 24 hours.

b. The floor vibrations in Exercise 1.2 are recorded with a sensitive accelerometer and found to have five main frequencies whose amplitudes each vary with time during the day. To model this signal it is represented as a summation of five sinusoids whose amplitudes may change every hour.

c. Instantaneous heart rate (IHR) is defined as the reciprocal of the duration of a beat. The heart rate of a subject in the Intensive Care Unit (ICU) is being monitored by averaging the IHR for all beats during each 5-minute time interval to obtain an index of the mean heart rate.

1.4   Consider the following signals and decide which components of each are information-bearing biomedical signals or noise. State your reasons for your choices:

a. A new biomedical sensor continuously measures the glucose level in the blood.

b. A set of ECG electrodes records the electrical activity of the heart and the electrical activity of respiratory muscles of the chest.

c. An ultrasound beam is detected by an ultrasound microphone after the beam reflects off a solid tumor.

d. A microelectrode implanted in the motor cortex of a monkey records action potentials from many neurons, but especially from one neuron that becomes active during a reaching movement with the right hand.

1.5   Classify each of the following signals as CT or DT signals and specify either an appropriate unit of time for each signal or the independent variable (if it is not time):

a. The instantaneous velocity of the left heel during a long jump.

b. The concentration of calcium inside a muscle cell.

c. The amount of blood ejected from the left ventricle with each heartbeat.

d. The number of red blood cells passing through a pulmonary capillary each second.

e. The average velocity of red blood cells in a pulmonary capillary.

f. The concentration of oxytocin in 5 ml samples of arterial blood taken every hour.

g. Pressure inside the eyeball.

h. The number of nerve cells in a thin slice of the brainstem, where each slice is taken from a different experimental animal (but from the same location).

i. The brightness of a light that is supplied with a current proportional to the work done by a subject peddling an exercise bicycle.

1.6 Classify each of these signals as deterministic, stochastic, fractal, or chaotic, and explain your reasoning. (There may be more than one correct answer for some.)

a. The signal from a blood glucose sensor after it is inserted into an artery.

b. The signal from a blood glucose sensor before it is inserted into an artery.

c. The signal from a pH meter whose electrode is in contact with gastric contents.

d. The heart rate signal of Fig. 1.2(e).

e. The intensity signal of Fig. 1.2(f).

f. The EMG signal of Fig. 1.2(b).

g. The three-dimensional coordinates of the path traced by the movements of a molecule of oxygen deep in the lung.

h. $x(t) = 0.5 \cos(6\Omega t) + 14 t\, u(t)$.

i. The voltage across the membrane of a muscle cell of the heart.

1.7 Consider the set of all functions which can be constructed by linear summations of scaled and delayed versions of the three basis functions shown in Fig. 1.12. Any such function can be represented in the form $f(t) = a\, x(t - \tau_1) + b\, y(t - \tau_2) + c\, z(t - \tau_3) + d$, where a, b, c, and d are constants. Evaluate these parameters (a, b, c, d, τ_1, τ_2, τ_3) for each of the three functions graphed in Fig. 1.12.

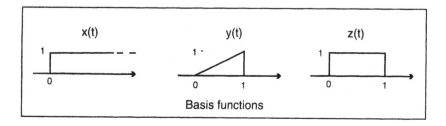

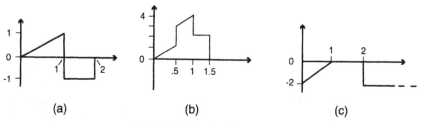

FIGURE 1.12 See Exercise 1.7.

1.8   In the steady state the liters per minute that a subject inhales during breathing (i.e., ventilation, symbolized as $\dot{V}_I$) is linearly related to the partial pressure of carbon dioxide in the arterial blood (symbolized as $PaCO_2$). To calculate this relationship a physician measures ventilation and $PaCO_2$ from a subject and plots the former versus the latter (Fig. 1.13). Because these data are noisy and do not fall on a perfectly straight line, he uses linear regression to fit a line through these data points. Draw an analysis model of this signal processing step, identifying the input signal and output(s), and describe the transformation processes which are occurring in the system. If you know about the procedure of linear regression, you should be able to give the equations of the transformation.

1.9   One tool for assessing the presence of lung disease is to measure the maximum expiratory airflow effort that a subject can make, starting from his maximum inspired volume. Fig. 1.14(a,b) shows simulated data from such a maximal expiratory flow maneuver. Theoretically the maximum flow is a function of the fluid mechanical properties of the lung and it is a decreasing function of lung volume. Therefore it is usual to plot airflow versus expired volume, as shown in Fig. 1.14(c), and to compare such flow–volume curves to typical curves from normal subjects. A common problem is the noise near the peak of the flow signal (Fig. 1.14(a)), which is "spread out" over a large initial part of the flow–volume curve. Furthermore, it is often difficult to filter this noise from the flow signal without also removing too much of the desired flow signal. Let's consider whether it is possible to filter the flow–volume curve directly.

a. If I digitize the flow and volume signals during the measurement (in this case, at 25 samples per second) and make a "discrete-time" plot from these data, I obtain the graph in Fig. 1.14(d). Explain why the data samples are not spaced uniformly along the volume axis.

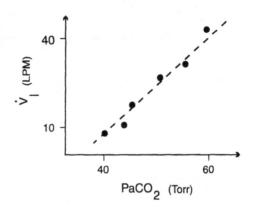

FIGURE 1.13   Steady-state ventilation of a human subject vs. partial pressure of CO_2 in arterial blood ($PaCO_2$). Solid circles: measured data. Dashed line: best fit line obtained from linear regression.

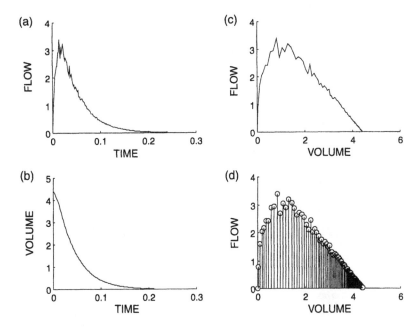

FIGURE 1.14 Simulated data from a maximum expiratory flow maneuver. (a) Air flow vs. time; (b) Lung volume vs. time; (c) Air flow vs. volume of air expired (CT); (d) Air flow vs. volume of air expired (sampled signals).

b. In theory, I could use interpolation of the data of Fig. 1.14(d) to generate samples of flow at uniformly spaced increments along the volume axis. Roughly sketch the "discrete" flow–volume plot obtained by interpolating at every 0.1-liter increment. Have I obtained a digitized signal that could be processed like any other "discrete-time" signal? Explain your answer.

2

MEMORY AND CORRELATION

2.1 INTRODUCTION

This chapter introduces the intimately related concepts of memory in physical systems and correlation in signals. The concept of *memory* in a physical system reflects the degree to which the present behavior (or output) is influenced by (or "remembers") past behaviors (inputs or outputs). Many physical systems exhibit memory. For example, if a pendulum is held at an angle to the vertical then released, it will oscillate transiently instead of dropping immediately to its rest position. The transient behavior reflects some memory of its initial potential energy. Memory in a physical system is an expression of the inability of the system to dissipate or redistribute its energy instantaneously. If the system is well understood, often one can relate this memory to specific parameters of the physical system. Thus there is potentially a great benefit from deriving a quantitative measure of memory based on observations of the output signal of a system, because such a measure might provide quantitative information about the parameters of an unknown system. Furthermore, as will be evident in Chapter 3, the input–output properties of a linear system can be almost completely described in terms of the memory properties of the system. On the other hand, virtually all methods of linear signal processing can be derived from the concept of correlation.

Correlation in a signal relates to the degree to which the signal at the present time reflects its values in the past. The similarity between these heuristic definitions of memory and correlation suggest the possibility of strong linkages between correlation properties of a signal and memory properties of the system from which the signal arises. Indeed, analysis of correlation in signals from physical systems provides a quantitative measure of memory, from which important properties of a physical system can be inferred.

This chapter begins by defining some basic properties of signal transformations and continues with an example and a discussion of memory in physical systems. Later the concept of correlation is defined mathematically and correlation analyses of several examples are presented. In this chapter the discussion will address deterministic signals only but the extension to stochastic signals in Chapter 9 will be straightforward.

2.2 PROPERTIES OF OPERATORS AND TRANSFORMATIONS

In Chapter 1 the input–output properties of a system were described by a transformation operation, $T[.]$, such that a system with input $x(t)$ and output $y(t)$ could be described by the relationship

$$y(t) = T[x(t)],$$

assuming that the system has no initial energy before x(t) is applied at the input. (This assumption, which is equivalent to assuming that the output and all internal variables are initially zero, is critical. The reason is that for a given $x(t)$ every different set of initial values of the system variables will produce a different output, $y(t)$.) All systems can be represented by transformation operators and we shall assume that any transformation operation represents a system. Consequently the terms "transformation operation" and "system" will be used interchangeably. It was also emphasized in Chapter 1 that signal processing could be viewed in this same "systems" framework—that is, the signal being "processed" would be considered the input to a transformation operation and the information or new signal derived from the signal processing procedure would be the output. Therefore both the analysis and the synthesis frameworks for signal processing can be expressed as transformation operations. Because transformation operations are so common, it is necessary to discuss some of their basic properties.

Transformations generally comprise one or more operations, many of which can be represented by mathematical operators. An *operator* expresses a mathematically well-defined action on a signal or group of signals. Operators may represent simple actions, such as the addition of two numbers, or complex actions, such as the integration of a function or a mapping of an n-dimensional signal from one function space to another. A familiar operator is the derivative operator, D. The representation "$Dx(t)$" is shorthand for dx/dt. The distinction between operators and transformations is not absolute, but often one envisions transformations as involving an ordered sequence of operators. To be more specific we consider an example.

Example 2.1 Transformation operator for a physical system   The act of breathing involves the reciprocating actions of bringing air into the lungs (inspiration) then expelling it (expiration). These actions can be described from a biomechanical perspective, as shown in Fig. 2.1(a). In a relaxed state with no airflow the pressure inside the lungs is atmospheric pressure, P_B. During inspiration the respiratory muscles—that is, the diaphragm and the rib cage muscles—contract and lower the pressure ($P_{pl}(t)$, or pleural pressure) around the lungs inside the chest.

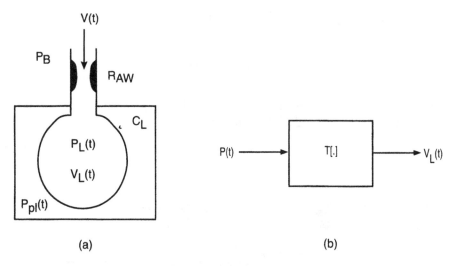

FIGURE 2.1. (a) Simple diagram of biomechanics of airflow into and out of the lungs. R_{AW} = airway flow resistance; C_L, lung compliance; P_B, atmospheric pressure; $P_L(t)$, $P_{pl}(t)$, pressures in the lungs and pleural space; $V_L(t)$, lung volume; $\dot{V}(t)$, airflow. (b) System representation of the dependence of lung volume, $V_L(t)$, on transrespiratory pressure, $P(t) = P_B - P_{pl}(t)$.

Because of their elasticity (described by their compliance, C_L, the inverse of elasticity), the transmural pressure drop across the lungs causes the lungs to expand, momentarily decreasing the pressure inside the lungs (represented as $P_L(t)$) and producing a pressure drop across the fluid flow resistance of the airways, R_{AW}, which results in airflow through the airways and into the lungs. As long as the muscles contract with progressively increasing vigor, this process repeats and inspiration (i.e., airflow into the lungs) continues. At the end of inspiration the expiratory muscles of the abdomen and rib cage may contract, forcing $P_{pl}(t)$ to be greater than P_B, which drives air out of the lungs. (In resting breathing the expiratory muscles do not contract, but relaxation of the inspiratory muscles allows $P_{pl}(t)$ to increase above P_B.)

This biomechanical process can be described mathematically from the physical properties of the structures involved. Like an electrical resistor, flow through the airway depends on airway resistance and the pressure drop across this resistance, so that

$$P_B - P_L(t) = \dot{V}(t)R_{AW},$$

where $\dot{V}(t)$ is the flow rate. Like an electrical capacitor, the compliance of the lungs relates the lung volume to the transmural pressure, as

$$P_L(t) - P_{pl}(t) = \frac{V(t)}{C_L}.$$

Note that airflow and lung volume are related by

$$V(t) = V(0) + \int_0^t \dot{V}(s)ds.$$

Since it is a constant, we can assume that $V(0) = 0$. The first two equations can be combined to yield a first-order differential equation

$$P_B - P_{pl}(t) = \dot{V}(t)R_{AW} + \frac{V(t)}{C_L}, \qquad t \geq 0.$$

For simplicity, let $P(t) = P_B - P_{pl}(t)$. Then the general solution of this differential equation (with $V(0) = 0$) is

$$V(t) = \frac{1}{R_{AW}} \int_0^t e^{-(t-s)/R_{AW}C_L} P(s)ds. \tag{2.1}$$

One can visualize this process from a systems viewpoint (Fig. 2.1(b)), with the output $V(t)$ resulting from the processing of the input signal $P(t)$, as shown in the above equation. In transformation notation,

$$V(t) = T[P(t)], \qquad \text{where } T[x(t)] = \frac{1}{R_{AW}} \int_0^t e^{-(t-s)/R_{AW}C_L}[x(s)]ds.$$

From the preceding example it is clear that system transformations may operate on functions and produce other functions as their output. A transformation, $T[.]$, is said to be *additive* if it satisfies the condition

$$T[x_1(t) + x_2(t)] = T[x_1(t)] + T[x_2(t)]$$

for all functions $x_1(t)$ and $x_2(t)$. $T[.]$ is said to be *homogeneous* if it satisfies the condition

$$T[ax(t)] = aT[x(t)]$$

for all values of the real scalar a. The transformation $T[.]$ is a *linear transformation* *if it is both additive and homogeneous*. A system is linear if its transformation operation is linear. Consequently, if $T[.]$ represents the input–output relationship of a system (under the assumption that all variables of the system are initially zero), then the criterion for linearity of the system is that

$$T[ax_1(t) + bx_2(t)] = aT[x_1(t)] + bT[x_2(t)] \tag{2.2}$$

for all real scalars (a, b) and all functions $x_1(t)$ and $x_2(t)$. Equation (2.2) is often termed the *superposition condition*. Note that if we let $a = -b$ and $x_2(t) = x_1(t)$, then $T[0] = 0$ for a linear system. That is, a linear system cannot have a persisting nonzero output if its input is a function that always has a value of zero.

Example 2.2 The system described by $y(t) = 10\,x(t-2) + 1$ is not linear. To prove this result it is sufficient to find one violation of the additivity or homogeneity conditions. Scaling $x(t)$ by 2, for example, does not scale $y(t)$ by 2 because of the "+1" term. Consider the system given by $y(t) = t\,x(t)$. Even though this system has a time-varying coefficient multiplying $x(t)$, one can show that it satisfies both of the conditions necessary for linearity.

Example 2.3 The biomechanical system for ventilating the lungs, described above, is linear. To show this result, choose two arbitrary input signals, $P_1(t)$ and $P_2(t)$, and let $P(t)$ be the weighted sum of these two signals, $P(t) = aP_1(t) + bP_2(t)$. By Eq. (2.1) the output $V(t)$ is

$$V(t) = \frac{1}{R_{AW}} \int_0^t e^{-(t-s)/R_{AW}C_L}[aP_1(s) + bP_2(s)]\,ds$$

$$= a\frac{1}{R_{AW}} \int_0^t e^{-(t-s)/R_{AW}C_L}P_1(s)\,ds + b\frac{1}{R_{AW}} \int_0^t e^{-(t-s)/R_{AW}C_L}P_2(s)\,ds$$

$$= aV_1(t) + bV_2(t),$$

where $V_1(t)$ and $V_2(t)$ are the outputs for the individual inputs $P_1(t)$ and $P_2(t)$. For systems such as this one that are described by differential equations, the system is linear (in the sense given above) if and only if the differential equation is linear.

Example 2.4 A common method for smoothing a discrete-time (DT) signal is to pass it through a filter that averages several consecutive values of the input signal. Let the input signal be $x[n]$ and the output $y[n]$. In the case of three-point smoothing, this process can be described by the difference equation

$$y[n] = \frac{1}{3}\{x[n-2] + x[n-1] + x[n]\}.$$

This equation is already in the form $y[n] = T[x[n]]$ and it is easy to show that this system is both additive and homogeneous, and therefore linear. Thus

$$T[ax_1[n] + bx_2[n]] = \tfrac{1}{3}\{ax_1[n-2] + ax_1[n-1] + ax_1[n] + bx_2[n-2] + bx_2[n-1]$$

$$+ bx_2[n]\} = aT[x_1[n]] + bT[x_2[n]].$$

(At several points in this book other properties of such filters, known as "moving averagers," will be analyzed.)

A transformation operation, $y(t) = T[x(t)]$, is *memoryless* if for all times t_1, $y(t_1)$ depends only on $x(t_1)$ and not on $x(t)$ at any other time. If a transformation is not memoryless, it is said to have *memory*. For example, the system represented by the relationship $y(t) = 5\,x(t) + 4$ is memoryless; however, the lung biomechanical system above has memory (because the output is determined by integration over past values of the input). Of course, a moving average filter also has memory.

A transformation is *time-invariant* if shifting of the input in time by an amount t_0 only has the effect of shifting the output in time by the same amount. That is, if $T[x(t)] = y(t)$, then $T[x(t - t_0)] = y(t - t_0)$. Equivalently, one can describe a time-invariant system by saying that the input–output relationship is independent of the chosen time origin. For example, the system $y(t) = x(t) - y(0)$ is neither time-invariant nor linear (unless $y(0) = 0$).

A transformation (and the system it represents) is *causal* if the output at any time depends on the input at that time or at earlier times but not on the input at future times. Specifically, if $y(t) = T[x(t)]$, then a transformation is causal if for any chosen time, t_0, $y(t_0)$ depends on $x(t)$, $t \le t_0$ only. Loosely speaking, for a causal system there can be no response to an input before that input occurs. Equivalently, the output cannot anticipate the input. The condition of causality is important for physical realizability of a system but is less important for digital signal processing systems that have all of the input signal stored in memory and do not operate in "real" time. The lung biomechanical system above is an example of a causal system because the integration to determine $V_L(t)$ involves the input signal up to and including, but not beyond, time equal to t. The two-point moving-average system described by the difference equation

$$y[n] = \tfrac{1}{2}(x[n] + x[n + 1])$$

is not causal.

Many causal systems can be described by linear differential equations of the form

$$a_N \frac{d^N y}{dt^N} + a_{N-1} \frac{d^{N-1} y}{dt^{N-1}} + \cdots + a_1 \frac{dy}{dt} + a_0 y(t) = b_0 x(t) + b_1 \frac{dx}{dt} + \cdots + b_M \frac{d^M x}{dt^M}, \quad (2.3)$$

where $\{a_0, a_1, \ldots, a_N, b_0, \ldots, b_M\}$ are scalar constants and $M < N$. If an equation of the form of Eq. (2.3) completely describes the input–output relationship of a system, and if it is not possible to do so after removing the N-th derivative term on the left-hand side, then the system is said to have an *order* equal to N. Similarly, the smallest value of N for which the input–output relationship of a discrete-time system can be written in the form

$$y[n + N] = \sum_{k=0}^{N-1} a_k y[n + k] + \sum_{m=0}^{M} b_m x[n + m] \quad (2.4)$$

is the order of the system.

2.3 MEMORY IN A PHYSICAL SYSTEM

With a basic understanding of fundamental properties of transformations and systems, we will now explore in depth the concept of memory as it applies to physical systems and mathematical descriptions of such systems.

Thought Problem: Braking an Automobile

Consider an automobile moving with an initial velocity, v_0, to which a constant braking force, f_b, is applied beginning at time $t = 0$. The braking force is maintained until the car stops. (It is assumed that the driver also releases the accelerator pedal so that the engine force, f_e, falls to zero immediately at $t = 0$.) A simplified diagram of this situation is presented in Fig. 2.2. Analysis of the motion of the car can be cast in a systems framework, in which the velocity of the car is related to the braking force by the transformation

$$\dot{x}(t) = T[f_b],$$

by writing and solving the differential equation of motion. The equation of motion for this system is

$$M\ddot{x} = -C_f\dot{x} - f_b, \tag{2.5}$$

where M is the mass of the car, C_f is the coefficient of friction, and f_b is the braking force.

Letting $y = \dot{x}$, $t_0 = 0$, $\dot{x}(0) = v_0$, one can write the above equation as

$$\dot{y} = -\frac{C_f}{M}y - \frac{1}{M}f_b. \tag{2.6}$$

Substituting $\tau = M/C_f$, which is the time constant of this system, the general solution for this first-order differential equation may be written as

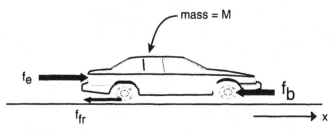

FIGURE 2.2. Simplified scheme of forces acting on a car. f_b, braking force; f_e, engine force; f_{fr}, friction force.

$$y(t) = v_0 e^{-t/\tau} + \int_0^t \left(\frac{-f_b}{M} \right) e^{-(t-s)/\tau} ds, \qquad t \geq 0$$

$$= v_0 e^{-t/\tau} - \frac{1}{M} f_b \left[\frac{M}{C_f} \right] e^{-C_f t/M} [e^{C_f s/M} |_0^t]$$

$$= v_0 e^{-C_f t/M} - \frac{f_b}{C_f} [1 - e^{-C_f t/M}], \qquad t \geq 0. \tag{2.7}$$

Consider the case when the engine force is removed but no braking force is applied—that is, $f_b = 0$. Then from Eq. (2.7) the velocity as a function of time is

$$\dot{x}(t) = y(t) = v_0 e^{-C_f t/M}, \qquad t \geq 0. \tag{2.8}$$

Note that for any time $t_1 > 0$, the velocity relative to the initial velocity is given by

$$\frac{\dot{x}(t_1)}{\dot{x}(0)} = e^{-C_f t_1/M} \doteq D(t_1). \tag{2.9}$$

That is, at $t = t_1$ the "residual" velocity is a constant fraction of the initial velocity, independent of the initial velocity, and that fraction is a function of the system parameters, M and C_f. Furthermore, this "memory" of the initial velocity decreases as time increases. A more remarkable expression of this property is apparent from taking the ratio of the velocities at any two times, t_1 and t_2:

$$\frac{\dot{x}(t_2)}{\dot{x}(t_1)} = e^{-C_f(t_2-t_1)/M} = D(t_2 - t_1). \tag{2.10}$$

This ratio depends only on the time difference and not on the actual times. That is, for a given *time difference* (i.e., $t_2 - t_1$) the memory of the past velocity is a constant fraction of the past velocity independent of the actual times, t_1 and t_2. One concludes that this memory of the past velocity must be an intrinsic property of the system because it applies equally to the initial velocity and all subsequent velocities. What does this property reveal about the physical system? Apparently the velocity of the physical system cannot be changed to an arbitrary value (in the absence of an externally applied force such as a braking force). To understand this statement consider that the velocity is one measure of the state of the system, where the *system state* is given by the set of simultaneously sampled values of all the system variables. The system state is determined by the distribution of energy in the system. In the absence of external forces, in order for the system state to change, either some energy has to be dissipated or the energy in the system has to be redistributed. In the present example all of the energy is present as kinetic energy and after the engine force is removed, it is being dissipated by the friction. This dissipation can not occur at an arbitrary rate (which would be necessary to change velocity instantly to an arbitrary value). The rate of dissipation has a simple time dependence that can be derived from the properties of the system and the basic definition of energy:

$$E = \int f\,dx = \int (M\ddot{x})(\dot{x}\,dt).$$

For the present example the energy of the car, $E(t)$, as a function of time is its initial energy at time t_0 (i.e., E_0) plus the change in energy as it is dissipated by friction. Using Eq. (2.8), the preceding equation for energy becomes

$$E(t) = E_0 + \int_0^t M\left(\frac{-C_f}{M}v_0e^{-C_ft/M}\right)(v_0e^{-C_ft/M})dt$$

$$= E_0 + (-C_fv_0^2)\int_0^t e^{-2C_ft/M}dt = \frac{1}{2}Mv_0^2[1-1+e^{-2C_ft/M}]$$

$$= \frac{1}{2}Mv_0^2e^{-2C_ft/M}, \qquad \text{where } E_0 = \frac{1}{2}Mv_0^2. \tag{2.11}$$

Therefore,

$$E(t) = E_0e^{-2C_ft/M} = E_0D(t).$$

The first term on the right side of the first equation in Eq. (2.11) is the initial kinetic energy of the automobile, and the second term is the cumulative energy dissipation (which will be negative). On a semilog plot (Fig. 2.3(a)) it is apparent that $\ln E(t)$ decreases linearly with time. Note from Eq. (2.11) that the magnitude of the

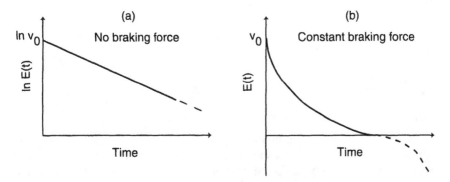

FIGURE 2.3. Log of energy vs. time when the car of Fig. 2.2 decelerates (a) without any braking force; and (b) with a constant braking force. Dashed curve in (b) indicates energy of reverse motion that would result if the constant "braking force" continued beyond the time when the velocity reached zero in the +x direction.

slope is proportional to the reciprocal of the system time constant. This relationship is a single straight line because this system has only one mechanism for storing energy—the kinetic energy of the car—and one mechanism for dissipating energy—the friction of the wheels. If there were two energy storage mechanisms (e.g., if the car were towing a trailer connected to the car by a viscously damped hitch), then $E(t)$ would contain two exponential terms and $\ln E(t)$ would not be linear but would approach linearity asymptotically (with different slopes) at $t = 0$ and infinity. This latter system would exhibit a more complicated form of memory that would reflect both energy dissipation and energy transfer between the car and trailer, and the velocity would not exhibit the single decaying exponential relationship of Eq. (2.8). *Thus memory in a physical system is simply an expression of the inability of the system to dissipate or redistribute its energy instantaneously.* This memory (sometimes called a "trace") of the energy distribution of the past is reflected in the time courses of the variables of the system.

Formally, a system is said to be *memoryless* if its output at any time t depends only on the value of its input(s) at the same time. In other words, there is no memory of the previous state of, or inputs to, the system. Systems which are not memoryless have memory. A system will have no memory if it has no mechanisms which store energy. In fact, it is difficult to design a physical system that has no memory, although any system describable by algebraic (and not differential) equations is memoryless. In practice, some systems may respond so quickly compared to others under study that the former can be considered to be memoryless on the time scale of study. For example, responses of the nervous system are so fast compared to body movements during locomotion that one often considers the neural components of locomotor control systems to be memoryless in the sense defined above. (The irony of assuming the nervous system to be memoryless should be obvious!)

Example 2.5 Do the following systems have memory?

(a) $y = x^2 + x + 1$;

(b) $z(t) = \dfrac{1}{C} \displaystyle\int_0^t i(t)dt$;

(c) $\dfrac{dy}{dt} = -ay(t) + bu(t)$;

(d) $y[n] = aw[n] + bw[n-1]$.

System (a) is algebraic and has no memory. $z(t)$ in system (b) depends explicitly on a summation (integration) of past values of $i(t)$ and therefore has memory. In system (c) the derivative cannot be evaluated from only the current value of $y(t)$. One needs also the value of $y(t - \delta)$ for some small δ. Therefore, a system described by a differential equation has memory. Finally, the output of system (d)—that is, $y[n]$—depends explicitly on the immediate past value of the input, $w[n - 1]$, as well as its current value, $w[n]$, and thus has memory.

Memory in the Presence of External Inputs

So far we have considered memory in the context of the dissipation or redistribution of preexisting energy in a system. How does memory affect the addition (or removal) of energy from a system by external inputs? Consider the general solution to the braking problem when f_b is constant (Eq. (2.7)):

$$\dot{x}(t) = v_0 e^{-C_f t/M} - \frac{f_b}{C_f}[1 - e^{-C_f t/M}]$$

over the interval $0 \leq t \leq t_z$, where t_z is the time at which the velocity of the car reaches zero, at which time we assume the driver ceases braking and f_b becomes zero. It is easy to show that

$$t_z = \frac{M}{C_f} \ln\left[1 + \frac{v_0 C_f}{f_b}\right].$$

The first term on the right of Eq. (2.7) is the *zero-input response*, which we have already discussed as the "memory" function $D(t)$. The zero-input response expresses the output due to initial conditions when the input signal is zero for all time. The specific zero-input response depends on the chosen initial conditions but, for a linear system, it always can be expressed as a weighted sum of (possibly complex) exponential functions that are invariant for a given system. It is a direct expression of the memory of the system. The remaining terms in Eq. (2.7) constitute the *zero-state response*—that is, the response of the system to an external input when its initial conditions are all zero. The zero-state response is an expression of the transformation operation, $T[.]$, of this system which (see Eq. (2.7)) would be expressed, for an arbitrary input force $f_b(t)$, as

$$\dot{x}(t) = T[f_b(t)] = \int_0^t \left(-\frac{f_b(s)}{M}\right) e^{-(t-s)/\tau} ds.$$

Both the amplitude and the form of the zero-state response depend on the specific input signal applied. It is apparent that this response also depends on the memory properties of the system and that the memory properties are expressed in both $D(t)$ and $T[f_b(t)]$ through the time constant, $\tau = M/C_f$.

The braking force in this example acts to dissipate energy in the same manner as the friction force—by causing the system to utilize some of its energy to move the opposing force through a distance "dx". One can solve for the energy of the car as was done above but using the general solution for d^2x/dt^2 obtained by differentiating Eq. (2.7). As shown in Fig. 2.3(b), $E(t)$ falls more quickly, as expected, but the decline is not a mono-exponential. The dotted line represents the mathematical solution for the theoretically realizable (but impossible) case in which the "braking force" continues to be applied after the velocity reaches zero. In this situation the car would start to move backwards and would achieve a steady-state velocity at which f_b is exactly counteracted by the force of friction. The magnitude of the energy becomes negative by convention because the force is acting along the negative direction of the x-axis.

It is apparent, then, that memory not only influences the dissipation of energy in the zero-input case, but also influences the rate at which energy in the system is altered by external forces. This result is heuristically tenable for the present example because in both cases physical objects have to be accelerated or decelerated and it is intuitive that such actions cannot occur instantaneously in the absence of infinite forces. Furthermore, *it is important to note that this result can be generalized to energy considerations for non-mechanical physical systems—for example, electrical circuits, chemical reactions.* In these latter cases one utilizes analogous definitions of energies and forces. In electrical circuits, for example, voltage can be analogous to force and charge to displacement. Thus electrical energy becomes the integral of voltage times the differential of charge, or

$$E(t) = \int v \, dQ = \int v(t) i(t) \, dt, \tag{2.12}$$

where $v(t)$ is voltage and $i(t)$ is current. Note also that by defining power to be the time rate of change of energy, we obtain the familiar result that instantaneous power equals voltage multiplied by current.

Example 2.6   Memory in a DT system   Consider the discrete-time system described by the input–output difference equation

$$y[n] - ay[n-1] = x[n], \qquad |a| < 1, \qquad y[-1] = y_0$$

To find the memory function, $D[n]$, of this system, one first determines the general solution for $y[n]$. Thus

$$y[0] = ay[-1] + x[0] = ay_0 + x[0]$$
$$y[1] = ay[0] + x[1] = a^2 y_0 + ax[0] + x[1],$$
$$y[2] = ay[1] + x[2] = a^3 y_0 + a^2 x[0] + ax[1] + x[2],$$
$$\vdots$$

from which it follows that

$$y[n] = a^{n+1} y_0 + \sum_{k=0}^{n} a^{n-k} x[k]. \tag{2.13}$$

As in Eq. (2.7), Eq. (2.13) contains both a zero-input response and a zero-state response. The zero-input response is $y_{zi}[n] = a^{n+1} y_0$, which has the form $y[n] =$

$D[n]y[0]$, where $D[n] = a^{n+1}$. Note that since $D[n]$ is determined by the zero-input response of a system, one only needs to solve for the output with $x[n] = 0$ (but with a nonzero initial condition on the output) in order to determine $D[n]$.

It may be apparent that the memory function $D(t)$ (or $D[n]$) is just the homogeneous solution of the differential (or difference) equation of a linear system. That is, the memory function expresses the influence of nonzero conditions at $t = 0$ on the future course of the system output. Although the exact function depends on the specific initial conditions, for a given linear system the memory function can always be expressed as a linear combination of the same (possibly complex) exponential functions for any set of initial conditions. The following example illustrates this point.

Example 2.7   Memory in a continuous-time (CT) system   A linear second-order system can be described by a differential equation of the form

$$\ddot{y}(t) + a\dot{y}(t) + by(t) = cx(t),$$

where (a, b, c) are arbitrary, but real, fixed parameters. The homogeneous solution can be found by setting the input to zero and assuming the form of the solution is

$$y_h(t) = C_1 e^{S_1 t} + C_2 e^{S_2 t}.$$

The parameters (C_1, C_2, S_1, S_2) can be determined by substituting the expression for $y_h(t)$ into the original differential equation and collecting all terms in S_1 (or S_2) and setting their sum equal to zero. The C_1 (and C_2) terms cancel and for both S_1 and S_2 one obtains the same equation:

$$S_i^2 + aS_i + b = 0, \qquad \text{for } i = 1, 2.$$

Thus,

$$S_i = -\frac{a}{2} \pm \frac{1}{2}\sqrt{a^2 - 4b}.$$

S_1 and S_2 are called the *eigenvalues* of the system. C_1 and C_2 are found by matching $y_h(t)$ and $dy_h(t)/dt$ to the given initial conditions, $y(0)$ and $\dot{y}(0)$. Thus, as asserted above, for any initial conditions the homogeneous solution for this system is a linear combination of (the same) two complex exponential functions.

Because of its dependence on the specific initial conditions, the homogeneous solution (i.e., the zero-input response) is not optimal as a quantitative measure of system memory. A more useful measure, the impulse response of a system, will be introduced in the next chapter. It is based on the observation noted earlier that the zero-state response also reflects the memory properties of the system.

2.4   ENERGY AND POWER SIGNALS

A force, $f(t)$, acting on a body imparts energy (which may be negative) to that body. Does it make sense to ask, "What is the energy of $f(t)$?" To ponder this question, consider again the car of Fig. 2.2. Let the car be initially at rest, with $E_0 = 0$, and let the engine produce a constant force, $f_e(t) = F$, $t \geq 0$. For simplicity, assume there is no friction. The energy imparted to the car by the engine is

$$E(t) = \int_0^t f_e(t)\dot{x}(t)dt = \int_0^t F\left(\frac{Ft}{M}\right)dt = \frac{F^2 t^2}{2M} \quad \text{since } \ddot{x}(t) = \frac{F}{M}. \quad (2.14)$$

$E(t)$ is proportional to $f_e^2(t)$ and, in the general case, if $f_e(t)$ is time-varying but is approximated as having a constant value over each small time interval, Δt, then ΔE during that time interval will be proportional to the square of the force. But from Eq. (2.14) the energy that $f_e(t)$ can impart to the car depends also on the mass, M, and therefore is not unique. Consequently, to assign a unique "energy" value to a force function, $f(t)$, it is customary to assume that $M = 1$. This same line of reasoning can be applied to electrical circuits to motivate the definition of electrical energy of a voltage signal to be the energy imparted to a 1 Ω resistor when the voltage is applied across the resistor. Generalizing from this reasoning, the *total energy* of a signal, $x(t)$, is defined as

$$E = \int_{-\infty}^{\infty} |x(t)|^2 dt = \int_{-\infty}^{\infty} x^*(t)x(t)dt \quad (2.15)$$

or, in the case of a real-valued signal,

$$E = \int_{-\infty}^{\infty} x^2(t)dt. \quad (2.16)$$

A signal $x(t)$ is said to be an *energy signal* if its total energy is finite and non-zero—that is,

$$0 < E = \int_{-\infty}^{\infty} |x(t)|^2 dt < \infty. \quad (2.17)$$

For example, the signal $x(t) = e^{-2t} u(t)$ is an energy signal since

$$E = \int_0^{\infty} e^{-4t}dt = \frac{1}{4}.$$

On the other hand, the signal $z(t) = e^{-4t}$, $-\infty < t < \infty$, is not an energy signal. However, any signal that is bounded and is nonzero only for a finite time is an energy signal.

A signal $x(t)$ is called a *power signal* if its average power is finite and nonzero, where average power, P, is defined as

$$P_x = \lim_{T_0 \to \infty} \left[\frac{1}{T_0} \int_{-T_0/2}^{T_0/2} |x(t)|^2 dt \right]. \tag{2.18}$$

Energy signals are not power signals because their average power is always zero. Likewise, power signals cannot be energy signals. Some signals, such as $z(t)$ above, are neither energy signals nor power signals. By convention, however, bounded signals that have finite duration (although they are energy signals) are also considered power signals because their average power *during their finite interval of observation* is nonzero and finite. In essence, one is assuming that the observed portion of the signal represents its behavior for all time, and therefore one may calculate power without letting T_0 approach infinity. For a bounded signal $x(t)$ observed on the interval (t_1, t_2), average power is defined as

$$P_x = \frac{1}{t_2 - t_1} \int_{t_1}^{t_2} |x(t)|^2 dt. \tag{2.19}$$

Periodic signals can be considered power signals if their average power determined over one cycle is nonzero and finite. Thus, for a periodic signal $x(t)$ with period T_1,

$$P_x = \lim_{T \to \infty} \frac{1}{T} \int_{-T/2}^{T/2} |x(t)|^2 dt = \frac{1}{T_1} \int_0^{T_1} |x(t)|^2 dt. \tag{2.20}$$

Note that Eq. (2.20) will be finite for any bounded periodic signal.

Equivalent definitions for energy and power signals exist for discrete-time (DT) signals. Thus a DT signal $x[n]$ is an energy signal if its energy, E, is nonzero and finite—that is,

$$0 < E = \sum_{n=-\infty}^{\infty} |x[n]|^2 < \infty. \tag{2.21}$$

For example, the signal $x[n] = a^n u[n]$, where $0 < |a| < 1$, is an energy signal because

$$E = \sum_{n=-\infty}^{\infty} |x[n]|^2 = \sum_{n=0}^{\infty} a^{2n} = \frac{1}{1 - a^2} < \infty.$$

Similarly, a DT signal $x[n]$ is a power signal if

$$0 < P_x = \lim_{N \to \infty} \frac{1}{2N + 1} \sum_{n=-N}^{N} |x[n]|^2 < \infty. \tag{2.22}$$

DT signals can be periodic and in such cases power is defined analogously to Eq. (2.20) as the summation of the squared magnitude of the signal over one period divided by the length of the period.

Example 2.8   Consider the periodic signal $x(t) = 10 \cos(4\pi t + \pi/16)$, which has a period $T_1 = 0.5$. Letting $A = 10$, the average power of this signal is

$$P_x = \frac{1}{T_1} \int_0^{T_1} A^2 \cos^2(4\pi t + \pi/16)dt = \frac{A^2}{4\pi T_1} \int_0^{2\pi} \cos^2(\theta)d\theta = \frac{A^2}{2} = 50.$$

Note that P_x is independent of the frequency of the cosine signal.

Example 2.9   To find the average power of $z[n] = \cos[9\pi n]$, first we must determine the period of $z[n]$. The period, N_1, must satisfy the condition that $9\pi N_1 = 2\pi k$, where both N_1 and k are integers. That is, $N_1 = 2k/9$, which has the solution $N_1 = 2$ when $k = 9$. Therefore the average power of $z[n]$ is

$$P_x = \frac{1}{N_1} \sum_{k=0}^{N_1-1} |z[k]^2| = \frac{1}{2}(\cos^2[0] + \cos^2[9\pi]) = \frac{1}{2}(1 + 1) = 1.$$

2.5   THE CONCEPT OF AUTOCORRELATION

The Deterministic Autocorrelation Function

The concept of memory was derived principally from the result in Eq. (2.10), which expresses the dependence of velocity at time t_2 on that at a preceding time t_1. Here we shall derive a more generalizable form of this expression starting from this previous result. Using Eq. 2.10 and letting $y(t)$ be an arbitrary system variable, one can write that $y(t_2) = D(t_2 - t_1)y(t_1)$. Multiplying by $y(t_1)$ we obtain

$$y(t_2)y(t_1) = D(t_2 - t_1)y(t_1)y(t_1). \tag{2.23}$$

Assume that, for practical situations, a signal is observed only for a finite length of time, T_0. Letting the time t_1 be an arbitrary time t and setting $s = t_2 - t$, one obtains

$$y(t + s)y(t) = D(s)y^2(t).$$

Now divide both sides of this equation by T_0 and integrate from $t = 0$ to $t = T_0$. Thus

$$\frac{1}{T_0} \int_0^{T_0} y(t + s)y(t)dt = D(s)\left[\frac{1}{T_0} \int_0^{T_0} y^2(t)dt \right]. \tag{2.24}$$

The term in the brackets on the right of Eq. (2.24) is the mean square value of $y(t)$, represented as msv(y). The left-hand side of Eq. (2.24) is defined as the *deterministic autocorrelation function* of y, denoted as $R_y(s)$. Note that the dependence on t is integrated out of this equation by the definite integral, so that the autocorrelation is a function of s only. The argument, s, of the autocorrelation function is usually referred to as the *lag*, since it represents the amount of time that $y(t)$ "lags behind" $y(t + s)$. Since $D(0) = 1$, $R_y(0) = \text{msv}(y)$ and

$$R_y(s) = D(s)R_y(0). \tag{2.25}$$

The interpretation of Eq. (2.24) and, therefore, of $R_y(s)$, is straightforward. The term on the left-hand side is the average of the product of y at time t and y at time $t + s$, taken over all of the time interval of observation. (The observant reader will recognize that $y(t + s)$ will not be defined for $t > T_0 - s$. We will assume that $y = 0$ whenever its argument lies outside the interval $[0, T_0]$.) $R_y(s)$ therefore expresses the degree to which $y(t)$ and $y(t + s)$ change in a similar fashion, on average. If, for example, for a given value of s, $y(t + s)$ always lies above zero whenever $y(t)$ is positive and always lies below zero when $y(t)$ is negative, then their product is always positive and their average product will be relatively large and positive. If the opposite is true, then their average product will tend to be large and negative. If, on the other hand, $y(t + s)$ is equally likely to be either positive or negative for any value of $y(t)$, then the integration of their product will tend toward a value of zero. Therefore, knowledge of $R_y(s)$ for many values of s provides considerable insight into the properties of $y(t)$. It should be noted, however, that although each $y(t)$ gives rise to a unique $R_y(s)$ through Eq. (2.24), there may be many functions, $y(t)$, that produce the same $R_y(s)$. Therefore, the inverse mapping from $R_y(s)$ back to $y(t)$ is not unique.

Since the term in brackets on the right-hand side of Eq. (2.24) is a constant, remarkably we find that the dependence of $R_y(s)$ on s is determined solely by $D(s)$, the function that expresses the memory of the system. That is, the autocorrelation function of $y(t)$ quantifies the memory of the system from which $y(t)$ arose. This result is intuitively compatible with the interpretation of $R_y(s)$, discussed in the preceding paragraph. Furthermore, this result is quite general and will be extended to many types of signals. In general, one does not know the function $D(s)$ of a system except for simple cases; however, one always can calculate $R_y(s)$ for any finite-length signal $y(t)$, $0 \le t \le T_0$, from the relationship

$$R_y(s) = \frac{1}{T_0} \int_0^{T_0} y(t + s)y(t)dt. \tag{2.26}$$

After calculating $R_y(0)$, one could determine $D(s)$ from Eq. (2.25). (For a system of second order or higher, for these results to be directly applicable one must evaluate $D(s)$ by setting the initial condition on the output to one and other initial conditions to zero.) Usually one omits this last step and works with $R_y(s)$ directly, since this function includes additional information about $\text{msv}(y)$

If $y(t)$ is the output of a system and $R_y(s)$ is nonzero for some $s > 0$, one cannot presume that this indication of memory in $y(t)$ necessarily reflects properties of the system. If the input to the system exhibits memory, this feature of the input is likely to propagate to the output. Memory in the input signal, however, is likely to be modified by the system. Therefore, memory in the output signal can be due both to system properties and to memory in the input signal. In Chapters 8 and 9 we will discuss important methods of system identification that are based on the relationship of output memory to input memory and the dependence of this relationship on system properties.

Properties of Deterministic Autocorrelation Functions

$R_y(s)$ has three important properties:

1. $R_y(-s) = R_y(s)$. That is, it is an even function of s. This result can be derived from Eq. (2.26) by first substituting $-s$ for s, then replacing t by $t + s$.
2. $|R_y(s)| \le |R_y(0)|$. This property is slightly more difficult to establish. Consider first that for any real values of the constants (a,b) the following must be true:

$$[ay(t + s) + by(t)]^2 \ge 0.$$

Expand this equation and divide by b^2. One obtains

$$\left(\frac{a}{b}\right)^2 y^2(t + s) + 2\left(\frac{a}{b}\right) y(t + s)y(t) + y^2(t) \ge 0.$$

Now, after dividing all terms by T_0 and integrating both sides from $t = 0$ to $t = T_0$, we get

$$\left(\frac{a}{b}\right)^2 R_y(0) + 2\left(\frac{a}{b}\right) R_y(s) + R_y(0) \ge 0,$$

which is a quadratic equation in the variable a/b with coefficients $R_y(0)$ and $R_y(s)$. Let $c = a/b$. Since a and b can have any values, so can c. That is, for any value of c this quadratic equation must be greater than or equal to zero. One can show that in order to meet this condition the equation cannot have two nonidentical real roots for c, or otherwise there will be some extent of the c-axis for which the value of the equation is negative. The equation will not have two nonidentical real roots if the discriminant is less than or equal to zero, or

$$4R_y^2(s) - 4R_y(0)R_y(0) \le 0.$$

Bringing the second term to the right-hand side and taking the square root establishes the above property.

3. The autocorrelation operation is not necessarily linear. That is, if $z(t) = x(t) + y(t)$, then $R_z(s) \ne R_x(s) + R_y(s)$ necessarily. An intuitive understanding of this property can be obtained by noting that in applying Eq. (2.26) to $z(t)$ there will be a term involving the product $x(t)y(t)$ which will not appear in either $R_x(s)$ or $R_y(s)$. If this term happens to evaluate to zero (and it does in some important cases), then the autocorrelation function of $z(t)$ will be the sum of the autocorrelation functions of $x(t)$ and $y(t)$.

Although most practical signals will be observed for a finite time, many theoretical signals are not so constrained. In this case, one defines the deterministic auto-correlation function as a limit, as

$$R_y(s) = \lim_{T_0 \to \infty} \left[\frac{1}{T_0} \int_{-T_0/2}^{T_0/2} y(t + s)y(t)dt \right]. \qquad (2.27)$$

For $s = 0$, the right-hand side of Eq. (2.27) (if it converges) is exactly the definition of the power of a signal. Therefore, from property 2 above, autocorrelation functions exist for power signals. Periodic signals are special cases of power signals and, analogously to Eq. (2.20), the autocorrelation function for a periodic signal, $y(t)$, with period P is

$$R_y(s) = \frac{1}{P} \int_0^P y(t + s)y(t)dt. \qquad (2.28)$$

In this case we do not assume that $y(t)$ is zero outside of the interval $[0, P]$, since it is given that $y(t)$ is periodic. Therefore, $R_y[s]$ can be calculated for $s > P$, and consequently periodic signals have periodic autocorrelation functions.

For energy signals the limit in Eq. (2.27) will be zero for all s. It is sometimes convenient to define an autocorrelation function for such signals which omits the division by T_0 in the bracketed term of Eq. (2.27) (see Exercise 2.8). While this definition of the autocorrelation function also is valid, one must be consistent about which definition is used in any given application.

From the right-hand side of Eq. (2.24) it is apparent that $R_y(s)$ is determined by two independent measures, msv(y) and $D(s)$. Since it is the variation of $R_y(s)$ with s that provides insight regarding the influences of memory processes on a signal, it is common to discuss the normalized autocorrelation function $r_y(s)$, where $r_y(s) = R_y(s)/\text{msv}(y)$. Since msv($y$) = $R_y(0)$, property 2 above implies that $|r_y(s)|$ is less than or equal to one.

Example 2.10   Autocorrelation functions   Determine the autocorrelation functions for the following signals: (a) $x(t) = u(t) - u(t - 3)$, $0 \le t \le 9$; (b) $z(t) = e^{-2t} u(t)$.

(a) $x(t)$ is shown in Fig. 2.4(a). By definition,

$$R_x(s) = \frac{1}{3} \int_0^3 x(t + s)x(t)dt$$

and for $s = 0$, $R_x(0) = 1$. For $0 \le t \le 3$, Fig. 2.4(a) indicates that $x(t + s)$ and $x(t)$ overlap for $0 \le t \le 3 - s$. Therefore,

$$R_x(s) = \frac{1}{3} \int_0^{3-s} (1)(1)dt = \frac{3-s}{3} = 1 - \frac{s}{3}.$$

For $s > 3$, $R_x(s) = 0$. Recall that $R_x(-s) = R_x(s)$. Finally, $R_x(s)$ is plotted in Fig. 2.4(b).

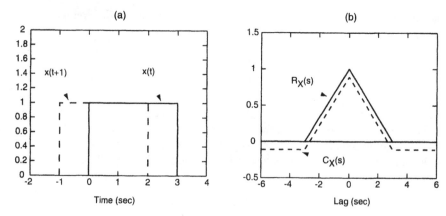

FIGURE 2.4. (a) $x(t)$ and $x(t + 1)$ for the function $x(t) = u(t) - u(t - 3)$; (b) Autocorrelation and autocovariance functions for $x(t)$.

(b) Since $z(t)$ is defined for all $t \geq 0$, $R_Z(s)$ is determined from Eq. (2.27). But since $z(t)$ has a finite value when integrated from zero to infinity, $R_Z(s) = 0$ for all s. If we use the alternative definition (i.e., without division by T_0), then $R_Z(s) = \frac{1}{4}e^{-2s}$, $s \geq 0$.

The Autocorrelation Function of Zero-Mean White Noise

Consider a signal, $w(t)$, that has the following properties: (1) its average value (for sufficiently large T_0) is zero; (2) its value at each point in time is random; (3) its value at each point in time is independent of its value at any other time. Such a signal is called *zero-mean white noise*. To evaluate $R_w(s)$, $s > 0$, first note the following:

$$R_w(s) = \frac{1}{T_0} \int_0^{T_0} w(t + s)w(t)dt = \frac{1}{T_0} \int_0^{T_0-s} w(t + s)w(t)dt, \qquad (2.29)$$

since we assume that $w(t) = 0$ outside of the interval of observation. Since $w(t)$ is random, for a fixed s, $w(t + s)$ is equally likely to be greater than or less than zero, and when one integrates $w(t + s)w(t)$ over a sufficiently large interval, $T_0 - s$, that integral is very likely to be zero. So for large (including infinite) intervals, zero-mean white noise has an autocorrelation of zero (except when $s = 0$, of course, when $R_w(0) = \text{msv}(w)$). The question of how large an interval is necessary in order to assure that the calculated autocorrelation function will be close to zero is considered later in this chapter.

The Deterministic Autocovariance Function

The autocorrelation function has one complicating feature when the mean level of the signal is not zero. To envision this complication, consider the simplest such signal, $x(t) = A$, where A is a constant, and $x(t)$ is nonzero only on the interval $[0, T_0]$.

Thus, by Eq. (2.26), $R_x(s) = A^2$, for $s \ll T_0$. Similarly, if $x(t) = A + f(t)$, where $f(t)$ is an arbitrary function with a mean level of zero, then

$$R_x(s) = \frac{1}{T_0} \int_0^{T_0} [A + f(t + s)][A + f(t)]dt = A^2 + R_f(s), \qquad \text{for } s \ll T_0. \quad (2.30)$$

That is, the autocorrelation of $x(t)$ is the square of its mean level plus the autocorrelation of the deviations of $x(t)$ about its mean level. Since it is easy to calculate the mean level of a signal, since a sufficiently large A^2 will obscure the dependence of $R_x(s)$ on s, and since the mean level of a signal contributes only an offset to the autocorrelation, therefore it is common to remove the mean level of a signal before determining its autocorrelation function. To distinguish this calculation from the autocorrelation function, it is referred to as the *deterministic autocovariance function*, $C_x(s)$. That is, the autocovariance function is a measure of the memory in the *deviations of x(t) about its mean level*. For a signal $x(t)$ observed on the interval $[0, T_0]$,

$$C_x(s) = \frac{1}{T_0} \int_0^{T_0} [x(t + s) - \bar{x}][x(t) - \bar{x}]dt, \qquad (2.31)$$

where

$$\bar{x} = \frac{1}{T_0} \int_0^{T_0} x(t)dt \triangleq \langle x(t) \rangle$$

is the mean level of $x(t)$. $C_x(s)$ and $R_x(s)$ are related by

$$R_x(s) = C_x(s) + \bar{x}^2. \qquad (2.32)$$

If $x(t)$ has a non-zero mean level and $y(t)$ is defined as $y(t) = x(t) - \bar{x}$, then

$$R_y(s) = C_y(s) = C_x(s) = R_x(s) - \bar{x}^2.$$

Example 2.11   To determine the autocovariance function for the example of Fig. 2.4 one can subtract the squared mean of the signal (i.e., 1/9) from the autocovariance function. The autocovariance function is shown as the dashed line in Fig. 2.4(b).

Example 2.12   The mean level of $x(t) = A \sin(\Omega_0 t)$ is zero and $R_x(s) = C_x(s)$. Let $P = 2\pi/\Omega_0$. Then

$$R_x(s) = \frac{1}{P} \int_0^P A^2 \sin(\Omega_0(t + s)) \sin(\Omega_0 t)dt$$

$$= \frac{A^2}{P} \int_0^P [\sin(\Omega_0 t) \cos(\Omega_0 s) \sin(\Omega_0 t) + \cos(\Omega_0 t) \sin(\Omega_0 s) \sin(\Omega_0 t)]dt$$

$$= \frac{A^2}{P} \frac{1}{\Omega_0} \int_0^{2\pi} [\sin^2(\theta) \cos(\Omega_0 s) + \cos(\theta) \sin(\theta) \sin(\Omega_0 s)]d\theta$$

Therefore,

$$R_x(s) = \frac{A^2}{2}\cos(\Omega_0 s).\tag{2.33}$$

Note that the autocorrelation and autocovariance functions of a sinusoid are cosines of the same frequency as the sinusoid, having an amplitude that is one-half of the squared amplitude of the original signal. Since $\text{msv}(x) = R_x(0)$, the mean square value of a sine wave is $A^2/2$. Previously we determined that the power of a zero-mean sinusoid of amplitude A is also $A^2/2$. Therefore, $R_x(0)$ (which equals $C_x(0)$) equals the power of a zero-mean sine or cosine wave. This important result will be generalized later.

The three properties described above for $R_x(s)$ also apply to $C_x(s)$. One can also define a normalized autocovariance function, $c_x(s)$, such that

$$c_x(s) = C_x(s)/C_x(0),\tag{2.34}$$

with the result that $|c_x(s)| \le 1$.

2.6 AUTOCOVARIANCE AND AUTOCORRELATION FOR DT SIGNALS

The concepts of autocovariance and autocorrelation are easily extended to DT signals. Let $x[n]$ be a bounded DT signal observed on the interval $[0, N_0 - 1]$. Its deterministic autocorrelation function is

$$R_x[m] = \frac{1}{N_0}\sum_{k=0}^{N_0-1} x[k+m]x[k]\tag{2.35}$$

and its autocovariance function is

$$C_x[m] = \frac{1}{N_0}\sum_{k=0}^{N_0-1} \{x[k+m]-\bar{x}\}\{x[k]-\bar{x}\},\tag{2.36}$$

where

$$\bar{x} = \frac{1}{N_0}\sum_{k=0}^{N_0-1} x[k].$$

As in the CT case, we assume $x[n] = 0$ whenever its argument is outside of the interval $[0, N_0 - 1]$. Consequently, the upper limit on the summations could be replaced by $N_0 - 1 - m$. The three properties discussed for $R_x(s)$ and the relationship of Eq. (2.32), apply also in the DT case. $C_x[m]$ and $R_x[m]$ will exist for discrete-time power signals but not for energy signals, as discussed above. Both functions will be periodic with period N_1 if $x[n]$ is periodic with period N_1.

Example 2.13 Autocorrelation function of a DT signal The signal $x[n] = n\{u[n] - u[n-4]\}$ is a truncated ramp which is nonzero for $0 \le n \le 3$ (Fig. 2.5). Therefore, $N_0 = 4$. Its mean level is

$$\bar{x} = \frac{1}{N_0} \sum_{k=0}^{3} x[k] = \frac{1}{4} \sum_{k=0}^{3} n = 1.5$$

and

$$R_x[m] = \frac{1}{N_0} \sum_{k=0}^{3} x[k+m]x[k] = \frac{1}{4}[x[m]x[0] + x[m+1]x[1] + x[m+2]x[2]$$

$$+ x[m+3]x[3]]$$

$$= \frac{1}{4}[x[m+1] + 2x[m+2] + 3x[m+3]], \qquad m \ge 0$$

Substituting specific values for m we find

$$R_x[0] = \tfrac{1}{4}(1 + 4 + 9) = 3.5$$

$$R_x[1] = \tfrac{1}{4}(2 + 6 + 0) = 2$$

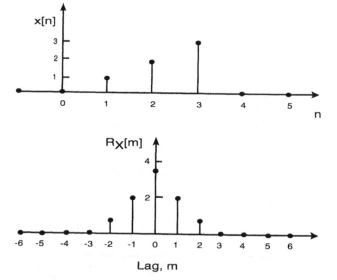

FIGURE 2.5. The function $x[n] = n\{u[n] - u[n-4]\}$ and its autocorrelation function.

$$R_x[2] = \tfrac{1}{4}(3 + 0 + 0) = 3/4$$

$$R_x[m] = 0, \qquad m \geq 2.$$

Recalling that $R_x[-m] = R_x[m]$ allows one to plot the autocorrelation function (Fig. 2.5). This same result can be obtained from a graphical approach. The method is shown in Fig. 2.6 for the case $m = 2$. First one plots $x[m + k]$ and $x[k]$ aligned vertically, then these two functions are multiplied to generate the function $x[m + k] x[k]$. This last function is summed from $k = 0$ to $k = N_0 - 1$ and the sum is divided by N_0. By visualizing this process for each value of m it is possible to sketch the general shape of $R_x[m]$ without actually calculating its values. Often knowledge of this shape is an important starting point of an analysis and *the student is strongly urged to practice this semiquantitative, graphical approach until it becomes second nature* (see Exercises 2.13 and 2.14).

Example 2.14 Autocovariance functions of white noise sequences   The MAT-LAB function randn was used to generate three sequences of a discrete-time white noise signal, $w[n]$, having lengths $N_0 = 64$, 512, and 4096. The normalized autocovariance functions of these sequences were obtained via Eq. (2.36) and are plotted in Fig. 2.7. From the longest sequence it is apparent that $c_w[m] = 0$ for $m > 0$, except for small random variations (whose amplitudes would diminish progressively as N_0 increases above 4096). For $N_0 = 64$ there are too few data points for the averaging in Eq. (2.36) to produce a result that is consistently close to zero. Even for $N_0 = 512$ the averaging is insufficient, although on a visual basis one might conclude that the

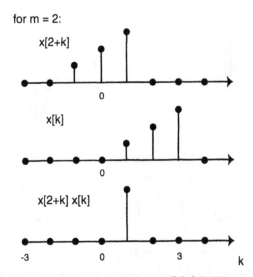

FIGURE 2.6. Graphical approach for evaluating $R_x[m]$, in this case for $m = 2$.

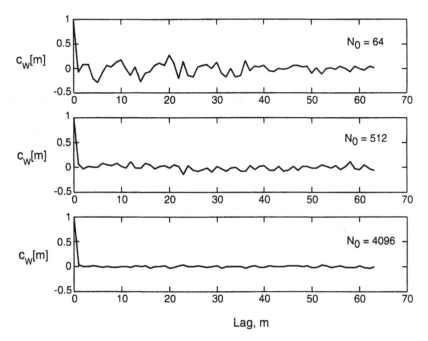

FIGURE 2.7. Autocovariance functions for three sample records of zero-mean white noise. Record lengths were $N_0 = 64$, 512, and 4096 data points, respectively.

autocovariance "is expected to be" zero for $m > 0$. Compare this example with the example of Fig. 9.4, in which the mean value of white noise is estimated from sequences of various lengths.

Example 2.15 Biomedical example   Autocovariance functions have an important role in the study of biomedical signals and systems. Consider, for example, the question of whether each breath that you take is independent of other breaths. This question is important to investigators who are attempting to define the neural networks in the brainstem which interact to generate the oscillatory rhythm of breathing. Besides studying the neurophysiology of respiratory neurons in the brainstem, the usual approach is to implement a mathematical model of a neural network that simulates the physiological neurons and their hypothesized interactions. If the model does not produce an oscillation like the respiratory rhythm, then it is rejected as an appropriate model of the physiological system. Even if it oscillates, to be acceptable the model must exhibit other behaviors that can be measured physiologically. All such models to date have exhibited the property that each breath is very nearly independent of other breaths. Does the physiological system behave this way?

To answer this question define a discrete-time variable $v[n]$ which is the volume which a subject inspires on the n-th breath in a sequence of N breaths. A typical time series of such data from an awake, resting human subject is shown in Fig. 2.8(a). It is apparent that $v[n]$ varies from breath to breath. To determine whether the value of v on breath n is independent of v from breaths $n-1, n-2, n-3, \ldots$, the MATLAB routine xcov was used (with the "biased" option) to calculate $C_v[m]$, which is plotted in Fig. 2.8(b). Several features can be noted. First the maximum of $C_v[m]$ is at zero lag, as expected. Second, the autocovariance is positive for lags up to approximately 25. Third, for lags > 25, $C_v[m]$ is negative. Fourth, since the length of the data sequence is $N = 120$, $C_v[m]$ is zero for $|m| > 120$. Note also that the summation (Eq. 2.36) involves only $N_0 - 1 - m$ product terms, whereas the multiplier in this equation, $1/N_0$, is independent of m; therefore as m approaches N_0, $C_x[m]$ will necessarily approach zero. From the plotted autocovariance function one concludes that if one breath volume is above the mean breath volume, then the next (approximately) 25 breaths are more likely, on average, to be above the mean breath volume also. This conclusion contradicts the hypothesis that each breath is generated independently of every other breath. Therefore, it is likely that the mathematical models still lack the physiologically important features that are responsible for the nonzero autocovariance observed in this data set.

Aside: To address the above problem one could have defined a new variable having a mean level of zero—for example, $z[n] = v[n] - \frac{1}{N}\sum_{k=0}^{N-1}v[n]$—and then calculated $R_z[m]$. $R_z[m]$ would equal the autocovariance function, $C_v[m]$, plotted in Fig. 2.8.

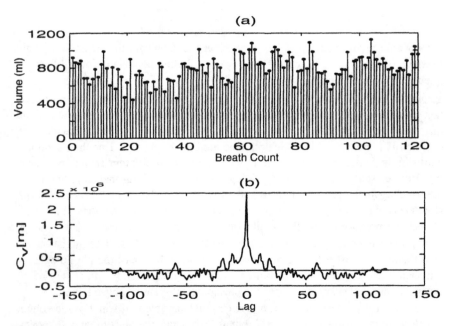

FIGURE 2.8. (a) Breath volumes of 120 consecutive breaths from a human subject; (b) Autocovariance function of the breath volume data considered as a DT signal. *Note:* Points at discrete lags in this graph are connected by straight lines.

2.7 SUMMARY

This chapter has introduced the concepts of memory in physical systems and correlation in signals. Memory is a system characteristic that reflects the ability of the system to increase, decrease, or redistribute its energy. It is dependent on the number (and size) of energy storage mechanisms in the system and on the mechanisms by which energy may be dissipated, absorbed, or transferred from one energy store to another. The time constants of the system are direct measures of the temporal properties of energy redistribution in a system. Because the rate of energy redistribution is limited by the system time constants, system variables (which are determined by the distribution of energy in the system) reflect memory properties in their time courses.

Energy signals are signals that have finite, nonzero energy and power signals are ones that have finite, nonzero power. Some signals are neither energy signals nor power signals. Bounded periodic signals are always power signals whose power is defined as their average squared value over one cycle. The deterministic autocorrelation function always exists for a power signal. In cases where a signal is observed (and is assumed to be nonzero) only for a finite time interval, the signal can be considered a power signal, even though its power is zero by the formal definition, as long as power is calculated only over the interval of observation.

The autocorrelation function, $R_x(s)$ and $R_x[m]$, (and the related measure, the autocovariance function) represents the average degree to which a signal varies in relation to its past values, and is a function of the time lag into the past. A positive autocorrelation at a particular lag, s, implies that an increase in the signal at time t is more likely to be followed by an increase at time $t + s$ than by a decrease. (But keep in mind that autocorrelation is a measure of the *average* relationship and "more likely" does not imply "always".) The opposite is true for a negative autocorrelation value. The autocorrelation of a signal contains information about the memory in the system from which the signal arose.

Autocovariance, $C_x(s)$ or $C_x[m]$, is the autocorrelation function that is determined after removing the mean level from a signal. It represents the correlation in a signal due to the deviations of the signal from its mean level. In fact, $C_x(s)$ and $C_x[m]$ represent all of the correlation and memory properties in a signal.

There is a close relationship between the correlation properties of a signal and the memory properties of the system that generated the signal and one may use the autocovariance (or autocorrelation) function as a quantitative measure of system memory. For many physical signals autocovariance and autocorrelation approach zero for very large lags, reflecting the property that memory in a physical system often does not persist forever. Determination of autocorrelation and autocovariance functions is a fundamental tool for biomedical signal processing.

EXERCISES

2.1 Classify the following systems as memoryless or having memory. In all cases, x is the input and y is the output.

a. $y(t) = x(t - 1)$
b. $y(t) = x^2(t) + 0.3x(t)$
c. $y[n] - 0.1\, y[n - 1] = x[n]$
d. $y[n] = \cos[0.6\, |x[n]|]$
e. $y[n] = nx[n]$

2.2 Derive Eq. (2.7) as the solution to Eq. (2.6).

2.3 Show that $z(t) = e^{-4t}$ is not an energy signal.

2.4 Determine whether each of the following signals is an energy signal, a power signal, or neither.

a. $x[n] = u[n]$
b. $x[n] = (-0.3)^n\, u[n]$
c. $x(t) = ae^{j\Omega_0 t}$
d. $x(t) = t[u(t) - u(t - 2)],\ 0 < t < 2$
e. $x[n] = \dfrac{n^2}{2}\{u[n] - u[n - 45]\}\ 0 \le n \le 44$, otherwise $x[n] = 0$

2.5 Determine the memory function (i.e., the homogeneous solution) for the systems described by the following input–output differential or difference equations:

a. $5\dfrac{dg(t)}{dt} + 2g(t) = u(t)$

b. $0.2\ddot{z}(t) + 0.01\dot{z}(t) + z(t) = 2x(t)$
c. $q[n] - 0.8\, q[n - 1] + w[n] = 0$
d. $c[n] = 0.1\, x[n - 2] + 0.2\, x[n - 1] + 0.4\, x[n] + 0.2\, x[n + 1] + 0.1\, x[n + 2]$

2.6 Calculate the power of the following periodic signals:

a. $s(t) = 0.2\sin(31\pi t) + 0.5\sin(62\pi t) + 0.1\sin(124\pi t)$

b. $z[n] = \displaystyle\sum_{k=1}^{2} a_k \cos\left[\left(\dfrac{2\pi k}{200}\right) n\right]$, where a_k, $k = 1, 2$, are constants.

2.7 Find the fundamental period of each of the cosine functions of Exercise 2.6(b).

2.8 Determine the autocorrelation function for $y(t) = Be^{-bt}u(t)$ using the alternative definition (i.e., without dividing by T_0 and taking the limit).

2.9 Prove Eq. (2.30).

2.10 Show that property 2 for the autocorrelation function also is true for the autocovariance function.

2.11 Referring to the description in this chapter of the biomechanical properties of the lungs, a common situation is the need to ventilate a patient artificially. Some ventilators work by applying a constant positive (i.e., greater than P_B) pressure at the airway opening to force air into the lungs (inflation) to mimic an inspiration, then open the airway to the room (via an electronically controlled valve) to allow the lungs to empty (deflation) without assistance, mimicking expiration. Figure 2.9 shows a simple diagram and biomechanical model. The model is similar to that of Fig. 2.1, except that the compliance, C, now represents the compliance of the lungs and chest together. Typical values for a normal subject are: $R = 4$ cm $H_2O/L/s$, $C = 0.10$ L/cm H_2O.

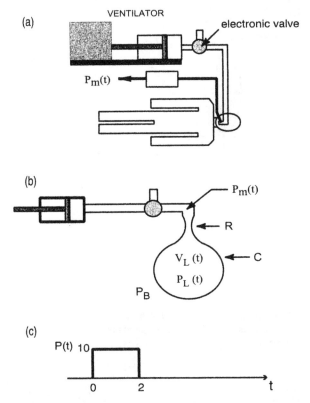

FIGURE 2.9. (a) Simplified diagram of a patient connected to a ventilator by way of a tube that is inserted into the airway (i.e., the trachea). An electronic valve connects the patient's airway either to the ventilator to permit inflation of the lungs, or to the room to permit the inflated lungs to deflate while the ventilator resets itself for another inflation; (b) Simple biomechanical diagram of the ventilator and patient. Symbols the same as in Fig. 2.1, except: C, total compliance of lungs and chest; $P_m(t)$, pressure at the site of the tracheal tube; (c) Pressure pulse applied by ventilator; $P(t)$ is a measure of the amount by which $P_m(t)$, produced by the ventilator, exceeds atmospheric pressure.

   a. Write a differential equation to describe the dependence of $V_L(t)$ on $P_m(t)$ when the lungs are being inflated by the ventilator. Let $P(t) = P_m(t) - P_B$ and rewrite this equation in terms of transrespiratory pressure, $P(t)$. Describe the memory properties of this system.
   b. During deflation $P_m(t) = P_B$. Assuming that the lungs reached a final volume V_T during inflation, write a differential equation to describe $V_L(t)$ during deflation.
   c. Assume that the single pressure pulse shown in Fig. 2.9(c) as $P(t)$ is applied by the ventilator. Solve for $V_L(t)$, $0 < t < 4$ s.
   d. Sketch $V_L(t)$, $0 < t < 12$, if the above pressure pulse is applied: (i) every 4 s; (ii) every 3 s.

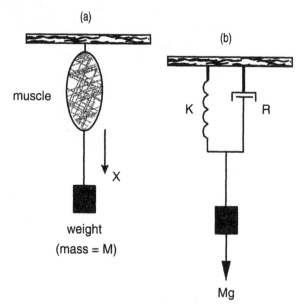

FIGURE 2.10. (a) A frog muscle is suspended from a rigid support and a mass is suspended from the muscle by a thread; (b) Simple biomechanical model of the muscle and weight. The muscle is modeled as a spring with spring constant, K, in parallel with a viscous resistance, R.

2.12 An excised frog muscle is suspended from a rigid support (Fig. 2.10). A weight with mass M is attached to the muscle and suddenly released. Consider this situation as a system with the input being the constant gravitational force, Mg, and the output being the position of the weight, x. Biomechanically the passive (i.e., not contracting) muscle can be represented as a spring in parallel with a viscous resistance.

a. By considering the balance of forces on the mass, M, write a differential equation whose solution will be the position, x, as a function of time.

b. Evaluate the two components of the homogeneous solution of the above differential equation in terms of the parameters R, K, and M. Discuss the memory properties of this system.

2.13 By visualizing $x(t)$ and $x(t + s)$ for various values of s, draw a rough sketch of $R_X(s)$ for each of the following functions. Indicate approximate values of s at which minima or maxima of $R_X(s)$ occur.

a. $x(t) = 2u(t) - 2u(t - 3)$

b. $x(t) = \begin{cases} t, \ 0 \le t < 2 \\ 0, \ \text{otherwise} \end{cases}$

c. $x(t) = 4 \cos(20\pi t)$

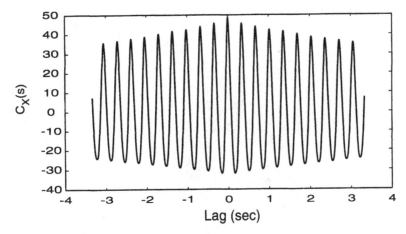

FIGURE 2.11. Autocovariance function of the first respiratory airflow signal from Fig. 9.6(a).

2.14 By visualizing $x[n]$ and $x[n + m]$ for various values of m, draw a rough sketch of $R_x[m]$ for each of the following functions. Indicate approximate values of m at which minima or maxima of $R_x[m]$ occur. Finally, calculate the exact autocorrelation functions to check your answers.

a. $x[n] = u[n] - u[n - 2]$
b. $x[n] = 2\,u[n] - u[n - 1] - u[n - 2]$
c. $x[n] = n^2\,\{u[n] - u[n - 4]\}$.

2.15 Fig. 9.6(a) presents eight examples of recordings of respiratory airflow patterns. The autocovariance function of the first signal of Fig. 9.6(a) is presented in Fig. 2.11. Interpret this result relative to possible memory properties of the physiological system that controls respiratory airflow. Estimate quantitative measures of memory from this graph.

2.16 Figure 2.12 shows the autocovariance functions for the first, second, sixth, and eighth examples of respiratory airflow recordings from the file ranproc2.mat (see Fig. 9.6(a)). Recalling that each example is from a different subject, interpret these results relative to possible memory processes in the physiological system that controls respiratory airflow and their similarities in different subjects. Give quantitative measures of memory.

2.17 The file hrv1.mat contains a DT signal representing the beat-by-beat instantaneous heart rate in beats per minute (i.e., 60/beat duration) obtained from a resting human subject. This heart rate signal and its autocovariance function are shown in Fig. 2.13. Interpret the autocovariance function with respect to possible memory properties of the physiological system controlling the heart rate.

2.18 The file hrv.mat contains 10 records (organized in a matrix of 10 columns with 256 entries per column) of heart rate versus heart beat number from 10 differ-

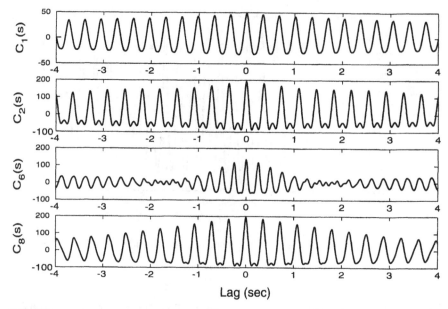

FIGURE 2.12. Autocovariance functions of four respiratory airflow signals from Fig. 9.6(a), which are included in the file `ranproc2.mat`.

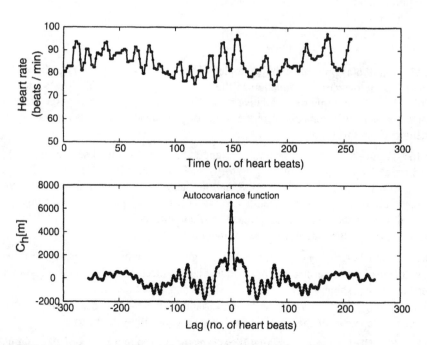

FIGURE 2.13. A heart rate signal from `hrv1.mat` and its autocovariance function. Both are DT functions that are graphed as discrete points joined by straight lines.

ent observations from the same human subject. Use the MATLAB functions xcov and xcorr to calculate the autocovariance and autocorrelation functions for the signals in the file hrv.mat. Compare these results and interpret these functions with respect to possible memory properties of the physiological system controlling the heart rate.

3

THE IMPULSE RESPONSE

3.1 INTRODUCTION

In Chapter 1 we discussed the fact that the transformation operator of a system describes the relationship between the input and output signals of the system. That chapter also described the fundamental theme of this text that any signal processing algorithm can be viewed as a transformation operator. The input to the transformation is the signal we wish to "process" and the output provides either the information that we wish to extract from the signal or an alternative representation of the signal. It is not necessary that transformations representing signal processing algorithms satisfy the conditions for linearity, but very many do. For example, any type of linear filtering can be framed as a signal processing operation that generates an alternative representation of the signal—for example, one with reduced noise components. Consequently, the study of linear systems and their properties is fundamental to much of signal processing.

In Chapter 2 it was stated that the input–output relationship for any linear, time-invariant system can be represented as a finite-order differential (or difference) equation. Assuming that the initial conditions of the system are zero, *then the closed-form equation for the zero-state solution of the differential (or difference) equation can be considered the transformation operator for the signal processing operation that the system is performing.* To understand this assertion, consider that the zero-state solution is simply the means of specifying how the system is to generate a modified, or alternative, representation of the input signal via the mathematical operations that appear in the equations of the zero-state solution. That is, where the control systems engineer sees a system as a black box whose output is "driven by" its input, the signal processing engineer sees the same system as an ordered set of mathematical operations which generate an alternative representation of the input signal. Given this framework, a question naturally arises: Is it possible to determine the output of this linear signal processing system for an arbitrary but well-specified input by some method other than directly solving the zero-state equation? The answer is yes and the method to be developed in this chapter has the advantage that it provides detailed insights into how the transformation operator acts to modify (or "process") the time-domain input signal.

Unless otherwise stated, we will consider only linear time-invariant (LTI) systems (or, in the case of DT systems, linear shift-invariant (LSI) systems). In these cases, if for a given input $x(t)$ one knows the output, $T[x(t)]$, then because of linearity one also knows the output for any scaled version of the input, $x_1(t) = ax(t)$. Similarly, one also knows the output for any time-shifted input of the form $x_2(t) = x(t - t_0)$. Conceivably one could construct a table of such known inputs and outputs. From this table one could determine the output for any input that can be represented as a sum of these inputs. Such a process would be cumbersome, of course, and generally is impractical. In contrast, the concept to be developed below, the impulse response of a system, will provide a compact and elegant solution to the problem of determining the output response to an arbitrary input.

3.2   THOUGHT EXPERIMENT AND COMPUTER EXERCISE: GLUCOSE CONTROL

The purpose of this exercise is to probe more deeply into the question of how to determine the output of a system in response to an arbitrary input, especially in the common biomedical situation in which the system transformation operator is not known. This exercise is based on the SIMULINK model glucos3.mdl, which is a simulation of the regulation of blood glucose concentration by insulin in a diabetic adult human (Fig. 3.1). The SIMULINK model assumes that the patient has no capability to secrete insulin and therefore it is necessary for the patient to receive injections or intravenous infusions of insulin at appropriate times in order to maintain blood glucose near the normal level of 4.5 mmol/L. For convenience, we will assume that the patient has a programmable insulin pump (an infusion controller) that is connected to an intravenous catheter. Three questions need to be answered:

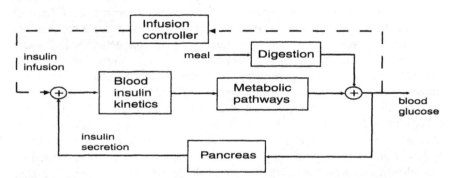

FIGURE 3.1. A simple model illustrating the control of blood glucose concentration by pancreatic insulin secretion. When the feedback loop fails in disease, an artificial feedback loop, shown by the dashed line, may be instituted. Equations for the feed-foreward loop are given in the text. Based on the model in Furler et al. (1985).

(1) What is the steady-state rate of insulin infusion that should be delivered in order to maintain a normal blood glucose level in the absence of disturbances? (2) If a blood glucose reading is found to be abnormal, how long must one wait to determine the effect of the new infusion rate that the physician will request? (3) If one can anticipate a disturbance, such as a meal, can one design an insulin infusion protocol to minimize the effect of the disturbance on glucose concentration?

The SIMULINK model implements five differential equations:

1. $$\frac{dG}{dt} = [P_1 - X(t)]G(t) - P_1 G_0 + \frac{F_G}{V_G};$$

2. $$\frac{dX}{dt} = P_2 X(t) + P_3 [I_F(t) - I_0];$$

3. $$\frac{dI_F}{dt} = \frac{F_1(t)}{V_1} - nI_F(t) - \frac{dI_{B1}}{dt} - \frac{dI_{B2}}{dt};$$

4. $$\frac{dI_{B1}}{dt} = k_{a1}I_F[C_{T1} - I_{B1}] - k_{d1}I_{B1}; \text{ and}$$

5. $$\frac{dI_{B2}}{dt} = k_{a2}I_F[C_{T2} - I_{B2}] - k_{d2}I_{B2}, \qquad (3.1)$$

where

G = blood (plasma) glucose concentration;
X = insulin concentration in a "remote" compartment;
I_F = free plasma insulin concentration;
I_{B1}, I_{B2} = concentration of insulin bound to antibodies with high and low affinity for insulin, respectively;
F_G = exogenous glucose load—for example, due to eating a meal;
F_I = rate of infusion of insulin;
I_0 = baseline insulin level (set to zero in this simulation).

Other entries in the equations are various parameters of insulin and glucose kinetics. Their values in glucos3.mdl have been adjusted to reflect a diabetic patient with no natural insulin secretion who is also partially "glucose resistant." Additional information and instructions for use are included in the file glucos3.mdl and the reader is urged to reproduce the simulations discussed below. This simulation is based largely on the model of Furler et al. (1985).

By running the simulation with the insulin infusion rate, F_I, set to zero, we observe that the steady-state value of G with no insulin control is approximately 10 mmol/L, or about twice the desired level. Through trial and error one can establish that a steady insulin infusion rate, F_I, of 0.9 U/hr will bring the glucose concentration to the normal level (Fig. 3.2). Note from the figure that the return of G to normal takes about four hours after the infusion is begun. Consuming a meal has

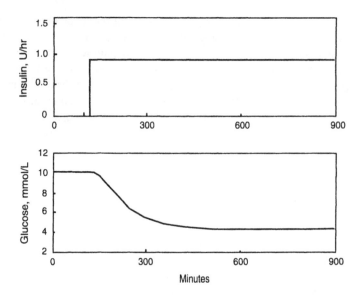

FIGURE 3.2. Response of glucose model to a step insulin infusion rate of 0.9 U/hr.

a substantial effffect on blood glucose level (Fig. 3.3), even in the presence of the baseline insulin infusion rate of 0.9 U/hr. Ideally, one would increase the insulin infusion rate at the start of a meal and then readjust it to the baseline level as the glucose concentration approaches normal after the meal. This type of control should be possible when reliable, rapidly responding glucose sensors are available but the design of control algorithms for programmable infusion pumps now depends heavily on the ability to predict the glucose response to a change in insulin infusion rate.

If F_I is constrained to be pulses of varying amplitude and duration, then perhaps one can predict the response of G from knowledge of the step response shown in Fig. 3.2. This idea was tested by comparing the actual model response to an input pulse with that predicted from the step response. Thus, in Fig. 3.2 let $t = 0$ refer to the time of onset of the step function. Let the input step be $u_1(t) = 0.9\ u(t)$ and the resulting *change* in G from its initial level (Fig. 3.2, bottom) be $G_1(t)$. Now select another input given by $u_2(t) = 0.2\ [u(t) - u(t - 200)]$. Assuming linearity, the predicted response to this input is

$$\hat{G}_1(t) = \frac{0.2}{0.9}[G_1(t) - G_1(t - 200)].$$

Figure 3.4 plots the actual response of the model (as the change in G from its initial level) to $u_2(t)$ as well as $\hat{G}_2(t)$. The prediction is very close to the actual response. If this process is repeated for an input $u_3(t) = 1.0\ [u(t) - u(t - 200)]$, the predicted and actual responses are quite different (Fig. 3.5), implying that the system is

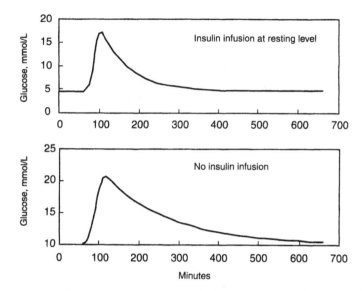

FIGURE 3.3. Response of glucose model to a simulated meal with a steady insulin infusion rate of 0.9 U/hr (top) and with no insulin infusion (bottom).

not truly linear. However, *for pulses of sufficiently small amplitude*, the glucose response to a pulsatile change in insulin infusion rate is very nearly linear and predictable from knowledge of the step response.

Although one cannot avoid the fact that the glucose system becomes nonlinear for larger inputs, it is possible to satisfy the conditions for linearity if one considers only small changes of F_I and G about their resting levels. Nonetheless, restricting the inputs to have the form of steps or pulses seriously constrains the types of con-

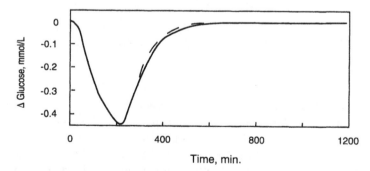

FIGURE 3.4. Change in output of glucose model to a 0.2 U/hr pulse of insulin lasting 200 minute (dashed). Response to the same input predicted by extrapolating from the step response of Fig. 3.2 (solid).

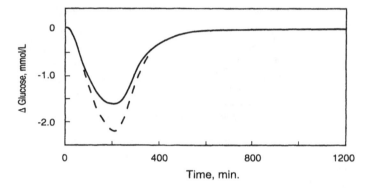

FIGURE 3.5. Change in output of glucose model to a 1.0 U/hr pulse of insulin lasting 200 minute (solid). Response to the same input predicted by extrapolating from the step response of Fig. 3.2 (dashed).

trolling signals that can be utilized. There would be a significant advantage to being able to predict the glucose response to an arbitrary input because then the control algorithm could be designed to fulfill some criterion such as compensating as rapidly as possible for disturbances to G. This chapter will develop such a method and then return to this application.

3.3 CONVOLUTION FORM OF AN LSI SYSTEM

Unit-Pulse Response of an LSI System

Consider a linear, shift-invariant, causal (LSIC) DT system described by the transformation operator $F[.]$ (Fig. 3.6(a)). Let its input $x[k]$ be the unit-pulse function

$$x[k] = \delta[k] = \begin{cases} 1 & k = 0 \\ 0 & k \neq 0. \end{cases} \tag{3.2}$$

Assuming zero initial conditions, its output is then given by

$$y[k] = F[x[k]] = F[\delta[k]] \triangleq h[k], \tag{3.3}$$

where $h[k]$ is defined as the *unit-pulse response* (also known as the *impulse response*) of the system. For a causal system one property of $h[k]$ is immediately apparent. Since $\delta[k] = 0$ for all $k < 0$, and since the output of a causal system can depend only on the past and current values of the input, $h[k]$ for a causal system must equal zero for all $k < 0$. Discrete-time systems are classified into two groups based on their impulse responses. A DT system is called a *finite impulse response (FIR) system* if $h[k] = 0 \; \forall k > K$, where $|K| < \infty$. Otherwise the system is an *infinite impulse response (IIR) system*. Note that the time-invariance and linearity

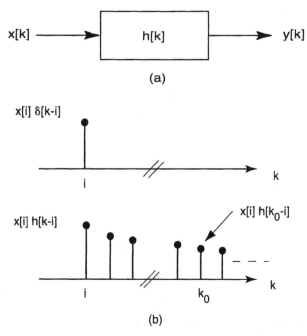

x[k] ———→ | h[k] | ———→ y[k]

(a)

x[i] δ[k-i]

i ———————————— k

x[i] h[k-i]

x[i] h[k₀-i]

i ——————— k₀ ———→ k

(b)

FIGURE 3.6. (a) An LSIC system. (b) Interpetation of a single term of the summation in Eq. (3.6).

conditions imply that the response of the system of Fig. 3.6(a) to the input $x[k] = \delta[k - i]$ will be $y[k] = h[k - i]$ and its response to the input $x[k] = a\delta[k]$ will be $y[k] = ah[k]$.

Consider now the response of this system to an arbitrary input $x[k]$ which may be represented as

$$x[k] = \ldots + x[-2]\delta[k + 2] + x[-1]\delta[k + 1] + x[0]\delta[k] + x[1]\delta[k - 1]$$
$$+ x[2]\delta[k - 2] + = \ldots$$
$$= \sum_{i=-\infty}^{\infty} x[i]\delta[k - i]. \tag{3.4}$$

Now $y[k] = F[x[k]]$ and by linearity one can write

$$y[k] = \ldots F[x[-2]\delta[k + 2]] + F[x[-1]\delta[k + 1]] + F[x[0]\delta[k]] + F[x[1]\delta[k - 1]]$$
$$+ F[x[2]\delta[k - 2]] + = \ldots$$
$$= \sum_{i=-\infty}^{\infty} x[i]F[\delta[k - i]]. \tag{3.5}$$

and from the definition of $h[k]$

$$y[k] = \sum_{i=-\infty}^{\infty} x[i]h[k-i].\qquad(3.6)$$

Equation 3.6 is a fundamental expression of the input–output relationship of an LSI discrete-time system known as the *convolution sum*. The indicated summation represents the convolution operation, or simply convolution, and it is symbolized as

$$y[k] = x[k] * h[k].\qquad(3.7)$$

Note that by a change of variables the convolution sum also can be written as

$$y[k] = \sum_{i=-\infty}^{\infty} h[i]x[k-i].\qquad(3.8)$$

If $h[k]$ represents a causal system, the upper limit of the summation in Eq. (3.6) can be replaced by k and the lower limit in Eq. (3.8) by zero. Similarly if $x[k] = 0$ for all $k < 0$, then the lower limit in Eq. (3.6) can be set to zero and the upper limit in Eq. (3.8) can be replaced by k.

The convolution sum relationship establishes the primary importance of the unit-pulse response of a DT system: Knowing the unit-pulse response, $h[k]$, one can calculate the system output for any arbitrary input without having to solve the difference equation of the system. How does one determine this important function for an LSI system? If the unit-pulse response is unknown but the difference equation of the system is known, one can solve the difference equation using a unit-pulse function as the input. The output then will be the unit-pulse response. If one is evaluating a physical system that has an accessible input, it is conceptually possible to apply a unit-pulse input to the system and measure $h[k]$ as the output. Often it is simpler to apply a unit-step function, $u[k]$, as the input and measure the unit-step response, $g[k]$. Since $\delta[k] = u[k] - u[k-1]$, then $h[k] = g[k] - g[k-1]$.

The interpretation of the convolution sum can be developed from an understanding of a single term in the summation of Eq. (3.6), $x[i] \, h[k-i]$, for specific values of k and i. Keeping in mind that since $x[i]$ in Eqs. (3.4)–(3.6) refers to the *specific value* of the function $x[k]$ when $k = i$, then the i-th term of the input signal as expressed in the r.h.s. of Eq. (3.4) is the function $x[i]\delta[k-i]$, which is shown in Fig. 3.6(b). Now pick a specific value of k, say $k = k_0$, at which to evaluate the output $y[k_0]$. In the figure a specific $h[k]$ has been chosen for purposes of illustration. The output response to the i-th input term is $x[i] \, h[k-i]$, as shown, which is also the i-th term in the r.h.s. of the equation (Eq. (3.6)) for the output, $y[k]$. At time $k = k_0$, the output response of Fig. 3.6(b) contributes a value $x[i] \, h[k_0 - i]$ to the summation for $y[k_0]$ (obtained by setting $k = k_0$ in Eq. (3.6)). Therefore for any value of k the convolution sum expresses the output, $y[k]$, as the

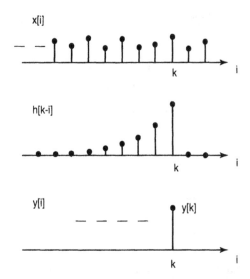

FIGURE 3.7. Visualization of the convolution sum as the multiplication of the function $x[i]$ (top) by the function $h[k-i]$ (middle) to calculate the result, $y[k]$ (bottom).

weighted sum of past values of the input, where each weighting factor is the value of the impulse response at a time equal to the difference between the current time, k, and the time, i, at which the input occurred.

A fruitful visualization of the convolution sum is presented in Fig. 3.7. Consider $y[k]$ as given in Eq. (3.6) to be a time function, $y[i]$, that is to be evaluated at $i = k$. First plot $x[i]$ and identify the time $i = k$ (Fig. 3.7, top). Then immediately below $x[i]$ make a graph of $h[k-i]$, which is $h[i]$ reversed in time and shifted to the right by k units of time (Fig. 3.7, middle). Now according to Eq. (3.6) one calculates $y[i]$ for $i = k$ by multiplying $x[i]$ and $h[k-i]$ point-by-point in time and summing all of the products. Now note that to calculate the next value of $y[i]$ (i.e., $y[k+1]$), one simply shifts the time-reversed $h[i]$ in Fig. 3.7 (middle) one unit of time to the right and repeats the multiplying and summing. Thus the convolution sum can be visualized by slowly shifting the time-reversed $h[i]$ to the right, stopping at each time point k only long enough to multiply $x[i]$ by $h[k-i]$ and to sum these products to obtain $y[k]$. For simple unit-pulse responses and simple input signals it is not difficult to calculate the exact output function by this graphical process. Furthermore, this process is very useful for approximating the convolution of two functions and the reader should practice until it is natural.

Example 3.1 Determine the unit-pulse response of the system described by the difference equation

$$y[n] + 0.4\, y[n-1] = 3\, x[n].$$

Solution: Let $x[n] = \delta[n]$ and solve for $y[n]$, $n = 0, 1, 2, \ldots$, assuming $y[-1] = 0$. Thus

$$y[0] = 3x[0] - 0.4y[-1] = 3$$
$$y[1] = 0 - 0.4(3) = -1.2$$
$$y[2] = 0 - 0.4(-1.2) = 0.48$$
$$\vdots$$
$$y[k] = (-0.4)^k(3) \quad k \geq 0$$

Since $x[n]$ is a unit-pulse function,

$$h[n] = y[n] = 3(-0.4)^n u[n].$$

Example 3.2   The response, $y[n]$, of an LSIC system to the input

$$x[n] = \begin{cases} 1, & n \geq 0 \\ 0, & n < 0 \end{cases}$$

is shown in Fig. 3.8(a). All initial conditions were zero.

(a) What is the unit-pulse response of this system? (b) What is its response to the input $x[n] = u[n] - u[n - 3]$?

Solution: (a) The input was a unit-step function, $u[n]$. Thus the output is the unit-step response, $g[n]$, and $h[n] = g[n] - g[n-1]$. Therefore, from the graph of Fig. 3.8(a),

$$h[0] = g[0] - g[-1] = 0$$
$$h[1] = g[1] - g[0] = 1 - 0 = 1$$
$$h[2] = g[2] - g[1] = 2 - 1 = 1$$
$$h[3] = g[3] - g[2] = 3 - 2 = 1$$
$$h[n] = 0, \, n > 3$$

A more general approach is to express $y[n]$ as a convolution sum and expand this equation. Thus, using Eq. (3.8),

$$y[k] = \sum_{i=0}^{k} h[i]x[k - i].$$

Writing this equation for $k = 0, 1, 2, \ldots$, and using $x[n]$ as defined above,

$$y[0] = 0 = h[0]$$
$$y[1] = 1 = h[0] + h[1] \Rightarrow h[1] = 1$$
$$y[2] = 2 = h[0] + h[1] + h[2] \Rightarrow h[2] = 1, \qquad \text{etc.}$$

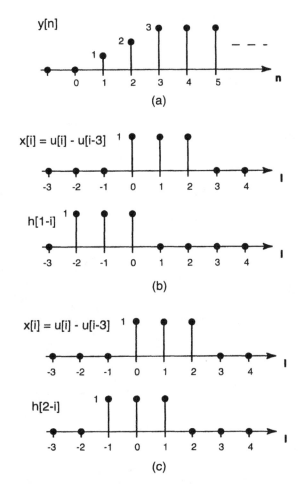

FIGURE 3.8. Visualization of the convolution of Example 3.2. (a) The unit-step response $y[n]$, of the system. (b) Calculating $y[1]$ in response to the input $x[n]=u[n] - u[n - 3]$. (c) Calculating $y[2]$ in response to this input.

(b) $y[k] = (u[k] - u[k - 3]) * h[k]$. Since $h[0] = 0, y[0] = 0$. Fig. 3.8(b,c) exhibits the next two steps in calculating this convolution graphically. Figure 3.8(b) shows the functions $x[i] = u[i] - u[i - 3]$ and $h[1 - i]$. Multiplying these two functions point-by-point in time and adding the products gives a value for $y[1]$ of 1. Similarly, Fig. 3.8(c) shows $u[i] - u[i - 3]$ and $h[2 - i]$. Multiplying these functions and adding the products gives $y[2] = 2$. Evaluating $y[k]$ for $k > 2$ is left as an exercise for the reader.

Example 3.3   An LSIC system has an impulse response given by $h[k] = a^k u[k]$, with $|a| < 1$. Determine its zero-state response to the input

$$x[k] = \begin{cases} k, 0 \le k \le 5 \\ 0, \text{ otherwise} \end{cases}.$$

Solution: By the convolution sum,

$$y[k] = \sum_{i=0}^{k} x[i]h[k-i] = \sum_{i=0}^{\min(k,5)} ia^i.$$

From direct calculation

$y[0] = 0$

$y[1] = h[0] = 1$

$y[2] = h[1] + 2h[0] = a + 2$

$y[3] = h[2] + 2h[1] + 3h[0] = a^2 + 2a + 3$

$y[4] = h[3] + 2h[2] + 3h[1] + 4h[0] = a^3 + 2a^2 + 3a + 4$

$y[5] = h[4] + 2h[3] + 3h[2] + 4h[1] + 5h[0] = a^4 + 2a^3 + 3a^2 + 4a + 5.$

For $k > 5$, $x[i]$ and $h[k-i]$ overlap for $i = 0{:}5$ (see Fig. 3.9, for $k = 7$) and the general solution for $y[k]$ is found directly from the convolution sum to be

$$y[k] = a^{k-1} + 2a^{k-2} + 3a^{k-3} + 4a^{k-4} + 5a^{k-5}.$$

Example 3.4   A *noncausal* LSI system has an impulse reponse given by

$$h[k] = \delta[k+1] - \delta[k-1].$$

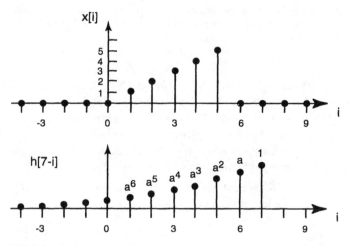

FIGURE 3.9. Calculating y[7] by graphical convolution for Example 3.3.

(a) Determine its response to the input signal $x[k] = \cos[Pk]$. The system is initially at rest.

(b) Calculate the output when $P = 4$; when $P = 2\pi$.

Solution:

(a) Call the output $y[k]$. Thus

$$y[k] = x[k] * h[k] = \sum_{i=-\infty}^{\infty} x[i]h[k-i] = -x[k-1] + x[k+1].$$

This system could be called a "two-step differencing filter" because the output at time k is the difference between input values separated by two steps in time—that is, the next input to come and the immediate past input. For the specified $x[n]$,

$$y[k] = \cos[P(k+1)] - \cos[P(k-1)]$$

$$= \cos[Pk]\cos[P] - \sin[Pk]\sin[P] - (\cos[Pk]\cos[-P] - \sin[Pk]\sin[-P])$$

$$= 2\sin[P]\sin[Pk].$$

(b) Substituting into the above solution, for $P = 4$, $y[k] = -1.5136 \sin[4k]$. For $P = 2\pi$, $y[k] = 2\sin[2\pi]\sin[2\pi k] = 0 \ \forall k$.

Properties of Discrete-Time Convolution

Although the convolution operation was derived as a representation of the input–output properties of a system, convolution is a general mathematical operation which can be applied to any two functions. Indeed, convolution arises naturally in many signal processing contexts. Consequently it is important to recognize some basic properties of the discrete-time convolution operator. This operator is:

- *Associative:* Given the bounded DT signals $x[k]$, $v[k]$, and $w[k]$, then $x[k] * (v[k] * w[k]) = (x[k] * v[k]) * w[k]$.
- *Commutative:* $x[k] * v[k] = v[k] * x[k]$.
- *Distributive:* $x[k] * (v[k] + w[k]) = x[k] * v[k] + x[k] * w[k]$.

These properties are readily provable from the basic definition of convolution. Thus to prove commutativity, one starts from the definition

$$x[k] * h[k] = \sum_{i=-\infty}^{\infty} x[i]h[k-i].$$

Substituting $l = K - i$,

$$x[k] * h[k] = \sum_{i=-\infty}^{\infty} x[k-l]h[l] = h[k] * x[k].$$

Another important property is the shifting property of convolution, which leads to the sifting property of the unit-pulse function. Note that if $y[k] = x[k] * h[k]$, then for an integer constant q, $y[k-q] = x[k-q] * h[k] = x[k] * h[k-q]$. The first equality is a basic property of convolution, provable from the definition; the second equality results from a change of variables—for example, let $w = k - q$, then let $k = w$ and $q = -q$. Next, consider convolution of a function $x[k]$ with a unit-pulse function:

$$x[k] * \delta[k] = \sum_{i=-\infty}^{\infty} x[i]\delta[k-i] = x[k].$$

Thus, by the shifting property,

$$x[k] * \delta[k-q] = x[k-q] * \delta[k] = x[k-q].$$

The consequence of convolving a function with a time-delayed unit-pulse function is just to "sift out" the value of the original function at that time delay (the *sifting* property of the unit-pulse function). Equivalently, to define a system which produces a pure time delay, q, its unit-pulse response should be $\delta[k-q]$.

Finally, the previously stated relationship between $g[k]$, the unit-step response of a system, and $h[k]$ of an LSIC system can be shown more rigorously. The unit-step response is

$$g[k] = u[k] * h[k] = \sum_{i=0}^{k} h[i].$$

and

$$g[k] - g[k-1] = \sum_{i=0}^{k} h[i] - \sum_{i=0}^{k-1} h[i] = h[k]. \tag{3.9}$$

The Impulse Response and Memory

Because the unit-pulse response can be interpreted as a set of weights that determine the contribution of past inputs to the current output, this response is a direct expression of memory in the system. Recall that the autocorrelation function of the system output, $y[k]$, is a measure of memory in $y[k]$ and that this memory may be due to memory processes in the system that generated $y[k]$ or to memory (i.e., autocorrelation) that was already present in the input to the system. Thus, for a given system, to determine the relationship between $h[k]$ and $R_y[m]$, it is necessary to excite the system with an input that has no signal memory. It is trivial to show that the autocorrelation function of a unit-pulse function is also a unit-pulse function, implying that $\delta[k]$ is uncorrelated and has no signal memory. Let us use this input signal and assume either that the response of the system to a unit-pulse input is observed for a very long time, N, or that $h[k] = 0$ for $k > K$, and $K \ll N$. Then for a unit-pulse input the autocorrelation function of the output, $y[k]$, is

$$R_y[m] = \frac{1}{N} \sum_{k=0}^{N} y[k] y[k + m] = \frac{1}{N} \sum_{k=0}^{N} h[k] h[k + m]. \tag{3.10}$$

That is, the memory the system adds to the output signal is completely describable from the unit-pulse response. That this should be true seems obvious since $h[k]$ represents the weights, or "gains" of the system, through which previous values of the input influence the current output. From Eq. (3.8) it may be seen that $R_y[m] = h[k] * h[-k]$. It may seem that one also could calculate the unit-pulse response of an unknown system indirectly from the autocorrelation function of the system output if it is certain that the input signal has zero autocorrelation everywhere (except at $m = 0$). For an FIR system with $K \ll N$ it is possible to write a set of equations based on Eq. (3.8), for $0 \le m \le k$ whose unknown r.h.s. terms are all functions of $h[k]$, $0 \le k \le K$. These equations, however, are nonlinear and their solution in the presence of typical noise levels is usually subject to considerable error.

Relation to Signal Processing

Applying a signal to the input of an LSI system (Fig. 3.6(a)) produces an output that is a modified form of the input signal, as indicated by the convolution sum equation (3.6). Our interpretation of this process is that the values of the input signal are weighted according to the values of the unit-pulse response of the system and summed to generate the output. This interpretation provides a convenient framework for visualizing signal processing in the time domain. Consider the unit-pulse response, $h_1[k]$, presented in Fig. 3.10(a). By examining the function $h_1[k - i]$ it is seen that this system adds up five input values—the current and four previous input values—all weighted by one-fifth, to generate $y[k]$. Such a system is known as a *moving averager* because the selection of data points being averaged "moves along" the time axis as k increases. It is apparent that this system will tend to reduce the influence of isolated events in the input on the output and that rapid changes in the input signal will be blurred in the output due to the averaging. In fact, one can readily show that a step change in the input signal will not be reflected completely in the output until four time units later. The system represented by the impulse response of Fig. 3.10(b) averages three previous input signal values but with unequal weightings. Now the fact that $h_2[2]$ is larger than other values of $h_2[k]$ implies that a sudden change in the input signal will be better represented in the output than was the case for the moving averager; however, it will be delayed by two samples. Yet we would still expect that rapid changes in the input would be attenuated by the weighted summation implicit in $h_2[k]$. Thus both of these examples could be classed as *lowpass filters*, meaning that they attenuate rapid changes in the input more than slow ones. Consider the unit-pulse response of Fig. 3.10(c). Because the action of this impulse response is to calculate the difference between the input at time $k - 2$

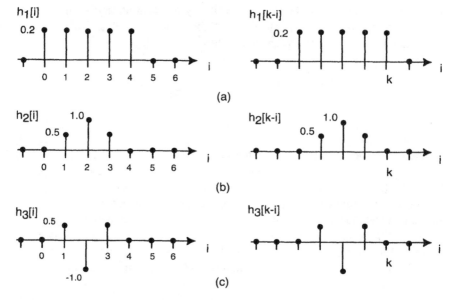

FIGURE 3.10. Three examples of interpreting the signal processing properties of a system by plotting the impulse response reversed in time. Left: impulse responses; right: impulse responses reversed in time and shifted to time point k.

and one-half of the sum of the input values at times $k-1$ and $k-3$, a sudden change in the input will be relatively unattenuated (although delayed by two samples). The output for a slowly varying input, however, will be close to zero because $x[k-3]$, $x[k-2]$, and $x[k-1]$ will be approximately equal. This system is a *highpass filter*.

Example 3.5   A DT filter   A causal, three-point moving average filter has the unit-pulse response

$$h[n] = \tfrac{1}{3}(u[n] - u[n-3]).$$

Determine the response of this filter to the signal $x[n] = a^n u[n]$ and specify whether the filter is highpass or lowpass.
    Solution. By convolution,

$$y[n] = \sum_{i=-\infty}^{\infty} x[n-i]h[i] = \frac{1}{3}\sum_{i=0}^{n} a^{n-i}\,(u[i] - u[i-3])$$

$$= \frac{1}{3}\sum_{i=0}^{n} a^n a^{-i} - \frac{1}{3}\sum_{i=3}^{n} a^n a^{-i}.$$

Note that

$$\sum_{i=3}^{n} a^n a^{-i} = \sum_{i=0}^{n} a^n a^{-i} - a^n (a^{-0} + a^{-1} + a^{-2}).$$

Therefore,

$$y[n] = \frac{a^n}{3}(a^{-0} + a^{-1} + a^{-2}) = \frac{1}{3}(a^n + a^{n-1} + a^{n-2}), \qquad n \geq 3.$$

By direct calculation from the convolution sum equation,

$$y[0] = \tfrac{1}{3}, \qquad y[1] = \tfrac{1}{3}(1 + a), \qquad y[2] = \tfrac{1}{3}(1 + a + a^2).$$

Note that if $a = 1$, then $x[n] = u[n]$. In this case the step change in the input signal becomes a slower, ramp change in the output (over the range $n = [0:3]$). Thus this system attenuates rapid changes in its input and is a lowpass filter.

Aside. The following closed-form expressions for summations are often useful in problems like the preceding one:

$$\sum_{r=0}^{\infty} a^r = \frac{1}{1-a}, \, |a| < 1 \qquad \sum_{r=0}^{N-1} a^r = \frac{1-a^N}{1-a}, \, \forall \, a.$$

3.4 CONVOLUTION FOR CONTINUOUS-TIME SYSTEMS

Define the CT *impulse function*, $\delta(t)$, by the relationships

$$\delta(t) = 0, t \neq 0, \qquad \text{and} \qquad \int_{-\epsilon}^{\epsilon} \delta(t)dt = 1 \; \forall \epsilon > 0. \qquad (3.11)$$

For a linear, time-invariant continuous-time system specified by the transformation operation $y(t) = F[x(t)]$, its *impulse response* is defined as

$$h(t) = F[\delta(t)]. \qquad (3.12)$$

Remember that the notation in Eq. (3.12) assumes that all initial conditions on the system are zero. As was the case for DT systems, if an LTI system is causal, then $h(t) = 0$, for all $t < 0$. Also, by linearity, if $x(t) = a\delta(t - t_0)$, then $y(t) = ah(t - t_0)$.

Since it is not possible to apply a true impulse function to a physical system, $h(t)$ must be evaluated by indirect means. If the differential equation of the system is known, one may solve it for the zero-state response using $\delta(t)$ as the input signal. For example, consider the problem of determining airflow into the lungs from

Chapter 2. Recall that the zero-state solution for the volume change of the lungs, $V(t)$, in response to pressure, $P(t)$, applied starting at $t = 0$ was

$$V(t) = \int_0^t e^{-(t-\lambda)/R_{AW}C_L}\left(\frac{P(\lambda)}{R_{AW}}\right)d\lambda.$$

Letting $P(t)$ be an impulse function,

$$V(t) = \int_0^t e^{-(t-\lambda)/R_{AW}C_L}\left(\frac{\delta(\lambda)}{R_{AW}}\right)d\lambda \triangleq h(t).$$

Therefore,

$$h(t) = \frac{1}{R_{AW}}e^{-t/R_{AW}C_L}\,u(t).$$

From the impulse response, one may evaluate the response of an LTI system to an arbitrary input. To prove this result, consider again the system specified by $y(t) = F[x(t)]$, having impulse response $h(t)$. Let $x(t)$ be an arbitrary but bounded input signal. Note that

$$\int_{-\infty}^{\infty} x(\lambda)\delta(t-\lambda)d\lambda = x(t),$$

which is an expression of the sifting property of CT convolution. Now,

$$y(t) = F[x(t)] = F\left[\int_{-\infty}^{\infty} x(\lambda)\delta(t-\lambda)d\lambda\right]$$

$$= \int_{-\infty}^{\infty} x(\lambda)F[\delta(t-\lambda)]d\lambda = \int_{-\infty}^{\infty} x(\lambda)h(t-\lambda)d\lambda. \qquad (3.13)$$

The final integral in Eq. (3.13) is called the *convolution integral* and it is another definition of the input–output relationship of an LTI system. This relationship is symbolized as

$$y(t) = \int_{-\infty}^{\infty} x(\lambda)h(t-\lambda)d\lambda = x(t) * h(t). \qquad (3.14)$$

By a change of variables it is easy to show that

$$x(t) * h(t) = \int_{-\infty}^{\infty} x(\lambda)h(t-\lambda)d\lambda = \int_{-\infty}^{\infty} x(t-\lambda)h(\lambda)d\lambda = h(t) * x(t). \qquad (3.15)$$

For a causal LTI system whose input, $x(t)$, is zero for $t < 0$, the lower and upper limits on the integrals in Eq. (3.13) can be replaced by 0 and t, respectively. The interpretation of CT convolution is the same as that for DT convolution, as seen in the following example.

Example 3.6 An LTIC system has the impulse response $h(t) = u(t) - u(t - 1)$. The input $x(t) = u(t) - 2u(t - 1) + u(t - 2)$ is applied to its input. Determine the output, $y(t)$.

Solution. Figure 3.11 shows the two functions, $h(t)$ and $x(t)$, as functions of the dummy variable λ. Since $x(t) = 0$ for $t < 0$, and since the system is causal, then

$$y(t) = x(t) * h(t) = \int_0^t x(\lambda)h(t - \lambda)d\lambda.$$

The figure indicates the functions inside the integral for $t = 0$ and for $t = 1.5$. For $t = 0$, $h(0 - \lambda)$ and $x(\lambda)$ do not overlap. Therefore the integral evaluates to zero and $y(0) = 0$. For $0 < t < 1$, these two functions will overlap in the interval $[0, t]$, and $y(t) = \int_0^t (1)(1) \, d\lambda = t$.

For $1 < t < 2$ the interval of overlap will be $[t - 1, t]$, and

$$y(t) = \int_{t-1}^1 (1)(1)d\lambda + \int_1^t (-1)(1)d\lambda = 3 - 2t.$$

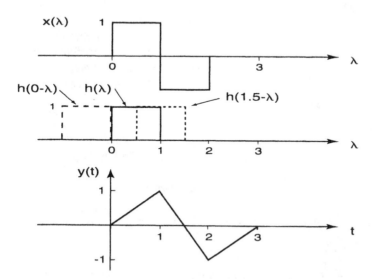

FIGURE 3.11. Graphical convolution for the CT example. Top: input signal; middle: impulse response of the system (solid) and time-reversed impulse response shifted to $t = 0$ (long dashes), and to $t = 1.5$ (short dashes); bottom: output signal, $y(t)$.

The figure indicates the overlap between $x(\lambda)$ and $h(1.5 - \lambda)$. For $2 < t < 3$,

$$y(t) = \int_{t-1}^{2} (-1)(1)d\lambda = t - 3.$$

Finally, for $t > 3$, $y(t) = 0$. These results are plotted in Fig. 3.11.

It is apparent from Fig. 3.11 that $h(t)$ expresses the weighting factor by which a previous input at time $t_0 - t$ contributes to the output, $y(t)$, calculated at time t_0, and that $y(t_0)$ comprises the integration (i.e., summation) of all past inputs multiplied by their weighting factors. To emphasize this point for an LTIC system one may rewrite Eq. (3.15), after a change of variables, as

$$y(t_0) = \int_{0}^{t_0} x(t)h(t_0 - t)dt, \qquad \text{where we have let } t \to t_0, \lambda \to t.$$

In this form the weighting of past values of the input by the time-reversed and shifted $h(t)$ is obvious. Thus the interpretation of CT convolution is equivalent to that of DT convolution even though the input $x(t)$ now is defined for all instants of time. Furthermore, the graphical method of calculating the convolution sum applies also to the convolution integral as long as we integrate where previously we summed. As was observed for $h[n]$, $h(t)$ also expresses the effect of system memory processes on the output.

Aside. Probably we sometimes unconsciously test physical systems with a pseudo-impulse input without thinking in terms of an impulse response. Consider a basketball player who bounces the ball in order to "get a feel" for its mechanical behavior. Notice too that a very young child who is presented with a new food may tap it with his spoon as part of exploring it. It is a common experience also that a driver wishing to evaluate the mechanical response of a new car will apply a stepwise depression of the accelerator pedal and monitor the change in velocity. This latter action is related to the impulse response by way of the step response, as discussed after the next example.

Example 3.7   CT convolution as a weighted moving average   The LTIC system of Fig. 3.12 has the impulse response $h(t) = e^{-2t}u(t)$. Find its output when

$$x(t) = [1 - \cos(2\pi t)]u(t).$$

Solution.   Figure 3.12 shows plots of $x(t)$ and $h(t)$. Note that $x(t)$ starts at $t = 0$. By convolution,

$$y(t) = \int_{0}^{t} e^{-2(t-\lambda)}[1 - \cos(2\pi\lambda)]d\lambda = e^{-2t}\int_{0}^{t} [e^{2\lambda} - e^{2\lambda}\cos(2\pi\lambda)]d\lambda$$

$$= \frac{1}{2}e^{-2t}[e^{2t} - 1] - e^{-2t}\left[\frac{e^{2\lambda}}{4 + 4\pi^2}(2\cos(2\pi\lambda) + 2\pi\sin(2\pi\lambda))\right]\Bigg|_{0}^{t}$$

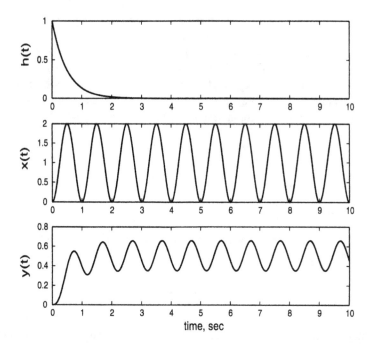

FIGURE 3.12. Response of a system with an exponentially decaying impulse response to a sinusoidal input having a nonzero mean level. Top: impulse response of the system; middle: input signal; bottom: output signal.

$$= \frac{1}{2}[1 - e^{-2t}] - \frac{e^{-t}}{4 + 4\pi^2}[e^{2t}(2\cos(2\pi t) + 2\pi\sin(2\pi t) - 2)]$$

$$y(t) = \frac{1}{2}[1 - e^{-2t}] - \frac{1}{4 + 4\pi^2}[2\cos(2\pi t) + 2\pi\sin(2\pi t) - 2e^{-2t}].$$

This function is also plotted in Fig. 3.12. Note in $y(t)$ the blurring of the abrupt on-set of the input signal due to the weighted averaging by $h(t)$ of inputs which were zero for $t < 0$.

Given the LTIC system specified by $y(t) = F[x(t)]$, the response of this system to a unit-step input, $u(t)$, is $y(t) = F[u(t)] = g(t)$, where $g(t)$ is the unit-step response. But since

$$u(t) = \int_{-\infty}^{t} \delta(\eta)d\eta = 1, \qquad t \ge 0,$$

then

$$g(t) = u(t) * h(t) = \int_{0}^{t}\left[\int_{0}^{t-\lambda}\delta(\eta)d\eta\right]h(\lambda)d\lambda = \int_{0}^{t}h(\lambda)d\lambda. \qquad (3.16)$$

That is, the unit-step response is the integral of the impulse response. (Similarly, for an LTIC DT system, we saw that the unit-step response is the summation of the unit-pulse response—i.e., $g[n] = \sum_{k=0}^{n} h[k]$.) This result provides a more practical alternative for evaluating the impulse response of an unknown physical system because often one can apply a step input, whereas a true impulse input is impossible to generate physically. Therefore, to determine $h(t)$ one excites the system with a scaled version of a unit-step input, $Au(t)$, and records the output, $y(t) = Ag(t)$. Then

$$h(t) = \frac{1}{A} \frac{dy(t)}{dt}.$$

Properties of CT Convolution

Continuous-time convolution is a general operator so that for any two bounded, integrable functions $w(t)$ and $v(t)$ their convolution is defined by

$$w(t) * v(t) = \int_{-\infty}^{\infty} w(t - \lambda)v(\lambda)d\lambda \qquad (3.17)$$

if the integral exists. This operator satisfies many of the same properties as the DT convolution operator. Thus CT convolution is:

Associative: $(x(t) * w(t)) * v(t) = x(t) * (w(t) * v(t))$;

Commutative: $w(t) * v(t) = v(t) * w(t)$;

Distributive: $x(t) * (v(t) + w(t)) = x(t) * w(t) + x(t) * v(t)$.

In addition, if $x(t) = w(t) * v(t)$, then $x(t - t_0) = w(t - t_0) * v(t) = w(t) * v(t - t_0)$.
Convolution of $x(t)$ with the unit impulse function just produces $x(t)$ again. Finally, a useful property of convolution is given by

$$\frac{d}{dt}(w(t) * v(t)) = \frac{dw(t)}{dt} * v(t). \qquad (3.18)$$

From this latter property one may also derive the relationship of Eq. (3.16) in the form

$$h(t) = \frac{dg(t)}{dt}. \qquad (3.19)$$

All of the above properties are readily proven from the basic definition of CT convolution.

Before addressing some examples of convolution, one further property will be noted. Consider the cascaded LTIC systems with impulse responses $h_1(t)$ and $h_2(t)$ that are shown in Fig. 3.13. Let $h(t)$ be the overall impulse response of the cascaded system, such that $y(t) = x(t) * h(t)$. To determine $h(t)$ note first that $x_1(t) = x(t) *$

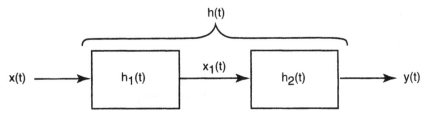

FIGURE 3.13. Two LTIC systems cascaded (i.e., connected in series). $h(t)$ is the equivalent overall impulse response of the cascaded systems.

$h_1(t)$. Therefore $y(t) = h_2(t) * x_1(t) = h_2(t) * (x(t) * h_1(t)) = (h_2(t) * h_1(t)) * x(t)$. Consequently, $h(t) = h_2(t) * h_1(t)$. If there are N such systems cascaded, then the overall impulse response of the cascade is

$$h(t) = h_N(t) * h_{N-1}(t) * \cdots * h_2(t) * h_1(t). \qquad (3.20)$$

An identical relationship applies for a cascade of DT systems.

Example 3.8 Biomedical application Consider once again the example of air-flow into the lungs. This system can be represented by the transformation $V(t) = F[P(t)]$. The impulse response of this system was given above as

$$h(t) = \frac{1}{R_{AW}} e^{-t/R_{AW}C_L} u(t).$$

Therefore, the system may also be represented by the convolution equation $V(t) = P(t) * h(t)$. Assume the following values for the parameters of a patient: $R_{AW} = 4$ cm $H_2O/L/s$, $C_L = 0.10$ L/cm H_2O. (These values are typical for a subject without respiratory disease.) The impulse response for this subject,

$$h(t) = \tfrac{1}{4} e^{-t/0.4} u(t),$$

is plotted in Fig. 3.14(a). Let $P(t)$ be the single pulse shown in Fig. 3.14(b) and estimate the response of $V(t)$ using graphical convolution. Confirm this estimate by calculating the exact solution.

Solution. The plot of Fig. 3.14(c) shows $P(\lambda)$ and $h(t - \lambda)$ for $t = 0, 1, 2, 3$. From this figure the convolution result, $V(t) = P(t) * h(t)$, can be estimated graphically. $V(t)$, $t < 0$, is zero because $P(\lambda) = 0$ for $t < 0$. $V(0) = 0$ since $h(0 - \lambda)$ and $P(\lambda)$ do not overlap. As t increases above zero (but $t < 2$) there is increasing overlap between $P(\lambda)$ and $h(t - \lambda)$ and the value of the convolution integral

$$V(t) = \int_0^t P(\lambda) h(t - \lambda) d\lambda$$

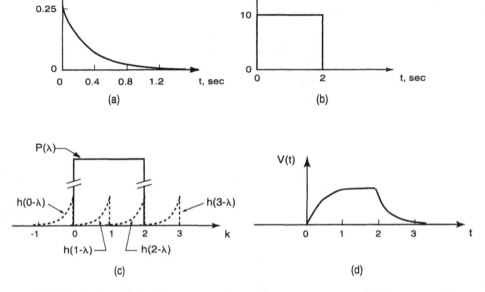

FIGURE 3.14. Determining the response of lung volume to a pressure pulse from a ventilator using convolution. (a) Impulse response of volume to a pressure input using average values for biomechanical parameters from normal subjects. (b) The pressure pulse from the ventilator. (c) Graphical convolution, showing the pressure pulse and the time-reversed impulse response at four different time shifts. (d) Sketch showing the expected shape of the resulting $V(t)$ waveform as estimated from visual analysis of the graphical convolution.

increases progressively. But because of the decaying shape of $h(t - \lambda)$, the increment added to this integral for each Δt increment of t decreases as t increases. That is, the derivative of $V(t)$ is continuously decreasing. Furthermore for $t = 1$, $h(t - \lambda)$ is almost zero for $\lambda < 0$. Therefore, the integral is almost at a maximum value which changes only slightly for $1 < t < 2$. For $t > 2$, there is decreasing overlap of between $P(\lambda)$ and $h(t - \lambda)$ as t increases and $V(t)$ falls progressively to zero. The rate of fall is largest near $t = 2$ and from the time constant of $h(t)$ one estimates that $V(t)$ reaches zero around $t = 3.2$ s. The plot of Fig. 3.14(d) indicates the inferred shape of $V(t)$. The peak value of $V(t)$ occurs at $t = 2$ and can be estimated as

$$V(2) = \int_0^2 (10)(0.25e^{-(t-\lambda)/0.4})d\lambda \approx 2.5\int_0^\infty e^{-\lambda/0.4}d\lambda = (2.5)(0.4) = 1 \text{ liter.}$$

This result can be confirmed by direct calculation of the convolution integral. For $0 \le t \le 2$:

$$V(t) = P(t) * h(t) = \int_0^t (10)(0.25e^{-(t-\lambda)/0.4})d\lambda$$

$$= 2.5e^{-t/0.4}\int_0^t e^{\lambda/0.4}d\lambda$$

$$= 2.5(0.4)\, e^{-t/0.4}[e^{t/0.4} - 1] = 1 - e^{-t/0.4}.$$

Thus $V(2) = 1 - e^{-5} = 0.9933$ liters. Similarly, one can show that for $2 < t$, $V(t) = V(2)\, e^{-(t-2)/0.4}$.

3.5   CONVOLUTION AS SIGNAL PROCESSING

The importance of convolution relative to signal processing derives from the understanding that convolution of the impulse response of a system with its input signal is a means of performing a weighted average of past values of the input signal. This weighted averaging, which slides through time, is a form of signal processing and we can gain insight into the nature of the output from our understanding of convolution. For example, if the input signal contains an impulse-like event, then for some time thereafter the output will "look like" the impulse response of the system. Similarly, if the input signal contains a step change in level, then for some time thereafter the output will "look like" the integral of the impulse response. Thus, if the impulse response is an exponentially decaying function, the response to a step change in the input will not be a step but will be a slower exponential rise. Figure 3.15 presents six examples using graphical convolution to estimate the effect on a signal of passing it through a system with a specified impulse response. The reader is urged to prove the qualitative validity of the right-hand column of plots using graphical convolution.

3.6   RELATION OF IMPULSE RESPONSE TO DIFFERENTIAL EQUATION

Since the impulse response is the zero-state solution of the differential equation of a system when the input is an impulse function, the parameters of the impulse response are determined by the coefficients of the differential equation and one can infer certain properties of the impulse response from these coefficients. This relationship is most useful for first- and second-order systems. Consider a first-order differential equation of the form

$$k_1 \frac{dy(t)}{dt} + \frac{1}{k_2}y(t) = x(t),$$

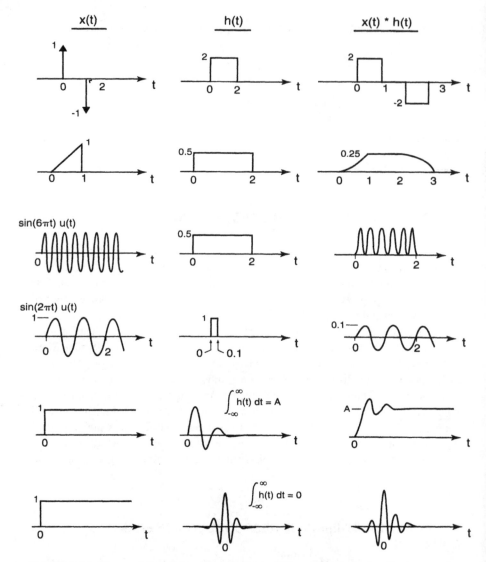

FIGURE 3.15. Six examples of signal processing by LTI systems. In the top three examples the system calculates a uniformly weighted summation of past inputs for 2 seconds. In the fourth example the summation is over 0.1 seconds; therefore, for low-frequency signals the system mainly reduces the signal amplitude as shown. The fifth example is a second-order lowpass filter while the last example is a second-order highpass filter. The reader should be able to predict the shapes of x(t)·h(t) for these examples using graphical convolution.

which can be rearranged into a standardized form as

$$\frac{dy(t)}{dt} + \frac{1}{k_1 k_2} y(t) = \frac{1}{k_1} x(t). \tag{3.21}$$

Letting $x(t) = \delta(t)$ and defining $\tau = k_1 k_2$, solving for $y(t)$ yields

$$y(t) = h(t) = \frac{1}{k_1} e^{-t/\tau} u(t).$$

Therefore the impulse response function is an exponential decay, as in Fig. 3.14(a), whose rate of decay is determined by the time constant, $\tau = k_1 k_2$. For larger values of τ the decay is slower, implying a greater weighting of past values of the input for a longer time. Consequently, thinking in terms of the graphical convolution of $h(t)$ with an input signal, for large values of the time constant a sudden change in the input signal will be severely attenuated. On the other hand, in the same situation high-frequency noise in the input will be removed from the output.

The situation is slightly more complicated for second-order systems. A standardized form for the differential equation of a second-order system is

$$\ddot{y}(t) + a\dot{y}(t) + by(t) = kx(t). \tag{3.22}$$

It is perhaps easiest to determine the impulse response of this system by calculating the unit-step response, then taking its derivative as indicated by Eq. (3.16). For $x(t) = u(t)$, $y(t)$ has two parts—the homogeneous solution, $y_h(t)$, and the particular solution, $y_p(t)$. If the particular solution is thought of as the steady-state response to $x(t)$, then the homogeneous solution provides the transient change from the initial conditions (assumed to be zero) to the mean level of the steady-state response to the input signal. For a step input $y_p(t)$ is just a constant level, c_0. Substituting this solution for $y(t)$ in Eq. (3.22) when $x(t) = u(t)$ yields $c_0 = k/b$. To find the homogeneous solution for a second-order system note that $y_h(t)$ has the form

$$y_h(t) = c_1 e^{s_1 t} + c_2 e^{s_2 t}. \tag{3.23}$$

The parameters s_1 and s_2 are determined by solving the homogeneous form of the differential equation—that is, by substituting Eq. (3.23) for $y(t)$ in Eq. (3.22) and letting $x(t) = 0$. Thus

$$s_{1,2} = -\frac{a}{2} \pm \frac{\sqrt{a^2 - 4b}}{2}. \tag{3.24}$$

Consequently, the step-response is

$$y(t) = c_1 e^{s_1 t} + c_2 e^{s_2 t} + c_0$$

and taking the derivative yields

$$h(t) = \frac{dy(t)}{dt} = c_1 s_1 e^{s_1 t} + c_2 s_2 e^{s_2 t} = d_1 e^{s_1 t} + d_2 e^{s_2 t}, \qquad t \geq 0, \qquad (3.25)$$

where d_1 and d_2 are determined by the initial conditions that

$$y(0) = \dot{y}(0) = 0.$$

Clearly, the impulse response of a second-order system is specified by the parameters s_1 and s_2 that, in turn, are determined by the coefficients of the differential equation of the system. Three different situations can be identified:

1. $a^2 > 4b$: The term inside the square root in Eq. (3.24) is positive and s_1 and s_2 are both real. Therefore, the impulse response comprises the sum of two exponentially decaying terms. For BIBO stability, $h(t)$ must be bounded (otherwise there will exist some inputs for which $y(t)$ will not be bounded). Consequently, both s_1 and s_2 must be negative. Since $h(0) = 0$, then $d_2 = -d_1$ and $h(t)$ has the general shape depicted in Fig. 3.16(a). For reasons that will become apparent imminently, these impulse responses are called *overdamped*. In terms of signal processing, an overdamped second-order system will modify input signals qualitatively like a first-order system, although there are some potentially important differences. First, the peak of $h(t)$ is not at $t = 0$ as it is for a first-order system. Therefore, the recognition of input events in the output is slightly delayed. Second, for the second-order system there is a much greater variety of possible shapes for $h(t)$. Consequently, the filtering effects of convolution can be adjusted more subtly in order to remove or emphasize certain features of the input signal. This latter effect is more striking in the next case below.

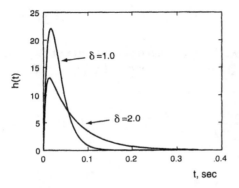 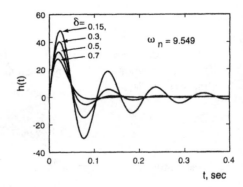

FIGURE 3.16. Impulse responses of second-order lowpass systems with various damping factors. (a) Overdamped responses. (b) Underdamped responses.

2. $a^2 < 4b$: The term inside the square root is negative and s_1 and s_2 are complex conjugates. Thus

$$s_1 = \sigma + j\Omega,\ s_2 = \sigma - j\Omega, \qquad \text{where } \sigma = -\frac{a}{2}, \qquad \Omega = \frac{\sqrt{4b - a^2}}{2} \qquad (3.26)$$

and

$$h(t) = d_1 e^{\sigma t} e^{j\Omega t} + d_2 e^{\sigma t} e^{-j\Omega t}$$
$$= d_1 e^{\sigma t}[\cos(\Omega t) + j\sin(\Omega t)] + d_2 e^{\sigma t}[\cos(\Omega t) - j\sin(\Omega t)].$$

To make $h(0) = 0$ we must have $d_2 = -d_1$, as discussed above. Consequently, the cosine terms cancel. After defining $d = jd_1$, we have

$$h(t) = d e^{\sigma t} \sin(\Omega t) u(t). \qquad (3.27)$$

That is, $h(t)$ is an exponentially damped sine wave. We can use a trick to find d in terms of the system parameters. Since the zero-state response to a unit-step input is the integral of $h(t)$, we can express the final value of the unit-step response as

$$g(t)_{t\to\infty} = \int_0^\infty h(t)dt = d \int_0^\infty e^{\sigma\lambda} \sin(\Omega\lambda) d\lambda$$
$$= \frac{d}{\sigma^2 + \Omega^2} [e^{\sigma\lambda}(\sigma \sin(\Omega\lambda) + \Omega \cos(\Omega\lambda))]|_0^\infty$$
$$= \frac{\Omega d}{\Omega_n^2},$$

where Ω_n is the value of Ω when $a = 0$ (i.e., when there is no term in the differential equation related to dy/dt, meaning that the differential equation represents a pure oscillation) and is called the *natural frequency*. If $a = 0$, then

$$\sigma = 0 \qquad \text{and} \qquad h(t) = d\sin(\Omega t), \qquad \text{where } \Omega = \frac{\sqrt{4b}}{2} = \sqrt{b} \triangleq \Omega_n. (3.28)$$

That is, when $a = 0$, $h(t)$ is an undamped oscillation having a frequency defined as the natural frequency of the system. Note that the natural frequency is readily identifiable from the differential equation since $\Omega_n^2 = b$.

Now, from the previous expression for the unit-step response, its final value is c_0. Since c_0 must satisfy the original differential equation, Eq. (3.22), when $x(t) = u(t)$, then $c_0 = k/b$. Using the fact that $b = \Omega_n^2$, equating the two expressions for the final value of the step response gives

$$\frac{\Omega d}{\Omega_n^2} = \frac{k}{\Omega_n^2}, \qquad \text{from which } d = \frac{k}{\Omega}.$$

Finally, the impulse response has the general form observable in Fig. 3.16(b). This important case will be discussed in more detail shortly.

3. $a^2 = 4b$: In this case $s_1 = s_2 = -a/2$ and

$$h(t) = d_1 e^{\sigma t} + d_2 t e^{\sigma t}, \qquad t \geq 0. \tag{3.29}$$

The general form of this impulse response is given in Fig. 3.16(a), labeled "$\delta = 1.0$."

Consider again the situation when s_1 and s_2 (known as the *eigenvalues* of the system) are complex conjugates. Let's explore how the impulse response changes as we vary the parameter a while keeping the natural frequency (and b) constant. The X's in Fig. 3.17 depict the locations of s_1 and s_2 on the complex plane for some value of a such that $0 < a^2 < 4b$. Notice first that at these locations the square of the length of the vector connecting s_1 (or s_2) to the origin is

$$l^2 = \sigma^2 + \Omega^2 = \left(\frac{-a}{2}\right)^2 + \left(\frac{4b - a^2}{4}\right) = b = \Omega_n^2. \tag{3.30}$$

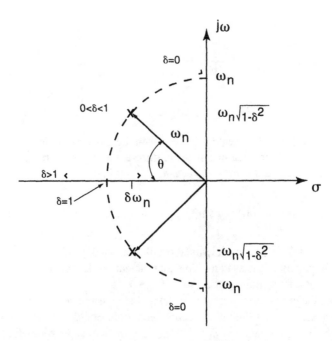

FIGURE 3.17. Locations in the complex plane of the eigenvalues of a second-order system (i.e., the exponents of the exponential terms in the homogeneous response) as a function of damping factor. For $\delta = 0$ the eigenvalues are at $\pm j\Omega_n$. For $0 < \delta < 1$, they lie on a semicircle of radius Ω_n, reaching the negative real axis simultaneously when the damping factor equals one. When $\delta \geq 1$, the eigenvalues are on the negative real axis.

That is,

$$l = \sqrt{\sigma^2 + \Omega^2} = \Omega_n. \tag{3.31}$$

Now, define a *damping factor*, δ, as the ratio of the magnitude of the exponential decay coefficient to the natural frequency. In other words,

$$\delta = \frac{|\sigma|}{\Omega_n} = \cos(\theta), \tag{3.32}$$

where θ is shown in the figure. For a given natural frequency the impulse response can be characterized by its damping factor. When $\delta = 0$, $\sigma = 0$ and we previously ascertained that $h(t)$ is an undamped sinusoid. In the complex plane of Fig. 3.17, s_1 and s_2 will lie on the imaginary axis. As the damping factor increases in the range $0 < \delta < 1$, s_1 and s_2 will move along the semicircle of radius Ω_n centered at the origin. They will reach the real axis simultaneously when $\delta = 1$. Thereafter, one eigenvalue will move left along the real axis and one right (but always to the left of the origin) with further increase of the damping factor.

The location of the two eigenvalues reveals much regarding the structure of $h(t)$. Note that these parameters can be expressed as

$$s_{1,2} = -\delta\Omega_n \pm j\Omega_n\sqrt{1 - \delta^2}. \tag{3.33}$$

Then $h(t)$ can be written in the form

$$h(t) = \frac{k}{\Omega_n\sqrt{1 - \delta^2}} e^{-\delta\Omega_n t} \sin(\Omega_n\sqrt{1 - \delta^2}t) \tag{3.34}$$

and $h(t)$ can be graphed for various values of its parameters. When $0 < \delta < 1$, $h(t)$ is an *underdamped* oscillation as shown in Fig. 3.16(b), and when $\delta > 1$, $h(t)$ is *overdamped*. When $\delta = 1$, $h(t)$ is said to be *critically damped*. Of course, when $\delta > 1$, we have the case discussed above when s_1 and s_2 are both real.

Thus one can readily sketch the impulse response of a second-order system from knowledge of the coefficients of the differential equation of that system. The signal processing implications of these impulse responses depends on the value of the damping factor. For δ near 1 the interpretation is similar to that when the impulse response has two real exponential terms. But as δ decreases, $h(t)$ assumes a more oscillatory nature. This oscillatory behavior will "add on" to sudden changes in the input signal and possibly obscure them. On the other hand, if the input signal contains an oscillation near the frequency of oscillation in $h(t)$, it should be apparent that the convolution process will amplify that input signal component while also attenuating higher frequencies. Therefore, an underdamped second-order system can be a sensitive detector of oscillation at a known frequency. This behavior, in which a system (that is not a true oscillator) oscillates at some preferred frequency in the presence of a weak input at that frequency, is known as *resonance* and the damped oscillatory impulse response of a resonant system often is said to exhibit *ringing*.

Example 3.9   Second-order systems   Describe the impulse response of the LTIC systems described by the following differential equations:

$$\text{(a) } \ddot{y}(t) + 6\dot{y}(t) + 10y(t) = 0.15x(t);$$

$$\text{(b) } \ddot{y}(t) + 10\dot{y}(t) + 6y(t) = 0.15x(t).$$

Solutions. (a) Solve for s_1 and s_2. Thus

$$s_{1,2} = -\frac{a}{2} \pm \frac{\sqrt{a^2 - 4b}}{2}, \qquad \text{where } a = 6, b = 10. \text{ Substituting,}$$

$$s_{1,2} = -3 \pm \tfrac{1}{2}\sqrt{36 - 100} = -3 \pm j8.$$

Therefore, $h(t) = d\, e^{-3t} \sin(8t)\, u(t)$. This impulse response is an exponentially decaying sinusoid having a frequency of approximately 1.27 Hz. Its amplitude reaches zero in approximately three time constants—that is, one second.

(b) In this case $s_{1,2} = -5 \pm 8 = -13, -3$. Thus $h(t)$ has two real roots and $h(t) = d(e^{-13t} - e^{-3t})u(t)$.

3.7   CONVOLUTION AS A FILTERING PROCESS

The process of convolving the impulse response of an LTI system with an input signal is the time-domain representation of *filtering*. Consequently, convolution is a key concept for understanding the modification of signals by filters. Does this concept also have a role in understanding how physical biomedical systems respond to stimuli? Very definitely! As long as the assumption of linearity is valid, then knowledge of the response of an unknown system to a step unit will permit at least qualitative insight into its impulse response, from which the response to any other stimulus may be estimated. It is important to recognize that it is *not* necessary to know the differential equation of the system in order to predict its behavior if one can determine its impulse (or, equivalently, its step) response. In fact, in many applications a semiquantitative knowledge of the impulse or step response is sufficient. For example, if I know that the response of a blood pressure transducer to a step increase in pressure has an approximately exponential rise with a time constant of 500 msec, then I would not want to use that transducer to measure blood pressure fluctuations between heartbeats because the impulse response will perform a weighted average of past blood pressure values over a time interval of about three time constants, or 1.5 s, which is larger than the typical intervals between heartbeats.

Example 3.10   Using convolution to predict the effect of filtering on a biomedical signal   Action potentials recorded from a nerve axon via an extracellular monopolar electrode have amplitudes on the order of microvolts. Thus it is necessary to amplify these signals in order to display or store them (Fig. 3.18(a)). This example will explore some of the necessary properties for electronic amplifiers designed to amplify action potential signals. A simple model of a train of action potentials is a sequence of pulses, each lasting a few milliseconds (Fig. 3.18(e)). An average "firing rate" of 10 pulses per second characterizes the behavior of many types of receptors and motor neurons.

Second-order electronic amplifiers are easy to build and there exist standard implementations for such electronic filters (discussed further in Chapter 6). First,

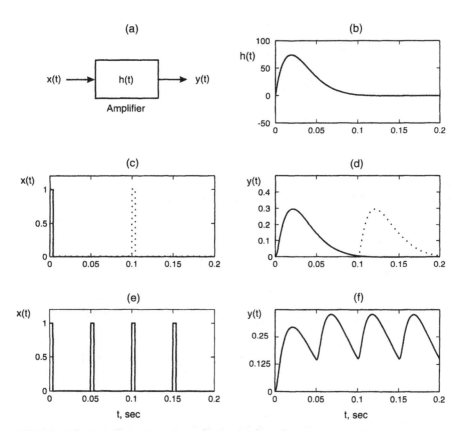

FIGURE 3.18. Determining the response of an amplifier to a train of simulated action potentials. (a) Model of the amplifier as an LTIC system. (b) Impulse response of the amplifier for the specified parameters, $\delta = 0.75$, $\Omega_n = 20\pi$, gain $= 10^4$. (c,d) Input (c) and output (d) for a signal containing one action potential (solid) or two action potentials separated by 100 msec (dotted). (e) Simulated action potential train with 20 pulses per second. (f) Output in response to the signal of (e).

however, one must specify the desired gain, natural frequency, and damping factor for the filter. In the present application a gain of at least 10,000 is necessary. Since it seems that having a strong oscillation in the output in response to a pulse-like input is not desirable (because action potentials are pulse-like), we specify an under-damped system with a damping factor of 0.75. Finally, one might be tempted to specify a natural frequency on the order of the expected highest frequency of action potentials—in our case, 10 per second. Therefore, $\Omega_n = 2\pi(10) = 20\pi$ rad/sec.

In order to estimate the effect of the amplifier on action potential signals, first find the output response to a single (simulated) action potential at the input. This response can be derived by convolution. Using the specified parameter values,

$$h(t) = \frac{10^4}{20\pi\sqrt{1-0.75^2}} e^{-0.75(20\pi)t} \sin(20\pi\sqrt{1-0.75^2}t)$$

$$= 368.3\, e^{-15\pi t} \sin(8.75\pi t). \tag{3.35}$$

$h(t)$ is plotted in Fig. 3.18(b). The single simulated action potential is shown as the solid graph of Fig. 3.18(c). Its width is 4 msec. Assuming an LTIC system, the response of this system to the simulated action potential input is given by the convolution integral as

$$y(t) = \int_0^t x(t-\lambda)h(\lambda)d\lambda = \int_0^t P_{0.004}(t-0.002)[(368.3)e^{-15\pi\lambda} \sin(8.75\pi\lambda)]d\lambda, \tag{3.36}$$

where $P_a(t-t_1)$ is a unit-amplitude pulse of width a that is centered at $t = t_1$. This integral can be evaluated using the general integration formula

$$\int e^{ax} \sin(bx)dx = \frac{e^{ax}}{a^2+b^2}(a\sin(bx) - b\cos(bx)).$$

For $t < 0.004$,

$$y(t) = 368.3\int_0^t e^{-15\pi\lambda} \sin(8.75\pi\lambda)d\lambda$$

$$= \frac{368.3}{(15\pi)^2 + (8.75\pi)^2}[e^{15\pi\lambda}(-15\pi\sin(8.75\pi\lambda) - 8.75\pi\cos(8.75\pi\lambda))]\big|_0^t.$$

Simplifying, we obtain

$$y(t) = 0.3839[8.75\pi - e^{-15\pi t}(15\pi\sin(8.75\pi t) + 8.75\pi\cos(8.75\pi t))], \quad t \le 0.004. \tag{3.37}$$

To evaluate the convolution integral for $t > 0.004$, notice that in Eq. (3.36) this integral is written in the alternate manner of Eq. (3.15), in which $x(t)$ is time-reversed instead of $h(t)$. Now at any time t, $x(t - \lambda)$ is nonzero on the interval $[t - 0.004, t]$ only. Therefore, the limits on the convolution integral become $t - 0.004$ and t, and we may calculate $y(t)$ as

$$y(t) = 368.3 \int_{t-0.004}^{t} e^{-15\pi\lambda} \sin(8.75\pi\lambda)d\lambda$$

$$= \frac{368.3}{(15\pi)^2 + (8.75\pi)^2} [e^{15\pi\lambda}(-15\pi \sin(8.75\pi\lambda) - 8.75\pi \cos(8.75\pi\lambda))]|_{t-0.004}^{t}.$$

Simplifying, we obtain

$$y(t) = 0.3839e^{15\pi t}[15\pi e^{0.06\pi}\sin(8.75\pi[t - 0.004]) + 8.75\pi e^{0.06\pi}\cos(8.75\pi[t$$
$$- 0.004]) - 15\pi \sin(8.75\pi t) - 8.75\pi \cos(8.75\pi t)]. \tag{3.38}$$

The output, $y(t)$, is graphed as the solid line in Fig. 3.18(d). This result also can be obtained using the MATLAB command conv(x,h) in which h is defined using Eq. (3.35) over a sufficiently fine time sampling—for example, $\Delta t = 0.00025$ s— and x is a single pulse defined using the stepfun function. Note that a single pulse-like event such as an action potential signal causes a slow rise and a slower fall in the output signal, which considerably outlasts the original event. The dotted graphs in Fig. 3.18(c,d) depict the input and output for two simulated action potentials separated by 100 msec. In this (and the following) case the output was generated using conv(x,h). Basically one can obtain an indication of the presence of each action potential but the signal is so distorted that the original action potential shape (which may be informative with respect to pathological changes in the neuron) is essentially unrecoverable. Finally, Fig. 3.18(e,f) show the input and response of the amplifier to a train of four action potentials occurring at a rate of 20 per second. Although one has an indication of the occurrence of each action potential, this signal is now so distorted that it must be deemed unreliable as a record of the action potential signal. An important conclusion is that designing a second-order system to achieve an acceptable level of distortion of a signal is not a trivial task. The reader may wish to use the m-file secondo52.m to generate second-order systems having various parameter values and try to determine "best" parameter values for the present application.

3.8 IMPULSE RESPONSES FOR NONLINEAR SYSTEMS

Consider a system given by the transformation operation $y(t) = S[x(t)]$ for which $S[.]$ does not satisfy both the additivity and homogeneity conditions for linearity. From a mathematical viewpoint one may define an impulse response, $h(t) = S[\delta(t)]$, but will knowledge of $h(t)$ allow the calculation of $y(t)$ for any arbitrary $x(t)$? To address this question, consider the simple nonlinear system, $y(t) = x^2(t)$, which has the impulse response $h(t) = \delta^2(t)$. This system fulfills the additivity criterion under many conditions. For example, let $x_1(t) = \delta(t)$ and $x_2(t) = 3\delta(t - t_0)$, yielding the outputs $y_1(t)$ and $y_2(t)$ respectively. If we now apply the input $x(t) = x_1(t) + x_2(t)$, the output will be $y(t) = y_1(t) + y_2(t)$ *except* when $t_0 = 0$. In the latter case, to satisfy additivity $y(t)$ apparently should be $10\delta^2(t)$, but it will actually be $16\delta^2(t)$. Thus knowledge of the response to the input $x_1(t) = \delta(t)$ does not permit one to correctly predict the response to the input $x(t) = 4\delta(t)$. In fact, the situation is more complicated because the homogeneity condition specifies that the response to $x(t)$ should be $4\delta^2(t)$, not $10\delta^2(t)$!

Although the system $S[x(t)] = x^2(t)$ is not linear, it would still be possible to determine its output for an arbitrary input if one knew both $h(t)$ and a "correction factor" that could be applied to compensate for the system not being linear. To evaluate this correction factor, consider again the response to $x(t) = \delta(t) + 3\delta(t)$. The response expected by linearity is $y_L(t) = (1 + 3)\delta^2(t)$, whereas the actual response is $y(t) = (1 + 3)^2\delta^2(t)$. The difference due to nonlinearity is $y_N(t) = [(1 + 3)^2 - (1 + 3)]\delta^2(t) = 12\delta^2(t)$. Similarly, for any input of the form

$$x(t) = \delta(t) + A\delta(t), \tag{3.39}$$

the output will be

$$y(t) = (1 + A)^2\delta^2(t) = (1 + A)\delta^2(t) + [(1 + A)^2 - (1 + A)]\delta^2(t)$$
$$= (1 + A)\delta^2(t) + (A^2 + A)\delta^2(t). \tag{3.40}$$

The term $(1 + A)\delta^2(t)$ is the linear response and the term $(A^2 + A)\delta^2(t)$ is the correction factor. For an input occurring at any arbitrary time, t_1, the situation is similar. Since $S[.]$ is memoryless, the output at time t_1 depends only on the input at that time. But since $x(t_1)$ can always be expressed as

$$x(t_1) = \delta(t - t_1) + [x(t_1) - 1]\delta(t - t_1) = \delta(t - t_1) + A\delta(t - t_1),$$

the output $y(t_1)$ can always be determined from Eq. (3.40). In this example, therefore, it is always possible to specify the output response to an arbitrary input if one knows $h(t)$ and the "correction factor" equation $y_N(t) = (A^2 + A)\delta^2(t)$. (Of course, in this simple example it is easier to calculate $x^2(t)$ directly!)

This approach can be extended to all time-invariant, analytic nonlinear systems although determining "correction factor" equations becomes much more complicated if the nonlinearity is of higher order than quadratic and especially if the system has memory. In the latter case an input at time t_1 will contribute to the output at times $t_2 > t_1$ until $t_2 - t_1$ exceeds the memory of the system. Therefore, although $x(t)$ can still be expressed by Eq. (3.39), $y_N(t)$ is a function of the amplitude nonlinearity (as given in Eq. (3.40)) plus the linear and nonlinear effects of past inputs. Volterra has shown that even in this general case it is possible to define an input–output relationship based on generalized impulse response functions.

The response of a time-invariant, analytic, nonlinear system, $S[.]$, to an arbitrary input $x(t)$ can be expressed as

$$y(t) = h_0 + \int_{-\infty}^{\infty} x(\lambda)h_1(t - \lambda)d\lambda + \int_{-\infty}^{\infty} \int_{-\infty}^{\infty} x(\lambda_1, \lambda_2)h_2(t - \lambda_1, t - \lambda_2)d\lambda_1 d\lambda_2$$

$$+ \int_{-\infty}^{\infty} \int_{-\infty}^{\infty} \int_{-\infty}^{\infty} x(\lambda_1, \lambda_2, \lambda_3)h(t - \lambda_1, t - \lambda_2, t - \lambda_3)d\lambda_1 d\lambda_2 d\lambda_3 + \dots . \quad (3.41)$$

Equation (3.41) is called the *Volterra expansion* of the system and the $h(\dots)$ functions are the *Volterra kernels*. Although it looks as if $h_1(t)$ is the impulse response of the system, $h_2(t_1,t_2)$ is the system response to two impulses, and so on, this simple interpretation is incorrect. Consider the impulse response of a system described by Eq. (3.41). When $x(t) = \delta(t)$, every integral term contributes to the output, with the result that

$$y(t) = h_0 + h_1(t) + h_2(t, t) + h_3(t, t, t) + \dots .$$

In other words, the impulse response comprises the diagonal parts of all of the Volterra kernels.

In general, it is difficult to determine the Volterra kernels for a nonlinear system, in part because the different terms in Eq. (3.41) are not orthogonal. Wiener has derived a similar expansion for the input–output properties of a nonlinear system when it is excited by Gaussian white noise, in which the terms are orthogonal. Identification of the Weiner kernels of a nonlinear system is generally easier than identification of Volterra kernels and numerous examples of applications of the Weiner method to nonlinear biomedical systems are in the literature. Estimation of Weiner kernels is not trivial, however, and usually requires large amounts of data for accurate identification. The interested reader should consult an advanced textbook (e.g., Marmorelis and Marmorelis, 1978).

3.9 THE GLUCOSE CONTROL PROBLEM, REVISITED

At the beginning of this chapter we discussed the problem of predicting the response of plasma glucose concentration to an arbitrary infusion of insulin and decided that the biological system (or at least our model of the biological system) is nonlinear. On the other hand, for sufficiently small inputs the difference between the actual response and that predicted from a known response to a pulsatile insulin infusion by assuming linearity was negligible (Fig. 3.4). Often in biomedical applications it is the case that linearity can be assumed for some restricted range of input amplitude and then convolution can be applied to predict the response of the system. To finish this exercise the reader first should determine the impulse response of the model using the SIMULINK file glucos3.mdl. In actual practice, one cannot test a patient with a true "impulse function" input, so it is necessary to utilize a more practical input from which one can derive the impulse response. Specify such an input. Furthermore, because the glucose level is not zero when the insulin input is zero, one must define the impulse response relative to the change from some baseline level. Specify a baseline level and utilizing your input defined above, determine $h(t)$. Then predict (using the MATLAB command conv) the responses of glucose to the various inputs discussed previously (see Figs. 3.2–3.5). Consider whether these predictions are sufficiently accurate by comparing them with the actual responses of the model to these inputs. In addition, determine the impulse response measured at a different baseline level. For example, if your initial baseline level was the glucose level with zero insulin infusion, choose a new baseline corresponding to a steady insulin infusion rate of 0.9 U/hr. Compare the original and new impulse responses and explain any differences.

3.10 SUMMARY

This chapter has introduced the related concepts of impulse response and convolution. The impulse response of an LTI (or LSI) system is its zero-state response to a unit impulse function (or a DT unit-pulse function). Systems with memory have impulse (or unit-pulse) responses which are nonzero for some finite (but nonzero) range of t (or n). A DT system is classified as a finite impulse response (FIR) system if its unit-pulse response, $h[n]$, reaches zero at a finite n and remains at zero for larger n. Otherwise it is an infinite impulse response (IIR) system. The impulse response expresses the degree to which past inputs contribute to the current output; this contribution can be evaluated using the convolution sum or convolution integral formulas. A graphical interpretation of convolution was developed in order to visualize the process of multiplying $x(t)$ (or $x[k]$) point-by-point in time by $h(t_0 - t)$ (or $h[k_0 - k]$), and then integrating (or summing) the product to calculate $y(t_0)$ (or $y[k_0]$). The convolution operation is associative, commutative, and distributive.

By viewing any linear system as a filter, one can visualize the signal processing effect of the system on an arbitrary input in the time domain through graphical convolution. Thus signal processing interpetations were developed for first-order and second-order systems. Because the impulse response of a system is determined by the coefficients of the differential equation of the system, we were able to develop insights into the filtering behaviors of first- and second-order systems from their differential equations. Second-order systems, in particular, can exhibit varying types of filtering effects, depending on whether they are overdamped or underdamped.

The concept of the impulse response can be extended to generalized impulse responses for nonlinear systems. Often, however, it is possible to utilize a linear system approximation to a nonlinear system, based on its impulse response, over a restricted amplitude range for its input signal. Usually a linear system approximation to a nonlinear system is valid only for small deviations around specific mean levels of the input and output signals.

EXERCISES

3.1 Use graphical convolution to sketch the results of convolving each pair of functions in Fig. 3.19.

3.2 Use graphical convolution to sketch the results of convolving each pair of functions below, then calculate the exact answers and compare to your sketches:
  a. $x(t) = 2\, u(t)$, $h(t) = e^{-t/5}\, u(t)$;
  b. $w[n] = 2\, u[n] - 4\, u[n-2] + 2\, u[n-4]$, $v[n] = \sin[2\pi n/4]\, u[n]$;
  c. $x[n] = u[n-3]$, $h[n] = -0.2\delta[n-1] + \delta[n] - 0.2\delta[n+1]$.

3.3 Two discrete-time systems are connected in series as shown in Fig. 3.20. Their input–output difference equations are:

$$T_1[.]\!: w[k] = 0.25\, x[k-2], \qquad T_2[.]\!: y[k] = 0.5\, w[k-1] + 0.5\, w[k-2].$$

  a. Determine the overall input–output difference equation relating $y[k]$ to $x[k]$.
  b. Determine the impulse response of the overall system.

3.4 Two systems are connected together as shown in Fig. 3.21 The input–output differential equations for these systems are:

$$T_2[.]\!: w(t) = 0.5y(t) \qquad T_1[.]\!: \ddot{y}(t) = -\dot{y}(t) - 2y(t) + v(t),$$

  a. What is the overall input–output differential equation relating $y(t)$ to $x(t)$?
  b. Describe the qualitative behavior of the overall impulse response.

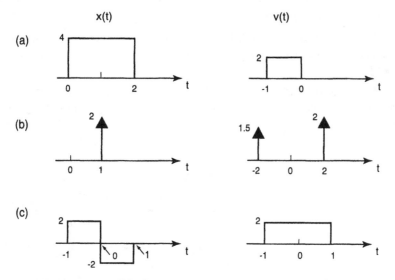

FIGURE 3.19. See Exercise 3.1.

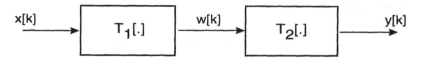

FIGURE 3.20. Two LSI systems connected in series. See Exercise 3.3.

c. Compare the impulse response of the system with feedback to that of $T_1[.]$ by itself.

3.5 A system has an impulse response $h(t) = e^{-t}u(t)$. Calculate its output when its input is $x(t) = \cos(2\pi t)$.

3.6 A discrete-time LTIC system has the impulse response $h[n] = 2u[n - 1] - 2u[n - 3]$. Graph $h[n]$. Calculate the output for the discrete exponential input $x[n] = 0.5^n u[n]$. Discuss the signal processing characteristics of a system having the unit-pulse response $h[n]$.

3.7 Calculate the convolution of $x(t) = e^{-at}u(t)$ with $h(t) = e^{-bt}u(t)$. Sketch the result for the case when a and b are both positive and $b < a$.

3.8 The impulse response of a system is given by $h(t) = e^{-5t} \sin(20\pi t)$. Using graphical convolution sketch the system response to the following inputs:
   a. $x(t) = u(t) - u(t - .00001)$;
   b. $x(t) = 10\ u(t)$.

3.9 Let $x(t) = u(t + 2) - u(t - 3)$ and $h(t) = \delta(t + 2) + \delta(t - 1)$. If $x(t)$ is the input to an LTI system and $h(t)$ is the impulse response of the system, calculate and graph the output. Discuss the signal processing characteristics of a system having the impulse response $h(t)$. Also calculate the output if the roles of $x(t)$ and $h(t)$ are reversed and discuss the signal processing characteristics of this new system.

3.10 If $x[n]$ is the input to an LSI system having the unit sample response, $h[n]$, determine the output $y[n]$. $x[n] = a^n u(n)$, $h[n] = b^n u[n]$, $a \neq b$. Sketch $x[n]$, $h[n]$, and $y[n]$.

3.11 Write the differential equation for the electrical circuit shown in Fig. 3.22 with $v(t)$ as the input and $e(t)$ as the output. Describe the signal processing characteristics of this filter.

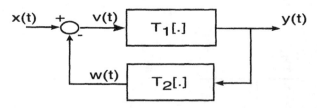

FIGURE 3.21. Feedback connection of two LTI systems. See Exercise 3.4.

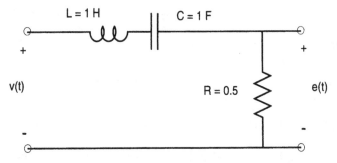

FIGURE 3.22. An electrical filter with input v(t) and output e(t). See Exercise 3.11.

3.12   A DT differencing filter has a unit pulse response given by

$$h[n] = \begin{cases} 1, & n = 0 \\ 0, & n = 1 \\ -1, & n = 2 \\ 0, & \text{otherwise} \end{cases}$$

Calculate the response of this filter to a square wave $x[n]$ described as follows:

$$x[n] \text{ has a period of 8,} \qquad x[n] = \begin{cases} 1, 0 \le n \le 3 \\ -1, 4 \le n \le 7 \end{cases}.$$

3.13   Draw the function that results from convolving $f(t)$ and $g(t)$ of Fig. 3.23 when: (a) $P = 2$; (b) $P = 1$.

3.14   A common method of smoothing discrete-time data is to pass it through a weighted moving average filter. A filter of this type with 3-point Bartlett weighting is defined by the difference equation $y[n] = 0.25\,x[n] + 0.50\,x[n-1] + 0.25\,x[n-2]$. Determine the unit-pulse response of this filter. Determine its response to $x[n] = u[n]$.

3.15   A discrete-time system is described by the difference equation $y[k + 1] + 0.5y[k] = 2x[k]$. What is the impulse response sequence of this system? Discuss the signal processing characteristics of this system.

FIGURE 3.23. See Exercise 3.13.

3.16   An LSIC system has a unit-step response $g[k]$. The system receives the input $x[k] = \sum_{i=0}^{N-1} c_i \{u[k - k_i] - u\{k - k_{i-1}]\}$, where $k_0 < k_1 < \ldots < k_{N-1}$ and the c_i are constants. Express the resulting output in terms of $g[k]$ and time shifts of $g[k]$. Assume the system starts from the zero-state.

3.17   A system comprises two cascaded subsystems as shown in Fig. 3.20. The difference equations for these two systems are:

$$T_1[.]: w[k] + 0.6\ w[k - 1] = x[k]; \qquad T_2[.]: y[k] = w[k] - 0.5\ w[k - 1] - 0.5\ w[k - 2].$$

a. Find one difference equation that expresses the overall input–output relation for these cascaded systems.

b. Use the result from part (a) to calculate the overall unit pulse response of the cascaded systems.

c. Find the unit pulse response for each of the two subsystems.

d. Calculate the overall unit pulse response for a cascaded system by convolving the two individual responses found in part (c) and show that it is the same as your answer to part (b).

3.18   An electrophysiological recording table is subject to floor vibrations. When the tip of a recording electrode is inside a nerve cell whose diameter may be tens of microns, very little table vibration can be tolerated. Often each leg of such a table is placed on a "damper" consisting of a dashpot and spring in parallel. Assume that a dashpot and a spring are placed under *each* leg of a recording table, which has a total mass M, so that the *combined* damping resistance is R and the *combined* spring constant is K. A mechanical diagram representing the table and dampers is presented in Fig. 3.24.

a. Write the differential equation for this mechanical system. The input is the up-and-down displacement of the floor and the output is the table displacement in the same direction. (Ignore lateral motions.)

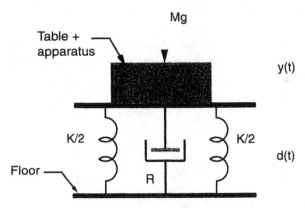

FIGURE 3.24. Simple mechanical model of a microscope table subjected to vertical floor vibrations. M = mass of table and apparatus; K = spring constant; R = viscous damping resistance. $d(t)$ is the floor displacement and $y(t)$ is the resulting table displacement.

b. Let $M = 25$ kg, $K = 0.1$ N/mm, and $R = 100$ N/m/s. Describe the characteristics of the impulse response of this mechanical system.

3.19 To control blood pressure in a patient, a physician gives him a bolus injection of a vasoactive drug every hour. Consider each injection as a discrete event, $x[n]$. A patient who has received no drug suddenly starts to receive the same dose, $x[n] = K$, every hour and the change in his blood pressure (from its value before starting the injections), $p[n]$, is measured immediately after each injection. Assume the first injection is given at $n = 0$. $p[n]$ is found to be: $p[n] = 5K\{1 - 0.7(0.5)^n - 0.3(0.2)^n\}$, $n \geq 0$.

a. Determine an equation for the impulse response, $h[n]$, of blood pressure to this drug and sketch $h[n]$.

b. If instead of giving the patient a bolus of size K every hour, the patient is given a bolus of size $6K$ every 6 hours, how will his blood pressure vary over a 24-hour time period ? To solve this part, use MATLAB (assume $K = 1$) and convolve the impulse response with an appropriate input signal. Do you think it is better to give the drug every hour or every 6 hours? Why?

3.20 To control the position of a robotic arm it is desirable to design a controller that compensates for the time constants of the arm and provides a simple relationship between the input signal and the position of the arm. Assume the controller in Fig. 3.25 is well designed and provides the position response, $P(t)$, for the input signal, $R(t)$, as shown.

a. What is the overall impulse response of arm position to the controlling input signal? Let $K = 10$ inches, $A = 10$.

b. Using convolution, calculate the position response to the input signal of Fig. 3.25(c).

3.21 (MATLAB exercise) Referring to the robotic arm of Exercise 3.20, when a mass load is placed on the arm the position response degrades so that in response to the input signal of Fig. 3.25(c) the arm position is

$$P(t) = 8[1 - e^{-4t}], \quad 0 \leq t \leq T$$
$$= 8e^{-4(t-T)}, \quad T < t \leq 2T$$
$$= 0, t > 2T.$$

a. Calculate the impulse response of the mass-loaded system.

b. Develop a SIMULINK model of the mass-loaded system. Test your model by exciting it with the signal of Fig. 3.25(c) and comparing the output of the model with the response specified above.

c. It is possible to measure $P(t)$ continuously in real time using a displacement transducer. Assume this measurement is available and that the measurement or a filtered form of it can be fed back and added to the control input signal, as depicted in Fig. 3.21, to attempt to compensate for the degradation of the position response. What characteristics should the impulse response of $T_2[.]$ possess in order to effect this compensation? Add this feedback path to your SIMULINK model and test your suggestions.

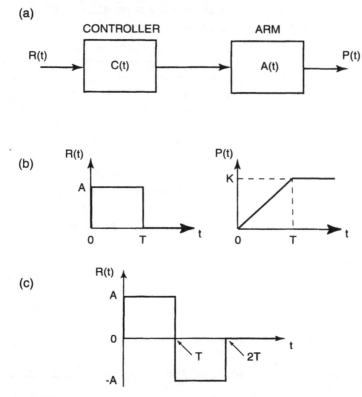

FIGURE 3.25. See Exercise 3.20. (a) Model of robotic arm driven by an electronic controller. (b) Input, $R(t)$, and arm position output, $P(t)$. (c) Test input.

3.22 (MATLAB exercise) Figure 3.26 shows a simple model for the ingestion and disposition of a drug in the body. The drug is taken orally into the gut, is absorbed into the blood, and is removed from the blood and excreted by the kidney. (Of course, many other physiological actions are possible, but this is a simple model!) A fluid model of this process is diagrammed in the figure. Let $x(t)$ be the rate of input of the drug to the gut, $q(t)$ be the total amount of drug in the gut, $y(t)$ be the amount of drug in the blood, and $z(t)$ be the rate of drug excretion. A usual approach with compartmental models such as this one is to assume that the *rate* of transfer of a substance between two consecutive compartments is proportional to the difference in amount of the substance in the two compartments. Thus, if k_1 and k_2 are constants, the following relationships apply:

$$z(t) = k_2 y(t) \qquad \text{and} \qquad w(t) = k_1[q(t) - y(t)].$$

a. Write the basic compartmental equations for the gut and blood compartments. Combine them into one second-order differential equation whose solution is $y(t)$, the amount of drug in the blood. (Usually the clinically effective dose is expressed

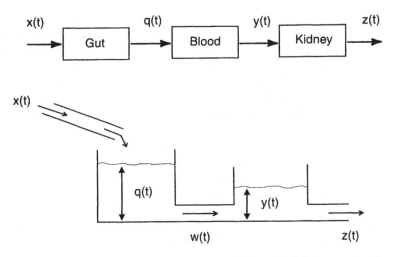

FIGURE 3.26. See Exercise 3.22. Top: simple conceptual model of the transport and eventual excretion of a drug that enters the gut at a rate $x(t)$. Bottom: fluid mechanical model of drug transport and excretion. This model ignores the removal of the drug from the blood via metabolism. $q(t)$ = amount of drug in gut; $y(t)$ = amount of drug in blood; $w(t)$ = rate of transfer of drug from gut to blood; $z(t)$ = rate of removal of drug from blood by excretion.

as the *concentration* of drug in the blood rather than the amount, but we have simplified the problem.)

b. Let $k_1 = 0.02$ mg/min/mg. Construct a plot of the damping factor of this system as a function of k_2. Let the range of k_2 be $[0.001, 0.10]$ mg/min/mg.

c. Let $k_2 = 0.01$ and $k_1 = 0.1$. Using MATLAB, determine the response of $y(t)$ to a drug dosage given by $x(t) = 30[u(t) - u(t-1)]$. (*Hint:* Determine the impulse response of the system from the parameters of the differential equation.) If the drug is clinically effective when $y(t) > 5$ mg, how long does it take after the initial dose before the drug is effective, and for how long is it effective?

3.23 (MATLAB exercise) Functional neuromuscular stimulation is the procedure by which natural activation of muscles is replaced by activation through electrical stimulation using trains of electrical pulses. For best control of the resulting force it is necessary to develop a model of the response of the muscle to electrical stimulation, then to use this model to calculate the "correct" stimulus to deliver. Recently Bernotas et al. determined models for two muscles of the cat hindlimb, the plantaris and the soleus. In their experimental preparation (and in the resulting models) the muscle was stimulated with a constant-frequency pulse train and the width of the pulse was varied. Therefore, the models represent the change in force caused by a change in pulse width from the baseline value of pulse width. Typical models were:

soleus: $y[k] = 0.585y[k-1] + 0.147y[k-2] + 0.011x[k-1]$;

plantaris: $y[k] = 0.653y[k-1] - 0.060y[k-2] + 0.018x[k-1]$, where in both cases $x[k]$ is the input pulse width and $y[k]$ is the output force. In both cases the DT model was created by sampling the input and output signals every time that an input stimulus pulse occurred. For soleus, a stimulus pulse occurred every 30 msec and for plantaris, every 108 msec.

a. Determine and plot the impulse responses of these two muscles.

b. Plot the impulse responses again but convert the time axis to continuous-time by explictly accounting for the frequency at which the data were sampled. From the impulse responses, can you tell which muscle is a fast-twitch muscle and which is a slow-twitch one?

c. To compare the muscle responses to the same sinusoidal input, it is necessary to specify the input in CT and sample it at the corresponding frequency for each muscle. Generate a unit-amplitude, 1-Hz sine wave that starts at $t = 0$ and persist for 5 cycles, and sample it at 33.3 Hz (for the plantaris) and at 9.26 Hz (for the soleus). Excite each model with the corresponding DT input signal and determine the force output. Which muscle more closely follows this input signal?

4

FREQUENCY RESPONSE

4.1 INTRODUCTION

From a biomedical engineering perspective there are several general goals that require one to obtain a quantitative description of the transformation properties of a system. One such goal is simply the measurement of system properties for classification. The auditory evoked potential (AEP), for example, is a wave that is recorded in an EEG lead in response to an auditory click. It may be viewed as an approximate impulse response of the auditory system. The AEP has several peaks and troughs that are associated with various subsystems in the central nervous system that process auditory signals. By measuring the amplitudes and latencies of these features the neurologist can classify the auditory system of the patient as normal or abnormal and suggest the site of an abnormality. Another example is the use of inert gas washout to determine the ventilation to perfusion ratio(s) in the lungs. In the MIGET technique the patient breathes a mixture of several inert gases, such as helium and argon, and the decay of the concentrations of these gases in the lungs is measured after resumption of air breathing. The "time constants" of these decays are indices of the transformation properties of the lung mechanisms for gas transport. From these time constants one may calculate ventilation/perfusion distributions and categorize them as normal or abnormal.

The need to quantify the transformation properties of a system arises also in the context of replacing a natural controlling signal. This type of application is exemplified by the use of insulin infusions to control blood glucose concentration, as discussed in the previous chapter. A third type of application requiring quantitative information about system transformations is probably the most ubiquitous—the need to determine whether a measuring device distorts the signal being measured. For example, previously we discussed the suitability of a certain blood pressure transducer for measuring inter-beat pressure fluctuations. This issue may arise repeatedly during use of a transducer even when the initial performance is acceptable, as in the case of pH sensors, which can become progressively impaired by surface deposition of proteins when used in blood.

In previous chapters two methods were discussed for deriving quantitative descriptions of the transformation of an input signal by a linear, time-invariant system: solving the differential equation of the system subject to the input of interest, and convolving the input of interest with the impulse response of the system. Both of these methods are mathematically complete. Furthermore, the process of graphical convolution permits semiquantitative insights into the filtering properties of a system without requiring rigorous solution of differential or integral equations. Nonetheless, for certain classes of input signals one can obtain deep insights into the transformation properties of a system from alternative descriptions of the system. This chapter addresses one such alternative description, based on the transformations of *periodic* input signals effected by LTI (and LSI) systems. Although few if any biological processes are truly periodic, often it is possible to approximate a biomedical signal as a periodic signal and obtain useful insights about processes that respond to (or, filter) the signal. Furthermore, and more importantly, developing an understanding of how linear systems manipulate periodic signals will lay the foundation for understanding how these systems manipulate nonperiodic signals.

The previous chapter also established the principle that any system having an input and an output can be viewed as a filter. That is, for (virtually) any system the output signal will differ from the input signal because of the properties of the system, as reflected in its impulse response. In that chapter graphical convolution was utilized to assess the manner in which the weighting of past input values by the time-reversed impulse response function could reduce or accentuate time-dependent features of the input signal. Thus, in addition to the obvious example of an electronic amplifier, any transducer or measuring device is itself a filter. Also the reader should be comfortable with the notion that any biological system can be analyzed as a filter for which the input is the stimulus or controlling signal of the biological system and the output is the response of the system. Consequently, we will adopt the shorthand of referring to the transformation properties of a system as its "filtering" properties and will use the terms "system" and "filter" interchangeably.

4.2 BIOMEDICAL EXAMPLE (TRANSDUCERS FOR MEASURING KNEE ANGLE)

In order to assist paraplegic subjects to stand or walk, electrical stimuli are applied to various muscles of the legs in an attempt to mimic the natural use of these muscles. It is desirable to adjust the timing and intensity of stimulation in order to produce the smoothest motion having the least energetic demands on the muscles. To do so requires knowledge of the exact position of each leg. An important measure of leg position is the angle of rotation between the lower and upper legs, the knee angle. Various types of transducers may be used to measure this angle; this example compares two of these types. The first is a tilt sensor that measures angle of rotation by sensing the movement of a thin, semicircular column of fluid. The second is an electromagnetic transducer that senses the local strength of an electromagnetic field produced by a companion device located at a fixed site. The strength of the field varies reciprocally with the distance of the sensor from the source. To compare the performance of these two sensors, investigators attached them both to a wooden bar that was hinged at one end so that it hung vertically. They then repeatedly rotated the free end of the bar from the vertical position to a nearly horizontal position at different frequencies of rotation and compared the outputs of the two sensors. Ideally, the readings from the two sensors should always be equal.

Figure 4.1 presents data from these tests. The dashed curves in Fig. 4.1(a,b) are outputs from the electromagnetic transducer and the solid curves are from the tilt sensor. By plotting one signal against the other it is apparent that the two signals are very similar during the low-frequency movement (Fig. 4.1(c)). On the other hand, clearly they are different during the faster movement (Fig. 4.1(d)). We are forced to conclude that the behavior of at least one of the sensors (and perhaps both) depends on the frequency at which it is stimulated. The obvious explanation for the frequency-dependent behavior of the tilt sensor is that its output depends on actual movement of a small fluid column and inertia will slow this movement. At higher frequencies of motion inertial slowing will cause a delay that becomes significant relative to the period of the motion. Furthermore, when direction changes at the peak of the motion, there may be some "sloshing" of the fluid that produces irregular "bumps" in the signal.

Frequency-dependent behavior is a universal phenomenon in physical systems and is another indication of the filtering properties of a system. Therefore, there exists a close relationship between frequency-dependent behavior and the impulse response that will be developed in this chapter. It will become apparent also that fre-

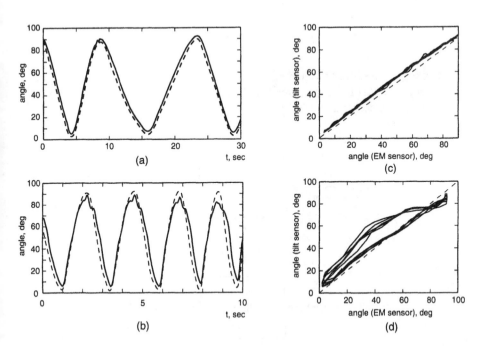

FIGURE 4.1. (a,b) Angle measured by a tilt sensor (solid) and an electromagnetic (EM) sensor (dashed) during rotation of a straight wooden board that is hinged at its upper end. (c,d) Tilt sensor output versus EM sensor output from "a" and "b," respectively. Courtesy of J. Abbas.

quency-dependent behavior does not always alter an input signal at high frequencies, as in the preceding example, but may have its effect over any range of frequencies.

4.3 SINUSOIDAL INPUTS TO LTIC SYSTEMS

Consider our standard LTIC system with input $x(t)$, output $y(t)$, impulse response $h(t)$, and transformation operator $F[.]$ (Fig. 4.2(a)). This system can be described by the general linear differential equation

$$a_n \frac{d^n y}{dt^n} + a_{n-1} \frac{d^{n-1} y}{dt^{n-1}} + \ldots + a_1 \frac{dy}{dt} + a_0 y = b_0 x + b_1 \frac{dx}{dt} + \ldots + b_m \frac{d^m x}{dt^m}. \quad (4.1)$$

Let $x(t) = C \cos(\Omega_0 t)$, $\forall t$, where Ω_0 is an arbitrary frequency, and assume the system is in a zero state initially. Consider the r.h.s. of Eq. 4.1 subject to this input. Note that

$$\frac{dx}{dt} = -\Omega_0 C \sin(\Omega_0 t) = \Omega_0 C \cos(\Omega_0 t - 3\pi/2).$$

Likewise,

$$\frac{d^i x}{dt^i} = (\Omega_0 C)^i \cos(\Omega_0 t - 3\pi i/2).$$

Since a weighted sum of cosine functions, all of the same frequency, is another cosine function at that frequency, then the r.h.s. of Eq. (4.1) simplifies to $B \cos(\Omega_0 t + \varphi)$, where B is a constant and φ is the net phase angle.

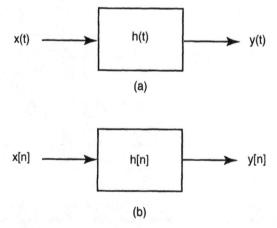

FIGURE 4.2. Generic CT (a) and DT (b) linear systems.

Now consider the l.h.s. of Eq. (4.1). Does $y(t)$ have the form $y(t) = D\cos(\Omega_0 t + \theta)$? Either all of the terms of $y(t)$ that are not of this form must cancel at all times (since there is no other form on the r.h.s.) or all terms of $y(t)$ have this form. The homogeneous solution of the differential equation is not likely to be purely sinusoidal but it *will cancel* (i.e., sum to zero over all of the l.h.s. terms) at all times. In the steady state, however, the homogeneous solution will have decayed to zero and then $y(t)$ must be a cosine function. That is, after the homogeneous part of the response has fallen to zero, the output of an LTIC system having a steady-state sinusoidal input must be a sinusoid of the same frequency as the input signal. Furthermore, there can be no other steady-state component in the output signal. Of course, the amplitude and phase of $y(t)$ are almost certain to differ from those of $x(t)$.

The above result can be expressed in terms of the system transformation operator as

$$\lim_{t\to\infty} y(t) = \lim_{t\to\infty} F[x(t)] = F_{ss}[C\cos(\Omega_0 t)] = C \cdot G(\Omega_0) \cdot \cos(\Omega_0 t + \theta(\Omega_0)), \quad (4.2)$$

where $F_{ss}[.]$ represents the steady-state response, and where $G(\Omega_0)$ is the *steady-state gain of the system for a sinusoidal input* having frequency Ω_0, and $\theta(\Omega_0)$ is the *steady-state phase shift* of the system for the same sinusoidal input. Note that $G(\Omega_0)$ and $\theta(\Omega_0)$ together contain all the necessary information to specify the *steady-state* response of the system to a cosine (or sine) input signal of frequency Ω_0. In other words, these two functions of frequency describe precisely the filtering properties of the system with respect to sinusoidal inputs. Together $G(\Omega_0)$ and $\theta(\Omega_0)$ specify the *(steady-state) frequency response of the system.*

This analytical approach is readily extended to situations where $x(t)$ comprises a sum of two or more sinusoidal signals. For example, let $x(t) = a_1\cos(\Omega_1 t) + a_2\cos(\Omega_2 t)$. By linearity, in the steady state,

$$y(t) = a_1 F_{ss}[\cos(\Omega_1 t)] + a_2 F_{ss}[\cos(\Omega_2 t)]$$
$$= a_1 G(\Omega_1)\cos(\Omega_1 t + \theta(\Omega_1)) + a_2 G(\Omega_2)\cos(\Omega_2 t + \theta(\Omega_2)).$$

Completely equivalent results can be demonstrated for DT systems (Fig. 4.2(b)) starting from the general difference equation for an LSIC system. Thus, if $x[n]$ is the input and $y[n]$ the output of an LSIC system, the steady-state response to the input $x[n] = C\cos[\omega_0 n]$ is $y[n] = C \cdot G(\omega_0) \cdot \cos[\omega_0 n + \theta(\omega_0)]$, where $G(\omega_0)$ and $\theta(\omega_0)$ are the steady-state gain and phase shift, respectively, of the DT system for sinusoidal inputs of frequency ω_0.

Example 4.1   Steady-state response of a first-order system to a sinusoidal input
To determine the steady-state output of the system $\dot{y}(t) + 1000y(t) = 1000x(t)$ to the input $x(t) = 5\cos(1000t)$, it is necessary to find the frequency response of this system at 1000 rad/s. Let $G = G(1000)$, $\theta = \theta(1000)$. In the steady state $y(t)$ still must

satisfy the differential equation of the system. Using Eq. (4.2) for $y(t)$ in the steady state and $x(t)$ as given above, the system differential equation becomes

$$-5G(1000)\sin(1000t + \theta) + 5(1000)G\cos(1000t + \theta) = 1000(5\cos(1000t)).$$

This equation may be expanded using standard formulas for $\sin(a + b)$ and $\cos(a + b)$ to yield

$$5000G[-\sin(1000t)\cos\theta - \cos(1000t)\sin\theta + \cos(1000t)\cos\theta - \sin(1000t)\sin\theta]$$

$$= 5000\cos(1000t),$$

from which, by equating the $\sin(\Omega_0 t)$ and $\cos(\Omega_0 t)$ terms on both sides of the equals sign, one obtains two relationships:

$$\cos\theta - \sin\theta = \frac{1}{G}, \qquad \cos\theta + \sin\theta = 0.$$

Using the result that $\sin^2 + \cos^2 = 1$, these equations may be solved to find

$$G = \frac{1}{\sqrt{2}}, \qquad \theta = -\frac{\pi}{4}\text{ rad.}$$

Therefore, the steady-state output is

$$y(t) = \frac{5}{\sqrt{2}}\cos\left(1000t - \frac{\pi}{4}\right).$$

Example 4.2 Frequency response of a biological system   The blood flow through a vascular bed (Fig. 4.3(a)) is determined by the pressure drop across the bed and the biomechanical properties of the vessels in the bed. A simple model (Fig. 4.3(b)) assumes that the inflow artery can be described by lumped parameters representing flow resistance, R_a, and wall compliance, C_a. The capillaries of the vascular bed are assumed to be noncompliant; therefore, the bed is described by a net flow resistance, R_c. Relative to the arterial pressure, P, pressure in the venus outflow is assumed to be zero. Now arterial pressure is the sum of a mean pressure plus the variations about the mean due to the heartbeat. A simple approximation is

$$P = P_{mean} + \Delta P(t) = P_{mean} + P_0\cos(\Omega_0 t). \tag{4.3}$$

The blood flow through the bed is the sum of the mean flow and the fluctuations in flow due to the heartbeat (i.e., the cosine term). The mean blood flow through the vascular bed is

$$\dot{Q}_{mean} = \frac{P_{mean}}{R_a + R_c}.$$

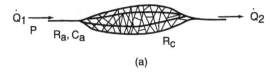

(a)

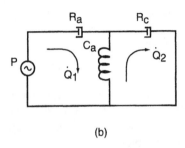

(b)

FIGURE 4.3. (a) Simplified schematic of a capillary plexus with arterial inflow and venous out-flow. P, instantaneous arterial pressure at the inflow side; $\dot{Q}_1$, $\dot{Q}_2$, instantaneous blood flow into and out of the capillary bed; R_a, R_c, flow resistances of the artery and capillary bed; C_a, compliance of the artery. (b) Electrical analog of the pressure-flow relationships of the capillary network above.

To determine the steady-state amplitude and phase of the fluctuations in blood flow through the bed due to the heartbeat requires finding the frequency response of this system at $\Omega = \Omega_0$. One approach is to ascertain the differential equation relating blood flow in the vascular bed to arterial pressure and then solve it for the cosine input term of Eq. (4.3). Begin by writing equations for the two loops of the model (Fig. 4.3(b)), considering only the $\Delta P(t)$ part of the input:

$$\Delta P = R_a \dot{Q}_1 + \frac{Q}{C_a}; \tag{4.4a}$$

$$0 = R_c \dot{Q}_2 - \frac{Q}{C_a}, \tag{4.4b}$$

where

$$Q = Q_1 - Q_2 = \int_0^t \dot{Q}_1 dt - \int_0^t \dot{Q}_2 dt. \tag{4.4c}$$

Note that $\dot{Q}_2(t)$ is the blood flow through the vascular (i.e., capillary) bed. Differentiating Eq. (4.4b),

$$0 = R_c \ddot{Q}_2 - \frac{\dot{Q}_1 - \dot{Q}_2}{C_a} \Rightarrow \dot{Q}_1 = \dot{Q}_2 + C_a R_c \ddot{Q}_2.$$

Substituting the above result in Eq. (4.4a), and then substituting for Q/C_a from Eq. (4.4b) yields

$$\Delta P = (R_a + R_c)\dot{Q}_2 + R_a C_a R_c \ddot{Q}_2. \tag{4.5}$$

It becomes apparent that Eq. (4.5) describes a first-order system if we let $X(t) = \dot{Q}_2(t)$. Thus, after making this substitution and rearranging,

$$\dot{X}(t) + \left(\frac{R_a + R_c}{R_a R_c}\right)\left(\frac{1}{C_a}\right)X(t) = \left[\frac{1}{R_a R_c C_a}\right]\Delta P(t). \tag{4.6}$$

For purposes of illustration, take specific values for the parameters of the problem. Thus let $P_{mean} = 100$ Torr, $P_0 = 9.6$ Torr, $C_a = 1.2$ ml/Torr, $R_a = 0.4$ Torr/ml/s, and $R_c = 2.0$ Torr/ml/s. Finally, assume the heart rate is one per second. Therefore, $\Omega_0 = 2\pi(1) = 2\pi$. Eq. (4.6) becomes

$$\dot{X}(t) + 2.5X(t) = \left[\frac{1}{0.96}\right]9.6 \cos(2\pi t). \tag{4.7}$$

There are numerous methods for solving Eq. (4.7). One approach is to use the knowledge that the steady-state solution must also be a cosine wave of the same frequency as the input signal. Thus, assuming $X(t) = A \cos(2\pi t + \theta)$, where A and θ are unknowns, and substituting into Eq. (4.7) yields

$$-2\pi A \sin(2\pi t + \theta) + 2.5A \cos(2\pi t + \theta) = 10 \cos(2\pi t).$$

Expanding the sine and cosine terms and grouping gives

$$[-2\pi A \cos(\theta) - 2.5A \sin(\theta)] \sin(2\pi t) + [-2\pi A \sin(\theta) - 2.5A \cos(\theta)] \cos(2\pi t)$$
$$= 10 \cos(2\pi t).$$

Since there is no sine term on the r.h.s., the coefficient of the $\sin(2\pi t)$ term on the l.h.s. must equal zero. Thus

$$-2\pi A \cos(\theta) = 2.5A \sin(\theta) \Rightarrow \tan(\theta) = \frac{-2\pi}{2.5} \Rightarrow \theta = -1.19 \text{ rad.}$$

Equating the coefficients of the cosine terms now permits evaluation of A. Thus

$$-2\pi A \sin(\theta) + 2.5A \cos(\theta) = 10 \Rightarrow [-2\pi(-0.929) + 2.5(0.370)] A = 10.$$

Solving, $A = 1.48$ ml/s. Therefore, in the steady state the fluctuation in blood flow through the capillary bed due to the heartbeat (under the assumption that the pressure fluctuations are cosinusoidal) is

$$\dot{Q}_2(t) = X(t) = 1.48 \cos(2\pi t - 1.19) \text{ ml/s.}$$

The mean level of the blood flow is found from the static relationship for series resistances presented above as

$$\dot{Q}_{mean} = \frac{P_{mean}}{R_a + R_c} = \frac{100}{2.4} = 41.7 \text{ ml/s}.$$

and the total steady-state capillary blood flow is the sum, $\dot{Q}_2(t) + \dot{Q}_{mean}$.

If the heart rate were to change, it would be necessary to evaluate the frequency response at the new frequency. Alternatively, one may determine the frequency response of flow through the capillary bed as a function of Ω by repeating the above solution for an arbitrary frequency, Ω. It is sufficient to let the input amplitude be unity since the system is linear and in the general case the output amplitude will be scaled in direct proportion to the input amplitude. Substituting $\Delta P(t) = \cos(\Omega t)$, by Eq. (4.2) the steady-state solution to Eq. (4.6) has the form $X(t) = G(\Omega) \cos(\Omega t + \theta(\Omega))$. Substituting these results into Eq. (4.6) and proceeding as above, we find

$$\theta(\Omega) = \tan^{-1}\left(\frac{-\Omega}{2.5}\right), \qquad G(\Omega) = \frac{1}{0.96\sqrt{6.25 + \Omega^2}}. \qquad (4.8)$$

These two functions are graphed in Fig. 4.4. Thus the *steady-state* blood flow response to a cosine pressure wave having arbitrary amplitude, $\Delta P(t) = P_0 \cos(\Omega t)$, is given by

$$\dot{Q}_2(t) = G(\Omega)P_0 \cos(\Omega t + \theta(\Omega)), \qquad (4.9)$$

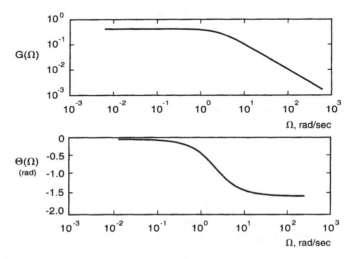

FIGURE 4.4. Magnitude (a) and phase shift (b) of the frequency response of the capillary bed model relating flow through the bed to arterial pressure.

where $G(\Omega)$ is the frequency-dependent *gain* of the system and $\theta(\Omega)$ is its frequency-dependent *phase shift*.

4.4 GENERALIZED FREQUENCY RESPONSE

Eigenfunctions and Generalized Frequency Response Function

Consider the LTI system with impulse response $h(t)$, input $x(t)$, and output $y(t)$. Let the input be a complex exponential function of the form $x(t) = ae^{j\Omega t}$, $\forall t$. By convolution one can express the resulting output as

$$y(t) = \int_{-\infty}^{\infty} h(\lambda)x(t - \lambda)d\lambda = \int_{-\infty}^{\infty} ah(\lambda)e^{j\Omega(t-\lambda)}d\lambda = ae^{j\Omega t}\int_{-\infty}^{\infty} h(\lambda)e^{-j\Omega\lambda}d\lambda$$

$$= ae^{j\Omega t}H(\Omega), \qquad \text{where } H(\Omega) \triangleq \int_{-\infty}^{\infty} h(\lambda)e^{-j\Omega\lambda}d\lambda. \tag{4.10}$$

For this *special case*, note from the definition of $x(t)$ that $y(t) = x(t)H(\Omega)$, where $H(\Omega)$ is a (possibly complex-valued) constant. Therefore $y(t)$ is a scaled version of the input. Any input signal for which the output of an LTIC system is simply a scaled version of the input is known as an *eigenfunction* of the system. Furthermore, because a complex exponential function is a generalized sinusoidal function, $H(\Omega)$ is regarded as the *generalized frequency response* of the LTI system. Since the generalized frequency response may be complex-valued, the response of the system to a complex exponential input may be written as $y(t) = a|H(\Omega)|e^{j(\Omega t + \arg(H(\Omega)))}$. (Keep in mind that this specific result applies *only* to eigenfunctions.) If one prefers to work with Herz instead of rad/s, the generalized frequency response can be defined equivalently as

$$H(f) = \int_{-\infty}^{\infty} h(t)e^{-j2\pi ft}dt. \tag{4.11}$$

If the LTI system is also causal, then $h(t) = 0$, $t < 0$, and the generalized frequency response is

$$H(\Omega) = \int_{0}^{\infty} h(\lambda)e^{-j\Omega\lambda}d\lambda. \tag{4.12}$$

The generalized frequency response of an LTI system and the frequency response defined previously are equivalent. To prove this assertion, consider the response of the above system to a cosine input. That is, let

$$x(t) = A\cos(\Omega_0 t) = \frac{A}{2}[e^{j\Omega_0 t} + e^{-j\Omega_0 t}], \qquad \forall t. \tag{4.13}$$

From the definition of the generalized frequency response, the steady-state output $y(t)$ can be expressed as

$$y(t) = \frac{A}{2}[H(\Omega_0)e^{j\Omega_0 t} + H(-\Omega_0)e^{-j\Omega_0 t}]$$

$$= \frac{A}{2}|H(\Omega_0)|[e^{j\Omega_0 t}e^{j\arg(H(\Omega_0))} + e^{-j\Omega_0 t}e^{-j\arg(H(\Omega_0))}], \qquad (4.14)$$

since it is easily proven that $H(-\Omega_0) = H^*(\Omega_0)$. Then by Euler's identity,

$$y(t) = A|H(\Omega_0)| \cos(\Omega_0 t + \arg(H(\Omega_0))). \qquad (4.15)$$

Equation (4.15) is equivalent to Eq. (4.2) if we let $G(\Omega_0) = |H(\Omega_0)|$ and $\theta(\Omega_0) = \arg(H(\Omega_0))$. *Therefore it is justifiable to refer to* $H(\Omega)$ *simply as "the frequency response" of the system,* while remembering that the frequency response summarizes the *steady-state* gain and phase shift imposed by the system on a sine or cosine input waveform as a function of frequency. That is, the frequency response quantifies the steady-state filtering effect of the system on sinusoidal (and complex exponential) inputs.

Note that one may express the frequency response function as

$$H(\Omega) = \int_{-\infty}^{\infty} h(\lambda)e^{-j\Omega\lambda}d\lambda = \int_{-\infty}^{\infty} h(\lambda)\cos(\Omega\lambda)d\lambda - j\int_{-\infty}^{\infty} h(\lambda)\sin(\Omega\lambda)d\lambda. \qquad (4.16)$$

Therefore, the real part of the frequency response is an even function of frequency, whereas the imaginary part is an odd function of frequency. Since the square of either term is an even function, $|H(\Omega)|$ is also an even function. The argument of $H(\Omega)$ depends on the ratio of the imaginary to the real part and is an odd function of frequency.

Equation (4.10) suggests another method for determining the frequency response of a system whose differential equation is known: let $x(t) = e^{j\Omega t}$ and, assuming that the transient response has decayed to zero, let $y(t) = H(\Omega)e^{j\Omega t}$. Substitute both of these relationships into the differential equation and solve for $H(\Omega)$.

Output Power

In the previous chapter the power, P_x, of a sinusoid was defined as the energy calculated over one cycle divided by the duration of the cycle. The power of the signal $x(t) = A\cos(\Omega t)$ is $A^2/2$. If $x(t)$ is the input to an LTI system with frequency response $H(\Omega)$, then the magnitude of the steady-state output will be $A|H(\Omega)|$. Consequently the steady-state output power will be $P_y = A^2|H(\Omega)|^2/2$. The *power gain* of the system is *the ratio of (sinusoidal) output power to input power in the steady state* and it equals $|H(\Omega)|^2$. The formal definition of the gain of a system in dB is

$$|H(\Omega)|_{dB} = 10\log(\text{power gain}) = 10\log(|H(\Omega)|^2) = 20\log(|H(\Omega)|). \qquad (4.17)$$

Frequency Response of a First-Order System

If the LTIC system shown in Fig. 4.2(a) is a first-order system, it can be described by the general first-order differential equation

$$\dot{y}(t) + \frac{1}{\tau}y(t) = bx(t),$$

with $h(t) = be^{-t/\tau}u(t)$.

From Eq. (4.12),

$$H(\Omega) = \int_0^\infty h(\lambda)e^{-j\Omega\lambda}d\lambda = \int_0^\infty be^{-\lambda/\tau}e^{-j\Omega\lambda}d\lambda = \frac{-b}{\frac{1}{\tau}+j\Omega}[e^{-(1/\tau+j\Omega)\lambda}\big|_0^\infty] = \frac{b}{j\Omega + \frac{1}{\tau}}. \quad (4.18)$$

Therefore,

$$H(\Omega) = \frac{|b|}{\sqrt{\Omega^2 + \frac{1}{\tau^2}}} = \frac{|b\tau|}{\sqrt{\Omega^2\tau^2 + 1}}. \quad (4.19)$$

Note that

$$|H(0)| = |b\tau| \text{ and } \left|H\left(\frac{1}{\tau}\right)\right| = \frac{|b\tau|}{\sqrt{2}} = \frac{1}{\sqrt{2}}|H(0)|.$$

The gain function for a first-order system is graphed in Fig. 4.5(a). Because the gain decreases as frequency increases, this system is referred to as a *lowpass filter*. The time constant defines the frequency, $1/\tau$, at which the gain is $1/\sqrt{2} \approx 0.7071$ of that at zero frequency. This frequency is called the *cutoff frequency*, Ω_c, although it is clear that input sinusoids having frequencies greater than the cutoff frequency are not "cut off" completely from the output.

The phase shift of this filter is

$$\arg(H(\Omega)) = -\tan^{-1}\left(\frac{\Omega}{1/\tau}\right) = -\tan^{-1}(\Omega\tau). \quad (4.20)$$

Note that

$$\arg(H(0)) = 0 \text{ and } \arg\left(H\left(\frac{1}{\tau}\right)\right) = -\tan^{-1}(1) = -\frac{\pi}{4} \text{ rad.}$$

This function is graphed in Fig. 4.5(b). The maximum phase shift for a first-order system is $-90°$.

Since the gain of a filter in dB is defined as 20 times the log of the magnitude of the frequency response, at the cutoff frequency

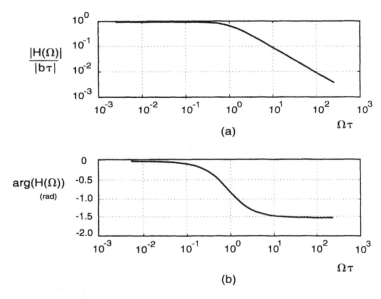

FIGURE 4.5. Magnitude (a) and phase shift (b) of the frequency response of a general first-order lowpass filter. Frequency is expressed as a ratio relative to the cutoff frequency.

$$\left| H\!\left(\frac{1}{\tau}\right) \right|_{dB} = 20 \log(|H(0)|) - 20 \log(\sqrt{2}), \qquad \text{and}$$

$$\left| H\!\left(\frac{1}{\tau}\right) \right|_{dB} - |H(0)|_{dB} = -20 \log(\sqrt{2}) \approx -3 \text{ dB}. \tag{4.21}$$

That is, at the cutoff frequency the gain is 3 dB less than the maximum gain. This result is applied to any lowpass filter by considering that the frequency at which gain (relative to the gain at zero frequency) is –3 dB is the cutoff frequency of the filter. For a lowpass filter it is common to refer to the frequency range below the cutoff frequency as the *passband* of the filter and the frequency range above the cutoff frequency as the *stopband* (although, of course, these terms are relative).

Example 4.3 *Frequency response of a first-order system*   Find the frequency response function of a system whose input–output differential equation is

$$\dot{y}(t) + 10y(t) = 2x(t)$$

and determine the steady-state filtering effect of this system on the input signal

$$x(t) + 3\cos(6\pi t) + \sin(\pi t)$$

Solution. $\tau = 0.1$, $b = 2$, and $h(t) = 2e^{-10t}u(t)$. Therefore, by Eqs. 4.18–4.20,

$$H(\Omega) = |H(\Omega)|e^{j\arg(H(\Omega))} = \frac{0.2}{\sqrt{0.01\Omega^2 + 1}}e^{-j\tan^{-1}(0.1\Omega)}.$$

Graphs of the magnitude and phase of this frequency response function would have the same shapes as those in Fig. 4.5. To determine the response to $x(t)$ it is necessary to evaluate the frequency response at the frequencies of the two sinusoids in the input signal. Thus

$$H(6\pi) = \frac{0.2}{2.133}e^{-j\tan^{-1}(0.6\pi)} = 0.0938e^{-j1.083}, \qquad H(\pi) = \frac{0.2}{1.048}e^{-j\tan^{-1}(0.1\pi)}$$

$$= 0.191e^{-j0.304}.$$

Applying Eq. (4.15) to each of the terms in $x(t)$, one obtains (in the steady state)

$$y(t) = 3(0.0938)\cos(6\pi t - 1.083) + 0.191\sin(\pi t - 0.304).$$

The filtering effect of this system on the input signal is severe. The output amplitudes of the two input components are reduced by 90.62% and 80.9%, respectively, and both components are phase-shifted. Figure 4.6 compares the input and output signals.

Frequency Response of a Second-Order Lowpass Filter

The general differential equation for a second-order LTIC system is

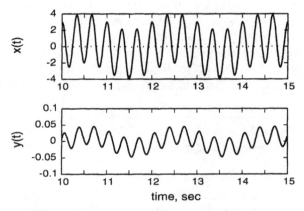

FIGURE 4.6. Steady-state input and output signals for Example 4.3. Note difference in the ordinate scales.

$$\ddot{y}(t) + a\dot{y}(t) + by(t) = cx(t). \tag{4.22}$$

Previously it was discussed that the character of the impulse response of this system depended on the values of the parameters a and b. In the case of two complex exponents in the homogeneous response, the impulse response was shown to be

$$h(t) = \frac{c}{\Omega}e^{\sigma t}\sin(\Omega t)u(t) = \frac{c}{\Omega_n\sqrt{1-\delta^2}}e^{-\delta\Omega_n t}\sin(\Omega_n\sqrt{1-\delta^2}\,t)u(t),$$

where the complex roots of the characteristic equation are

$$s_{1,2} = \sigma \pm j\Omega, \quad \text{and} \quad \Omega_n^2 = b, \quad \Omega = \sqrt{b - \frac{a^2}{4}}, \quad \delta = \frac{a}{2\sqrt{b}}, \quad \text{and} \quad \sigma = -\delta\Omega_n.$$

Therefore,

$$H(\Omega) = \int_0^\infty h(\lambda)e^{-j\Omega\lambda}d\lambda = \frac{c}{(j\Omega + \delta\Omega_n)^2 + \Omega_n^2(1-\delta^2)} = \frac{c}{\Omega_n^2 - \Omega^2 + j2\delta\Omega_n\Omega}. \tag{4.23}$$

The magnitude of the frequency response is

$$H(\Omega) = \frac{|c|}{\sqrt{(\Omega_n^2 - \Omega^2)^2 + 4\delta^2\Omega_n^2\Omega^2}}. \tag{4.24}$$

For a fixed value of the natural frequency (i.e., b), the frequency response is a function of the damping factor, δ, which is a function of the parameter a (Fig. 4.7). When $a = 0$, then

$$\delta = 0, \text{ and } H(\Omega) = \frac{|c|}{\Omega_n^2 - \Omega^2}.$$

For frequencies much greater than the natural frequency, the gain of this filter decreases monotonically, but at $\Omega = \Omega_n$, the gain is infinite. That is, when an input sinusoid at this frequency has zero amplitude, it appears that the output sinusoid can have nonzero amplitude—that is, the system can be an oscillator that produces an output in the absence of an input. For values of damping factor greater than zero but less than about 0.5, the magnitude of $H(\Omega)$ will exceed $|H(0)|$ in some range of frequencies, implying that frequencies in this range will be amplified to a greater extent than other frequencies. This effect, known as *resonance*, will be especially noticeable around $\Omega = \Omega_n$ for $\delta < 0.3$. It reflects the frequencies at which energy redistribution within the system occurs with the least loss.

When the damping factor is one, the gain function simplifies to

$$|H(\Omega)| = \frac{|c|}{\Omega_n^2 + \Omega^2}.$$

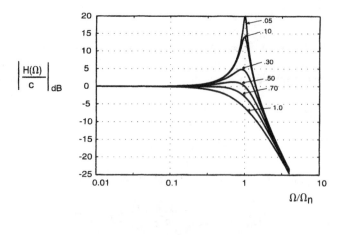

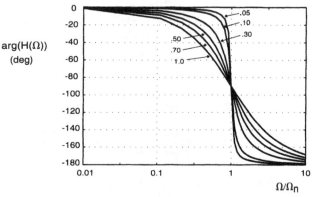

FIGURE 4.7. Gain (top) and phase shift (bottom) of the frequency response of a lowpass second-order filter as a function of damping factor. Frequency is expressed as a ratio relative to the natural frequency.

Now the gain is monotonically decreasing. In particular, when the input frequency equals the natural frequency, the gain is one-half of the gain at zero frequency—that is,

$$|H(\Omega_n)| = \frac{|c|}{2\Omega_n^2} = \frac{1}{2}|H(0)|.$$

The relative logarithmic gain in dB is

$$20 \log|H(0)| - 20 \log|H(\Omega_n)| = -20 \log(2) \approx -6dB, \qquad \text{when } \delta = 1.$$

When the damping factor is greater than one, the impulse response of a second-order system has two real exponential terms and the gain function is monotonically decreasing and everywhere below the gain function for $\delta = 1$ shown in Fig. 4.7. The maximum phase shift is $-180°$ and at $\Omega = \Omega_n$ the phase shift is $-90°$.

Highpass Filters

A highpass filter is one for which the passband includes all frequencies greater than or equal to some cutoff frequency. The frequency response functions for first-order and second-order highpass filters can be visualized by reflecting the frequency response of the corresponding lowpass filter about the line $\Omega = \Omega_c$ (first-order) or $\Omega = \Omega_n$ (second-order). Whereas lowpass filters often are utilized for removing high-frequency instrumentation noise from a signal, highpass filters are used to remove artifactual low-frequency variations in the baseline of the signal.

4.5   FREQUENCY RESPONSE OF DISCRETE-TIME SYSTEMS

The generalized frequency response of an LSI system with unit-pulse response, $h[n]$, and having the input $x[n]$ and output $y[n]$, is found in a manner parallel to that used for CT systems. Let $x[n]$ be a DT exponential signal of the form $x[n] = ae^{j\omega n}$, $\forall n$. The output is found by convolution to be

$$y[n] = \sum_{k=-\infty}^{\infty} h[k]x[n-k] = \sum_{k=-\infty}^{\infty} h[k]ae^{j\omega(n-k)} = ae^{j\omega n}\sum_{k=-\infty}^{\infty} h[k]e^{-j\omega k}.$$

Therefore, in this *special case*, $y[n] = x[n]H(e^{j\omega})$, where

$$H(e^{j\omega}) \triangleq \sum_{k=-\infty}^{\infty} h[k]e^{-j\omega k} \tag{4.25}$$

is the *(generalized) frequency response* of the DT system and $e^{j\omega n}$ is an eigenfunction of an LSI system.

$H(e^{j\omega})$ fulfills the same role for DT systems that $H(\Omega)$ serves for CT systems. Consider the steady-state response of an LSI system to the input $x[n] = A \cos[\omega n]$, $\forall n$. Using Eq. 4.25 and Euler's identity,

$$y[n] = \frac{A}{2}\{H(e^{j\omega})e^{j\omega n} + H(e^{-j\omega})e^{-j\omega n}\} = \frac{A}{2}|H(e^{j\omega})|\{e^{j(\omega n + \arg(H(e^{j\omega})))} + e^{-j(\omega n - \arg(H(e^{j\omega})))}\}$$

$$= A|H(e^{j\omega})| \cos[\omega n + \arg(H(e^{j\omega}))]. \tag{4.26}$$

$|H(e^{j\omega})|$ and $\arg(H(e^{j\omega}))$ are known as the *gain* and *phase shift* of the frequency response, respectively. Because Eq. (4.26) was derived under the assumption that the input signal was present for all time, this relationship applies only to the *steady-state* response to a sinusoidal input.

If the LSI system is also causal, then $h[n] = 0$, $n < 0$, and the generalized frequency response is

$$H(e^{j\omega}) = \sum_{k=0}^{\infty} h[k]e^{-j\omega k}. \qquad (4.27)$$

Note in addition that $e^{-j\omega k} = e^{-j(\omega + 2\pi)k}$. Consequently, $H(e^{j\omega})$ is *periodic with period* 2π. This result is necessary since the functions $\cos[\omega_0 n]$ and $\cos[(\omega_0 + 2\pi)n] = \cos[\omega_0 n + 2\pi n]$ are *identical DT functions*. Therefore, the system *must* filter the two functions identically. Note also that the magnitude of $H(e^{j\omega})$ is an even function of frequency (see Exercise 4.6), whereas the argument is an odd function.

Example 4.4   Frequency response of a DT system   An LSIC system has the unit-pulse response $h[n] = a^n u[n]$, where $|a| < 1$. Find its frequency response and discuss the steady-state filtering effects of this system on sinusoidal inputs.
   Solution.

$$H(e^{j\omega}) = \sum_{k=-\infty}^{\infty} h[k]e^{-j\omega k} = \sum_{k=0}^{\infty} a^k e^{-j\omega k} = \sum_{k=0}^{\infty}(ae^{-j\omega})^k = \frac{1}{1 - ae^{-j\omega}}. \qquad (4.28)$$

The gain and phase shift of this system are

$$|H(e^{j\omega})| = \left| \frac{1}{1 - a\cos(\omega) + ja\sin(\omega)} \right| = \left[\frac{1}{(1 - a\cos(\omega))^2 + (a\sin(\omega))^2} \right]^{1/2}$$

$$= \left[\frac{1}{1 + a^2 - 2a\cos(\omega)} \right]^{1/2}$$

$$\arg(H(e^{j\omega})) = -\tan^{-1}\left[\frac{a\sin(\omega)}{1 - a\cos(\omega)} \right]. \qquad (4.29)$$

These functions are plotted in Fig. 4.8 for $a = 0.70$. Since the frequency response has period 2π and the gain (phase) is an even (odd) function of frequency, only the range $0 \leq \omega \leq \pi$ is shown. The gain function is reminiscent of the gain function of a first-order, lowpass, CT system and one can define a cutoff frequency, ω_c, by the relationship $|H(e^{j\omega_c})| = (1/\sqrt{2})|H(e^{j0})|$. Here $\omega_c \simeq 1.192$ rad/s. (See Exercise 4.7.) The phase shift is unlike those of previous examples of CT systems but is not unusual for a DT system.

Example 4.5   An LSIC filter has the frequency response $H(e^{j\omega}) = 0.5\cos(\omega)e^{-j\omega/5}$. Find the steady-state output for the input signal

$$x[n] = 2\sin\left[\frac{2\pi}{10}n \right] - 3\cos[0.025n + 0.20].$$

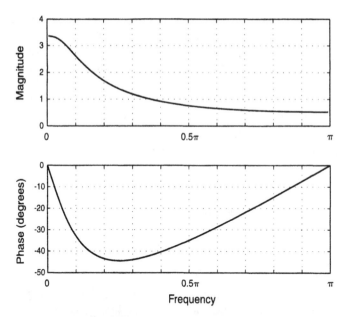

FIGURE 4.8. Magnitude (top) and phase shift (bottom) of the frequency response of a first-order DT lowpass filter with $a = 0.70$.

Solution. It is necessary to evaluate the frequency response at each of the two frequencies present in the input signal. Thus

$$H(e^{j0.2\pi}) = 0.5 \cos(0.2\pi)e^{-j0.04\pi} = 0.405e^{-j0.04\pi},$$

$$H(e^{j0.025}) = 0.5 \cos(0.025)e^{-j0.005} = 0.5e^{-j0.005}.$$

Taking the magnitude and phase of $H(e^{j\omega})$ at each frequency we find

$$y[n] = 2(0.405) \sin\left[\frac{2\pi}{10}n - 0.04\pi\right] - 3(0.5) \cos[0.025n + 0.20 - 0.005]$$

$$= 0.810 \sin\left[\frac{2\pi}{10}n - 0.04\pi\right] - 1.5 \cos[0.025n + 0.195].$$

Example 4.6 Determining the frequency response of an unknown system One way to determine the frequency response of an unknown physical system is to excite it with a sinusoidal input, then measure the amplitude and phase of the output after all transient responses have disappeared. The gain at the excitation frequency is the ratio of output amplitude to input amplitude and the phase shift is the phase difference between the output and input. This process can be repeated at many different frequencies covering the desired range of the frequency response function. Furthermore, if linearity can be assumed, then the input signal can comprise the sum of several sinusoids having a range of frequencies. The output signal will contain a summation of sinusoids at these same frequencies and selective filters can be applied to extract each output sinusoid individually. (Chapter 5 will discuss other methods for determining the amplitudes and phases of combined sinusoids.) By this method many measurements of the frequency response function at different frequencies can be obtained simultaneously. Figure 4.9 presents a diagram of this process.

As an example of the above process, consider an "unknown" system that is a two-point moving averager—that is, its impulse response is

$$h[n] = [0.5, 0.5].$$

Let this system be excited by the input signal

$$x[n] = \sum_{k=1}^{5} \frac{1}{k} \cos[0.15k\pi n]. \tag{4.30}$$

The resulting steady-state output signal may be analyzed by using filters which have zero gain everywhere except in a narrow range around one of the input frequencies, where the gain is unity. (Such filters are examples of *bandpass* filters.) Figure 4.10 shows input signal $x[n]$ defined by Eq. (4.30) and the steady-state output re-

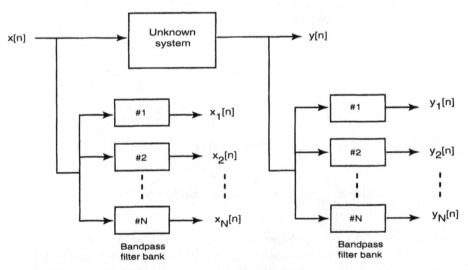

FIGURE 4.9. A method for identifying the frequency response function of an unknown linear system. The steady-state response, $y[n]$, is measured when the input, $x[n]$, comprises a sum of sine waves of N different frequencies. Both $x[n]$ and $y[n]$ are filtered by identical bandpass filter banks, each filter of which is tuned to pass only one of the N frequencies. Then estimates of the gain and phase of $H(e^{j\omega})$ at each frequency are determined as the relative amplitudes and phases of the outputs of corresponding filters.

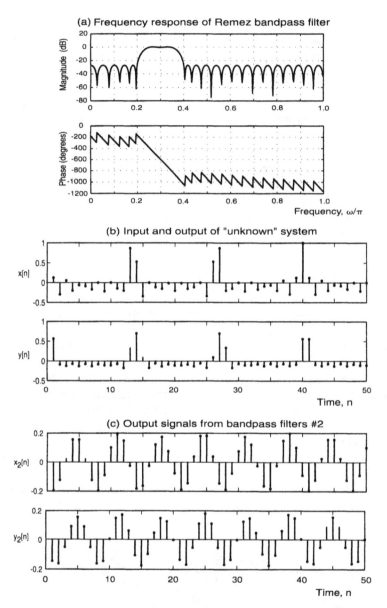

FIGURE 4.10. Identification of the frequency response of a two-point moving averager filter using an input that is the sum of five sinusoids (Example 4.6) having frequencies of 0.15π, 0.30π, 0.45π, 0.60π, 0.75π. (a) Frequency response of the bandpass filters tuned to pass signals at the second of the five frequencies ($\omega = 0.30\pi$). This filter was designed using the Remez algorithm (Chapter 8). (b) Steady-state input and output of the two-point moving average filter. (c) Steady-state outputs of the bandpass filter of (a) in response to the input signals $x[n]$ and $y[n]$.

sponse, $y[n]$. A bandpass filter was designed (see Chapter 8) to pass the signal components corresponding to $k = 2$ (i.e., $\omega = 0.3\pi$); the frequency response of this bandpass filter, and the filtered signals $x_2[n]$ and $y_2[n]$ are also shown in the figure. The bandpass filtered signals are sinusoids having peak-to-peak amplitudes of 0.400 and 0.358, respectively. Dividing the output amplitude by the input amplitude yields the estimated gain at this frequency of 0.895. The phase shift may be estimated by plotting $y_2[n]$ versus $x_2[n]$ and using Lissajou's formula. It is simple to calculate the actual frequency response of this filter and show that these estimates are reasonably accurate (see Exercise 4.8). This example is available in the m-file idxdemo.m.

The above unknown system is implemented also in the MATLAB function file unknown4.m. (Use help unknown4 to learn how to call this function.) It would be instructive to test this system at each of the individual frequencies in $x[n]$ (Eq. (4.30)) and demonstrate that one obtains the theoretical values for the gain and phase shift.

4.6 SERIES AND PARALLEL FILTER CASCADES

In many physical situations one encounters an ensemble of interconnected filters rather than one isolated system. For example, many measurement systems comprise a sensing device that provides an input signal to a high-gain amplifier whose output then is modified by a filtering stage, the output of the latter being presented to a display device for viewing. To assess the net effect of the measuring system on the signal being measured, one must know the frequency response of each intervening device or amplifier/filter stage. The net filtering effect depends on how the filters are interconnected. Although the following discusses CT filters, the equivalent results for DT systems are obvious.

Consider first the series connection of two filters shown in Fig. 4.11(a). Let $x(t) = A\cos(\Omega_0 t)$. Then in the steady state $v(t) = A|H_1(\Omega_0)|\cos(\Omega_0 t + \arg(H_1(\Omega_0)))$. Similarly, the steady-state output $y(t)$ is $y(t) = A|H_1(\Omega_0)||H_2(\Omega_0)|\cos(\Omega_0 t + \arg(H_1(\Omega_0)) + \arg(H_2(\Omega_0)))$. Therefore, the overall gain from $x(t)$ to $y(t)$ is the product of the in-

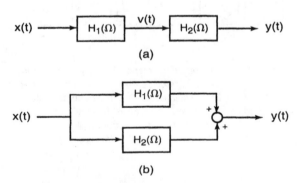

FIGURE 4.11. (a) Series and (b) parallel connections of two arbitrary linear filters.

dividual gains of the two filters, and the overall phase shift is the sum of the two phase shifts. Consequently, if $H(\Omega)$ is the overall frequency response of the cascaded filters, then

$$|H(\Omega)| = |H_1(\Omega)||H_2(\Omega)|, \quad \text{and} \quad \arg(H(\Omega)) = \arg(H_1(\Omega)) + \arg(H_2(\Omega)). \quad (4.31)$$

Note that the two parts of Eq. 4.31 are equivalent to the single relationship

$$H(\Omega) = H_1(\Omega)H_2(\Omega). \qquad (4.32)$$

If the cascade consists of N filters connected in series, then the net frequency response is the product of all of the individual frequency responses.

For the parallel connection of two filters shown in Fig. 4.11(b) it is apparent that the steady-state output in response to $x(t)$ given above is

$$y(t) = A|H_1(\Omega_0)| \cos(\Omega_0 t + \arg(H_1(\Omega_0))) + A|H_2(\Omega_0)| \cos(\Omega_0 t + \arg(H_2(\Omega_0))). \quad (4.33)$$

The terms in Eq. (4.33) can be combined to evaluate the overall frequency response. For simplicity, write $H_1(\Omega) = H_1 = |H_1|e^{j\angle H_1}$, $H_2(\Omega) = H_2 = |H_2|e^{j\angle H_2}$ If the overall frequency response is $H(\Omega) = H$, then it is clear that $H = H_1 + H_2$. Since H_1 and H_2 are complex numbers,

$$|H|^2 = [|H_1| \cos(\angle H_1) + |H_2| \cos(\angle H_2)]^2 + [|H_1| \sin(\angle H_1) + |H_2| \sin(\angle H_2)]^2.$$

After expanding, combining terms, and taking the square root,

$$|H(\Omega)| = [|H_1(\Omega)|^2 + |H_2(\Omega)|^2 + 2|H_1(\Omega)||H_2(\Omega)| \cos(\arg(H_1(\Omega)) - \arg(H_2(\Omega)))]^{1/2}.$$
$$(4.34)$$

Similarly, the phase shift is

$$\arg(H(\Omega)) = \tan^{-1}\left[\frac{|H_1(\Omega)| \sin(\arg(H_1(\Omega))) + |H_2(\Omega)| \sin(\arg(H_2(\Omega)))}{|H_1(\Omega)| \cos(\arg(H_1(\Omega))) + |H_2(\Omega)| \cos(\arg(H_2(\Omega)))} \right]. \quad (4.35)$$

If the two filters produce the same phase shift, then Eq. (4.42) indicates that the overall gain is the sum of the individual gains. Equality of phase, however, is more likely to be the exception than the rule.

If one is confronted with a more complex interconnection of filters than a simple series or parallel cascade, then it is necessary to successively combine series or parallel subsets of filters until the overall frequency response is found.

4.7 IDEAL FILTERS

A filter allows sinusoidal waveforms in some frequency range that are present in its input signal to "pass through" to the output signal relatively unchanged, whereas sinusoidal waveforms in other frequency ranges are relatively attenuated in (or "cut out" of) the output signal compared to their amplitudes in the input signal. In the steady state, the relative amplification or attenuation (and the phase shift) as a function of the frequency of the sinusoid is described by the frequency response of the filter. Earlier in this chapter the cutoff frequency of a filter was defined as the frequency at which the gain is approximately 70.71% of the maximum gain.

Despite the name, the cutoff frequency usually does not represent a frequency that absolutely demarcates between frequencies for which the gain is nonzero and frequencies for which the gain is zero because most practical filters do not exhibit such sharp changes in gain. In practical situations, however, one may be able to specify the range of important frequencies present in the signal of interest; therefore, an absolute cutoff of sinusoidal components having frequencies outside of that range may be desirable. Of course, it is also desirable that the filter not distort those important frequency components that are not cut off from the output. A filter meeting these criteria is known as an "ideal" filter, and it is common to evaluate a practical filter in terms of how closely it approximates the frequency response of an ideal filter.

The magnitude response of an ideal filter has a gain of one for frequencies that should be passed through it unattenuated and a gain of zero for all other frequencies. Figure 4.12 plots the frequency responses of the four types of ideal filters: (1) lowpass, (2) highpass, (3) bandpass, and (4) band reject. The frequency range for which the gain is 1 is the passband of the ideal filter.

In addition to complete attenuation of undesired frequency components, an ideal filter should not alter the timing relationships among the sinusoidal components that are not attenuated. This goal is achieved by specifying an ideal phase response that is a linear function of frequency. To show that timing relationships are not changed by this phase response, consider an input signal having two sinusoids:

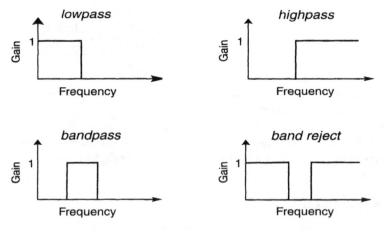

FIGURE 4.12. General representations of the gain functions for the four different types of ideal filters.

$x(t) = A \sin(\Omega_1 t) + B \sin(\Omega_2 t)$. Assume a filter frequency response $H(\Omega)$ such that the two frequencies lie in the passband of the filter and $\arg(H(\Omega)) = -\Omega d$, where d is a constant. The steady-state output of the filter is

$$y(t) = A \sin(\Omega_1 t - \Omega_1 d) + B \sin(\Omega_2 t - \Omega_1 d) = A \sin(\Omega_1(t - d)) + B \sin(\Omega_2(t - d)).$$

That is, the output is just the input signal delayed by d seconds and the timing relationship between the signal components is unaltered.

4.8 FREQUENCY RESPONSE AND NONLINEAR SYSTEMS

A simple example illustrates the difficulty of applying the concept of frequency response to a nonlinear system. Consider the algebraic nonlinear system given by $y[n] = F\{x[n]\} = x^2[n]$ and let $x[n] = A \sin[\omega_0 n]$. The output is

$$y[n] = A^2 \sin^2[\omega_0 n] = \frac{A^2}{2}(1 - \cos[2\omega_0 n]).$$

Using the relationship of Eq. (4.26) we find that the frequency response of this sytem, $F(e^{j\omega})$, has the apparent values $F(e^{j\omega_0}) = 0$, $F(e^{j2\omega_0}) = \infty$. It should be clear that for *any* choice of input frequency, the apparent gain at that frequency will be zero even though the apparent gain at that same frequency will be infinite if the input frequency is halved! That is, the gain depends on the input signal. One might suppose that the concept of frequency response could be "extended" for this system by defining the gain as the relationship between the input amplitude at ω_0 and the output amplitude at $2\omega_0$. That gain would be $A/2$ and it would depend on the amplitude of the input signal. Therefore, the concept of frequency response described for linear systems generally is not directly applicable to nonlinear systems.

For certain combinations of a simple nonlinear system with a linear system, however, it is possible to account for nonlinear interactions such as those identified above and to describe the output (in response to simple inputs like a sum of several sinusoids) in terms of the frequency response of the linear system and the parameters of the nonlinear system. Bendat (1990) illustrates these approaches extensively. Obviously, for complex nonlinear systems, the specification of a "frequency response" becomes more problematical. Typically, any such response is a function of the mean level, amplitude, and frequency content of the input signal.

In cases where the mean level of the input signal is constant and the amplitude of the input is small relative to the amplitude distortion produced by nonlinear gain, it often is possible to "linearize" the nonlinear system and approximate its behavior by the frequency response of a linear system. In the case of a static nonlinearity, such as the above example, the linearization can be accomplished by computing a Taylor series expansion of the nonlinear function about the mean level of the input signal. Thus, for the system given above, assuming the input has a mean level of one-half,

$$y[n] = \bar{x}^2 + (x[n] - \bar{x})\frac{dy}{dx}\Big|_{\bar{x}} + (x[n] - \bar{x})^2\frac{d^2y}{dx^2}\Big|_{\bar{x}} + \ldots \approx \bar{x}^2 + (x[n] - \bar{x})\frac{dy}{dx}\Big|_{\bar{x}}$$

$$= \bar{x}^2 + (x[n] - \bar{x}).$$

If we let $\bar{y} = \bar{x}^2$, $\Delta y[n] = y[n] - \bar{y}$, $\Delta x[n] = x[n] - \bar{x}$, then $\Delta y[n] = \Delta x[n]$ is the linear approximation to this system at the operating point $\bar{x} = 0.5$. The gain of a filter whose output equals its input is, of course, one at all frequencies. (What is its phase shift?) The accuracy of this approximation decreases rapidly for $x < 0.45$ or $x > 0.55$. For example, for $x = 0.55$, $y \approx \bar{x}^2 + \Delta x = 0.2500 + 0.0500 = 0.3000$. The correct value is $y = (0.55)^2 = 0.3025$. But for $x = 0.60$, the estimate and true value of y are 0.3100 and 0.3600, respectively. The linearized estimate is 16% too small.

4.9 OTHER BIOMEDICAL EXAMPLES

Frequency Response of Transducers

The frequency response functions of biomedical transducers and amplifiers may impose distortions on a measured signal and it is imperative that the user certify that his or her transducers and amplifiers do not introduce misleading distortions of the signals. For example, Hartford, et al. (1997) discuss the use of fluid-filled catheters for measuring respiratory variations in esophageal pressure of neonates, which is used for calculating lung compliance. For the immature lung of the neonate, lung compliance is an important index of the biomechanical state of the lungs from which one infers the effort required to breathe. The authors used a loudspeaker to apply pressures to a water-filled chamber and measured the frequency response of the pressure signal from various catheters placed in the chamber. The results obtained from a Ven FG-size 8 nasogastric tube implemented as a water-filled catheter are presented in Fig. 4.13. This frequency response function has a peak reminiscent of resonance in a second-order filter (see Exercise 4.20). Consequently, when using this catheter it may be necessary to filter the recorded signal with a filter having a

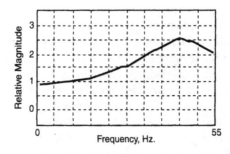

FIGURE 4.13. Frequency response of a Ven FG-size eight water-filled catheter (Hartford et al., 1997). © IEEE

frequency response that approximates the inverse of that of the catheter. Then the series cascade of the catheter and the filter can have an overall relative magnitude response that is more nearly constant over the whole frequency range shown.

Impedance of the Respiratory System

In Chapter 2, Fig. 2.1, we used a simple model of the biomechanical properties of the lungs in which the lungs were considered to be an elastic bag that fills and empties via a resistive flow channel. Because the actual structure of the lungs involves many branching airways and many compliant air spaces, a more complex model is necessary for a more complete description of pulmonary biomechanics; however, the large number of flow channels and air spaces overwhelms most modeling efforts. Consequently, the lungs are often modeled as a "black box" with the transpulmonary pressure (i.e., $P_B - P_{pl}$ in the figure) as its input and airflow as its output. This black box can be characterized by its frequency response and changes in the frequency response can be correlated with disease processes in the lungs and airways. Since measurement of P_{pl} is an invasive procedure, for clinical diagnosis the frequency response of the lungs and chest together is often measured. Generally in an intact subject one replaces the airway opening to the room with a mouthpiece connected to a pressure source such as a loudspeaker (Fig. 4.14(a)). The pressure source delivers sinusoidal pressure oscillations at one frequency at a time and the net pressure across the lungs and chest (i.e., pressure at the mouth, P_m, minus pressure around the chest, P_B) is measured as the input signal. The resulting flow into the respiratory system ($\dot{V}_m$) is the output signal, from which the gain and phase shift may be calculated. Instead of gain and phase shift, however, a more common approach is to calculate the reciprocal of the frequency response—that is, the pressure divided by the air flow. Since pressure is analgous to voltage and air flow to current, this relationship is a measure of generalized resistance known as *impedance*. The pressure-flow properties of the lungs and chest are characterized by measuring the magnitude of impedance and the phase shift (or, equivalently, the real and imaginary parts of impedance) at each frequency over a range of frequencies. In fact, contemporary methods for measuring pulmonary impedance utilize pressure signals that are sums of sinusoids or even random noise. The methods for separating such signals into their individual frequency components will be discussed in Chapter 5.

The average results from one such study are presented in Fig. 4.14(b) (Peslin, R., et al., 1985). The real (Resistance) and imaginary (Reactance) parts of the impedance are shown for 14 normal subjects (top) and 10 patients (bottom) diagnosed with chronic obstructive lung disease (COLD). Clearly the Reactance varies more with frequency than the Resistance. Note, however, that the Resistance at any frequency is higher in the patient group and the Reactance is shifted to the right in COLD patients. By determining the range of values for Resistance and Reactance in a large group of normal subjects, it is possible to identify a patient with abnormal pressure-flow biomechanical properties from such measurements of respiratory impedance.

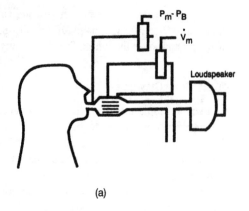

(a)

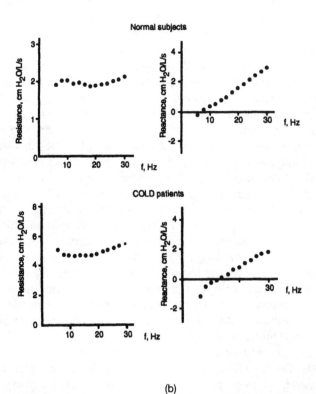

(b)

FIGURE 4.14. (a) Experimental setup for applying oscillatory pressures at the mouth. See text for details. (b) Average frequency response of the respiratory system to sinusoidal pressure applied at the mouth. Top: 14 normal subjects. Bottom: 10 COLD patients. Resistance: real part of impedance. Reactance: imaginary part of impedance (Peslin et al., 1985).

In order to determine the anatomical or physiological site of an abnormality, it is necessary to associate anatomical or physiological parameters with various features of the impedance graphs so that parameter values may be determined from the graphs. This objective is an area of active research.

Frequency Response of the Auditory System

To characterize overall performance of the auditory system, clinicians present pure sinusoidal tones (usually to one ear at a time) and determine the signal amplitude at which the sound is just audible to the subject. The frequency range of the sinusoids covers approximately 10 Hz to 16,000 Hz. If we consider the output to be a signal in the auditory cortex of "constant amplitude"—that is, a just-detectable signal—then a graph of the minimum input amplitude required for detection versus frequency is equivalent to the reciprocal of gain (as was the case for impedance above). Thus an *audibility* graph (Fig. 4.15(a)) is a type of frequency response plot whereas the *audiogram* (Fig. 4.15(b)) represents the difference from the audibility plot of an average normal subject. The figure depicts both the normal auditory sensitivity versus frequency and the loss of high-frequency sensitivity often observed in elderly subjects. Although useful for diagnosis, such graphs by themselves are not very insightful for diagnosing the anatomical site of a hearing deficit because they represent overall hearing performance. The auditory evoked potential waveform mentioned earlier in this chapter has features that have been associated with different sites of information processing in the nervous system and abnormalities in these features can aid in identifying specific sites of hearing deficit. There is much research intended to develop mathematical models of auditory neural information processing, from which it is hoped that specific diagnoses will be possible by combining modeling results with assessments of auditory performance. Another important application of audibility curves is in the tuning of a hearing aid for optimal performance for an individual subject.

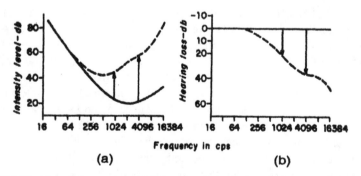

FIGURE 4.15. (a) Audibility plot (sound intensity at hearing threshold vs. frequency) of a normal young subject (solid) and a normal elderly subject (dashed). (b) Audiogram (difference of audibility plot from normal) for the elderly subject (Ruch and Patton, 1965).

Other Biomedical Applications of Frequency Response

Physiological systems for which the frequency of a stimulus is a relevant parameter often are characterized by frequency response graphs. Examples include touch receptors and vision. Assessing the response of a muscle to the frequency of variations of an applied electrical stimulus also is an important application of this concept. This information is needed for designing functional neuromuscular stimulation systems which apply artificial stimulation to a muscle in order to replace natural control by the nervous system that has been lost due to injury. In some other applications the frequency response is determined in order to ascertain the frequency of resonance or the conditions which lead to resonance behavior. Many studies aimed at understanding the causes of periodic breathing and sleep apnea utilize this latter concept. It will become apparent in the next chapter that many rhythmic biological behaviors have been described, to a first approximation, by signals that are sums of sinusoids. As a result, it is natural to consider the frequency responses of the systems generating these behaviors and of the functional mechanisms that are affected by the behaviors. The reader should be cautioned, however, that such "first approximations" often have developed because of familiarity with the methodology rather than on the basis of careful consideration of the most appropriate descriptors of a biological behavior.

4.10 SUMMARY

The frequency response of a filter describes the steady-state transformation that the filter applies to an input signal that is a sine (or cosine) function. The magnitude of the frequency response (the gain) represents the degree to which the amplitude of the sine wave is amplified and the angle of the frequency response (the phase shift) represents the phase change imposed on the sine wave by the filter. As a consequence of the close relationship between sinusoidal functions and complex exponential functions, the frequency response can be calculated from the impulse response of the filter. Thus, for an LTI filter, $H(\Omega) = \int_{-\infty}^{\infty} h[\lambda]e^{-j\Omega\lambda}d\lambda$, and for an LSI filter, $H(e^{j\omega}) = \sum_{n=-\infty}^{\infty} h[n]e^{-j\omega n}$. For a filter having the frequency response $H(\Omega)$ (or $H(e^{j\omega})$) and a steady-state input signal $x(t) = \sin(\Omega t)$ (or $x[n] = \sin[\omega n]$), the steady-state output signal will be $y(t) = |H(\Omega)| \sin(\Omega t + \arg(H(\Omega)))$ (or $y[n] = |H(e^{j\omega})| \sin[\omega n + \arg(H(e^{j\omega}))]$). The cutoff frequency of a lowpass (or highpass) filter is that frequency for which the gain is 3 dB less than the maximum gain at low (or high) frequencies.

The cutoff frequency of a first-order filter is the reciprocal of its time constant. The frequency response of a second-order filter depends on its natural frequency and damping factor. When the damping factor is smaller than 0.5, the gain at frequencies near the natural frequency increases (relative to gain at lower frequencies), a phenomenon known as *resonance*. The frequency response of a network of filters can be determined by considering them in subunits of series or parallel cascades. Ideal lowpass, highpass, bandpass, and band reject filters have a gain of unity in their passbands and a phase shift that is linear with frequency. Applying the concept of frequency response to nonlinear systems is complicated by the fact that the frequency response may depend on the mean operating point of the system or on the amplitude of the input signal.

The frequency response functions of biomedical transducers and amplifiers may impose distortions on any measured signal and it is imperative that the user certify that transducers and amplifiers do not introduce misleading distortions of signals. Frequency response measurements have been utilized to characterize physiological systems such as the mechanical state of the lungs and the state of information processing in the auditory nervous system. Diagnoses of abnormal function have been achieved on the basis of such measurements.

EXERCISES

4.1. An LTI system has the frequency response $H(\Omega) = |10 - \Omega|e^{-j\Omega}$, $|\Omega| \leq 10$. Sketch graphs of the magnitude and phase of the frequency response and calculate the steady-state output in response to the following input signals:

  a. $x(t) = \cos(3.5t) - 4\sin(6t)$;

  b. $x(t) = 3\cos(6t + \pi/6) + 2\cos(3t + \pi/3) + \cos(12t + \pi/12)$;

  c. $x(t) = 1 + \sum_{n=1}^{\infty} \frac{1}{n}\cos(4nt)$.

4.2. The frequency response of an LTI system is

$$|H(\Omega)| = \frac{5}{6 + \Omega}, \quad \arg(H(\Omega)) = -2\Omega,\ 0 \leq \Omega < 25 \quad \text{and} \quad |H(\Omega)| = 0,\ \Omega \geq 25.$$

Calculate the steady-state output in response to the following input signals:

  a. $x(t) = 2\cos(4t)$;

  b. $x(t) = 2\cos(10t - 1) - \sin(20t)$.

4.3. Consider a causal system for which the input $x[n]$ and output $y[n]$ satisfy the following linear, constant-coefficient difference equation:

$$y[n] - \tfrac{1}{2}y[n-1] = x[n] - \tfrac{1}{2}x[n-1].$$

  a. Determine the unit-pulse response of this system.

  b. Use convolution to determine the response to the complex sinusoidal input $x[n] = e^{j\omega n}$.

  c. Determine the frequency response of the system from its unit-pulse response. Is it the same as your answer to part b?

  d. Calculate the steady-state response to the input $x[n] = \cos\left[\dfrac{\pi n}{2} + \dfrac{\pi}{4}\right]$.

  e. Calculate the power gain of this filter as a function of ω.

4.4. The unit-pulse response of a three-point averager with Bartlett weighting is $h[n]$ $= \frac{1}{4}\delta[n + 1] + \frac{1}{2}\delta[n] + \frac{1}{4}\delta[n - 1]$. Determine the frequency response of this filter.

4.5. Calculate the frequency response for a second-order system described by the differential equation

$$\frac{d^2y}{dt^2} + 20\frac{dy}{dt} + 100y(t) = x(t).$$

Calculate the power gain of this filter.

4.6. Show that $|H(e^{j\omega})|$ is an even function of frequency.

4.7. Determine the cutoff frequency for the DT filter of Example 4.4.

4.8. Calculate the theoretical frequency response of a two-point moving averager and show that the estimated gain of 0.895 at $\omega = 0.3\pi$ (see Example 4.6) is an accurate estimate of the true gain at this frequency.

4.9. A DT second-order differencing filter has an impulse response given by $h[0] = 1$, $h[1] = 0$, $h[2] = -1$, and $h[n] = 0$ for $n > 2$. Calculate its frequency response and power gain.

4.10. A Butterworth filter of order N is the filter which has the maximally uniform gain in a given passband among all filters describable by linear differential equations of order N. For $N = 2$, a lowpass Butterworth filter is a second-order filter having a damping factor $\delta = 1/\sqrt{2}$ and unity gain at $\Omega = 0$. Specify and graph the frequency response of a second-order Butterworth lowpass filter with a natural frequency equivalent to 10 Hz. Show that the –3 dB cutoff frequency of this Butterworth filter equals the natural frequency.

4.11. In Fig. 4.11 let

$$H_1(\Omega) = \frac{10}{\Omega^2 + 5\Omega + 1}, \quad H_2(\Omega) = \frac{3}{\sqrt{2\Omega^2 + 1}}e^{-j\tan^{-1}(\sqrt{2}\Omega)}.$$

Determine the overall frequency responses of the series and parallel cascade networks shown in the figure.

4.12. A major limitation of pH electrodes is their slow response time. A typical "fast," commercially available, pH electrode has a time constant of one second. (That is, if the response of the electrode to a step change in pH is modeled as a first-order response, the time constant of the first-order system is 1.0 s.) In arterial blood the primary oscillatory component of pH is due to respiration. Assume that this component is exactly sinusoidal. If the above electrode is used to measure the pH of arterial blood *in vivo*, and if the subject is breathing at 0.5 Hz, in the steady state how much will the true amplitude of the oscillation in pH be attenuated at the output of the pH electrode?

4.13. Proper functioning of the neurons of the brain depends very much on the pH of brain tissue. Brain tissue pH has several oscillatory components. Assume that this pH, $p(t)$, can be described by the following equation:

$$p(t) = 7.037 + 0.010 \sin(2\pi(0.225)t) + 0.002 \sin(2\pi(0.9)t) + 0.005 \sin(2\pi(0.01)t).$$

If the pH electrode from the previous problem were used to measure this pH, what would the actual steady-state output from the pH electrode be? Which components of this measurement (if any) would have significant errors?

4.14. An ideal linear lowpass filter has the frequency response

$$H(\Omega) = \begin{cases} e^{-j\Omega}, & -2 \le \Omega \le 2 \\ 0, & \text{otherwise} \end{cases}.$$

Calculate its steady-state output when its input is

$$x(t) = \sum_{n=1}^{\infty} \frac{1}{n^2} \cos\left(\frac{n\pi}{2}t + \frac{\pi}{6}\right).$$

4.15. An LTIC system has the frequency response

$$H(\Omega) = \frac{1}{j\Omega + 1}.$$

Find its steady-state output for these inputs:
   a. $x(t) = \cos(t + 1.732)$;
   b. $x(t) = \cos(t) \sin(t)$.

4.16. An LSI system produces the following steady-state output, $y[n]$, when its input is $x[n]$. Calculate its frequency response at all frequencies for which there is sufficient information.

$$x[n] = 1 + 4\cos\left[\frac{2\pi}{6}n\right] + 8\sin\left[\frac{3\pi}{20}n - \frac{\pi}{4}\right], \qquad y[n] = 2 - 2\sin\left[\frac{2\pi}{6}n - \frac{\pi}{12}\right].$$

4.17. The input signal

$$x(t) = 1.5 + \sum_{n=1}^{\infty}\left\{\frac{1}{n\pi}\sin(n\pi t) + \frac{2}{n\pi}\cos(n\pi t)\right\}$$

is applied to an LTIC system. The resulting steady-state output is

$$y(t) = 0.5 + \sum_{n=1}^{\infty}\left\{\frac{2}{n\pi}\sin(n\pi t + 0.1n\pi) - \frac{1}{n\pi}\cos(n\pi t + 0.1n\pi)\right\}.$$

Calculate $H(\Omega)$ at all frequencies for which there is sufficient information.

4.18. The signal shown in Fig. 4.16 can be represented as

$$p(t) = \sum_{n=-\infty}^{\infty} 0.2 \cos(0.2n\pi)e^{-j0.2n\pi t}.$$

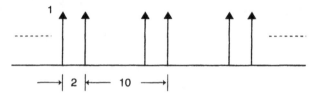

FIGURE 4.16. A train of impulses in which two impulses separated by 2 seconds occur every 10 seconds.

a. $p(t)$ is applied as the input to the filter of Exercise 4.5. Determine the steady-state output.

b. To visualize the effect of the filter, generate $p(t)$ and the filter in MATLAB, then calculate and plot the output signal.

4.19. To measure respiratory biomechanical properties in humans, pressure within the esophagous is measured as an estimate of pleural pressure. Esophageal pressure may be measured by way of a catheter introduced through the nose. Hartford et al. (1997) measured the frequency response (Fig. 4.13) of a water-filled catheter used for recording esophageal pressure waveforms in infants. The authors asked whether the common assumption that such catheters have a frequency response like that of a second-order system is correct. Read values from this graph about every 5 Hz and use MATLAB to try to fit a second-order frequency response to these data. (*Hint:* See the command ord2) For what value of damping factor and natural frequency do you obtain the best fit? Would you agree with the authors that a second-order model is not a good one for this catheter?

4.20. Kamen (1990) presents a model of the control of the angular position of a human eyeball in response to a sudden change in the position of a visual target (Fig. 4.17). His model is described by the equations

$$T_e \frac{d\theta_e}{dt} + \theta_e(t) = R(t), \qquad R(t) = b\theta_T(t - d) - b\theta_T(t - d - c) + \theta_T(t - d),$$

where $R(t)$ is the firing rate of action potentials in the nerve to the eye muscle, d is a time delay due to the central nervous sytem, and the other parameters are positive constants.

a. Find the impulse response of this model. (*Hint:* Find the step response, then differentiate.)

b. Determine the frequency response as a function of the unspecified parameters. Is this system first-order or second-order?

4.21. A researcher is studying muscle tremor during constant mean force (i.e., isotonic) contractions. She wishes to record the small-amplitude displacements of the tip of the thumb when the subject tries to produce an isotonic force by pressing against a force transducer. The researcher will attach a miniature accelerome-

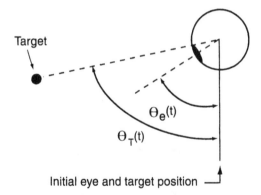

FIGURE 4.17. A model of eye movement in response to a jump in position of a visual target. $\Theta_e(t)$, $\Theta_T(t)$: angular position of eye and target (Kamen, 1990; pp. 180–181). (INTRODUCTION TO SIGNALS AND SYSTEMS by Kamen, © 1990, Reprinted by permission of Prentice-Hall, Inc., Upper Saddle River, NJ.)

ter to the tip of the thumb. The accelerometer has a constant gain from 1 Hz to 1000 Hz. It will be connected to a Validyne™ PA89 vibration amplifier that can be set to have a constant gain in one of four frequency ranges: 0–60, 0–200, 0–600, or 0–2000 Hz. From the literature the researcher has obtained some force tremor recordings from the thumb under similar experimental circumstances (Fig. 4.18). Based on these data, select the frequency range setting that she should use and explain your choice.

4.22. Schmid-Schoenbein and Fung (1978) describe a mechanical analysis of the response of the respiratory system to a small pressure disturbance such as that applied during a "forced oscillation" test (Fig. 4.14). Their final equation states that

$$E(t) = \frac{1}{K_{RS}}\xi + R_{RS}\dot{\xi} + L_{RS}\ddot{\xi},$$

where $E(t)$ is the applied pressure and $\xi(t)$ is the change in volume. They report typical values of the coefficients in normal human subjects as

$$\frac{1}{K_{RS}} = 32.5, R_{RS} = 2.4, L_{RS} = -0.35 \cdot 10^{-2}.$$

Calculate and graph the frequency response of the respiratory system based on this analysis.

4.23. Chen et al. (1998) report successfully detecting differences in control of saccadic eye movements in patients with Parkinsons disease compared to normal subjects. They measured horizontal saccades in response to sudden displacement

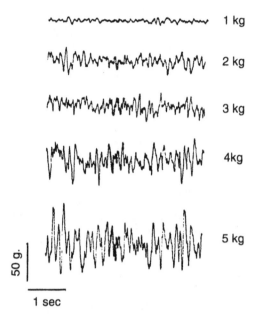

FIGURE 4.18. Tremor recorded from a thumb during contractions at the indicated force levels (Data from Allum et al., 1978).

of a visual target. (A lighted LED target directly in front of the subject was suddenly extinguished and an LED placed laterally was illuminated simultaneously.) The authors measured the angular rotation of the eye as it moved to the new target and normalized it by its final value. They then found the best damping factor and natural frequency to fit the *unit-step* response of a second-order filter to each measured response. Two of their responses are shown in Fig. 4.19. The authors found damping factor values of 0.5229 and 0.9740 for the normal and Parkinsons disease subject, respectively. Digitize these responses by reading 10 to 20 values from each graph. Use MATLAB to generate unit-step responses of various second-order systems and compare them to these data. (What should the gain at zero frequency be?) Choose some criterion for selecting the best values of δ and Ω_n to fit each response (e.g., minimize the mean-square difference between the digitized data and the unit-step response) and compare your "best" values of damping factor with those found by the authors. Are you confident that the response curve from the normal subject clearly cannot be fit using the damping factor found for the patient, and vice versa? How might you test or increase your confidence?

4.24. The file sinenoiz.mat contains a sine wave signal with added random noise. Plot this signal and estimate the frequency of the sine wave. Now try to find the best m-point moving average filter to reduce the noise contamination of this signal, where an m-point moving average filter has the impulse response $h = ones(m, 1)/m$. For several choices of m, use the MATLAB function freqz to create

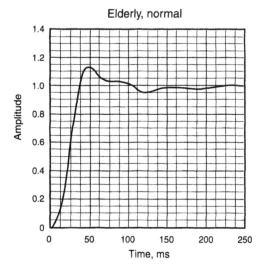

Elderly, normal

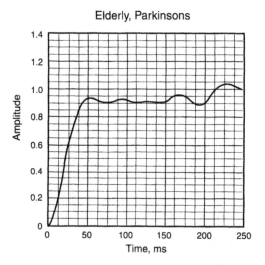

Elderly, Parkinsons

FIGURE 4.19. Saccadic position vs. time from a normal elderly subject (top) and an elderly subject with Parkinson's disease (bottom). The investigators' best-fit second-order models for these subjects produced damping factor estimates of 0.5229 and 0.9740, respectively (Chen et al., 1998). (Reprinted with permission from ANNALS OF BIOMEDICAL ENGINEERING, 1998, Biomedical Engineering Society.)

a graph of the frequency response of the filter and choose the value of m that you think will produce a filter that least attenuates the sine wave while best removing the noise. Filter the noisy signal and compare the filtered and unfiltered signals. How adequate is your filter? Can you suggest simple modifications to your chosen m-point filter that might produce a better filter?

5

MODELING CONTINUOUS-TIME
SIGNALS AS SUMS OF SINE WAVES

5.1 INTRODUCTION

In the previous chapter we found that transformations of steady-state sinusoidal signals by linear systems can be described compactly using frequency response functions. This result is a direct consequence of the fact that complex exponential functions are eigenfunctions of linear systems. This elegant theory, however, might seem almost inconsequential for biomedical applications since so few biomedical signals are truly sinusoidal. On the other hand there are biomedical signals that are almost periodic and one might model such signals as periodic signals corrupted by added noise. This type of model probably is most appropriate when biomedical processes are driven by periodic stimuli. For example, many physiological processes such as endocrine function, body metabolism, body temperature, breathing, and heart rate exhibit oscillations linked to the daily light-dark cycling (Fig. 5.1). Therapeutic interventions also might be periodic—for example, injection of drugs at regular intervals or artificial ventilation of a patient. In such cases one might derive diagnostic information from knowledge of the frequency and amplitude of a "best" periodic approximation to the fluctuations in a signal. Thus there are certain situations in which modeling a biomedical signal as a noise-corrupted sinusoid or a noise-corrupted periodic signal is a valid approach.

If a (possibly noise-corrupted) signal is represented as a sine wave, then it may be manipulated using the methods of the previous chapter. Are there similar advantages to representing a signal as a periodic signal that is not a simple sine wave? The answer is yes, but the advantages depend very much on the specific structure of the model that is chosen. Because sine and cosine signals are amenable to the analyses developed in the previous chapter, initially we will develop models of periodic signals that comprise only these functions. Furthermore, we will extend this development to signals that are not periodic although the models are based on sine and cosine functions. Even in these latter cases the methods of the previous chapter will be applicable.

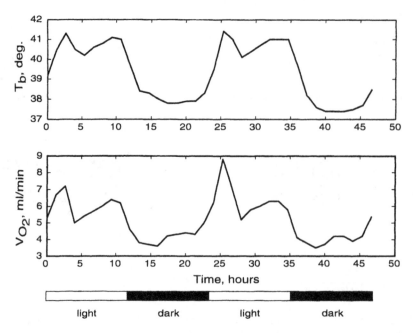

FIGURE 5.1. Body temperature (T_b) obtained by telemetry and oxygen consumption (V_{O2}) from a chronically instrumented pigeon over a 48-hour period (Berger and Phillips, 1988).

5.2 INTRODUCTORY EXAMPLE (ANALYSIS OF CIRCADIAN RHYTHM)

Many species, from bacteria to humans, maintain a daily rhythm of life by way of a circadian clock. This clock is not invariant. Its periodicity is modifiable by certain afferent inputs, but under controlled conditions its period is essentially fixed. Examples of the circadian variations in deep abdominal temperature (T_b) and oxygen consumption (V_{O2}) from a chronically instrumented pigeon in a controlled light-dark environment are shown in Fig. 5.1. Although both variables exhibit clear daily fluctuations, in neither case is the signal truly periodic. One could assume, however, that in the absence of noisy disturbances both signals would be periodic and therefore it would be appropriate to model each as a periodic function. Consider using one sine wave to represent the body temperature signal. Clearly this sine wave should have a period of 24 hours, but what should its amplitude and phase shift be? A typical approach to this question involves determining the combination of amplitude and phase that minimizes some measure of the error between the sine wave approximation and the data. In Fig. 5.1 there are 18 data points per day sampled at intervals of $\Delta t = 1.333$ hours. Refer to the temperature data as $T(n \cdot \Delta t)$, $0 \leq n \leq 35$. Let the approximating signal be given by $\hat{T}(t) = A \sin(2\pi t/24 + \theta) + \bar{T}$, where t has units of hours and T with an overbar indicates the average (or "DC") value of the temperature signal. Each data point can be predicted from the approximating signal in terms of the unknown amplitude, A, and phase, θ, as

$$\hat{T}(n \cdot \Delta t) = A \sin(2\pi n \cdot \Delta t/24 + \theta) + \overline{T}, \qquad 0 \le n \le 35. \qquad (5.1)$$

J, the mean square error (MSE) of the prediction, is defined as

$$J = \frac{1}{N} \sum_{n=0}^{N-1} [\hat{T}(n \cdot \Delta t) - T(n \cdot \Delta t)]^2, \qquad (5.2)$$

where $N = 36$ in this example. The task is to determine the optimal values of A and θ that together will minimize J. For this simple example it is reasonable to use a grid search method—that is, to select ranges for A and θ that seem likely to encompass the optimal values, then evaluate J for many combinations of the two parameters and choose the combination that yields the smallest value of J. Since the vertical range of the data is approximately $38°$ to $41.5°$ and $\overline{T} \approx 39.3556$, we choose to test values of A in the range $[1.00, 2.50]$. Similarly, since the oscillations in the data are nearly in phase with the light-dark cycling, we choose $-\pi/2 \le \theta \le \pi/2$. If we test 30 values in each range, then A can be resolved to the nearest 0.05 and phase to the nearest 6 degrees. After evaluating J for these 900 combinations, the values that minimize J are $A^* = 1.85$, $\theta^* = 0$, where * indicates the optimal value. (The reader may access these data from the file circadian.mat and test smaller intervals using the file fit-sine.m in order to define the optimal parameters more precisely.)

Figure 5.2(a) shows the original data and the approximating signal, $\hat{T}(t)$, using the optimal parameter values, while the dashed line in Fig. 5.2(b) shows the difference between the two (i.e., the error of the approximation). As expected from visual inspection, the approximating sine wave provides a rough guide to the circadian variations in body temperature but there are significant deviations of these data from a sine wave as evidenced by the error signal. Interestingly the deviations are similar in the two cycles shown and seem to have a frequency of three per day. If one were to use the same approach to approximate the *error* signal in Fig. 5.2(b) with a sine wave having a period of 8 hours, this second sine wave could be added to the first to achieve a better approximation of the data. When the same method was utilized to fit such a sine wave to the error signal, the optimal amplitude and phase were 1.0000 and 0.1047, respectively. The resulting approximating signal is depicted also in Fig. 5.2(b). Finally, the sum of these two approximating sine waves yields a better approximation of the original data (Fig. 5.2(c)).

One could find the difference between the two signals in Fig. 5.2(c) and determine whether another sine wave could usefully approximate this new error signal. Indeed this process could be repeated many times, although one has the impression that the error would soon become so small as to be negligible. In this way it would be possible to model the noisy periodic signal, $T_b(t)$, as a sum of sine waves. The remainder of this chapter is devoted to formalizing this procedure.

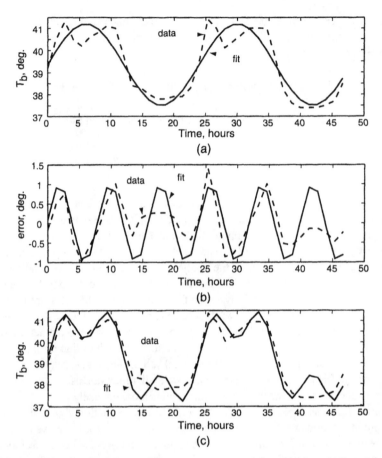

FIGURE 5.2. (a) T_b data and optimal fit of a 24-hour sine wave to these data. (b) Error from the fit in (a) and optimal fit of an 8-hour sine wave to the error. (c) Comparison of the sum of the optimal fit signals in (a) and (b) to the T_b data.

5.3   ORTHOGONAL FUNCTIONS

Consider a collection of functions $P = \{p_i(t), i = 1, 2, \ldots k_p\}$ and some operator $O[.]$. The functions in P are said to be *orthogonal under the operator* $O[.]$ if

$$O[p_i(t)p_j(t)] = \begin{cases} 0, & i \neq j \\ K_0, & i = j \end{cases}, \qquad 1 \leq i,j \leq k_p, \tag{5.3}$$

where K_0 is a constant. Depending on the specific choice of $O[.]$ the functions might have to satisfy conditions related to such properties as boundedness, continuity, or integrability. If one considers the $p_i(t)$'s to be analogous to vectors, then $O[.]$

is analogous to a dot (or inner) product and Eq. (5.3) is analogous to the definition of orthogonal vectors. This analogy is appropriate, as the following example shows. Let $p_1(t) \in P$, and let $x(t) = ap_1(t)$. Now

$$\frac{1}{K_0} O[x(t)p_j(t)] = \begin{cases} 0, j \neq 1 \\ a, j = 1 \end{cases}.$$

In other words, the result of the inner product of $x(t)$ with each function in the set P is a scalar that expresses "how much" of each function is present in $x(t)$. Just as one can expand arbitrary vectors as sums of scaled orthogonal vectors, it is possible to express arbitrary functions as sums of scaled orthogonal functions. The set of orthogonal functions is called a *basis set*, or simply a *basis*, and all of the functions that can be expressed as a weighted sum of these basis functions together comprise the *function space* that is spanned by the basis functions. Any function in this space can be expressed as

$$x(t) = \sum_{i=0}^{k_p} a_i p_i(t), \qquad \text{where } a_i = \frac{1}{K_0} O[x(t)p_i(t)]. \tag{5.4}$$

Consequently, if the a_i coefficients are unknown but the function $x(t)$ is known, it is possible to determine these coefficients uniquely using the second part of Eq. (5.4). Furthermore, because the basis functions are orthogonal, the value of any a_i is independent of the values of the other coefficients.

5.4 SINUSOIDAL BASIS FUNCTIONS

There is an infinite variety of basis sets but our current interest lies in representing arbitrary but periodic functions as a weighted sum of functions to which one may apply the concept of frequency response. Functions meeting the latter criterion are sine and cosine functions and one needs to consider under what operations such functions are orthogonal. For a periodic function it is natural to define operations that effectively average over one period. Therefore, let the first choice of basis set comprise the functions $p_1(t) = \sin(\Omega_0 t)$, $p_2(t) = \cos(\Omega_0 t)$. Note that $\int_0^{2\pi/\Omega_0} p_1(t)p_2(t)dt = 0$. Consequently, if one defines the inner product operator as $O[.] = \int_0^T [.]dt$, where $T = 2\pi/\Omega_0$, then the two functions are orthogonal. Unfortunately, relatively few functions could be expressed as weighted sums of $p_1(t)$ and $p_2(t)$. Fortunately, the following generalizations of the above results allow one to define a very large set of orthogonal trigonometric functions:

$$\int_0^T \sin(n\Omega_0 t) \sin(m\Omega_0 t)dt = \begin{cases} 0, & n \neq m, \\ 0, & n = m = 0 \\ T/2 = \pi/\Omega_0, & n = m \neq 0; \end{cases} \tag{5.5a}$$

$$\int_0^T \cos(n\Omega_0 t)\cos(m\Omega_0 t)dt = \begin{cases} 0, & n \neq m, \\ T, & n = m = 0 \\ T/2 = \pi/\Omega_0, & n = m \neq 0; \end{cases} \tag{5.5b}$$

$$\int_0^T \sin(n\Omega_0 t)\cos(m\Omega_0 t)dt = 0 \ \forall m, n. \tag{5.5c}$$

These results establish that the infinite collection of functions $P_F = \{\sin(n\Omega_0 t),$ $\cos(m\Omega_0 t), \forall m, n\}$ comprises an orthogonal function set with $K_0 = T/2$. (The only deviation from Eq. (5.3). occurs in Eq. (5.5b) when $n = m = 0$, in which case the integral of the squared cosine function equals $2K_0$.)

Establishing that functions in a set are orthogonal does not establish that this set may be used to approximate arbitrary functions. In fact, Fourier demonstrated the extremely powerful result that the set of trigonometric functions P_F can be used *not just to approximate, but to exactly represent*, most periodic functions. That is, P_F is a set of basis functions for a large class of periodic functions. Formally, any periodic function $x(t)$ with period T can be written as a weighted sum of the sine and cosine functions in the set of functions, P_F, if it satisfies the following conditions (known as the Dirichlet conditions):

1. $x(t)$ is absolutely integrable over one cycle—i.e., $\int_0^T |x(t)|dt < \infty$.
2. $x(t)$ has a finite number of maxima and minima in the interval $(0, T)$.
3. $x(t)$ has a finite number of discontinuities in the interval $(0, T)$.

It appears that many measurable real signals should meet these conditions. Therefore, most periodic signals likely to be encountered will be decomposable into a weighted sum of sine and cosine functions, and the analysis methods of the previous chapter will be applicable. The potential disadvantage is that P_F comprises an infinite number of functions so it may be necessary to utilize a very large number of sines and/or cosines to represent a given signal. On the other hand, in practice it often is possible to truncate the summation after a manageable number of terms without accumulating large errors.

5.5   THE FOURIER SERIES

Given a periodic function $x(t)$, with period T and frequency $\Omega_0 = 2\pi/T$, that exists for all t and satisfies the Dirichlet conditions, it may be represented as the following functional expansion known as a Fourier series:

$$x(t) = \sum_{n=0}^{\infty} [a_n \cos(n\Omega_0 t) + b_n \sin(n\Omega_0 t)]. \tag{5.6}$$

Note that because of the even and odd symmetries of the cosine and sine functions, respectively, it is not necessary to utilize the functions corresponding to negative values of m and n in P_F. Note also that it is assumed that T is the *smallest* interval that satisfies the criterion that $x(t + T) = x(t)$ for all t. Ω_0 is known as the *fundamental frequency* of the periodic function and the n-th term in the series is called the n-th *harmonic*.

To evaluate the weighting coefficients in the Fourier series (i.e., a_n and b_n), first substitute k for n in Eq. (5.6) then multiply both sides by $\cos(n\Omega_0 t)$ and integrate over one cycle. Thus

$$\int_0^T x(t) \cos(n\Omega_0 t)dt = \int_0^T \sum_{k=0}^{\infty} [a_k \cos(k\Omega_0 t) + b_k \sin(k\Omega_0 t)] \cos(n\Omega_0 t)dt.$$

Due to the orthogonality properties of these basis functions, only one term on the r.h.s. (i.e., when $k = n$) does not integrate to zero. Therefore,

$$\int_0^T x(t) \cos(n\Omega_0 t)dt = \int_0^T a_n \cos(n\Omega_0 t) \cos(n\Omega_0 t)dt = a_n \frac{T}{2}, \qquad n > 0$$

and, therefore,

$$a_n = \frac{2}{T} \int_0^T x(t) \cos(n\Omega_0 t)dt, \qquad n > 0. \tag{5.7}$$

From Eq. (5.5b) there is a factor of 2 difference when $n = 0$. Thus

$$a_0 = \frac{1}{T} \int_0^T x(t) \cos(0)dt = \frac{1}{T} \int_0^T x(t)dt. \tag{5.8}$$

Note that a_0 is the mean level of the signal over one cycle.

Repeating this procedure after multiplying both sides of Eq. 5.6 by $\sin(n\Omega_0 t)$, one finds

$$b_n = \frac{2}{T} \int_0^T x(t) \sin(n\Omega_0 t)dt, \; n \geq 0. \tag{5.9}$$

In particular, b_0 always evaluates to zero.

Since $\cos(0) = 1$ and $b_0 = 0$, it is common to write Eq. (5.6) in the form

$$x(t) = a_0 + \sum_{n=1}^{\infty} [a_n \cos(n\Omega_0 t) + b_n \sin(n\Omega_0 t)]. \tag{5.10}$$

The form of Eq. (5.10) makes it clear that $x(t)$ is being represented as the sum of its mean value over a cycle plus deviations from that mean value.

Example 5.1 Fourier series representation of a square wave   The signal $x(t)$ of Fig. 5.3(a) has period T and satisifes the Dirichlet conditions. Therefore it has a Fourier series representation. The Fourier series coefficients may be determined by substitution into Equations (5.7)–(5.9) using the relationship $\Omega_0 = 2\pi/T$:

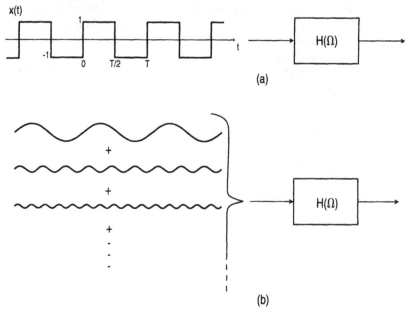

FIGURE 5.3. (a) A square-wave signal that is the input to an LTI system. (b) Representing the square wave as the sum of its Fourier series components.

$$a_0 = \frac{1}{T}\left[\int_0^{T/2}(1)dt + \int_{T/2}^T(-1)dt\right] = 0;$$

$$a_n = \frac{2}{T}\left[\int_0^{T/2}\cos(n\Omega_0 t)dt - \int_{T/2}^T \cos(n\Omega_0 t)dt\right] = 0, \qquad n > 0;$$

$$b_n = \frac{2}{T}\left[\int_0^{T/2}\sin(n\Omega_0 t)dt - \int_{T/2}^T \sin(n\Omega_0 t)dt\right] = \frac{2}{T}\frac{1}{n\Omega_0}[-2\cos(n\pi) + \cos(n\pi) + 1];$$

$$\therefore b_n = \begin{cases} 0, & n \text{ even} \\ \dfrac{4}{n\pi}, & n \text{ odd.} \end{cases} \qquad\qquad (5.11)$$

Therefore, the Fourier series for $x(t)$ is

$$x(t) = \frac{4}{\pi}\left[\sin(\Omega_0 t) + \frac{1}{3}\sin(3\Omega_0 t) + \frac{1}{5}\sin(5\Omega_0 t) + \ldots\right]$$

$$= \sum_{n=1}^{\infty}\frac{4}{(2n-1)\pi}\sin((2n-1)\,\Omega_0 t). \qquad\qquad (5.12)$$

Figure 5.3(b) indicates that if one is interested in how an LTI system will transform such a square wave, one may model the input as the sum of sinusoids given by Eq. (5.12) and use the frequency response of the system to determine the transformation at each frequency.

Equation (5.12) indicates that the Fourier series requires an infinite number of terms to accurately reproduce the square wave signal. (*Note:* At a discontinuity the Fourier series converges to the average value of the signal before and after the discontinuity.) If the series is truncated, the largest errors occur at the points of discontinuity of the square wave (Fig. 5.4, in which the series is truncated after 30 terms). This result, known as *Gibbs phenomenon*, illustrates that the higher-frequency harmonics of the Fourier series are most important at the times where the signal changes most rapidly. (The file gibbdemo.m provides a demonstration of Gibbs phenomenon.)

Finally consider the effect of time-shifting the square wave (Fig. 5.3(a)) so that $t = 0$ occurs in the middle of the positive half-cycle of the signal. This new signal, call it $v(t)$, equals $x(t + T/4)$. The reader may evaluate the Fourier series coefficients for $v(t)$ and show that

$$v(t) = \frac{4}{\pi}\left[\cos(\Omega_0 t) - \frac{1}{3}\cos(3\Omega_0 t) + \frac{1}{5}\cos(5\Omega_0 t) - \ldots\right]$$

$$= \sum_{n=1}^{\infty} \frac{4(-1)^{n+1}}{(2n-1)\pi}\cos((2n-1)\Omega_0 t). \tag{5.13}$$

It is apparent that the cosine terms in Eq. (5.13) could be rewritten as phase-shifted sines. Comparing this result to Eq. (5.12), therefore, the same frequency components are seen to be present in $v(t)$ as in $x(t)$, and they have the same absolute magnitude in both signals but different phase shifts. This result, that time-shifting a

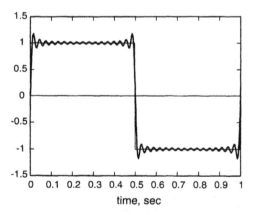

FIGURE 5.4. One cycle of a square wave compared to the sum of the first 30 components of its Fourier series. The presence of oscillations at discontinuities of the square wave is known as *Gibbs' phenomenon.*

signal merely changes the phase shifts of the frequency components of its Fourier series, is a general property of Fourier series.

Some Properties of Fourier Series

The following properties of Fourier series are easy to prove from the basic definition, Eq. (5.6):

1. If $x(t)$ is an even, periodic function (i.e., $x(-t) = x(t)$), then its Fourier series, if it exists, comprises only cosine terms. This property is self-evident because sine functions are odd and sums of odd functions are still odd; therefore, the Fourier series cannot contain sine waves.

2. Similarly, if $x(t)$ is an odd, periodic function (i.e., $x(-t) = -x(t)$), then its Fourier series, if it exists, comprises only sine terms.

3. If $x(t)$ has half-wave symmetry (i.e., $x(t + T/2) = -x(t)$), then its Fourier series coefficients are nonzero only for odd values of n. This property is not self-evident but is easily proven. (Note that if $x(t)$ satisfies the condition $x(t + T/2) = x(t)$, then its period should be considered to be $T/2$ rather than T!)

The value of the above properties is that one can greatly reduce the computational demands for calculating the Fourier series coefficients if one identifies symmetries in a signal.

Example 5.2   Half-wave rectification   To find the Fourier series for a half-wave rectified cosine wave (e.g., Fig. 5.8(a)), note that this signal is an even function; therefore, its Fourier series contains only cosine terms. Consequently, since $T = 2\pi/\Omega_0$,

$$a_0 = \frac{A\Omega_0}{2\pi} \int_{-\pi/\Omega_0}^{\pi/\Omega_0} \cos(\Omega_0 t)dt = \frac{A\Omega_0}{\pi} \int_0^{\pi/\Omega_0} \cos(\Omega_0 t)dt = \frac{A}{\pi} \sin(\Omega_0)\Big|_{t=0}^{t=\pi/2\Omega_0} = \frac{A}{\pi};$$

$$a_n = \frac{2\Omega_0}{\pi} \int_0^{\pi/2\Omega_0} A \cos(\Omega_0 t) \cos(n\Omega_0 t)dt, \ n \geq 1.$$

For $n \geq 2$,

$$a_n = \frac{A}{\pi}\left[\frac{\sin([n-1]\Omega_0 t)}{2(n-1)} + \frac{\sin([n+1]\Omega_0 t)}{2(n+1)}\right]_{t=0}^{t=\pi/2\Omega_0} = \frac{A}{\pi}\left(\frac{-2\cos(n\pi/2)}{n^2-1}\right),$$

from which we find

$$a_n = \begin{cases} 0, & n > 1 \text{ and odd} \\ \dfrac{2A}{\pi}\dfrac{(-1)^{(n/2)+1}}{n^2-1}, & n \text{ even} \end{cases}$$

For $n = 1$, $a_1 = (2\Omega_0/\pi) \int_0^{\pi/2\Omega_0} A \cos^2(\Omega_0)dt = (A/2)$.

The Fourier series also may be written in the form

$$x(t) = c_0 + \sum_{n=1}^{\infty} 2|c_n| \cos(n\Omega_0 t + \theta_n),$$ (5.14)

where $c_0 = a_0$ and

$$|c_n| = \frac{1}{2}\sqrt{a_n^2 + b_n^2}, \quad \theta_n = -\tan^{-1}\left[\frac{b_n}{a_n}\right] = \angle c_n, \quad a_n = 2|c_n|\cos(\theta_n), \quad b_n = -2|c_n|\sin(\theta_n).$$ (5.15)

This result is easy to prove by using the common expansion for $\cos(a + b)$ and equating like terms in Eq. (5.6) and in the expanded form of Eq. (5.14). (See Exercise 5.7.) In the form of Eq. (5.14) the Fourier series component at each frequency is represented by a magnitude and a phase shift. This form is easier to use than Eq. (5.6) if one wishes to calculate the effect of the frequency response of a system on a periodic input signal. Before illustrating this point we discuss a third form of the Fourier series.

Complex Exponential Form of the Fourier Series

Another form for the Fourier series becomes evident by substituting Euler's formulas into Eq. (5.10). Thus

$$x(t) = a_0 + \sum_{n=1}^{\infty}\left[a_n\left(\frac{e^{jn\Omega_0 t} + e^{-jn\Omega_0 t}}{2}\right) + b_n\left(\frac{e^{jn\Omega_0 t} - e^{-jn\Omega_0 t}}{2j}\right)\right]$$

$$= a_0 + \sum_{n=1}^{\infty}\left[\left(\frac{a_n - jb_n}{2}\right)e^{jn\Omega_0 t} + \left(\frac{a_n + jb_n}{2}\right)e^{-jn\Omega_0 t}\right]$$

$$= \sum_{n=-\infty}^{\infty} c_n e^{jn\Omega_0 t},$$ (5.16)

where

$$c_n = \frac{a_n - jb_n}{2}, \quad c_{-n} = \frac{a_n + jb_n}{2}, \quad c_0 = a_0, \quad |c_n| = \frac{1}{2}\sqrt{a_n^2 + b_n^2}, \quad \angle c_n = -\tan^{-1}\left[\frac{b_n}{a_n}\right].$$ (5.17)

Notice that the c_n here are the same coefficients that appear in Eq. (5.14). Note also the different lower limits on the summations in Eq. (5.14) and Eq. (5.16). Finally, note that $c_n = c_{-n}^*$, implying that $|c_n| = |c_{-n}|$ and $\angle c_{-n} = -\angle c_n$.

Given $x(t)$, it is not necessary to evaluate the Fourier series of Eq. (5.6) first in order to find the c_n coefficients. Proceeding as we did to determine a_n and b_n, multiply both sides of Eq. (5.16) by $e^{-jn\Omega_0 t}$ and integrate over one cycle. These complex exponential functions are orthogonal since the sines and cosines that comprise them are orthogonal, and again there is only one nonzero term on the r.h.s. Therefore,

$$\int_0^T x(t)e^{-jn\Omega_0 t}dt = \int_0^T c_n e^{jn\Omega_0 t}e^{-jn\Omega_0 t}dt = Tc_n$$

and

$$c_n = \frac{1}{T}\int_0^T x(t)e^{-jn\Omega_0 t}dt, \qquad \forall n. \tag{5.18}$$

(In the case of even or odd functions, there often is a computational advantage to using limits on the integral of $-T/2$ and $T/2$ instead of 0 and T.) Once the c_n's have been determined, if the form of the Fourier series in Eq. (5.6) is desired, it is readily found by noting that $a_n = 2Re\{c_n\}$ and $b_n = -2Im\{c_n\}$.

Example 5.3 *Exponential Fourier series for a pulse train* The signal of Fig. 5.5(a) is a pulse train with period T and duty cycle a/T. Its exponential Fourier series coefficients may be calculated using Eq. (5.18) as

$$c_n = \frac{1}{T}\int_{-a/2}^{a/2} A_0 e^{-jn\Omega_0 t}dt = \frac{A_0}{T}\left[\frac{-1}{jn\Omega_0}\right][e^{-jn\Omega_0 a/2} - e^{jn\Omega_0 a/2}] = \frac{A_0 a}{T}\frac{\sin(n\Omega_0 a/2)}{n\Omega_0 a/2}. \tag{5.19}$$

This equation is in the form $sinc(x) = (\sin(x)/x)$, $x = n\Omega_0 a/2 = n(2\pi/T)(a/2) = n\pi(a/T)$. Since $sinc(x) = 0$ whenever x is a nonzero integer multiple of π, c_n will be zero if the equation $n\pi(a/T) = k\pi$ has a solution (where k is any integer different from zero). For example, if $a = T/3$, then c_n equals zero for $n = 3k$—that is, for $n = 3$, 6, 9, ... These values of $n = 3k$ correspond to the frequencies $n\Omega_0 = 3k\Omega_0$ in Eq. (5.16), with

$$3k\Omega_0 = 3k\frac{2\pi}{T} = 3k\frac{2\pi}{3a} = k\frac{2k\pi}{a}.$$

Figure 5.5(b) shows a discrete plot of c_n versus $n\Omega_0 = 2n\pi/3a$ for the case $a = T/3$. The continuous sinc function, $(\sin(\Omega a/2)/\Omega a/2)$, that forms the envelope of the c_n coefficients, is also drawn. Notice that the frequencies at which zero-crossings occur (i.e., $2n\pi/a$) depend on 'a' but not on T. Likewise the spacing of the c_n's along the frequency axis (i.e., $2\pi/T$) depends on T but not on 'a'. Thus, if T is unchanged but 'a' decreases, then the point $\Omega = 2\pi/a$ moves to the right and the first zero coefficient occurs at $n > 3$. If 'a' is constant but T increases, then the frequencies $n\Omega_0$ move closer together and again the first zero coefficient occurs at $n > 3$. Although these Fourier series coefficients are all real, in general they may be complex. These important results will be used later in this chapter to obtain insights into the frequency content of signals for which $T \to \infty$.

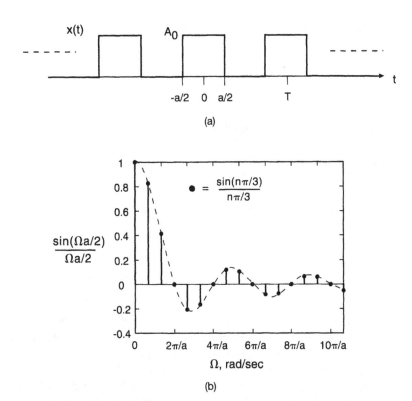

FIGURE 5.5. (a) A pulse train and (b) its Fourier series coefficients (normalized by $A_0 a/T$) versus frequency, for the case $a/T = 1/3$. Dashed line in (b) is the sinc function that determines the envelope of the Fourier series coefficients. Dots: $(\sin(n\pi/3)/n\pi/3)$ versus $n\Omega_0 = (2n\pi/3a)$.

Amplitude and Phase Spectra

From an analysis viewpoint, Eq. (5.14) can be considered as a decomposition of the periodic signal, $x(t)$, into an infinite summation of weighted, phase-shifted cosine functions. (Alternatively, from a synthesis viewpoint, one may view Eq. (5.14) as specifying the construction of $x(t)$ from weighted, phase-shifted cosines.) Given the fundamental frequency, Ω_0, the coefficients c_n embody all of the information necessary to specify this decomposition. A useful synopsis is achieved by constructing graphs of the magnitude and angle of c_n versus either n or $n\Omega_0$. The graph of $|c_n|$ versus $n\Omega_0$ is the *amplitude spectrum* of $x(t)$ whereas the graph of $\angle C_n$ vs. $n\Omega_0$ is the *phase spectrum* of $x(t)$.

Example 5.4 Calculating the exponential Fourier series
(a) Let $x(t) = 2 + 2\cos(\Omega_0 t) + \sin(\Omega_0 t)$. To find the complex exponential Fourier series for $x(t)$, note that $x(t)$ already is expressed in the standard form of Eq. (5.10). By

inspection, the only nonzero terms present correspond to $a_0 = 2$, $a_1 = 2$, and $b_1 = 1$. Therefore, $c_0 = a_0 = 2$. Also,

$$c_1 = \frac{a_1 - jb_1}{2} = \frac{2 - j}{2}, \qquad c_{-1} = \frac{2 + j}{2}, \qquad c_n = 0 \;\forall n \neq -1, 0, 1.$$

Thus, since $|c_1| = (\sqrt{5}/2)$, $\angle c_1 = -0.841$ rad,

$$x(t) = 2 + \left(1 - \frac{j}{2}\right)e^{j\Omega_0 t} + \left(1 + \frac{j}{2}\right)e^{-j\Omega_0 t} = 2 + \frac{\sqrt{5}}{2}e^{-j0.841}e^{j\Omega_0 t} + \frac{\sqrt{5}}{2}e^{j0.841}e^{-j\Omega_0 t}.$$

  (b) Find the exponential Fourier series for $x(t) = \cos(2t)\sin(3t)$ and graph its amplitude and phase spectra.

  Solution: By Euler's formulas,

$$x(t) = \left(\frac{e^{j2t} + e^{-j2t}}{2}\right)\left(\frac{e^{j3t} - e^{-j3t}}{2j}\right) = \frac{1}{4j}[e^{j5t} - e^{-j5t} - e^{-jt} + e^{jt}].$$

The largest common frequency in these complex exponential functions is $\Omega_0 = 1$. Since $x(t)$ is already in the complex exponential form, it remains to specify the coefficients at each frequency. Thus $c_0 = 0$, $c_1 = (1/4j) = c_5$, $c_{-1} = -(1/4j) = c_{-5}$. Otherwise $c_n = 0$. The corresponding spectra are plotted in Fig. 5.6.

  (c) Evaluate the exponential Fourier series coefficients for $y(t) = 5\sin^2(t)$.

  Solution: First note that $T_0 = \pi$, implying that $\Omega_0 = 2\pi/T = 2$ rad/s. Again using Euler's formulas,

$$y(t) = 5\left(\frac{e^{jt} - e^{-jt}}{2j}\right)^2 = -\frac{5}{4}(e^{j2t} - 2 + e^{-j2t}).$$

By inspection, $c_0 = 5/2$, $c_1 = -5/4$, $c_{-1} = -5/4$.

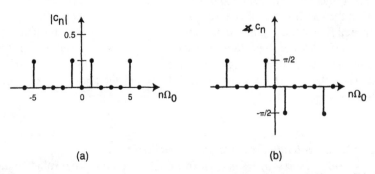

(a)   (b)

FIGURE 5.6. (a) Magnitude and (b) phase spectra of the Fourier series of $x(t) = \cos(2t)\sin(3t)$.

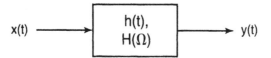

FIGURE 5.7. A general LTI system.

5.6 THE FREQUENCY RESPONSE AND NONSINUSOIDAL PERIODIC INPUTS

If the input, $x(t)$, to the LTIC system of Fig. 5.7 is a steady-state sine wave of frequency Ω_0, then the steady-state output, $y(t)$, may be found by multiplying the amplitude of the input sine wave by $|H(\Omega_0)|$ and adding $\angle H(\Omega_0)$ to the phase of $x(t)$. Because of linearity, if $x(t)$ comprises a sum of sine or cosine functions, then $y(t)$ (in the steady state) can be determined by applying this principle to every frequency component in $x(t)$ and summing the resulting signals. That is, if

$$x(t) = c_0^x + \sum_{n=1}^{\infty} 2|c_n^x| \cos(n\Omega_0 t + \angle c_n^x), \qquad (5.20)$$

then

$$y(t) = H(0)c_0^x + \sum_{n=1}^{\infty} 2|H(n\Omega_0)||c_n^x| \cos(n\Omega_0 t + \angle c_n^x + \angle H(n\Omega_0)). \qquad (5.21)$$

Note that since $H(0) = \int_{-\infty}^{\infty} h(t)e^{-j0}dt$, if $h(t)$ is real, then $H(0)$ must be real and its angle is zero or $-\pi$. Conversely, if $x(t)$ has the representation of Eq. (5.20) and the Fourier series of the output is known—that is,

$$y(t) = c_0^y + \sum_{n=1}^{\infty} 2|c_n^y| \cos(n\Omega_0 t + \angle c_n^y), \qquad (5.22)$$

then from Eqs. (5.21) and (5.22) the frequency response can be determined at multiples of Ω_0 as

$$|H(n\Omega_0)| = \frac{|c_n^y|}{|c_n^x|}, \qquad \angle H(n\Omega_0) = \angle c_n^y - \angle c_n^x, \qquad \text{or simply, } H(n\Omega_0) = \frac{c_n^y}{c_n^x}. \quad (5.23)$$

Example 5.5 *Steady-state output via the Fourier series* Determine the steady-state output of the system given by

$$|H(\Omega)| = \frac{5}{6 + (\Omega)}, \qquad \angle H(\Omega) = -2\Omega$$

in response to the periodic triangular signal satisfying

$$v(t) = \begin{cases} 4At/T, & 0 \le t \le T/2 \\ 4A(T-t)/T, & T/2 \le t \le T \end{cases},$$

where $T = 2.0$ s (Fig. 5.8(b)).

Solution: The reader may verify that the Fourier series for $v(t)$ is given by

$$v(t) = A - \frac{8A}{\pi^2} \sum_{n=1}^{\infty} \frac{1}{(2n-1)^2} \cos((2n-1)\Omega_0 t),$$

where $\Omega_0 = \pi$. Note that because $v(t)$ is an even function, the b_n coefficients are all zero. Call the output $y(t)$. By Eq. (5.21) and the specified frequency response,

$$y(t) = A\left(\frac{5}{6}\right) - \frac{8A}{\pi^2} \sum_{n=1}^{\infty} \frac{1}{(2n-1)^2}\left(\frac{5}{6+(2n-1)\pi}\right)\cos((2n-1)\pi t - (2n-1)\pi).$$

Although it is difficult to sketch $y(t)$, it is clear that the components in $v(t)$ at higher frequencies will be relatively more attenuated than those at lower frequencies because of the $5/(6 + (2n-1)\pi)$ term. Therefore, we should expect that sudden changes in $v(t)$, such as occur at its peaks and troughs, will be distorted.

Example 5.6 Determining frequency response using a periodic input   An alternative approach to Example 4.6 is now apparent. Instead of using a bank of filters,

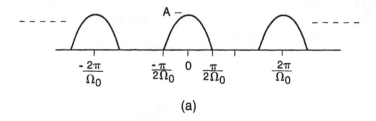

(a)

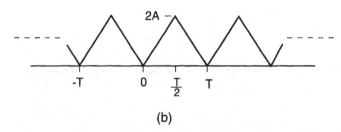

(b)

FIGURE 5.8. Two even periodic functions: (a) Half-wave rectified cosine. (b) Triangular wave.

one could analyze the input and output signals using Fourier series. To determine the a_n and b_n coefficients for an actual data signal, one could implement approximations to Eqs. (5.7)–(5.9) using numerical integration. For example, if a signal $x(t)$ is sampled every Δt seconds such that K samples are acquired, a simple approximation to Eq. (5.7) is

$$a_n^x \approx \frac{2}{T} \sum_{k=0}^{K-1} x(k \cdot \Delta t) \cos(n\Omega_0 \cdot k \cdot \Delta t)\Delta t, \, n > 0.$$

Because of the orthogonality of the trigonometric functions, this process is reasonably accurate if (1) $K \cdot \Delta t \gg T$, where T is the fundamental period of $x(t)$, and (2) $\Delta t \ll 2\pi/N\Omega_0$, where N is the largest multiple of the fundamental frequency that must be considered in order to represent $x(t)$ sufficiently closely. Often the "rule of thumb" that N should be 10 is adequate (but it should be confirmed for a given signal). In the next chapter a mathematically more elegant solution to this problem will be developed.

5.7 PARSEVAL'S RELATION FOR PERIODIC SIGNALS

Previously the power of a periodic signal having period of T was defined as the average energy per cycle, or $P_x = (1/T)\int_0^T x(t)x^*(t)dt$. If $x(t)$ has a Fourier series, one may express P as

$$P_x = \frac{1}{T}\int_0^T \sum_{n=-\infty}^{\infty} c_n e^{jn\Omega_0 t} \sum_{m=-\infty}^{\infty} c_m^* e^{-jm\Omega_0 t} dt = \sum_{n=-\infty}^{\infty} |c_n|^2. \qquad (5.24)$$

In other words, the power in the signal is distributed across the various frequency components in direct relation to the squared magnitudes of the Fourier series coefficients. This result is known as *Parseval's relation for periodic signals*. It is a direct consequence of the orthogonality of the functions in the Fourier series. The graph of $|c_n|^2$ vs. $n\Omega_0$ is the *power spectrum of a periodic signal*. Since power occurs only at discrete frequencies, this graph is known as a line spectrum.

Example 5.7 Contamination of a biomedical signal by 60-Hz interference EMG signals usually are recorded using differential amplifiers. If the impedances of the two electrodes are not well matched, or if shielding is inadequate, the EMG signal will contain a 60-Hz component due to electromagnetic radiation from power lines. Often EMG signals are quantified by passing them first through a full-wave rectifier, then through a lowpass filter whose output approximates the mean level of the rectified signal over the preceding T_{aver} msec, where T_{aver} might be 200 msec, for example. The frequency content of the rectified 60-Hz component will no longer be at 60-Hz. It may be determined by calculating the Fourier series for the rectified, 60-Hz signal $x(t) = A|\sin(2\pi(60)t)|$. Note that $2\pi \cdot 60 = 377$ and that the period of $x(t)$ is half that of the original 60-Hz signal, or $T = (1/2) \cdot (2\pi/377) = (\pi/377)$. Therefore,

$$c_0 = \frac{1}{T} \int_0^{\pi/377} A \sin(377t)dt = \frac{A}{377T}[-\cos(377t)|_0^{\pi/377}] = \frac{2A}{\pi}.$$

Since $x(t)$ is an even function with frequency $\Omega_0 = 2\pi(120) = 754$ rad/s,

$$c_n = \frac{1}{T} \int_0^{\pi/377} A \sin(377t)e^{-j754nt}dt = \frac{A}{T} \int_0^{\pi/377} \sin(377t) \cos(754nt)dt$$

$$= \frac{A}{T}\left[-\frac{\cos(2(0.5-n) \cdot 377t)}{2(0.5-n)(377)} - \frac{\cos(2(0.5+n) \cdot 377t)}{2(0.5+n)(377)} \right]\Bigg|_0^{\pi/377} = \frac{2A}{\pi}\left[\frac{1}{1-4n^2} \right].$$

That is, the spectral content of the rectified 60-Hz signal comprises a "DC" component plus all of the harmonics of 120 Hz.; however, the amplitudes of the third and higher harmonics are less than 5% of that of the original (unrectified) 60-Hz signal. The power of the original 60-Hz signal can be determined from its Fourier series, for which $c_1 = c_{-1} = A/2$; therefore, its power is $A^2/4$ at 60 and -60 Hz whereas the rectified signal has power $(4A^2/9\pi^2)$, or about 18% of the power of the 60-Hz signal, at 120 Hz (and -120 Hz).

5.8 THE CONTINUOUS-TIME FOURIER TRANSFORM (CTFT)

The Fourier series provides an alternative description of a periodic signal that is especially useful when one must evaluate the effects of filtering the signal. There is an obvious advantage if one also could model a non-periodic signal as a sum of trigonometric or complex exponential functions. In fact the derivation of a similar representation of a nonperiodic signal is straightforward if we consider the parallels between a nonperiodic signal and a periodic signal constructed by replicating the nonperiodic signal. Consider first a bounded, finite-duration, signal $x(t)$ that is nonzero only on the inteval $[-T_1, T_1]$ (Fig. 5.9). Now choose some $T > 2|T_1|$ and create a periodic signal

$$\tilde{x}(t), \qquad \ni \tilde{x}(t + nT) = x(t), \qquad n = \pm 1, \pm 2, \pm 3, \ldots .$$

Then $x(t) = \tilde{x}(t), -(T/2) \le t \le (T/2)$, and also $x(t) = \lim_{T \to \infty}[\tilde{x}(t)]$. The periodic signal has a Fourier series of the form

$$\tilde{x}(t) = \sum_{n=-\infty}^{\infty} c_n e^{jn\Omega_0 t}, \qquad \Omega_0 = 2\pi/T,$$

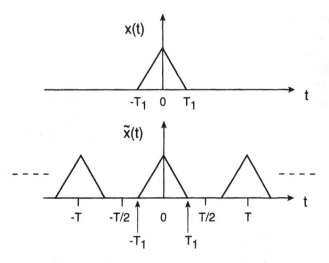

FIGURE 5.9. A bounded, finite-duration signal $x(t)$ (top) and a periodic extension of that signal (bottom).

where

$$c_n = \frac{1}{T} \int_{-T/2}^{T/2} \tilde{x}(t) e^{-jn\Omega_0 t} dt = \frac{1}{T} \int_{-\infty}^{\infty} x(t) e^{-jn\Omega_0 t} dt. \tag{5.25}$$

Defining

$$X(\Omega) \triangleq \int_{-\infty}^{\infty} x(t) e^{-j\Omega t} dt, \tag{5.26}$$

then one may write that

$$c_n = \frac{1}{T} X(n\Omega_0). \tag{5.27}$$

Substituting these values for c_n into the Fourier series equation,

$$\tilde{x}(t) = \sum_{n=-\infty}^{\infty} c_n e^{jn\Omega_0 t} = \sum_{n=-\infty}^{\infty} \frac{1}{T} X(n\Omega_0) e^{jn\Omega_0 t} = \frac{1}{2\pi} \sum_{n=-\infty}^{\infty} X(n\Omega_0) e^{jn\Omega_0 t} \Omega_0. \tag{5.28}$$

If we now take the limit as $T \to \infty$, then $\tilde{x}(t) \to x(t)$, $\Omega_0 \to d\Omega$, $n\Omega_0 \to \Omega$, and the summation becomes integration. Thus, applying this limit to Eq. (5.28) yields

$$x(t) = \frac{1}{2\pi} \int_{-\infty}^{\infty} X(\Omega) e^{j\Omega t} d\Omega. \tag{5.29}$$

Equation (5.29) describes a reconstruction of $x(t)$ based on a function of frequency, $X(\Omega)$, which is defined by Eq. (5.26). $X(\Omega)$ is the *Fourier transform* of $x(t)$ and the reconstruction equation, Eq. (5.29), is the *inverse Fourier transform* of $X(\Omega)$. Equation (5.27) expresses the relationship between the Fourier transform of $x(t)$ and the Fourier series coefficients of the periodic function $\tilde{x}(t)$ that is constructed by repeating $x(t)$ every T seconds. Regarding notation, the Fourier transform is symbolized as $X(\Omega) = \Im\{x(t)\}$ and the inverse Fourier transform as $x(t) = \Im^{-1}\{X(\Omega)\}$. The Fourier transform, $X(\Omega)$, is known as the frequency-domain representation of the time function, $x(t)$.

Viewing integration as a process of summation, one may interpret Eq. (5.29) as follows: The bounded, finite-duration signal $x(t)$ may be reconstructed via a weighted summation of an infinite collection of complex exponential functions. The weight at each frequency (or, more appropriately, for each $d\Omega$ range of frequencies) is given by $(1/2\pi)X(\Omega)d\Omega$. Thus one's understanding of the Fourier transform is not fundamentally different from that of the Fourier series. There are, however, two practical differences: (1) whereas a periodic signal can be reconstructed from a (possibly very large) number of discrete frequency components, reconstruction of a nonperiodic signal requires a continuous distribution of frequencies; (2) as a consequence, for a nonperiodic signal one does not talk about the weighting at each individual frequency but the weighting of a $d\Omega$ range of the continuous distribution of frequencies.

Although the above requirement that $x(t)$ be bounded and have finite duration is more restrictive than necessary, not every nonperiodic signal has a Fourier transform. The ordinary conditions for a signal $x(t)$ to have a Fourier transform are: (1) $x(t)$ must be absolutely integrable—that is, $\int_{-\infty}^{\infty}|x(t)|dt < \infty$, and; (2) $x(t)$ must have finite numbers of discontinuities, maxima, and minima in every finite time interval. Because energy signals are square-integrable, every energy signal satisfies the first condition and continuous energy signals are likely to meet condition 2 also and have a Fourier transform. Signals such as $x(t) = 1$ and $x(t) = u(t)$ are not absolutely integrable, however, and their ordinary Fourier transforms do not exist. Later in the chapter we will discuss a generalized Fourier transform which will be definable for some signals that do not meet the above integrability condition.

Example 5.8 Fourier transform of a pulse train Previously the Fourier series coefficients for a pulse train (Fig. 5.10(a)) were determined to be

$$c_n = \frac{A_0 a}{T} \frac{\sin(n\Omega_0 a/2)}{n\Omega_0 a/2}.$$

Now create a nonperiodic signal, $x(t)$, comprising only the pulse around $t = 0$ (Fig. 5.10(b)). The relationship between the Fourier series coefficients for the pulse train and the Fourier transform of $x(t)$ is given by Eq. (5.27), $X(n\Omega_0) = Tc_n$. Letting T approach infinity means that the pulse train converges to $x(t)$ and $n\Omega_0 \to \Omega$. Thus

$$X(\Omega) = \lim_{T\to\infty}[X(n\Omega_0)] = \lim_{n\Omega_0 \to \Omega}[Tc_n] = A_0 a \frac{\sin(\Omega a/2)}{\Omega a/2}. \qquad (5.30)$$

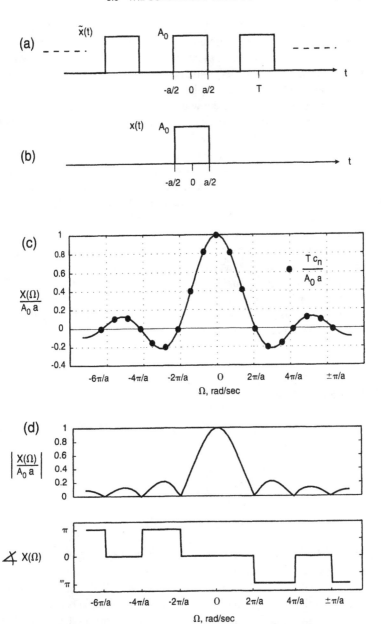

FIGURE 5.10. (a) A pulse train $\tilde{x}(t)$ and (b) a nonperiodic signal $x(t)$ obtained by taking one pulse from the pulse train. (c) Normalized Fourier transform of $x(t)$ (solid curve) and T times the Fourier series coefficients of $\tilde{x}(t)$ (dots) for the case $a/T = 1/3$. (d) Magnitude and phase of the normalized Fourier transform of $x(t)$.

Equation (5.30) has the form of sinc(x). It is plotted in Fig. 5.10(c) along with the points Tc_n for $a/T = 1/3$. In this example (but not always) $X(\Omega)$ is real-valued. Note that the zero-crossings of $X(\Omega)$ occur at integer multiples of $2\pi/a$ and, in particular, the first zero-crossing occurs at $2\pi/a$. That is, the frequency range that contains the complex exponential components of $x(t)$ having the largest amplitudes is limited by $|\Omega| < 2\pi/a$. It is common to quantify the width of this large central peak encompassing $\Omega = 0$ as half of the distance between the zero-crossings that mark the edges of the peak. In this case the width would be $2\pi/a$.

As the width, a, of the pulse decreases, the first zero-crossing of the sinc function moves further from $\Omega = 0$. In other words, the narrower the pulse, the greater the width of the central (main) lobe of $X(\Omega)$. Conversely if the pulse is very wide, the main lobe of $X(\Omega)$ is very narrow. This relationship between the time-domain extent of a pulse and the width of the main lobe of its Fourier transform assumes major importance in signal processing.

The intimate relationship between the Fourier series coefficients of the pulse train and the Fourier transform of the single pulse, $x(t)$, can be visualized by plotting $T \cdot c_n$ versus $n\Omega_0$ on the same graph with $X(\Omega)$. For illustration, let $a/T = 1/3$. Therefore, $\Omega_0 = 2\pi/T = 2\pi/3a$. By Eq. (5.27), $T \cdot c_n = X(\Omega)|_{\Omega=n\Omega_0}$; therefore, these points concide exactly with the plot of $X(\Omega)$ at the frequencies $n\Omega_0$ (Fig. 5.10(c)). Note that if $T/a = L$, where L is an integer, then $c_j = 0, j = L, 2L, 3L, \ldots$.

5.9 RELATIONSHIP OF FOURIER TRANSFORM TO FREQUENCY RESPONSE

The frequency response of a general LTIC system, such as that of Fig. 5.7, is defined as

$$H(\Omega) = \int_{-\infty}^{\infty} h(t)e^{-j\Omega t}dt.$$

This equation is equivalent to Eq. (5.26), from which it is clear that *the frequency response of an LTI system is exactly the Fourier transform of its impulse response.* Thus $H(\Omega)$ has two interpretations: (1) as a measure of the steady-state output of the system in response to unit-amplitude sinusoidal inputs, and (2) as the amplitudes of complex exponential signals (or, equivalently, phase-shifted cosine signals) that must be added together to construct $h(t)$. These two interpretations are quite consistent as we can demonstrate by showing that the second interpretation implies the first. Consider the response of the system to $x(t) = \sum_{n=-\infty}^{\infty}\delta(t - nT)$. The output, of course, is $y(t) = \sum_{n=-\infty}^{\infty}h(t - nT)$. The input and output are periodic signals. Now consider one pulse from the periodic input signal as we did for the pulse train above. Its Fourier transform is

$$\Im\{\delta(t)\} = \int_{-\infty}^{\infty}\delta(t)e^{-j\Omega t}dt = 1, \qquad \forall\Omega.$$

Therefore, by Eq. (5.27), the input comprises complex exponential functions such that $T \cdot c_n^x = 1$. Similarly, the amplitudes of the frequency components in the output signal are such that $T \cdot c_n^y = H(n\Omega_0)$, where $\Omega_0 = 2\pi/T$ and $H(\Omega)$ is the Fourier transform of $h(t)$. From Eq. (5.23) we know that the frequency response is obtained by dividing the Fourier series coefficients of the output by those of the input. Thus

$$\text{Frequency response} = \frac{c_n^y}{c_n^x} = \frac{T \cdot c_n^y}{T \cdot c_n^x} = \frac{H(n\Omega_0)}{1} = H(n\Omega_0).$$

By letting T approach infinity we obtain the general result that $H(\Omega)$, the Fourier transform of $h(t)$, is indeed the frequency response of the system.

5.10   PROPERTIES OF THE FOURIER TRANSFORM

Using Euler's formula the Fourier transform of a signal $x(t)$ may be written as

$$X(\Omega) = \int_{-\infty}^{\infty} x(t)e^{-j\Omega t}dt = \int_{-\infty}^{\infty} x(t)\cos(\Omega t)dt + j\left[-\int_{-\infty}^{\infty} x(t)\sin(\Omega t)dt\right] \triangleq R(\Omega) + jI(\Omega). \quad (5.31)$$

Therefore,

$$X(\Omega) = |X(\Omega)|e^{j\angle X(\Omega)} = \sqrt{R^2(\Omega) + I^2(\Omega)}\, e^{-j\tan^{-1}[I(\Omega)/R(\Omega)]}.$$

If $x(t)$ is real, then some special results apply. First, it is easy to show in this case that $X(-\Omega) = X^*(\Omega)$. Consequently, $|X(-\Omega)| = |X^*(\Omega)| = |X(\Omega)|$, implying that the magnitude of the Fourier transform is an even function of frequency. Similar consideration about the angle of $X(-\Omega)$ leads to the conclusion that the angle of the Fourier transform is an odd function of frequency (when $x(t)$ is real). A graph of $|X(\Omega)|$ versus Ω is known as the *amplitude* (or *magnitude*) *spectrum* of $x(t)$, while a graph of $\angle X(\Omega)$ vs. Ω is the *phase spectrum*.

The Fourier transform is a linear operator. That is, if $\Im\{x(t)\} = X(\Omega)$ and $\Im\{y(t)\} = Y(\Omega)$, then $\Im\{ax(t) + by(t)\} = aX(\Omega) + bY(\Omega)$. If $x(t)$ is an even function of time, then $I(\Omega) = 0$ (because the integral of an odd function between symmetric limits is zero) and the Fourier transform is entirely real. Similarly if $x(t)$ is an odd function, then $R(\Omega) = 0$ and the Fourier transform is entirely imaginary. There are numerous properties of the Fourier transform that either simplify its calculation or are important for linear filtering applications. Some of these properties are stated (without proof) in Table 5.1. Others that are important in signal processing are discussed below.

Time shift property: If $x(t)$ has the Fourier transform $X(\Omega)$, then the signal $x(t-d)$, where d is a constant, has the Fourier transform $e^{-j\Omega d}X(\Omega)$. To prove this result, let $z(t) = x(t-d)$. Then

$$Z(\Omega) = \int_{-\infty}^{\infty} x(t-d)e^{-j\Omega t}dt = \int_{-\infty}^{\infty} x(\lambda)e^{-j\Omega(\lambda+d)}d\lambda = e^{-j\Omega d}X(\Omega).$$

TABLE 5.1   Fourier Transform Properties

Property	Time domain	Frequency domain
Linearity	$ax_1(t) + bx_2(t)$	$aX_1(\Omega) + bX_2(\Omega)$
Conjugation	$x^*(t)$	$X^*(-\Omega)$
Time shift	$x(t - t_0)$	$e^{-j\Omega t_0}X(\Omega)$
Time scaling	$x(bt)$	$\dfrac{1}{\lvert b \rvert}X\left(\dfrac{\Omega}{b}\right)$
Time reversal	$x(-t)$	$X(-\Omega)$
Modulation	$e^{j\Omega t}x(t)$	$X(\Omega - \Omega_0)$
Duality	$X(t)$	$2\pi x(-\Omega)$
Time derivative	$\dfrac{d^n x(t)}{dt^n}$	$(j\Omega)^n X(\Omega)$
Frequency derivative	$-jtx(t)$	$\dfrac{dX(\Omega)}{d\Omega}$
Integration	$\displaystyle\int_{-\infty}^{t} x(\lambda)d\lambda$	$\dfrac{1}{j\Omega}X(\Omega) + \pi X(0)\delta(\Omega)$
Time convolution	$h(t) * x(t)$	$H(\Omega)X(\Omega)$
Time multiplication	$p(t)x(t)$	$\dfrac{1}{2\pi}P(\Omega) * X(\Omega)$
Amplitude modulation	$x(t)\cos(\Omega_0 t)$	$\dfrac{1}{2}[X(\Omega + \Omega_0) + X(\Omega - \Omega_0)]$
	$x(t)\sin(\Omega_0 t)$	$\dfrac{j}{2}[X(\Omega + \Omega_0) - X(\Omega - \Omega_0)]$

Example 5.9   Time shift property   Let $x(t)$ be the pulse of Fig. 5.10(b). (For simplicity, we shall represent a pulse of width 'a' as $P_a(t - c)$, where $T = c$ is at the center of the pulse. Therefore, $x(t) = A_0 P_a(t)$). Let $z(t)$ be the same pulse, delayed by $a/2$ seconds so that it now begins at $t = 0$ and ends at $t = a$—that is, $z(t) = A_0 P_a(t - a/2)$. By the time shift property,

$$Z(\Omega) = e^{-j\Omega a/2}X(\Omega) = e^{-j\Omega a/2}A_0 a\,\frac{\sin(\Omega a/2)}{\Omega a/2}.$$

Note that $|Z(\Omega)| = |e^{-j\Omega a/2}||X(\Omega)| = |X(\Omega)|$, and $\angle Z(\Omega) = \angle X(\Omega) - (a\Omega/2)$. Therefore time-shifting does not change the amplitude spectrum of a signal; however, it alters the phase spectrum by adding an angle that increases linearly with frequency in direct proportion to, and of the same sign as, the time shift.

In advanced applications involving fractal signals or wavelets it is necessary to consider signals that are scaled in time. That is, given a signal $x(t)$, one must also

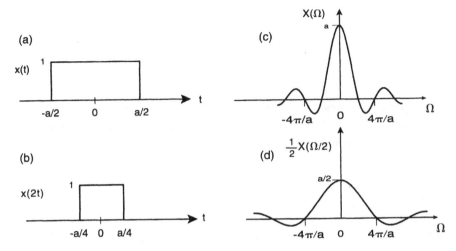

FIGURE 5.11. An example of the effect of time-scaling a signal (a, b) on its Fourier transform (c, d).

consider the signal $x(bt)$ where b is a constant. If $b > 1$, then $x(bt)$ is a time-compressed version of $x(t)$ (because the argument 'bt' increases faster than t). An example is shown in Fig. 5.11. Likewise, if $b < 1$, then $x(bt)$ is a time-expanded version of $x(t)$. Using the definition of the Fourier transform it is easily proven that if $x(t)$ has the transform $X(\Omega)$, then $x(bt)$ has the Fourier transform $(1/|b|)X(\Omega/b)$. In other words, compression in the time-domain produces broadening of the Fourier transform (Fig. 5.11). For the particular case of $x(t)$ real and $b = -1$,

$$\Im\{x(-t)\} = X(-\Omega) = X^*(\Omega).$$

Because complex exponential functions are so important in linear signal processing, often one encounters signals of the form $x(t)e^{j\Omega_0 t}$. It is easily shown from Eq. (5.26) that the Fourier transform of this signal is $X(\Omega - \Omega_0)$.

One often encounters derivatives of signals in signal processing. When $x(t)$ has a Fourier transform, the Fourier transform of its derivative is

$$\Im\left\{\frac{d^n x(t)}{dt^n}\right\} = (j\Omega)^n X(\Omega). \tag{5.32}$$

This property is provable by taking the derivative of Eq. (5.29).

The convolution operation is encountered often in signal processing, especially in the context of filtering. If two signals $x(t)$ and $v(t)$ have Fourier transforms $X(\Omega)$ and $V(\Omega)$, respectively, and if $y(t) = x(t) * v(t)$, the Fourier transform of $y(t)$ can be found as

$$Y(\Omega) = \int_{-\infty}^{\infty}\left[\int_{-\infty}^{\infty}x(t-\lambda)v(\lambda)d\lambda\right]e^{-j\Omega t}dt = \int_{-\infty}^{\infty}\int_{-\infty}^{\infty}x(t-\lambda)e^{-j\Omega(t-\lambda)}v(\lambda)e^{-j\Omega\lambda}d\lambda\,dt$$

$$= \left[\int_{-\infty}^{\infty}x(t-\lambda)e^{-j\Omega(t-\lambda)}dt\right]\left[\int_{-\infty}^{\infty}v(\lambda)e^{-j\Omega\lambda}d\lambda\right] = X(\Omega)V(\Omega). \qquad (5.33)$$

In other words, convolution in the time domain corresponds to multiplication in the frequency domain. Conversely, one can show that convolution in the frequency domain corresponds to multiplication in the time domain—that is,

$$\Im^{-1}\{X(\Omega) * V(\Omega)\} = 2\pi x(t)v(t). \qquad (5.34)$$

The most common application of Eq. (5.33) is to determine the Fourier transform of the output of a linear system given the input signal and the impulse response of the system (Fig. 5.7). Since $y(t) = h(t) * x(t)$, then

$$Y(\Omega) = H(\Omega)X(\Omega). \qquad (5.35)$$

That is, relative to a small frequency range, $d\Omega$, the amplitudes of the frequency components of the output signal are equal to the amplitudes of the corresponding frequency components in the input signal multiplied by the frequency response in that frequency range. This result implies the following two specific relationships:

$$|Y(\Omega)| = |H(\Omega)||X(\Omega)|, \qquad \measuredangle Y(\Omega) = \measuredangle X(\Omega) + \measuredangle H(\Omega).$$

The property of duality is often useful in theoretical derivations involving Fourier transforms. If $x(t)$ has the Fourier transform $X(\Omega)$, and if a time function is defined by substituting t for Ω in $X(\Omega)$, then $\Im\{X(t)\} = 2\pi x(-\Omega)$.

Example 5.10 Amplitude modulation An amplitude-modulated signal is the product of two other signals: $y(t) = a(t)\,x(t)$. In the case of amplitude modulation $a(t)$ usually varies much more slowly than $x(t)$ so that the amplitude of $y(t)$ is determined by $a(t)$. Sometimes the variations in time of a complex biomedical signal can be modeled as a basic signal whose amplitude is being modulated. For example, an EMG from a leg muscle varies cyclically throughout each cycle of walking. If one is most interested in the variations of the amplitude of the EMG signal (rather than its fine details), this signal might be modeled as amplitude-modulated random noise. Then the modulating signal is a measure of the variations of the amplitude of the EMG.

Another application of amplitude modulation occurs when a finite-duration signal is represented as an infinite-duration signal multiplied by a unit-amplitude pulse (Fig. 5.12). Since real-world signals always are observed for finite times, this model is encountered frequently in practice. The Fourier transform of the signal may be determined readily from those of the infinite-duration signal and the modulating signal. Consider the example of Fig 5.12, in which $y(t) = x(t)\sin(\Omega_0 t)$, and $x(t) = P_a(t)$. By Euler's identity,

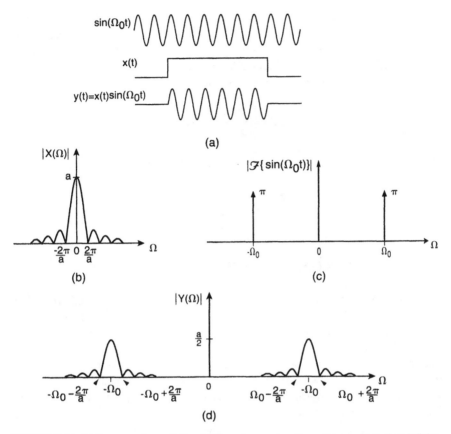

FIGURE 5.12. (a) Representing a finite-duration sine wave, $y(t)$, as the product of an infinite-duration sine wave and a unit-amplitude pulse. (b, c) Magnitude of the Fourier transform of the unit-amplitude pulse (b) and of the infinite-duration sine wave (c). (d) Magnitude of the Fourier transform of $y(t)$.

$$y(t) = x(t)\left[\frac{e^{j\Omega_0 t} - e^{-j\Omega_0 t}}{2j}\right] \Rightarrow \Im\{x(t)\sin(\Omega_0 t)\} = Y(\Omega) = \frac{j}{2}[X(\Omega + \Omega_0) - X(\Omega - \Omega_0)].$$

(5.36)

Using Eq. (5.30),

$$X(\Omega) = P_a(\Omega) = A_0 a \frac{\sin(\Omega a/2)}{\Omega a/2}.$$

Therefore,

$$Y(\Omega) = \frac{ja}{2}\left[\frac{\sin((\Omega + \Omega_0)a/2)}{(\Omega + \Omega_0)a/2} - \frac{\sin((\Omega - \Omega_0)a/2)}{(\Omega - \Omega_0)a/2}\right].$$

The magnitude spectra of $X(\Omega)$, $Y(\Omega)$ are graphed in Fig. 5.12(b,d). Note that in $Y(\Omega)$ the magnitude spectrum of the modulating pulse is shifted so that it is centered at $\pm\Omega_0$. This shifting is a direct consequence of the convolution of $X(\Omega)$ (Fig. 5.12(b)) with $\Im\{\sin(\Omega_0 t)\}$ (Fig. 5.12(c)) indicated in Eq. (5.34). For an amplitude-modulated cosine wave the equivalent result to Eq. (5.36) is

$$\Im\{x(t)\cos(\Omega_0 t)\} = \frac{1}{2}[X(\Omega + \Omega_0) + X(\Omega - \Omega_0)]. \tag{5.37}$$

This result may also be derived by expressing Eq. 5.34 in the general form

$$\Im\{x(t)v(t)\} = \frac{1}{2\pi}X(\Omega) * V(\Omega).$$

5.11   THE GENERALIZED FOURIER TRANSFORM

Previously we derived the Fourier transform of a unit-impulse function, $x(t) = \delta(t)$: $\Im\{\delta(t)\} = X(\Omega) = 1$, $\forall\Omega$. If one defines a time function $X(t) = 1$, $\forall t$, by the duality property its Fourier transform is $\Im\{X(t)\} = 2\pi\delta(-\Omega) = 2\pi\delta(\Omega)$. By this approach one is able to define the Fourier transform of a function $x(t) = 1$ which does not meet the strict conditions for the existence of its Fourier transform. This type of Fourier transform is known as a generalized Fourier transform and it permits one to define the Fourier transform for many other functions besides constant functions. For example, since by Eq. (5.37)

$$\Im\{x(t)\cos(\Omega_0 t)\} = \tfrac{1}{2}[X(\Omega + \Omega_0) + X(\Omega - \Omega_0)],$$

if we let $x(t) = 1$, then

$$\Im\{1 \cdot \cos(\Omega_0 t)\} = \pi[\delta(\Omega + \Omega_0) + \delta(\Omega - \Omega_0)].$$

Similarly, the Fourier transform of $\sin(\Omega_0 t)$ is $j\pi[\delta(\Omega + \Omega_0) - \delta(\Omega - \Omega_0)]$. Its magnitude spectrum is plotted in Fig. 5.12(c).

By similar reasoning, since $\Im\{x(t)e^{j\Omega_0 t}\} = X(\Omega - \Omega_0)$, $\Im\{1 \cdot e^{j\Omega_0 t}\} = 2\pi\delta(\Omega - \Omega_0)$. Based on this result, it is possible to write a generalized Fourier transform for any periodic function that has a Fourier series representation. Thus, if $x(t)$ with period T has the Fourier series $x(t) = \sum_{n=-\infty}^{\infty} c_n e^{jn\Omega_0 t}$, then $X(\Omega) = 2\pi\sum_{n=-\infty}^{\infty} c_n \delta(\Omega - n\Omega_0)$. This Fourier transform comprises a set of impulses located at $n\Omega_0$ along the frequency axis, each having the amplitude $2\pi c_n$.

The unit-step function, $u(t)$, is important in signal processing and one can define its generalized Fourier transform, $U(\Omega)$, as follows. First, let $x(t) = -0.5 + u(t)$. Thus $U(\Omega) = X(\Omega) + \Im\{0.5\}$. The derivative of $x(t)$ is an impulse function whose Fourier transform equals 1. But also by Eq. (5.32), $\Im\{dx/dt\} = j\Omega X(\Omega)$. Equating these two results implies $X(\Omega) = (1/j\Omega)$. Since $\Im\{-0.5\} = -\pi\delta(\Omega)$, then $U(\Omega) = (1/j\Omega) + \pi\delta(\Omega)$.

5.12 EXAMPLES OF FOURIER TRANSFORM CALCULATIONS

Example 5.11   The function $x(t) = \sin(\Omega_0 t + \theta)$ may be written as

$$x(t) = \frac{e^{j\Omega_0 t}e^{j\theta} - e^{-j\Omega_0 t}e^{-j\theta}}{2j} = \frac{j}{2}[e^{-j\theta}e^{-j\Omega_0 t} - e^{j\theta}e^{j\Omega_0 t}].$$

Therefore, $X(\Omega) = j\pi[e^{-j\theta}\delta(\Omega - n\Omega_0) - e^{j\theta}\delta(\Omega - n\Omega_0)$.

Example 5.12   The Fourier transform of $h(t) = e^{-at} u(t)$ may be calculated as

$$H(\Omega) = \int_{-\infty}^{\infty} h(t)e^{-j\Omega t}dt = \int_{0}^{\infty} e^{-at}e^{-j\Omega t}dt = \frac{1}{j\Omega + a}.$$

Example 5.13   Let $x(t)$ be a one-second segment of a cosine wave: $x(t) = \cos(\pi t)P_1(t)$. The Fourier transform of $x(t)$ may be determined from knowledge of that of $P_1(t)$. Thus

$$P_1(\Omega) = \frac{\sin(\Omega/2)}{\Omega/2}, \qquad X(\Omega) = \frac{1}{2}[P_1(\Omega + \pi) + P_1(\Omega - \pi)].$$

Therefore,

$$X(\Omega) = \frac{1}{2}\left[\frac{\sin\left(\frac{1}{2}(\Omega + \pi)\right)}{\frac{1}{2}(\Omega + \pi)} + \frac{\sin\left(\frac{1}{2}(\Omega - \pi)\right)}{\frac{1}{2}(\Omega - \pi)} \right].$$

Example 5.14   The Fourier transform of many signals that start abruptly at $t = 0$ can be determined easily. For example, let $x(t) = \sin(\Omega_0 t)u(t)$. Then

$$X(\Omega) = \frac{j}{2}[U(\Omega + \Omega_0) - U(\Omega - \Omega_0)]$$

$$= \frac{j}{2}\left[\frac{1}{j(\Omega + \Omega_0)} + \pi\delta(\Omega + \Omega_0) - \frac{1}{j(\Omega - \Omega_0)} - \pi\delta(\Omega - \Omega_0) \right]$$

$$= \left[\frac{\Omega_0}{\Omega_0^2 - \Omega^2} \right] + \frac{j\pi}{2}[\delta(\Omega + \Omega_0) - \delta(\Omega - \Omega_0)].$$

Example 5.15   Find $x(t)$ having the Fourier transform $X(\Omega) = P_4(\Omega) \cos(\pi\Omega/4)$. This function is shown in Fig. 5.13(a). One may evaluate the inverse Fourier transform directly:

$$x(t) = \frac{1}{2\pi} \int_{-\infty}^{\infty} X(\Omega)e^{j\Omega t}d\Omega = \frac{2}{2\pi} \int_0^2 \cos(\pi\Omega/4) \cos(\Omega t)d\Omega$$

$$= \frac{1}{\pi} \left[\frac{\sin(\Omega(\pi/4 - t))}{2(\pi/4 - t)} + \frac{\sin(\Omega(\pi/4 + t))}{2(\pi/4 + t)} \right]_0^2.$$

Evaluating the above expression,

$$x(t) = \frac{1}{\pi} \left[\frac{\sin(\pi/2 - 2t)}{\pi/2 - 2t} + \frac{\sin(\pi/2 + 2t)}{\pi/2 + 2t} \right].$$

This function is plotted in Fig. 5.13(b).

Example 5.16   Determine the Fourier transform of a pulse train having its minimum value at zero, its maximum at five, a pulse width of two seconds, and a period of four seconds. We may substitute the following values into Eq. (5.19): $a = 2$, $A_0 = 5$, $T = 4$, $\Omega_0 = 2\pi/4$; therefore, the Fourier series coefficients of $x(t)$ are given by

$$c_n = \frac{A_0 a}{T} \frac{\sin(n\Omega_0 a/2)}{n\Omega_0 a/2} = \frac{5 \cdot 2}{4} \frac{\sin(2\pi n/4)}{2\pi n/4}$$

and the Fourier series for $x(t)$ is $x(t) = \sum_{n=-\infty}^{\infty} (5/n\pi) \sin(n(\pi/2))e^{jn\pi t/2}$. Consequently, $X(\Omega) = \sum_{n=-\infty}^{\infty} (10/n) \sin(n(\pi/2))\delta(\Omega - n(\pi/2))$.

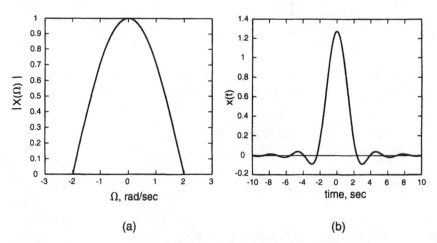

(a)   (b)

FIGURE 5.13. (a) Magnitude of the Fourier transform $X(\Omega) = P_4(\Omega) \cos(\pi\Omega/4)$. (b) Its inverse transform, $x(t)$.

5.13 PARSEVAL'S RELATION FOR NONPERIODIC SIGNALS

Let $x(t)$, $v(t)$ be arbitrary real or complex-valued functions that have Fourier transforms $X(\Omega)$, $V(\Omega)$ and let $y(t) = x^*(t)\, v(t)$. Then, after noting that $\Im\{x^*(t)\} = X^*(-\Omega)$,

$$Y(\Omega) = \frac{1}{2\pi} \int_{-\infty}^{\infty} x^*(t)v(t)e^{-j\Omega t}dt = \frac{1}{2\pi}\Im\{x^*(t)\} * V(\Omega) = \frac{1}{2\pi}\int_{-\infty}^{\infty} X^*(\lambda - \Omega)V(\lambda)d\lambda.$$

Since this result applies at all frequencies, one may let $\Omega = 0$. Thus

$$\int_{-\infty}^{\infty} x^*(t)v(t)dt = \frac{1}{2\pi}\int_{-\infty}^{\infty} X^*(\lambda)V(\lambda)d\lambda.$$

Changing the variable of integration from λ to Ω, we find that

$$\int_{-\infty}^{\infty} x^*(t)v(t)dt = \frac{1}{2\pi}\int_{-\infty}^{\infty} X^*(\Omega)V(\Omega)d\Omega.$$

Finally, let $v(t) = x(t)$. The above equation becomes

$$\int_{-\infty}^{\infty} |x(t)|^2 dt = \frac{1}{2\pi}\int_{-\infty}^{\infty} |X(\Omega)|^2 d\Omega. \tag{5.38a}$$

Note that the l.h.s. of Eq. (5.38) is the energy of $x(t)$. This equation is known as *Parseval's relation for nonperiodic signals*. It states that the energy in a signal, $x(t)$, that has a Fourier transform is distributed across its frequency components such that the energy in each frequency interval $(\Omega, \Omega + d\Omega)$ is proportional to the squared magnitude of $X(\Omega)$. Recalling the discussion of energy, memory, and correlation in Chapter 2, it is apparent that understanding the distribution of energy in a signal and how that distribution can be modified will provide insights about memory processes in the physical system that generated the signal. Parseval's relation provides a convenient tool, the Fourier transform, to dissect the distribution of energy in a signal.

The *energy spectral density* (or, simply, the *energy spectrum*) of the deterministic signal $x(t)$ is the graph $|X(\Omega)|^2$ versus Ω. This function is referred to as an energy "density" function because the actual energy in a $d\Omega$ band is proportional to $X(\Omega)d\Omega$. For a deterministic, finite-length signal observed on the interval $[0, T_0]$, the *power spectral density* (or, simply, the *power spectrum*) is the energy spectrum divided by T_0. That is,

$$P_x(\Omega) = \frac{|X(\Omega)|^2}{T_0}. \tag{5.38b}$$

Furthermore, we shall prove the very important result that

$$P_x(\Omega) = \Im\{R_x(s)\} \tag{5.38c}$$

Given $x(t)$, $0 \leq t \leq T_0$, by definition its autocorrelation function is

$$R_x(s) = \frac{1}{T_0} \int_0^{T_0} x(t + s)x(t)dt = \frac{1}{T_0} x(t) * x(-t).$$

The Fourier transform of $R_x(s)$ is

$$\Im\{R_x(s)\} = \frac{1}{T_0} X(\Omega)X^*(\Omega) = \frac{|X(\Omega)|^2}{T_0} = P_x(\Omega).$$

This result is a foundation of signal processing because it relates the distribution of power in a signal across frequency to temporal correlation in the signal. It also provides an alternative (and computationally fast) method for calculating $R_w(s)$ by first using a fast Fourier transform (FFT) method (see Chapter 7) to calculate $X(\Omega)$, then using the FFT again to calculate $\Im^{-1}\{P_x(\Omega)\}$.

Example 5.17 Power spectrum of infinite-length, zero-mean, white noise Consider a zero-mean, white-noise signal, $w(t)$. Previously we showed that $R_w(s) = msv(w) \, \delta(s)$. For simplicity, let $msv(w) = s_w^2$. Then the power spectrum of $w(t)$ is

$$P_w(\Omega) = \Im\{s_w^2 \, \delta(s)\} = s_w^2, \qquad \forall \Omega.$$

That is, the theoretical power spectrum of white noise is constant and has an amplitude equal to the mean-square value of $w(t)$.

5.14 FILTERING

Earlier in this chapter we established two related facts: (1) the frequency response of an LTI system is actually the Fourier transform of its impulse response, and; (2) the Fourier transform of the output of a linear system equals the Fourier transform of the input signal multiplied by the frequency response of the system. Consequently, the frequency response has importance that transcends its application to steady-state sinusoidal inputs: one may utilize the frequency response to obtain a measure of the alteration of the frequency content of any signal having a Fourier transform, such as nonperiodic signals and indeed even transient signals. Often this process is useful for removing noise. If a signal comprises a desired signal plus added high-frequency noise, and if one can estimate the frequency range for which the magnitude of the Fourier transform of the desired signal is significantly above that of the noise, then it may be possible to find a linear filter having a frequency response that is esssentially constant over this frequency range and nearly zero elsewhere. In theory, passing the recorded signal through this filter will allow the important frequency components of the desired signal to "pass through" to the output, whereas the frequency components associated with high-frequency noise will be very much attenuated in the output. A major limitation is that no real filter can have the sharp demarcation between the passband and the stopband described for ideal filters in Chapter 4. Therefore, if one chooses a frequency response that avoids any distortion of the desired signal, some noise components at nearby frequencies will not be much attenuated. The topic of filter design will be the focus of later chapters. Here we discuss some common types of CT (or analog) filters.

Example 5.18   60-Hz notch filter   A common application of noise removal is the use of a filter to remove a 60-Hz signal from a bioelectrical recording such as an electromyogram. This type of noise may be due to electromagnetic fields that are not blocked because of inadequate grounding or shielding. The design of notch filters whose gain has a null centered at 60 Hz can be quite sophisticated. An example of the frequency response of a 60-Hz notch filter is given in Fig. 5.14.

Impulse Response of an Ideal Lowpass Filter

The frequency response of an ideal, linear-phase, lowpass filter (Chapter 4) may be described compactly as

$$H(\Omega) = \begin{cases} e^{-j\Omega d}, & |\Omega| \le B \\ 0, & |\Omega| > B \end{cases}, \tag{5.39}$$

where B is its cutoff frequency. We can determine the impulse response, $h(t)$, of this filter using the duality property. Note that

$$H(\Omega) = P_{2B}(\Omega)\, e^{-j\Omega d}, \qquad \forall \Omega.$$

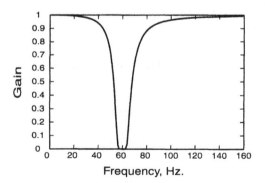

FIGURE 5.14. Magnitude of the frequency response of a tenth-order Bessel band-reject filter with cutoff frequencies of 55 Hz and 65 Hz.

Since by duality

$$\mathfrak{F}^{-1}\{P_{2B}(\Omega)\} = \frac{B}{\pi}\frac{\sin(Bt)}{Bt}, \text{ then } h(t) = \frac{B}{\pi}\frac{\sin(B(t-d))}{B(t-d)}, \forall t.$$

That is, the impulse response is a sinc function centered in time at $t = d$. Furthermore, $h(t)$ extends infinitely in both positive and negative time; therefore, this ideal filter is non-causal.

Common Practical (Causal) Filters

There are certain types of filters that are encountered often in practice because of their special properties. Butterworth filters are important because they meet a certain criterion for uniformity of gain in the passband. Among all linear filters of order N (i.e., filters for which their differential equations have order N), the Butterworth filter of order N has the most uniform gain in the passband (and its gain is said to have "maximally flat magnitude"). To achieve this goal the frequency response meets the criterion that the first $2N - 1$ derivatives of $|H(\Omega)|^2$ equal zero at $\Omega = 0$. For $N = 2$, the lowpass Butterworth filter is a second-order filter with $\delta = (1/\sqrt{2}) \approx 0.7071$. Referring to the previous chapter, the frequency response of such a second-order filter (assuming a desired passband gain of unity) will be

$$H(\Omega) = \frac{\Omega_n^2}{\Omega_n^2 - \Omega^2 + j\Omega\sqrt{2}\Omega_n}.$$

Its magnitude is

$$|H(\Omega)| = \frac{1}{\sqrt{1 + \left(\dfrac{\Omega}{\Omega_n}\right)^4}}. \tag{5.40}$$

The frequency response of this filter is graphed in Fig. 4.7.

A Chebyshev filter of order n has a sharper separation between its passband and stopband than is achieved by a Butterworth filter of the same order. On the other hand, Chebyshev filters exhibit strikingly nonuniform gain in either the passband or the stopband (but not both). The frequency response of an n-th order Chebyshev filter is specified by an n-th order Chebyshev polynomial function that is defined recursively as

$$T_0(\Omega) = 1, \qquad T_1(\Omega) = 2, \qquad T_n(\Omega) = 2\Omega\, T_{n-1}(\Omega) - T_{n-2}(\Omega). \tag{5.41}$$

The frequency response of the n-th order, Type 1, Chebyshev filter is

$$|H(\Omega)| = \frac{1}{\sqrt{1 + \varepsilon^2 T_n^2\left(\dfrac{\Omega}{\Omega_1}\right)}}. \tag{5.42}$$

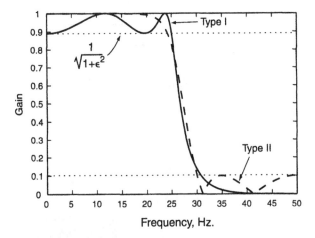

FIGURE 5.15. Magnitude of the frequency response of a Chebychev fourth-order Type I (solid) and Type II (dashed) lowpass filter.

Figure 5.15 shows this function for a second-order filter. (See also Exercise 5.18 regarding deriving this frequency response.) Ω_1 is a design parameter that does not equal the cutoff frequency of the filter unless $\varepsilon = 1$. Note that although the frequency response magnitude is variable in the passband, each fluctuation in gain traverses from the same minimum to the same maximum value, and these values are related to the design parameter, ε. This behavior is known as an "equiripple" response. Type 2 Chebyshev filters exhibit monotonically decreasing gain in the passband and gain fluctuations in the stopband.

5.15   OUTPUT RESPONSE VIA THE FOURIER TRANSFORM

On the basis of Eq. (5.35) it is possible to calculate the Fourier transform of the output signal from a linear system in response to many input signals besides sinusoids. If one then calculates the inverse Fourier transform, it is possible to determine the output for many nonperiodic and transient signals without directly solving the differential equation of the system! This last step may be difficult and typically one attempts to express the Fourier transform as a sum of terms, each of which has a form that is recognizable in a table of Fourier transforms (such as the brief list in Table 5.2). The decomposition usually is accomplished using the method of partial fraction expansion. An example will illustrate the procedure, as well as the determination of $H(\Omega)$ directly from the differential equation of a system.

Example 5.19   Determining the frequency response and output of a system using Fourier transforms   An exponential signal, $v_1(t) = V e^{-at} u(t)$, is applied to the circuit of Fig. 5.16. If $v_2(0) = 0$, what is the output voltage, $v_2(t)$?

TABLE 5.2 Common Fourier Transform Pairs

Property	Time domain	Frequency domain
Unit impulse	$\delta(t)$	1
Causal exponential	$e^{-at}u(t)$, $a > 0$	$1/(j\Omega + a)$
Complex sinusoid	$e^{j\Omega_0 t}$	$2\pi\delta(\Omega - \Omega_0)$
Unit step	$u(t)$	$\dfrac{1}{j\Omega} + \pi\delta(\Omega)$
Time-weighted exponential	$te^{-at}u(t)$, $a > 0$	$\dfrac{1}{(j\Omega + a)^2}$
Sine wave	$\sin(\Omega_0 t)$	$j\pi[\delta(\Omega + \Omega_0) - \delta(\Omega - \Omega_0)]$
Cosine wave	$\cos(\Omega_0 t)$	$\pi[\delta(\Omega + \Omega_0) + \delta(\Omega - \Omega_0)]$
Constant	A_0	$2\pi A_0\delta(\Omega)$
Rectangular pulse	$P_a(t)$	$a\,\text{sinc}(\Omega a/2)$
Impulse train	$\displaystyle\sum_{n=-\infty}^{\infty} \delta(t - nT)$	$\dfrac{2\pi}{T}\displaystyle\sum_{n=-\infty}^{\infty} \delta\left(\Omega - n\dfrac{2\pi}{T}\right)$
Signum	$\dfrac{t}{|t|}$	$\dfrac{2}{j\Omega}$

Solution: Consider the "system" to be the part of the circuit outlined by the dashed box in the figure. Its input is $v_1(t)$ and its output is $v_2(t)$. For any $v_1(t)$ having a Fourier transform one may find the output as $v_2(t) = \mathfrak{J}^{-1}\{H(\Omega)V_1(\Omega)\}$, where $H(\Omega)$ is the frequency response of the "system". To find the frequency response, write the differential equation(s) of the system:

$$i(t) = \frac{1}{R}[v_1(t) - v_2(t)];$$

$$v_2(t) = \frac{1}{C}\int_{-\infty}^{t} i(t)dt.$$

Taking the derivative,

$$\frac{dv_2(t)}{dt} = \frac{i(t)}{C} = \frac{1}{RC}[v_1(t) - v_2(t)].$$

Taking the Fourier transform of both sides, assuming initial conditions are zero,

$$j\Omega V_2(\Omega) = \frac{1}{RC}V_1(\Omega) - \frac{1}{RC}V_2(\Omega).$$

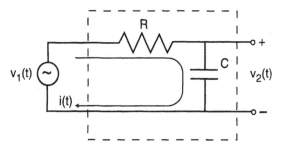

FIGURE 5.16. A simple electrical circuit viewed as a system having the frequency response $H(\Omega) = V_2(\Omega)/V_1(\Omega)$.

Thus

$$H(\Omega) = \frac{V_2(\Omega)}{V_1(\Omega)} = \frac{1}{RC} \frac{1}{j\Omega + \dfrac{1}{RC}}. \tag{5.43}$$

Since the Fourier transform of the input is $V_1(\Omega) = V(1/(j\Omega + a))$, then

$$V_2(\Omega) = \frac{V}{RC} \frac{1}{j\Omega + \dfrac{1}{RC}} \frac{1}{j\Omega + a} = \frac{K_1}{j\Omega + \dfrac{1}{RC}} + \frac{K_2}{j\Omega + a}. \tag{5.44}$$

Equation (5.44) expresses the partial fraction expansion of the Fourier transform. The constants K_1 and K_2 are evaluated as

$$K_1 = [V_2(\Omega)(j\Omega + 1/RC)]|_{j\Omega = -1/RC}, \qquad K_2 = [V_2(\Omega)(j\Omega + a)]|_{j\Omega = -a}.$$

Thus

$$K_1 = \frac{V}{RC} \frac{1}{j\Omega + a}\bigg|_{j\Omega = -1/RC} = \frac{V}{RC\left(a - \dfrac{1}{RC}\right)},$$

$$K_2 = \frac{V}{RC} \frac{1}{j\Omega + \dfrac{1}{RC}}\bigg|_{j\Omega = -a} = \frac{-V}{RC\left(a - \dfrac{1}{RC}\right)}.$$

The inverse transforms of the terms on the r.h.s. of Eq. 5.44 can be determined by inspection; thus,

$$v_2(t) = \frac{V}{RC\left(a - \dfrac{1}{RC}\right)}[e^{-t/RC} - e^{-at}]u(t).$$

The above example illustrates both the process of determining the frequency response of a system directly from its differential equation(s) and the process of solving for the output response using Fourier transforms. The former process is especially important in biomedical signal processing when physical processes are describable by (linear) differential equations. By determining the corresponding frequency response (either from the differential equations or by direct experimental measurement), one achieves insight about the response of the system to arbitrary inputs (i.e., stimuli). For example, one can at least qualitatively anticipate an unobserved system response to a stimulus that may have been observed during monitoring or that one is planning to apply. (See Exercises 5.14 and 5.20.) Conversely, if one observes a response from the system, it is possible to infer the characteristics of the input giving rise to that response. (See Exercises 5.21 and 5.23.) These benefits should not be considered trivial because in most biomedical applications it is difficult to monitor many variables simultaneously. Often just basic knowledge, such as the passband of a system, permits one to infer whether a stimulus will be effective in eliciting a response—for example, whether an injected drug will achieve an efficacious concentration at a distant site in a desired time.

Example 5.20   Filtering to remove 60-Hz signals   A signal that has features approximately like an electrocardiogram signal with added 60-Hz noise is

$$e(t) = 0.4 + \frac{1.731}{\pi}[\cos(8t) - 0.5\cos(16t + \pi) + 0.25\cos(40t)$$

$$+ 0.33\cos(32t + \pi/6)] + 0.2\sin(120\pi t).$$

One could attempt to remove the 60-Hz signal by passing $e(t)$ through a lowpass filter. The simplest such filter is a first-order filter having the frequency response

$$|H(\Omega)| = \frac{|b\tau|}{\sqrt{1 + \Omega^2\tau^2}},$$

where $1/\tau$ is the cutoff frequency. Letting $b = 1/\tau$ so that $H(0) = 1$, choose a cutoff frequency above 40 rad/s but below 60 Hz (which is equivalent to 120π rad/s). Choosing 100 rad/s, for example, the gain of the filter at the various frequency components in $e(t)$ is: 0.9968, 0.9874, 0.9524, 0.9285, and 0.2564 at the frequencies 8, 16, 32, 40, and 120π rad/s, respectively. For each of the non-noise components of $e(t)$ the gain of the filter is nearly unity although at 32 and 40 rad/s the gain is about 5% and 7% below unity. Probably this degree of alteration of the desired signal components is acceptable. But at 60-Hz the gain is still 0.2564, implying that the amplitude of the corresponding frequency component will be reduced only by 75%. Practically, this degree of attenuation is probably too little. One could use a lower cutoff frequency to reduce the gain at 60 Hz but the desired signal would be further distorted. One concludes that a simple lowpass filter is inadequate for this application.

5.16 SUMMARY

Periodic functions that satisfy the Dirichlet conditions can be represented by an orthogonal function expansion known as a Fourier series. The Fourier series is an infinite summation of sine and cosine functions having frequencies that are integer multiples (i.e., harmonics) of the frequency of the original function. The weighting coefficient for each term in the series can be determined by integrating the product of the original signal and the corresponding sine or cosine term (or equivalently, the corresponding complex exponential function) over one cycle. In general, the higher harmonic terms are more important when the original signal has abrupt changes. A pulse train is a model for regularly recurring events. The amplitudes of its Fourier series coefficients vary as a sinc function. If a periodic signal is applied to the input of an LTI system, then by expressing the input as a Fourier series, one may utilize the frequency response of the system to determine the steady-state output. Parseval's Theorem for periodic signals expresses the power of the signal as the summations of the squared magnitudes of its (exponential form) Fourier series coefficients, c_n. The graph of $|c_n|^2$ versus $n\Omega_0$ is the power spectrum of a periodic, deterministic signal.

The continuous-time Fourier transform (CTFT) extends the concept of the Fourier series model to nonperiodic signals. For a signal $x(t)$ that meets certain conditions (the foremost of which is absolute integrability), its Fourier transform is defined as

$$\Im\{x(t)\} = \int_{-\infty}^{\infty} x(t)e^{-j\Omega t}dt = X(\Omega).$$

From the inverse Fourier transform,

$$x(t) = \Im^{-1}\{X(\Omega)\} = \frac{1}{2\pi}\int_{-\infty}^{\infty} X(\Omega)e^{j\Omega t}d\Omega,$$

one obtains the interpretation that $x(t)$ can be expressed as a summation of complex exponentials (or sines and cosines) of all frequencies. In each range of frequencies, $[\Omega, \Omega + d\Omega]$, the weighting factor for this summation is proportional to $X(\Omega)$. There is a close relationship between the Fourier transform of a nonperiodic signal, $x(t)$, and the Fourier series coefficients of a periodic extension of this signal, $\tilde{x}(t)$. The generalized Fourier transform uses the duality property to extend this transform to some signals that are not absolutely integrable, such as the unit step function and sine and cosine functions.

The frequency response of an LTI system is equal to the Fourier transform of its impulse response. Consequently the Fourier transform of the output signal of an LTI system equals the product of the Fourier transforms of the input signal and the impulse response. This result permits calculation of the system response to nonperiodic and transient input signals without directly solving the differential equations of the system. Conversely one can determine the frequency response from the Fourier transforms of the differential equations of the system. These results provide the foundation for an enhanced understanding of filtering. The basic concept of filtering developed in Chapter 4 can be applied to nonperiodic as well as periodic signals and to finite-length as well as infinitely long signals.

Parseval's relation for nonperiodic signals states that the energy of a signal is proportional to the integral over all frequencies of the squared magnitude of its Fourier transform. This result permits one to define an energy spectral density function, $|X(\Omega)|^2$ vs. Ω, and a power spectral density function (power spectrum), $|X(\Omega)|^2/T_0$ vs. Ω, for a signal observed on $[0, T_0]$.

Although most biomedical signals are not truly periodic, situations in which stimuli are presented at regular intervals can be analyzed using Fourier series methods. A great number of biomedical signals meet the conditions for the existence of their Fourier transforms, and Fourier transforms have important roles both in filtering of biomedical signals and in analysis of biomedical systems.

EXERCISES

5.1 Find the Fourier series in both the complex exponential and trigonometric (i.e., having sines and cosines) forms for $f(t) = \cos(4t) + \sin(5t)$.

5.2 Consider the basic signal $f(t) = 2t - 1, -0.5 \le t < 0.5$.

a. Construct a periodic signal, $x(t)$, by making a periodic extension of $f(t)$ and determine its Fourier series.

b. Find the Fourier transform of the signal

$$y(t) = \begin{cases} f(t), & -0.5 \le t \le 0.5 \\ 0, & \text{otherwise} \end{cases}.$$

5.3 Find the Fourier series in trigonometric form for $f(t) = |\sin(\pi t)|$. Graph its power spectrum.

5.4 Exercise 4.14 presented a model of the periodic oscillations found in a recording of the pH of brain tissue. Determine the Fourier transform of this pH signal.

5.5 A periodic function $p(t)$ with period P is also periodic with period $2P$. In computing the Fourier series of $p(t)$ I mistakenly use $\Omega_0' = 2\pi/2P$ rather than $\Omega_0 = 2\pi/P$ as the fundamental frequency. Explain how it is possible to determine the correct Fourier series coefficients from the coefficients I calculated.

5.6 Consider the periodic signal $x(t)$ consisting of unit-amplitude delta-functions, as shown in Fig. 5.17. Calculate its exponential Fourier series. Let $t = 0$ occur in the middle of two of the impulses that are close together. Graph its power spectrum.

5.7 Derive Eq. (5.14) from Eq. (5.6) and prove the relationships given in Eq. (5.15).

5.8 A signal $x(t)$ is periodic with period $T = 4$. Its Fourier series coefficients are given by:

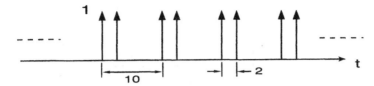

FIGURE 5.17. A periodic pulse train

$$c_n = \begin{cases} 0, & n = 0 \\ -1/n, & n = even. \\ 1/n, & n = odd \end{cases}$$

Write its Fourier series as a sum of phase-shifted cosines.

5.9   Consider the signal $g(t) = [1 + \cos(\Omega_0 t)]\cos(\Omega_1 t)$, $\Omega_0 \gg \Omega_1$.

a. Find the magnitude of the frequency component of $g(t)$ at $\Omega = \Omega_1$.

b. $g(t)$ is applied to the input of a Butterworth filter having the frequency response

$$H(\Omega) = \frac{\Omega_1^2}{-\Omega^2 + 1.414j\Omega_1\Omega + \Omega_1^2}.$$

What is the steady-state output at $\Omega = \Omega_1$?

5.10   Consider the following coupled differential equations which describe a system with input $x(t)$ and output $y_1(t)$:

$$\frac{dy_1}{dt} = y_2, \qquad \frac{dy_2}{dt} = -600\, y_2 - 10^6\, y_1 + x(t).$$

Find $H(\Omega)$ for this system and calculate $y_1(t)$ when $x(t) = 2 \cos(200\pi t)u(t)$.

5.11   Calculate the frequency response of the filter whose impulse response is $h(t) = (t - 1)^2 e^{-2(t-1)}u(t - 1)$.

5.12   The pulse train shown in Fig. 5.18 as $x(t)$ is given as the input to an LTIC system and the output response is shown as $y(t)$. For $0 \le t < 1$, $y(t)$ is given by the equation $y(t) = e^{-t/0.15}$.

a. Calculate the Fourier series of $y(t)$ (in complex exponential form).

b. The Fourier series coefficients for $x(t)$ are given by

$$c_n = 0.001[\sin(0.0005n\Omega_0)/\, 0.0005n\Omega_0]e^{-j0.0005n\Omega_0},$$

where Ω_0 is the same as you determined in part a. Specify Ω_0 and calculate $H(n\Omega_0)$.

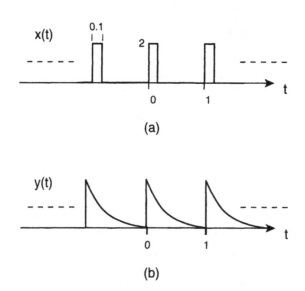

FIGURE 5.18. (a) A pulse train. (b) Response of a system to this pulse train.

5.13   In Exercise 3.18 you determined the differential equation describing the displacement of a microscope table in response to floor vibrations.

a. Find the frequency response of this system.

b. Three motors in an adjacent room each cause a sinusoidal movement of the floor. Their amplitudes and frequencies (in the steady state) are: 0.01 mm at 60 Hz, 0.03 mm at 30 Hz, and 0.05 mm at 10 Hz. Assuming there is no phase difference between these movements, calculate the resulting motion of the table.

5.14   "Stretch" receptors in the airways of the lungs respond to stretching the wall of the airway. These receptors have been modeled as second-order systems whose input is the applied stretch, $x(t)$, and output is the instantaneous firing frequency of the receptor, $y(t)$. Assume that one such receptor can be described by the differential equation

$$\frac{d^2y}{dt^2} + 2\frac{dy}{dt} + 4y(t) = 10x(t) + \frac{dx}{dt}.$$

a. Determine the frequency response of this receptor to stretching.

b. There are devices that can apply sinusoidal stretch to a small segment of tissue. Calculate the steady-state response, $y(t)$, to stretching an airway segment according to $x(t) = 2 \cos(\pi t)u(t)$.

5.15   A very sensitive, variable reluctance displacement transducer is being used to measure the compression of an excised vertebral disc in response to a sudden, impulse-like, compressive force applied to its top surface (Fig. 5.19). With this type of transducer an electronic circuit (known as a transducer coupler) applies a high-

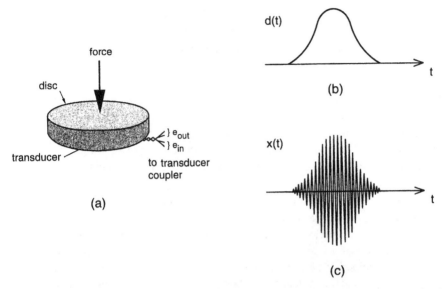

FIGURE 5.19. (a) Diagram of a force applied to a vertebral disc. (b) The displacement, $d(t)$, of the disc. (c) Output of the transducer.

frequency sine wave, e_{in}, to the electrical input of the transducer and the output, e_{out}, is a sine wave of the same frequency whose instantaneous amplitude is directly proportional to the displacement. For a compressive impulse applied at $t = 0$, assume the actual displacement (i.e., the amount of shortening of the height of the vertebral disc) is found to be $d(t) = 0.1[1 - \cos(50\pi t)]$ mm, $0 \leq t \leq 40$ msec. This results in the e_{out} signal $x(t)$. This signal has the same frequency as the input, e_{in}, but its amplitude varies in direct proportion to $d(t)$. (Assume a proportionality constant of unity.)

a. Let e_{in} be a sine wave of frequency 1000 Hz and amplitude one. What is the Fourier transform of $x(t)$?

b. I want to digitize $x(t)$. I should lowpass filter $x(t)$ before sampling it in order to separate the signal components related to $x(t)$ from any electrical noise. Specify an appropriate cutoff frequency so that this filter will pass all important components of $x(t)$ with little distortion of them.

c. Specify the parameters of a second-order Butterworth filter (i.e., $N = 2$) to achieve the cutoff frequency you specified in part b, and write its equation for $H(\Omega)$.

5.16 Use the duality property to calculate the Fourier transform of $h(t) = (1/t)$, $-\infty < t < \infty$. A system having this impulse response is known as a Hilbert transformer. Determine its response to the signal $x(t) = A_0 \cos(\Omega_0 t)$, $\forall t$.

5.17 The signal $P_{2a}(t - a)$ is applied to an LTIC system, resulting in an output signal whose Fourier transform is $Y(\Omega) = 4a^2 A e^{-j2a\Omega}(\sin^2(\Omega a)/(\Omega a)^2)$. Determine the frequency response of this system.

5.18   Determine the magnitude of the frequency response of a Chebyshev Type I lowpass filter as a function of the parameters ε, Ω_1 for the particular case $n = 2$.

5.19   Determine the Fourier transform of a sequence of three pulses, each of amplitude B and width $b/2$, that begin at times $t = -0.75\ b$, $-0.25\ b$, and $0.75\ b$, respectively.

5.20   In Example 4.2 a model of a capillary bed was formulated and differential equations to describe this model were developed. Starting from the differential equations and using the parameter values given in that example, determine the frequency response function of capillary flow to the blood pressure input. Graph the magnitude and phase of the frequency response. Does the frequency response have a lowpass, highpass, or bandpass character?

5.21   Figure 5.20 shows a simple model of a train of action potentials from a single nerve cell that are recorded using bipolar electrodes placed near the axon of the cell. (Assume the train is periodic and infinitely long.) Consider each action potential to be a single cycle of a sine wave, with period $T_1 = 4$ msec. Let the repetition period of the action potentials be $T_2 = 100$ msec. The peak amplitude of each action potential is 10 mv. The action potentials are being amplified by a second-order linear filter with a frequency response specified by: $\Omega_n/2\pi = 150\ Hz$, $\delta = 0.75$, gain at zero frequency $= 10,000$.

    a. Find an appropriate Fourier representation for the action potential train.

    b. Write the equations for the magnitude and phase of the frequency response of the amplifier, and sketch these functions.

    c. Calculate the corresponding Fourier representation of the output of the amplifier when the action potential train is the input.

    d. Indicate in a drawing how the action potentials in the output signal differ from those in the input signal (if they do). If there are differences, suggest new values for the properties of the amplifier (still being second order) to reduce these differences. Show mathematically why your suggestion(s) should work.

5.22   Amplitude-modulated signals can be useful as test signals to test the frequency response of electronic amplifiers. One such signal is described by the equation $z(t) = \sin(\Omega_1 t)\cos(\Omega_0 t)[P_{\pi/\Omega_0}(t)]$, where $\Omega_1 \gg \Omega_0$, the signal is non-zero

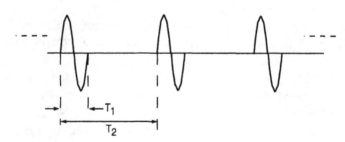

FIGURE 5.20. Model of a train of action potentials in which each action potential is represented as one cycle of a sine wave.

for $-T/4 \le t \le T/4$, and $T = 2\pi/\Omega_0$. Sketch this signal and calculate its Fourier transform.

5.23   Variations in lung volume during breathing cause variations in heart rate, an effect known as *respiratory sinus arrhythmia*. The file rsa.mat contains two signals measured from a human subject: instantaneous lung volume and heart rate at three different frequencies of breathing. Use the grid search method to fit a sine wave to each signal and determine the frequency response of heart rate to lung volume at each frequency. Does the frequency response change with frequency?

5.24   The signal $x(t)$ (Fig. 5.18) is the input to an LTIC system, yielding the output

$$y(t) = \sum_{n=1}^{\infty} [8 \sin(0.1n\pi)] \cos(2n\pi t + \pi/2).$$

Show that this system is an ideal differentiator having the frequency response $H(\Omega) = \Omega e^{-j\pi/2}$.

6

RESPONSES OF LINEAR CONTINUOUS-TIME FILTERS TO ARBITRARY INPUTS

6.1 INTRODUCTION

Up to this point we have focused on the frequency response of a linear system as the primary direct measure of its filtering properties. Thus, in Chapter 4 the frequency response was developed as a means of determining the system response to steady-state sinusoids and in Chapter 5 it was recognized that the frequency response may be utilized with any input having a Fourier transform, including transient signals. Practically speaking, though, calculation of the output of a linear system by way of the inverse Fourier transform was seen to be problematic for all but the simplest cases. One of the difficulties is that many real signals do not extend in time to negative infinity but have a defined onset time, and the Fourier transforms of such signals often contain delta functions as well as rational polynomials in Ω. These problems are avoidable if one defines a new transform that expresses a signal as a weighted sum of waveforms that begin at $t = 0$. This chapter introduces one such transform, the (unilateral) Laplace transform. The Laplace transform models a signal as a sum of exponentially decaying sinusoids having various frequencies and rates of decay. It is especially useful for calculating the time response of a linear system to an arbitrary input. Thus the Laplace transform is an essential complement to frequency response analysis for signal processing in situations involving transient signals. Furthermore there is an intimate relationship between the frequency response of an LTIC system and the Laplace transform of its impulse response (known as the "transfer function). In most practical cases it is (usually) straightforward to determine the frequency response of a filter if the Laplace transform of its impulse response is known. This last relationship will provide a basis (in Chapter 8) for designing discrete-time filters whose frequency response properties are derived from those of continuous-time filters.

6.2   INTRODUCTORY EXAMPLE

In the delivery of therapeutic drugs, the speed with which the concentration of the drug reaches the therapeutic threshold in the target tissue is usually an important issue. Concentrations at sites remote from the delivery site can be predicted using models of mass transport and storage. Often the parameters of such models are calculated from measurements of the change in drug concentration in response to an experimental dose. Typically these measurements are noisy and one may wish to filter the data before estimating the parameters. One must consider, however, the extent to which any temporal distortion of the data signal by the filter will alter the time courses of such signals and introduce errors in estimates of the parameter values. In such a situation it may not be sufficient to choose a filter based on steady-state filtering properties because the experimental measurements are transient signals. Consider the simple example of Fig. 3.26 which comprises a site of entry of a drug, two storage compartments, and a route of elimination. The differential equation relating the concentration in the second compartment, $y(t)$, to the inflow rate, $x(t)$, was derived in Exercise 3.22 as

$$\ddot{y}(t) + a\dot{y}(t) + by(t) = k_1x(t), \qquad a = 2k_1 + k_2, \qquad b = k_1k_2. \tag{6.1}$$

In response to a rapid bolus input of $x(t)$ (which is equivalent to having a nonzero initial condition on $q(t)$), $y(t)$ initially will rise with a time course that reflects the flow "conductance" from the first to the second compartment, then fall at an eventual rate determined by the "conductance" of the elimination pathway. If we perform an experiment in which $x(t) = 0$ and $q(0) = q_0$ and measure $y(t)$, we can theoretically estimate the two parameters k_1 and k_2 from the early and late rates of change of $y(t)$. With this knowledge one could utilize the differential equation of the system (Eq. (6.1)) to predict the response in the second compartment to any time course of infusion of the drug, $x(t)$.

Estimation of the initial and final slopes of $y(t)$ is subject to considerable error in the presence of noise. A more reliable approach to estimating values for k_1 and k_2 is possible. Assume the signal $y(t)$ is available after performing the experiment just described. In concept, one could calculate the first and second derivatives of $y(t)$ and use some parameter search algorithm, such as that utilized in Chapter 4, to fit Eq. (6.1) to these three signals and estimate values for a and b, from which k_1 and k_2 may be calculated. If, however, $y(t)$ is noisy, we may need to filter it before calculating the derivatives. For example, consider the response of the SIMULINK model drugdisp.mdl, which implements the model of Fig. 3.26, when $x(t) = 0$ and $q_0 = 5$ mg (Fig. 6.1(a)). In this simulation $k_1 = 0.1$ mg/s and $k_2 = 0.025$ mg/s, and noise has been added to $y(t)$ to simulate noisy measurements. One may test the suggested method for estimating the values of k_1 and k_2 by numerically calculating $\dot{y}(t)$ and $\ddot{y}(t)$, performing the parameter search (as if the values were unknown), and comparing the estimates of k_1 and k_2 with the known values. It is desirable though to filter out the noise from $y(t)$ before calculating the derivatives. Since the peak of $y(t)$ oc-

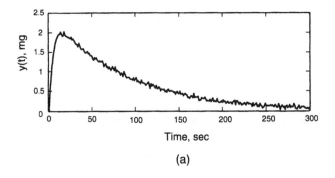

(a)

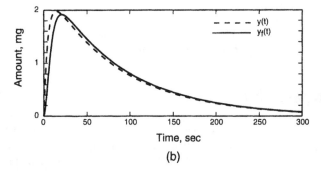

(b)

FIGURE 6.1. (a) Response of $y(t)$ of the model `drugdisp.mdl` when $x(t) = 0$ and $q(0) = 5$ mg. (b) Calculated response of the above model before (dashed) and after (solid) passing through a unity gain lowpass filter having a cutoff frequency of 0.04 Hz.

curs after approximately 25 seconds, we might be tempted to choose a unity-gain, first-order lowpass filter with a cutoff frequency of $1/25 = 0.04$ Hz. Will this filter cause an error in the estimated parameters? To answer this question we evaluate the exact, noise-free solution to Eq. (6.1), $y^*(t)$, and determine whether the filter would distort it significantly.

Referring to Chapter 4, the zero-input response of a second-order system is

$$y_{zi} = y^*(t) = c_1 e^{s_1 t} + c_2 e^{s_2 t}, \qquad t \geq 0,$$

where for the present example,

$$s_{1,2} = -\frac{a}{2} \pm \frac{\sqrt{a^2 - 4b}}{2} = -0.2133, -0.0117.$$

Now $y^*(0) = 0$, but from the original mass balance equations of Exercise 3.22 we see that $\dot{y}^*(0) = k_1 q(0) = 0.5$. From these two conditions one finds that $c_1 = 2.480$ and $c_2 = -2.480$. The impulse response of the proposed noise filter is $h(t) =$

$0.08\pi e^{-0.08\pi t}u(t)$. Using convolution, after some algebra the output of the filter when $y^*(t)$ is the input is

$$y_f(t) = y^*(t) * h(t) = 2.60e^{-0.0117t} - 16.60e^{-0.2133t} + 14.0e^{-0.2513t}, \qquad t \geq 0. \tag{6.2}$$

Figure 6.1(b) plots both $y^*(t)$ and $y_f(t)$. It is clear that the noise filtering does indeed distort the initial rise of the response and would cause errors in estimating the parameters (especially k_1). To avoid this problem the cutoff frequency must be higher than 0.04 Hz. Of course, a higher cutoff frequency will permit more noise to pass, so the smallest acceptable cutoff frequency is best. The reader may wish to utilize the simulation to evaluate alternative filters.

In this example it was necessary to determine the actual temporal output of the filter, a process that required calculating a convolution. Later we will revisit this problem using Laplace transform methods and find it much easier to solve.

6.3 CONCEPTUAL BASIS OF THE LAPLACE TRANSFORM

The conceptual basis of the Laplace transform will be developed in two steps. Initially the development will parallel that presented for the Fourier transform and will culminate in defining the two-sided Laplace transform. This development will engender complications related to existence and uniqueness, so the second step will define the one-sided Laplace transform that is less burdensome so long as its use is restricted to signals that are zero for all $t < 0$.

Consider the usual LTIC system with input $x(t)$, impulse response $h(t)$, and output $y(t)$. Let the input be a sinusoid having an exponentially modulated amplitude of the form $x(t) = e^{\sigma t}e^{j\Omega t}$. The resulting output may be calculated using convolution as

$$y(t) = \int_{-\infty}^{\infty} h(\tau)e^{(\sigma+j\Omega)(t-\tau)}d\tau = e^{st}\int_{-\infty}^{\infty} h(\tau)e^{-s\tau}d\tau, \; s \triangleq \sigma + j\Omega. \tag{6.3}$$

It is tempting to define the integral on the r.h.s. of Eq. (6.3) as some function, $H(s)$, so that one may write $y(t) = e^{st}H(s) = H(s)x(t)$ as was done earlier when discussing frequency response. Caution is required, however, because this integral may exist for some values of σ but not for others. Loosely speaking, the function $h(\tau)$ must grow more slowly than the function $e^{\sigma\tau}$ so that the integrand, $h(\tau)e^{-s\tau}$, does not grow without bound. Consequently, one indeed may define a transform $H(s)$, where

$$H(s) = \int_{-\infty}^{\infty} h(t)e^{-st}dt, \tag{6.4}$$

but it is necessary to specify also the range of σ for which $H(s)$ exists.

Equation (6.4) is the *two-sided Laplace transform* of $h(t)$. The interpretation of $H(s)$ is more apparent if we consider the function $h'(t) = h(t)e^{-\sigma t}$, in which case Eq.

(6.4) may be considered the Fourier transform of $h'(t)$. This Fourier transform may not exist for all values of σ but for any value for which it does exist one may use the inverse Fourier transform to show that

$$h(t) = e^{\sigma t}h'(t) = \frac{1}{2\pi}e^{\sigma t}\int_{-\infty}^{\infty}H(\sigma+j\Omega)e^{j\Omega t}d\Omega = \frac{1}{2\pi}\int_{-\infty}^{\infty}H(\sigma+j\Omega)e^{(\sigma+j\Omega)t}d\Omega.$$

But since $s = \sigma + j\Omega$, $d\Omega = ds/j$ and

$$h(t) = \frac{1}{2\pi j}\int_{\sigma-j\infty}^{\sigma+j\infty}H(s)e^{st}ds. \tag{6.5}$$

Note that Eq. (6.5) is true for any value of σ for which Eq. (6.4) converges and therefore as long as σ is in the region of convergence, the determination of $h(t)$ from $H(s)$ is independent of the actual value of σ. One may interpret Eq. (6.5) as follows: $h(t)$ may be represented as an infinite summation of exponentially modulated sinusoids, whose frequencies range over all possible values of Ω, where the weighting factors in the summation are proportional to $H(s)$. This interpretation leads us to define the two-sided Laplace transform for an arbitrary signal $x(t)$ as

$$X(s) = \int_{-\infty}^{\infty}x(t)e^{-st}dt. \tag{6.6}$$

$X(s)$ is defined for those values of σ for which it converges. In the region of convergence (ROC) the inverse transform relationship is given by

$$x(t) = \frac{1}{2\pi j}\int_{\sigma-j\infty}^{\sigma+j\infty}X(s)e^{st}ds. \tag{6.7}$$

Example 6.1   Two-sided Laplace transforms   The signal $x(t) = e^{-at}u(t)$ has the two-sided Laplace transform

$$X(s) = \int_{-\infty}^{\infty}e^{-at}u(t)e^{-st}dt = \int_{0}^{\infty}e^{-(\sigma+a)t}e^{-j\Omega t}dt.$$

This integral is identical to the Fourier transform of the function $e^{-(\sigma+a)t}u(t)$. From Chapter 5, the integral converges to

$$X(s) = \frac{1}{\sigma+a+j\Omega} = \frac{1}{s+a}$$

for $\sigma > -a$. The ROC is indicated in Fig. 6.2 on a plane defined by the real and imaginary parts of s. If one considers evaluating $X(s)$ for various values of s on this s-plane, it is clear that the Laplace transform is generally complex-valued. Further-

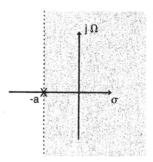

FIGURE 6.2. Region of convergence of the two-sided Laplace transform of $x(t) = e^{-at}u(t)$ is to the right of the dotted line.

more, when $s = -a$, $X(s)$ is unbounded. Consequently the point $s = -a$, which is known as a *pole* of $X(s)$, cannot be in the ROC of $X(s)$ (nor can any other pole location). The resemblance to the Fourier transform is more than accidental. Since the ROC includes $\sigma = 0$, then $X(s)|_{\sigma=0} = X(\Omega)$. That is, if the ROC strictly includes the entire imaginary axis, then the Fourier transform equals the Laplace transform evaluated along this axis.

Consider now the signal $y(t) = -e^{-at}u(-t)$ which equals zero for $t > 0$. Its Laplace transform is

$$Y(s) = -\int_{-\infty}^{0} e^{-at}e^{-st}dt = \frac{1}{s + a}$$

with the ROC specified by $\sigma < -a$. Thus, although $x(t)$ and $y(t)$ are quite different, $X(s)$ and $Y(s)$ differ only in their regions of convergence.

Because the relationships of Eqs. (6.6) and (6.7) are incomplete without specifying the region of convergence, the two-sided Laplace transform has limited usefulness for signal processing. One may show, however, that the Laplace transform (*if it exists*) of any function that is identically zero for $t < 0$ has an ROC that always lies to the right of its rightmost pole in the s-plane, as seen in the example of Fig. 6.2. Furthermore, a given $X(s)$ then corresponds to only one $x(t)$. Consequently, by adopting the convention that the Laplace transform will be applied only to such functions, it is unnecessary to specify the ROC. This previous statement may seem inconsistent with Eq. 6.7, where it is indicated that the inverse transform requires evaluating a contour integral inside the ROC. Indeed to evaluate Eq. (6.7) it is necessary to know the ROC but fortunately there is an alternative method (see later) for determining inverse Laplace transforms that does not require explicit knowledge of the ROC. *We shall adopt the convention that the Laplace transform will be applied only to functions, $x(t)$, for which $x(t) = 0$, $t < 0$,* and shall consider the formal definition of the Laplace transform to be a modified form of Eq. 6.6:

$$X(s) = \int_0^\infty x(t)e^{-st}dt \triangleq L\{x(t)\}. \tag{6.8}$$

Equation 6.8 defines the *one-sided* (or *unilateral*) *Laplace transform*. To ensure that this convention is met, sometimes we will explicitly multiply $x(t)$ by $u(t)$. When $x(t)$ is nonzero around $t = 0$, there is a question of exactly how to define the lower limit on the integral of Eq. (6.8). In part because of the importance of the delta-function in systems analysis, it is common to use a lower limit of $t = 0-$ so that the integral of the delta-function is preserved. This convention will be utilized here. The *inverse Laplace transform*, $L^{-1}\{X(s)\}$, is still given by Eq. (6.7).

Existence of the One-Sided Laplace Transform

In order for $X(s)$ to exist it is required that the integral of Eq. (6.8) should converge. The following set of conditions on $x(t)$ is sufficient to ensure convergence (McGillam and Cooper, 1991; p. 221): If $x(t)$ is integrable in every finite interval $a < t < b$, where $0 \le a < b < \infty$, and for some value c, $\lim_{t \to \infty} e^{-ct}|x(t)| < \infty$, then the integral of Eq. (6.8) converges absolutely and uniformly for $\text{Re}\{s\} > c$. Although many practical signals meet these conditions and therefore have a one-sided Laplace transform, signals that grow "rapidly" such as e^{t^2} do not satisfy the above limit and do not have a Laplace transform.

Relation of Fourier and Laplace Transforms

There are two issues that confound the relationship between Fourier and Laplace transforms. The first is simply a notational issue; the second is the theoretical consideration of convergence properties (and ROC of the Laplace transform). Regarding the first, one should note that despite their similarity in name $X(s)$ and $X(\Omega)$ are not referring to the same function using different variables as their arguments. This point is obvious from the following example: Previously we determined that $L\{e^{-at}u(t)\} = X(s) = 1/(s + a)$, whereas $\Im\{e^{-at}u(t)\} = X(\Omega) = 1/(j\Omega + a)$. Clearly $X(\Omega) \ne X(s)|_{s=\Omega}$. In this case, in fact,

$$X(\Omega) = X(s)|_{s=j\Omega} \text{ and } X(s) = X(\Omega)|_{\Omega=-js} \tag{6.9}$$

Unfortunately, the relationships of Eq. 6.9 are not universal but depend on the locations of poles in $X(s)$. Whenever a unilateral Laplace transform has poles *only* in the left half of the s-plane, then the imaginary axis will be in its ROC and the function will have a Fourier transform that has the relationships to its Laplace transform, as indicated in Eq. (6.9).

Conversely, when $X(s)$ is a valid unilateral Laplace transform that has poles in the right half plane, then $x(t)$ does not have a Fourier transform because the ROC of the (unilateral) Laplace transform is always to the right of the rightmost pole of $X(s)$.

If $X(s)$ has complex conjugate poles on the imaginary axis, a further complication arises because the signal $x(t)$ is then a pure sinusoid that starts at $t = 0$ and does not have a normal Fourier transform. Using a general Fourier transform, in Chapter 5 it was shown that

$$\Im\{\sin(\Omega_0 t)u(t)\} = \frac{\Omega_0}{\Omega_0^2 - \Omega^2} + j\frac{\pi}{2}[\delta(\Omega + \Omega_0) - \delta(\Omega - \Omega_0)]. \qquad (6.10)$$

The Laplace transform of this function is

$$L\{\sin(\Omega_0 t)u(t)\} = \int_0^\infty \sin(\Omega_0 t)e^{-st}dt = \int_0^\infty \left(\frac{e^{j\Omega_0 t} - e^{-j\Omega_0 t}}{2j}\right)e^{-st}dt$$

$$= \left[\frac{k_1}{s + j\Omega_0} - \frac{k_1}{s - j\Omega_0}\right], \qquad k_1 = \frac{j}{2}. \qquad (6.11)$$

That is, $X(s)$ has complex conjugate poles on the imaginary axis. Recognizing that the first term on the r.h.s. of Eq. (6.10) may be written as

$$\frac{\Omega_0}{\Omega_0^2 - \Omega^2} = \frac{\frac{j}{2}}{j\Omega + j\Omega_0} - \frac{\frac{j}{2}}{j\Omega - j\Omega_0},$$

it is clear that substituting $s = j\Omega$ into the Laplace transform gives the first term of the Fourier transform. However, the pair of imaginary-axis poles of the Laplace transform (Eq. (6.11)) also gives rise to a term $\pi k_1[\delta(\Omega + \Omega_0) - \delta(\Omega - \Omega_0)]$ in the Fourier transform (Eq. (6.10)). k_1 is known as the *residue* of the poles at $s = \pm j\Omega_0$. It may be evaluated using a partial fraction expansion (see later).

Example 6.2   Unilateral Laplace transforms
   a. Let $x(t) = u(t)$. Then

$$X(s) = \int_0^\infty e^{-st}dt = -\frac{1}{s}e^{-st}\Big|_0^\infty = \frac{1}{s}, \qquad \sigma > 0.$$

   b. Let $y(t) = \delta(t)$. Then $Y(s) = \int_0^\infty \delta(t)e^{-st}dt = 1 \ \forall s$.
   c. Let $z(t) = e^{-t/\tau}u(t)$. Then $Z(s) = 1/(s + 1/\tau) = \tau/(\tau s + 1), \qquad s > -1/\tau$.

6.4   PROPERTIES OF (UNILATERAL) LAPLACE TRANSFORMS

There are many important properties of the Laplace transform that facilitate its application to linear systems. For the following discussion of properties, let

$$L\{x(t)\} = X(s), \qquad L\{v(t)\} = V(s).$$

Linearity: The Laplace transform operation is linear. That is,

$$L\{ax(t) + bv(t)\} = aX(s) + bV(s).$$

This result is proven easily from the definition, Eq. (6.8).

Right shift in time (time delay): $L\{x(t - d)\} = e^{-sd}X(s)$. One may prove this result by substituting $t' = t - d$ in Eq. (6.8). Note that a comparable result for $x(t + d)$ does not exist, since we are assuming that $x(t) = 0$ for $t < 0$.

Example 6.3.   The unilateral Laplace transform of the function $x(t) = P_1(t)$ does not exist because this function is nonzero for some $t < 0$. However, the shifted function $x(t) = P_1(t - 0.5)$ does have a Laplace transform given by

$$L\{P_1(t - 0.5)\} = L\{u(t)\} - L\{u(t - 1)\} = \frac{1}{s} - e^{-s}\frac{1}{s} = \frac{1 - e^{-s}}{s}.$$

Time scaling: Scaling in the time domain scales the transform reciprocally. That is, $L\{x(at)\} = (1/a)X(s/a)$. Again, this property is easily demonstrated from Eq. (6.8).

Multiplication by t^n: $L\{t^n x(t)\} = (-1)^n d^n X(s)/ds^n$. This result is readily provable by induction.

Example 6.4.   The ramp function, $r(t) = t\, u(t)$, has the Laplace transform given by

$$R(s) = (-1)^1 \frac{d}{ds} U(s) = \frac{1}{s^2}.$$

Similarly, $L\{t^n u(t)\} = (n!/s^{n+1})$.

Exponential modulation: From the definition, Eq. (6.8), one may show that

$$L\{e^{at}x(t)\} = X(s - a).$$

Amplitude modulation of sinusoids: From Example 6.4 and the immediately preceding property, using Euler's formulas one may prove that

$$L\{\cos(\Omega_0 t)u(t)\} = L\left\{\frac{1}{2}[e^{j\Omega_0 t} + e^{-j\Omega_0 t}]u(t)\right\} = \frac{1}{2}[U(s + j\Omega_0) + U(s - j\Omega_0)]$$

$$= \frac{1}{2}\left[\frac{1}{s + j\Omega_0} + \frac{1}{s - j\Omega_0}\right] = \frac{s}{s^2 + \Omega_0^2}.$$

Likewise, if one replaces $u(t)$ by an arbitrary $x(t)$, a similar derivation yields the result that

$$L\{x(t)\cos(\Omega_0 t)\} = \tfrac{1}{2}[X(s + j\Omega_0) + X(s - j\Omega_0)].$$

The comparable result for an amplitude-modulated sine wave is given in Table 6.1.

Example 6.5   The previous two results may be combined to determine the Laplace transform of an exponentially damped cosine function. Thus, if $x(t) = \cos(\Omega_0 t)u(t)$,

$$L\{e^{-bt}\cos(\Omega_0 t)u(t)\} = X(s + b) = \frac{s + b}{(s + b)^2 + \Omega_0^2}.$$

One of the most useful properties of the Laplace transform relates to the convolution of two functions in the time domain. As we saw for the Fourier transform, convolution in time corresponds to multiplication of the Laplace transforms of the two functions. That is, if $x(t)$ and $v(t)$ both have Laplace transforms, then

$$L\{x(t) * v(t)\} = X(s)V(s). \tag{6.12}$$

TABLE 6.1   Properties of the Unilateral Laplace Transform

Property	Time Domain	Transform
Linearity	$ax(t) + by(t)$	$aX(s) + bY(s)$
Time scaling	$x(bt)$	$\dfrac{1}{b}X\left(\dfrac{s}{b}\right), \quad b > 0$
Convolution	$x(t) * y(t)$	$X(s)Y(s)$
Integration	$\displaystyle\int_0^t x(\tau)d\tau$	$\dfrac{1}{s}X(s)$
Differentiation	$\dot{x}(t)$	$sX(s) - x(0)$
	$\ddot{x}(t)$	$s^2 X(s) - sx(0) - \dot{x}(0)$
	$x^{(n)}(t)$	$s^n X(s) - \displaystyle\sum_{i=0}^{n-1} s^{n-1-i}x^{(i)}(0)$
Right shift	$x(t - d)u(t - d)$	$e^{-sd}X(s)$
Modulation by t^n	$t^n x(t)$	$(-1)^n \dfrac{d^n}{ds^n}X(s)$, n an integer
Modulation by e^{at}	$e^{at}x(t)$	$X(s - a)$
Modulation by sine	$x(t)\sin(\Omega t)$	$\tfrac{1}{2j}[X(s + j\Omega) - X(s - j\Omega)]$
Modulation by cosine	$x(t)\cos(\Omega t)$	$\tfrac{1}{2}[X(s + j\Omega) + X(s - j\Omega)]$

The proof of this result parallels the proof of the corresponding property of Fourier transforms in Chapter 5.

Example 6.6   *Calculating a convolution using Laplace transforms*   Let $x(t) = v(t) = P_1(t - 0.5)$. By Eq. (6.12) and Example 6.3,

$$L\{x(t) * v(t)\} = \left[\frac{1 - e^{-s}}{s}\right]^2 = \frac{1}{s^2} - 2\frac{e^{-s}}{s^2} + \frac{e^{-2s}}{s^2}. \tag{6.13}$$

Using the result (Table 6.2) that $L\{tu(t)\} = (1/s^2)$, and that multiplication by e^{-ds} corresponds to a delay of 'd', then the inverse Laplace transform of Eq. (6.13) is

$$x(t) * v(t) = tu(t) - 2(t - 1)u(t - 1) + (t - 2)u(t - 2).$$

One may compute this convolution graphically to verify this result.

TABLE 6.2   Some Basic Unilateral Laplace Transforms

Time Domain	Transform
$u(t)$	$1/s$
$\delta(t)$	1
$\delta^{(n)}(t)$	s^n
$P_{2a}(t - b)$	$[e^{-(b-a)s} - e^{-(b+a)s}]/s$
$t^n u(t)$	$\dfrac{n!}{s^{n+1}}$
$e^{-bt}u(t)$	$\dfrac{1}{s + b}$
$te^{-bt}u(t)$	$\dfrac{1}{(s + b)^2}$
$\cos(\Omega t)u(t)$	$\dfrac{s}{s^2 + \Omega^2}$
$\sin(\Omega t)u(t)$	$\dfrac{\Omega}{s^2 + \Omega^2}$
$e^{-bt}\cos(\Omega t)u(t)$	$\dfrac{s + b}{(s + b)^2 + \Omega^2}$
$e^{-bt}\sin(\Omega t)u(t)$	$\dfrac{\Omega}{(s + b)^2 + \Omega^2}$
$\cos^2(\Omega t)$	$\dfrac{s^2 + 2\Omega^2}{s(s^2 + 4\Omega^2)}$
$\sin^2(\Omega t)$	$\dfrac{2\Omega^2}{s(s^2 + 4\Omega^2)}$

Integration and differentiation are common time domain operations whose effects in the transform domain can be determined. For example, if $y(t) = x'(t)$, where $x'(t) = dx(t)/dt$, then

$$Y(s) = L\{x'(t)\} = sX(s) - x(0^-).$$ (6.14)

The reader is referred to basic signal processing textbooks (e.g., McGillem and Cooper, 1991), for the proof of this result and a discussion of the origin of the term $x(0^-)$. Eq. (6.14) may be applied repeatedly for higher derivatives. Thus

$$L\left\{\frac{d^2x(t)}{dt^2}\right\} = s^2X(s) - sx(0^-) - x'(0^-).$$ (6.15)

Form of the Laplace transform: Although there is no inherent restriction on the form of the unilateral Laplace transform, in very many cases it may be expressed as the ratio of two polynomials in s of the form

$$X(s) = \frac{N(s)}{D(s)} = \frac{b_m s^m + b_{m-1} s^{m-1} + b_{m-2} s^{m-2} + \ldots + b_1 s + b_0}{a_n s^n + a_{n-1} s^{n-1} + a_{n-2} s^{n-2} + \ldots + a_1 s + a_0}.$$ (6.16)

If the order of the numerator polynomial, m, is strictly less than the order of the denominator polynomial, n, then Eq. (6.16) is known as a proper rational function.

Initial and final value theorems: Equation (6.14) defines the Laplace transform of $x'(t)$. But one may also say that $L\{x'(t)\} = \int_{0^-}^{\infty} x'(t)e^{-st}dt$. Rewriting this result and equating it to the r.h.s. of Eq. 6.14 we note that

$$\int_{0^-}^{\infty} x'(t)e^{-st}dt = x(0^+) - x(0^-) + \int_{0^+}^{\infty} x'(t)e^{-st}dt = sX(s) - x(0^-).$$

Canceling the common $x(0^-)$ terms on either side of the second equality, and taking the limit as s approaches infinity, yields

$$\lim_{s \to \infty}\left[x(0^+) + \int_{0^+}^{\infty} x'(t)e^{-st}dt\right] = \lim_{s \to \infty}[sX(s)].$$

So long as t is greater than zero, the indicated limit of the integral is zero. Therefore,

$$\lim_{s \to \infty} x(0^+) + 0 = x(0^+) = \lim_{s \to \infty}[sX(s)].$$ (6.17)

This result is known as the *initial value theorem*, which is valid only if $X(s)$ is a proper rational function. A similar approach may be taken to derive the *final value theorem*, which states that if $X(s)$ is a proper rational function, then

$$\lim_{t \to \infty} x(t) = \lim_{s \to 0}[sX(s)].$$ (6.18)

6.5   THE INVERSE (UNILATERAL) LAPLACE TRANSFORM

It was stated previously that a given unilateral Laplace transform, $X(s)$, corresponds to a unique $x(t)$. In fact, two functions, $x_1(t)$ and $x_2(t)$, that have different values only at discontinuities can have the same Laplace transform as long as $\int_0^t [x_1(\tau) - x_2(\tau)] d\tau = 0$, $\forall t > 0$. Given this restriction, it is usual to assume that the differences between the two functions have little "practical" consequence and that the functions are equal. Therefore, the previous statement regarding uniqueness of the inverse transform applies and any mathematically valid operation may be used to determine the inverse transform (including of course the definition, Eq. (6.7)). The easiest and most common general approach is the method of partial fractions. The basis of this method is to expand a transform, $X(s)$, into a sum of terms in which the inverse transform of each term is recognizable from a table of common transforms (e.g., Table 6.2). By the linearity of the Laplace transform operation, the inverse transform of $X(s)$ is then the sum of the inverse transforms of the individual terms.

To utilize the method of partial fractions it is necessary that the Laplace transform, $X(s)$, can be written as a ratio of two polynomials as in Eq. (6.16) and that $m < n$. If $X(s)$ is not a proper rational function, then it is necessary to divide $N(s)$ by $D(s)$ until the remainder function is proper. In such cases $X(s)$ will have the form

$$X(s) = k_0 + k_1 s + \ldots + k_r s^r + \frac{N'(s)}{D'(s)}.$$

To evaluate the inverse Laplace transform of $X(s)$ one handles the ratio of polynomial terms using partial fractions, as described below. The inverse transform of the initial terms may be evaluated using the general result that $L\{\delta^{(n)}(t)\} = s^n$.

Given that $X(s)$ is already expressed as a proper rational function, then the polynomial functions $N(s)$ and $D(s)$ of Eq. (6.16) may be factored so that $X(s)$ may be rewritten as

$$X(s) = \frac{b_m(s - z_1)(s - z_2) \cdots (s - z_m)}{a_n(s - p_1)(s - p_2) \cdots (s - p_n)} = \frac{b_m \Pi_{j=1}^m (s - z_j)}{a_n \Pi_{j=1}^n (s - p_j)}. \tag{6.19}$$

Each z_j is known as a *zero* of $X(s)$ since $X(s)|_{s=z_j} = 0$. Similarly, each p_j is known as a *pole* of $X(s)$ since $X(s)|_{s=p_j} = \infty$. The initial step of the partial fraction method is to express the product of terms in Eq. (6.19) as a sum of terms. Depending on the specific values of the poles, the method for calculating the inverse transforms of the terms in the sum will differ. The various possibilities will be considered separately.

Partial Fractions: Simple, Nonrepeated Poles

Note from previous examples (or from Table 6.2) that $L^{-1}\{c/(s - p)\} = ce^{pt}u(t)$. Therefore, if it is possible to expand $X(s)$ into a sum of terms of this form, one can express the inverse transform as a sum of (causal) exponential functions. Assume such an expansion is possible. Then there must be one term for each pole. Thus

$$X(s) = \frac{b_m \Pi_{j=1}^m (s - z_j)}{a_n \Pi_{j=1}^n (s - p_j)} = \frac{c_1}{s - p_1} + \frac{c_2}{s - p_2} + \ldots + \frac{c_n}{s - p_n}. \tag{6.20}$$

To evaluate any constant, c_i, first multiply both sides of Eq. (6.20) by $(s - p_i)$ obtaining

$$(s - p_i)X(s) = \frac{c_1(s - p_i)}{s - p_1} + \frac{c_2(s - p_i)}{s - p_2} + \ldots + c_i + \ldots + \frac{c_n(s - p_i)}{s - p_n}.$$

Evaluating both sides of this equation at $s = p_i$ causes all of the r.h.s. terms to evaluate to zero except the term involving c_i. Therefore,

$$c_i = [(s - p_i)X(s)]|_{s=p_i}. \tag{6.21}$$

Once all of the c_i coefficients are determined, then the inverse transform of Eq. (6.20) is

$$x(t) = L^{-1}\{X(s)\} = c_1 e^{p_1 t} u(t) + c_2 e^{p_2 t} u(t) + \ldots + c_n e^{p_n t} u(t). \tag{6.22}$$

Example 6.7 Simple nonrepeated poles To find the inverse transform of

$$X(s) = \frac{s + 2}{s^2 + 5s + 4},$$

first the denominator is factored into $(s + 1)(s + 4)$. Then by partial fraction expansion,

$$X(s) = \frac{s + 2}{(s + 1)(s + 4)} = \frac{c_1}{s + 1} + \frac{c_2}{s + 4}.$$

To determine c_1:

$$c_1 = [(s + 1)X(s)]|_{s=-1} = \frac{(s + 2)}{(s + 4)}\bigg|_{s=-1} = \frac{1}{3}.$$

Likewise,

$$c_2 = [(s + 4)X(s)]|_{s=-4} = \frac{(s + 2)}{(s + 1)}\bigg|_{s=-4} = \frac{2}{3}.$$

Finally, the inverse transform is

$$x(t) = \tfrac{1}{3} e^{-t} u(t) + \tfrac{2}{3} e^{-4t} u(t).$$

Partial Fraction Expansion: Repeated Simple Poles

A given factor of $D(s)$ may appear more than once—e.g., r times. In this case Eq. (6.19) may be written as

$$X(s) = \frac{b_m \Pi_{j=1}^m (s - z_j)}{a_n (s - p_1)^r \Pi_{j=2}^{n-r+1}(s - p_j)}$$

$$= \frac{c_1}{s - p_1} + \frac{c_2}{(s - p_1)^2} + \ldots + \frac{c_r}{(s - p_1)^r} + \frac{c_{r+1}}{(s - p_2)} + \ldots + \frac{c_n}{(s - p_{n-r+1})}. \quad (6.23)$$

To determine c_{r+1} to c_n one proceeds as described above for simple, nonrepeated poles. Similarly it is apparent that

$$c_r = [(s - p_1)^r X(s)]|_{s=p_1}. \quad (6.24)$$

Note also that differentiating the function $(s - p_1)^r X(s)$ with respect to s leaves every term except the one involving c_{r-1} multiplied by some power of $(s - p_1)$. Consequently, if this derivative is evaluated at $s = p_1$, then one may determine c_{r-1}. By taking higher derivatives one may generalize this result to

$$c_{r-i} = \frac{1}{i!}\left[\frac{d^i}{ds^i}(s - p_1)^r X(s)\right]\Bigg|_{s=p_1}. \quad (6.25)$$

Example 6.8 Repeated simple pole   Let

$$X(s) = \frac{s^2}{(s + 1)^2(s + 2)} = \frac{c_1}{s + 1} + \frac{c_2}{(s + 1)^2} + \frac{c_3}{s + 2}.$$

One may calculate as above that $c_3 = 4$. Using Eqs. (6.24) and (6.25),

$$c_2 = \frac{s^2}{s + 2}\bigg|_{s=-1} = 1, \qquad c_1 = \frac{d}{ds}\left[\frac{s^2}{s + 2}\right]\bigg|_{s=-1} = \frac{s^2 + 4s}{(s + 2)^2}\bigg|_{s=-1} = -3.$$

To determine the inverse transform, note from Table 6.2 that $L\{te^{-at}u(t)\} = 1/(s + a)^2$. Therefore, $x(t) = -3e^{-t}u(t) + te^{-t}u(t) + 4e^{-2t}u(t)$.

Partial Fractions: Distinct Poles with Complex Roots

Suppose that one of the factors of $D(s)$ has a complex pole of the form $p_1 = \alpha + j\beta$. Then if all of the coefficients of $D(s)$ are real, there must also be a factor having the pole $p_2 = \alpha - j\beta = p_1^*$. In this case, c_1 and c_2 may be evaluated in the usual manner for simple nonrepeated poles, and it is easy to prove that $c_2 = c_1^*$. Thus $X(s)$ has the form

$$X(s) = \frac{c_1}{(s - p_1)} + \frac{c_1^*}{(s - p_1^*)} + \sum_{j=3}^{n} \frac{c_j}{(s - p_j)}. \tag{6.26}$$

Now let $X_1(s) = (c_1/(s - p_1)) + (c_1^*/(s - p_1^*))$. Then we can say that

$$x_1(t) = [c_1 e^{p_1 t} + c_1^* e^{p_1^* t}]u(t) = [c_1 e^{\alpha t} e^{j\beta t} + c_1^* e^{\alpha t} e^{-j\beta t}]u(t).$$

Recall from the discussion of the various forms of the Fourier series that $c_n e^{jn\Omega_0 t} + c_n^* e^{-jn\Omega_0 t} = 2|c_n|\cos(n\Omega_0 t + \measuredangle c_n)$. By direct comparison we conclude that

$$x_1(t) = 2|c_1| e^{\alpha t}\cos(\beta t + \measuredangle c_1). \tag{6.27}$$

Although there are other forms in which to express a transform having complex conjugate poles, their inverse transforms are all equivalent to Eq. (6.27) and we shall use this form.

Example 6.9 Complex poles   To determine the inverse transform of

$$X(s) = \frac{s^2 + 3s + 1}{s^3 + 3s^2 + 4s + 2},$$

first it is necessary to factor the denominator, $D(s)$. The result is

$$D(s) = (s^2 + 2s + 2)(s + 1) = (s - (-1 + j))(s - (-1 - j))(s + 1).$$

Therefore,

$$X(s) = \frac{c_1}{s - (-1 + j)} + \frac{c_1^*}{s - (-1 - j)} + \frac{c_3}{s + 1}.$$

Then

$$c_3 = \frac{s^2 + 3s + 1}{(s + 1 - j)(s + 1 + j)}\bigg|_{s=-1} = \frac{-1}{-j^2} = -1;$$

$$c_1 = \frac{s^2 + 3s + 1}{(s + 1 + j)(s + 1)}\bigg|_{s=-1+j} = 1 - \frac{j}{2}.$$

Consequently,

$$|c_1| = \frac{\sqrt{5}}{2}, \qquad \measuredangle c_1 = \tan^{-1}\left(-\frac{1}{2}\right) \approx -0.464 \text{ rad.}$$

Thus, $x(t) = \sqrt{5} e^{-t}\cos(t - 0.464) - e^{-t}, t \geq 0$.

An alternative method for such problems is to consider the factor $s^2 + 2s + 2$ in $D(s)$ as a separate term. Then one may write the partial fraction expansion as

$$X(s) = \frac{c_1 s + c_2}{s^2 + 2s + 2} + \frac{c_3}{s + 1}. \tag{6.28}$$

c_3 has the same value as before. Once c_3 is known, one may combine the terms on the r.h.s. of Eq. (6.28) and set the numerator of the resulting expression equal to the numerator polynomial of $X(s)$, then solve for c_1 and c_2 by equating coefficients of like powers of s.

When $X(s)$ contains repeated complex poles, the partial fraction expansion is written with multiple occurrences of the repeated pole just as in the case of a repeated simple pole. It may be easier, however, to utilize the form for a complex pole given in Eq. (6.28) and to solve for the coefficients using the alternative method discussed in the previous example.

6.6 TRANSFER FUNCTIONS

The convolution theorem for Laplace transforms provides one of the most useful results for signal processing applications. Consider an LTIC system having input $x(t)$, impulse response $h(t)$, and output $y(t)$. Let $h(t)$ have a Laplace transform, $H(s)$, where $H(s)$ is known as the *transfer function* of the system. If the system is initially at rest, then $y(t) = h(t) * x(t)$. Therefore, the convolution theorem implies that the Laplace transform of the output signal is the product of the Laplace transforms of the impulse response and the input signal. That is, $Y(s) = H(s)X(s)$. If we interpret $X(s)$ and $Y(s)$ as weighting functions for the reconstruction of $x(t)$ and $y(t)$ from an ensemble of exponentially damped sinusoids, then $H(s)$ is the (complex-valued) gain for each specific combination of frequency, Ω, and exponential-decay coefficient, σ. This interpretation is a generalization of the interpretation of the frequency response, $H(\Omega)$. The latter may be determinable from $H(s)$, as discussed earlier.

This result suggests a powerful method for determining the exact temporal response of a filter to any input signal $x(t)$ having a well-defined Laplace transform. If $H(s)$ is known, then $Y(s)$ may be determined as the product of $H(s)$ and $X(s)$. Finally, $y(t)$ may be found from $L^{-1}\{Y(s)\}$ using the partial fraction method. A major practical limitation of this approach is that the direct calculation of the Laplace transform for an arbitrary input signal via Eq. (6.8) may have no closed-form solution. Consequently it often is difficult to apply the Laplace transform to a real data signal. On the other hand it is relatively easy to determine the exact response of a filter having a known transfer function to a variety of test inputs or to analytical signals that approximate actual data signals. An example is presented following the next section.

The Transfer Function and the System Equation

The properties of the Laplace transform related to differentiation and integration provide a convenient method of determining the transfer function (and usually also the frequency response) of an LTIC system from its differential equation. Recall from Chapter 2 that any LTIC system may be represented by a differential equation of the form

$$a_n \frac{d^n y}{dt^n} + \sum_{i=0}^{n-1} a_i \frac{d^i y}{dt^i} = \sum_{i=0}^{m} b_i \frac{d^i x}{dt^i}.$$

Assume that $x(t)$ and all of its derivatives are zero at $t = 0-$. Taking the Laplace transform of this equation,

$$a_n s^n Y(s) - f_n(y(0), y^{(1)}(0), \ldots, y^{(n-1)}(0), s)$$

$$+ \sum_{i=0}^{n-1} [a_i s^i Y(s) - f_i(y(0), y^{(1)}(0), \ldots, y^{(i-1)}(0), s)] = \sum_{i=0}^{m} b_i s^i X(s),$$

where each $f_i(.)$ is a function of the initial conditions on $y(t)$ and its derivatives. Rearranging, and combining all of the $f_i(.)$ functions into one such function, $f(.)$, one obtains

$$Y(s) \left[a_n s^n + \sum_{i=0}^{n-1} a_i s^i \right] = f(y(0), y^{(1)}(0), \ldots, y^{(n-1)}(0), s) + \sum_{i=0}^{m} b_i s^i X(s).$$

Finally, solving for $Y(s)$,

$$Y(s) = \frac{f(y(0), y^{(1)}(0), \ldots, y^{(n-1)}(0), s)}{\sum_{i=0}^{n} a_i s^i} + \frac{\sum_{i=0}^{m} b_i s^i}{\sum_{i=0}^{n} a_i s^i} X(s). \qquad (6.29)$$

According to Eq. (6.29) the output response, $y(t)$ comprises two components. One component is due to the initial conditions on $y(t)$ and its derivatives and is called the *zero-input response*. The second component is the output due to the input signal when all initial conditions are zero and is called the *zero-state response*. But from the preceding section the zero-state response may be determined from the transfer function as $Y(s) = H(s)X(s)$. Therefore, setting $f(.) = 0$ in Eq. 6.29, we find that

$$H(s) = \frac{Y(s)}{X(s)} = \frac{\sum_{i=0}^{m} b_i s^i}{\sum_{i=0}^{n} a_i s^i} \qquad (6.30)$$

Eq. 6.30 establishes the fundamental relationship between the transfer function of a single-input, single-output, LTIC system and the coefficients of its input-output differential equation. Given either the transfer function or the differential equation, it is trivial to specify the other.

Example 6.10   Output response via the Laplace transform   In prior discussions of the frequency responses of sensors, emphasis was placed on assessing the steady-state filtering properties of the sensor. For example, if a displacement transducer has a cutoff frequency of 5 Hz, then it might not be adequate for measuring finger tremor because tremor frequencies may exceed 5 Hz. One might ask whether the same transducer could be employed successfully for measuring displacement of a finger in response to an abrupt, step-like activation of a finger flexion muscle. In this case it is the transient response of the sensor that is important and, although the transient and steady-state response characteristics are highly correlated, often the transient response to a specified input may not be directly apparent from knowledge of $H(\Omega)$. Thus one must calculate the output response.

Assume the displacement sensor can be represented as a first-order system with a cutoff frequency of 5 Hz. Thus its system equation has the form

$$\frac{dy(t)}{dt} + \frac{1}{\tau}y(t) = bx(t).$$

Assuming that all initial conditions are zero and taking Laplace transforms,

$$sY(s) + \frac{1}{\tau}Y(s) = bX(s),$$

from which one finds

$$H(s) = \frac{Y(s)}{X(s)} = \frac{b}{s + \dfrac{1}{\tau}}.$$

(This result could have been obtained also from Eq. (6.30)). Now to approximate the experimental situation, let $x(t) = u(t)$. Then

$$Y(s) = H(s)X(s) = \frac{b}{s + \dfrac{1}{\tau}}\frac{1}{s} = \frac{c_1}{s} + \frac{c_2}{s + \dfrac{1}{\tau}}.$$

It is easy to show that $c_1 = b\tau$, $c_2 = -b\tau$, and thus

$$y(t) = b\tau[1 - e^{-t/\tau}]u(t). \tag{6.31}$$

The cutoff frequency of 5 Hz corresponds to $\tau = 1/10\pi \approx 0.0318$ s. Let $b = 1/\tau$

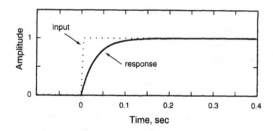

FIGURE 6.3. Theoretical step response of a displacement sensor modeled as a first-order system having a cutoff frequency of 5 Hz.

so that $|H(0)| = 1$. The input and output signals are plotted in Fig. 6.3. Note the slow rise of $y(t)$ due to the time constant of the transducer. One concludes that if resolving the trajectory over the initial 100 msec were critical for interpretation of the data, then this transducer would not be suitable because of the distortion of the output signal relative to the rapid change in the input.

Example 6.11   Step response of underdamped second-order system   The LTIC system described by the equation $6\ddot{y} + 0.6\dot{y} + 2.4y = x$ has the transfer function

$$H(s) = \frac{1}{6s^2 + 0.6s + 2.4}.$$

Its complex conjugate poles are $p_1 = -0.0500 + j0.631$, $p_1^* = -0.0500 - j0.631$. The response of this system to the unit step, $x(t) = u(t)$, may be determined using Laplace transforms. Thus,

$$Y(s) = H(s)X(s) = \frac{1}{6s^2 + 0.6s + 2.4}\frac{1}{s} = \frac{c_1}{s - p_1} + \frac{c_1^*}{s - p_1^*} + \frac{c_2}{s}, \qquad \text{where}$$

$$c_1 = \frac{1/6}{s(s + 0.0500 + j0.631)}\bigg|_{s=-0.0500+j0.631} = -0.2080 + j0.0165 = 0.2086e^{-j3.221},$$

and $c_2 = 0.417$. Therefore,

$$y(t) = 0.417e^{-0.05t}\cos(0.361t - 3.221)u(t) + 0.417u(t).$$

This step response is plotted in Fig. 6.4. Note the long-lasting damped oscillation characteristic of systems having complex conjugate poles near (but not on) the imaginary axis.

Example 6.12   System equation from transfer function   What is the differential equation of a system having the transfer function

$$H(s) = \frac{2}{s^2 + 6s + 8}?$$

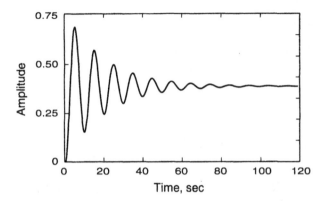

FIGURE 6.4. Step response of the underdamped second-order system of Example 6.11.

Set $H(s) = Y(s)/X(s)$ and cross-multiply, yielding

$$s^2 Y(s) + 6s Y(s) + 8Y(s) = 2X(s).$$

Since $H(s)$ is defined for a system initially at rest, assume that all initial conditions are zero. Taking inverse transforms we obtain

$$\frac{d^2 y(t)}{dt^2} + 6\frac{dy(t)}{dt} + 8y(t) = 2x(t).$$

Example 6.13   Transfer function of a filter   Figure 6.5 shows the circuit diagram of an electronic filter. We may determine its transfer function, $H(s) = E(s)/V(s)$, and frequency response by first writing the differential equations of the circuit. Using the current loops indicated in the figure and assuming all initial conditions are zero, these equations are

$$v(t) = \int_0^t \frac{1}{C_1} i_1(t) dt + \int_0^t \frac{1}{C_2} [i_1(t) - i_2(t)] dt; \qquad (6.32a)$$

$$0 = \int_0^t \frac{1}{C_2} [i_2(t) - i_1(t)] dt + Ri_2(t); \qquad (6.32b)$$

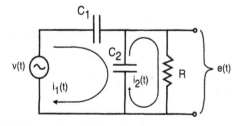

FIGURE 6.5. Circuit diagram of the electronic filter of Example 6.13.

$$e(t) = Ri_2(t). \tag{6.32c}$$

The Laplace transforms of Eqs. (6.32b) and (6.32a) yield

$$0 = I_2(s)\left[\frac{1}{sC_2} + R\right] - I_1(s)\frac{1}{sC_2}; \tag{6.32d}$$

$$V(s) = I_1(s)\left[\frac{1}{sC_1} + \frac{1}{sC_2}\right] - I_2(s)\frac{1}{sC_2}; \tag{6.32e}$$

and from Eq. (6.32c), $E(s) = R\,I_2(s)$. Combining Eqs. (6.32d) and (6.32e), after some algebra we obtain

$$V(s) = I_2(s)\left[\frac{1 + sRC_1 + sRC_2}{sC_1}\right],$$

from which we may solve for $I_2(s)/V(s)$ and then for $H(s)$:

$$H(s) = \frac{E(s)}{V(s)} = R\frac{I_2(s)}{V(s)} = \frac{sRC_1}{sR(C_1 + C_2) + 1}. \tag{6.33}$$

It is apparent that this filter is highpass with a zero at $\Omega = 0$ and a pole at $\Omega = -(1/R(C_1 + C_2))$. Since $H(s)$ has no complex conjugate poles on the imaginary axis and no poles in the right-half s-plane, the frequency response may be evaluated as

$$H(\Omega) = H(s)|_{s=j\Omega} = \frac{j\Omega RC_1}{j\Omega R(C_1 + C_2) + 1}.$$

Pole Locations and Stability

An important issue for signal processing is whether certain fundamental properties of the input signal will be retained in the output from a filter. For example, if the input signal is bounded, will the output also be bounded? A filter meeting the condition that the output signal will be bounded whenever the input signal is bounded is said to be *bounded-input, bounded output (BIBO) stable*. For an LTIC filter with a bounded input, $x(t)$, the output is

$$y(t) = h(t) * x(t) = \int_0^t h(\tau)x(t-\tau)d\tau \le \max(x(t))\int_0^t h(\tau)d\tau \le \max(x(t))\int_0^\infty |h(\tau)|d\tau. \tag{6.34}$$

In other words, a *sufficient* condition for BIBO stability is that $h(t)$ is absolutely integrable over the positive time axis. Then $y(t)$ will be bounded if $x(t)$ is bounded.

This criterion may be expressed in terms of the pole locations of $H(s)$ since, on the basis of a partial fraction expansion of $H(s)$, one may express $h(t)$ as the summa-

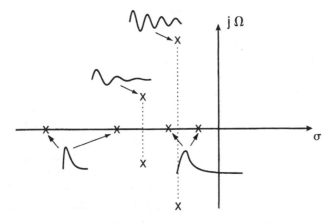

FIGURE 6.6. The various forms of the impulse responses of second-order lowpass systems have pairs of poles that are either (1) on the negative real axis, with one close to the origin and one not close to the origin, or (2) are complex conjugate poles that are both closer to or further away from the negative real axis.

tion of the time functions associated with the poles. Consider poles in the left half of the s-plane. For a pole on the negative real axis, the corresponding time function has the form $e^{-at}u(t)$, where $a > 0$. This function will be absolutely integrable as required for BIBO stability. For second-order systems the situation is qualitatively similar, as depicted in Fig. 6.6. For complex conjugate poles in the left half-plane, the corresponding time function has the form $e^{-at}\cos(bt + \theta)u(t)$ where $a, b > 0$. This function also is absolutely integrable. On the other hand, poles on the imaginary axis or in the right half-plane will correspond to time functions that are not absolutely integrable. Therefore, a sufficient condition for BIBO stability is that all poles of $H(s)$ lie in the left half of the s-plane.

6.7 FEEDBACK SYSTEMS

Feedback refers to the sensing of the output of a system and transmission of this signal back to the system input in order to influence the future output of the system (Fig. 6.7). Feedback is often utilized when it is desired to automatically adjust the input to a system if the output deviates from a desired output. Although conceptually with knowledge of the transfer function of the system one may accomplish this goal by modifying the input signal itself, the use of feedback provide more precise adjustment of the input to a system in direct relationship to unanticipated changes in the output. By adding feedback one modifies the overall transfer function between the input signal and the system output and, by judicious choice of the properties of the feedback path, one may (within limits) "redesign" the transfer function to suit one's objectives. Indeed there are many published algorithms for designing feedback paths to satisfy various criteria of optimality such as minimizing steady-state errors, maximizing speed of response, or minimizing the energy expenditure of the

system to achieve a specified output. Feedback paths also may be designed to provide compensation for changes in the properties of the original system so that the
overall control of the output by the input signal is not seriously degraded by such
changes (if they are small enough). Probably for this latter reason feedback paths
have evolved in natural biological systems for many of their control processes. Indeed it seems to be the exception when a physiological process is controlled without
feedback (called "open-loop" control), although there may be cases in which open-
loop control is utilized initially to respond to a disturbance, followed by control by
way of feedback (called "closed-loop" control) to achieve an appropriate steady-
state response. Feedback is important in signal processing because the presence of
feedback, or alteration of the properties of a feedback path, modifies the overall frequency response of a system. We will focus on this consequence of feedback, but it
should be recognized that entire textbooks are devoted to the analysis and design of
feedback systems.

As indicated in Fig. 6.7, the basic linear feedback system comprises a *feed-for-
ward system*, $G(s)$, a *feedback system*, $F(s)$, and a node which linearly combines the
input signal, $x(t)$, and the feedback signal, $y_f(t)$. If the node subtracts the feedback signal from the input signal, the overall system is called a *negative-feedback system*.
Conversely, if the node adds the two signals, it is a *positive-feedback system*. (In general, linear positive-feedback systems are unstable and therefore have less relevance
to normal biological systems, although positive feedback may be relevant to understanding certain disease processes. Positive-feedback systems can be physically realizable if their outputs are "clipped"—for example, by saturation phenomena—so that
they never require infinite energy.) Other names for $G(s)$ are the *open-loop system* or
the *plant*. The complete system with intact feedback path is called the *closed-loop
system*.

To determine the transfer function of a closed-loop, negative-feedback system,
note first that

$$Y(s) = G(s)[X(s) - Y_f(s)] = G(s)[X(s) - F(s)Y(s)].$$

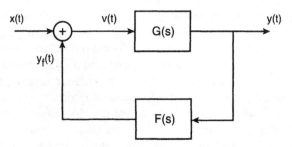

FIGURE 6.7. General schematic of a simple feedback system. Output of the linear combiner
(i.e., circle with plus sign) can be either $x(t) + y_f(t)$ (positive feedback) or $x(t) - y_f(t)$ (negative
feedback).

Solving for $Y(s)$ yields

$$Y(s) = \frac{G(s)}{1 + F(s)G(s)} X(s),$$

from which one concludes that

$$H(s) = \frac{Y(s)}{X(s)} = \frac{G(s)}{1 + F(s)G(s)}. \tag{6.35}$$

The potentially profound differences between $H(s)$ and $G(s)$ that may be accomplished through the introduction of the feedback system, $F(s)$, may not be evident from this seemingly simple equation. Consider a $G(s)$ that contains only one pole on the negative real axis. In the absence of feedback, that pole dominates all transient responses of $y(t)$. By introduction of an $F(s)$ also having a single pole, one bestows on $H(s)$ a second pole and a second component of the transient response that may be much faster or much slower than that of $G(s)$ by itself. One extreme occurs when $F(s) = s$. Then $y_f(t)$ is the derivative of $y(t)$ and when $y(t)$ is increasing rapidly, the feedback adds a large negative signal to the input and decreases the rate of rise of $y(t)$ (assuming that $G(0) > 0$). When $y(t)$ is constant, $y_f(t)$ is zero but as soon as $y(t)$ deviates from a constant value—for example, in response to a noisy disturbance to the system—then the feedback quickly responds to force $y(t)$ in a direction opposite to its initial change. Derivative feedback is a powerful means of keeping a system output near a constant steady-state level. One of its disadvantages, however, is that it may be too sensitive to small disturbances in $y(t)$ and actually amplify their effect. In some biological systems derivative feedback has been combined in parallel with feedback paths having other properties so that the overall feedback balances sensitivity to noise disturbances, steady-state output error, and speed of response.

Proportional feedback, in which $F(s) = k_f$, a constant, is common in natural physiological systems. The control of many basic physiological process, from heart rate to insulin secretion to stages of the cell cyle, involve some sort of proportional feedback. It should be noted, however, that in many real biological processes the proportional feedback also involves a nonlinear term such as the square of the feedback signal or the product of two signals.

One profound effect of the introduction of feedback into a system is that the right type and amount of feedback may cause a nonoscillatory system to become oscillatory. Assume that $G(s)$ is BIBO stable, so that all of its poles are in the left half-plane. Introducing negative feedback will cause the overall transfer function (Eq. (6.35)) to have the denominator $1 + F(s)G(s)$. If there is some value of s on the imaginary axis, say s_0, such that $s_0 = j\Omega_0$, and $1 + F(j\Omega_0)G(j\Omega_0) = 0$, then we can conclude that: (1) there is another value of s, specifically $s = s_0^*$, that also makes the denominator equal to zero; (2) s_0 and s_0^*, are poles of $H(s)$; (3) the denominator polynomial has factors $(s - s_0)(s - s_0^*) = (s^2 + |s_0|^2)$; (4) $h(t)$ has a term that is an undamped sinusoid. That is, if $F(s)G(s) = -1$ at some point on the imaginary axis, then the system produces an undamped, sustained oscillation in response to any input or nonzero initial condition. This statement is equivalent to the conditions: $|F(j\Omega_0)G(j\Omega_0)| = 1$, $\angle F(j\Omega_0)G(j\Omega_0) = -\pi$. Because the gain of the feedback system "around the loop" is $F(s)G(s)$, this criterion for producing an oscillation is often stated as "the loop gain must be equal to negative one".

Example 6.14   Determining H(s) for a feedback system   The system of Fig. 6.8 has nested feedback loops. To determine the overall transfer function, $H(s) = P(s)/R(s)$, it is necessary to proceed in steps. First we evaluate the transfer function from $x(t)$ to $y(t)$: $G_1(s) = Y(s)/X(s)$. By Eq. (6.35), the transfer function of this internal feedback loop is

$$G_1(s) = \frac{C(s)}{1 + B(s)C(s)}.$$

Now the "plant" transfer function for the outer feedback loop is the transfer function from $x(t)$ to $p(t)$, which is $G(s) = G_1(s)A(s)$. Thus the overall closed-loop transfer function is

$$H(s) = \frac{G(s)}{1 + F(s)G(s)} = \frac{A(s)\left[\dfrac{C(s)}{1 + B(s)C(s)}\right]}{1 + F(s)A(s)\left[\dfrac{C(s)}{1 + B(s)C(s)}\right]} = \frac{A(s)C(s)}{1 + C(s)[B(s) + F(s)A(s)]}.$$

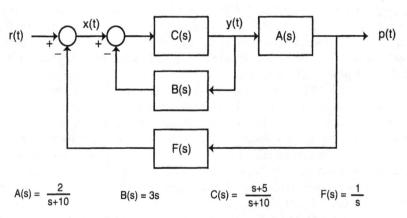

$$A(s) = \frac{2}{s+10} \qquad B(s) = 3s \qquad C(s) = \frac{s+5}{s+10} \qquad F(s) = \frac{1}{s}$$

FIGURE 6.8. A system having nested feedback loops.

Substituting the specified transfer functions of the four subsystems into this equation yields

$$H(s) = \frac{\left[\dfrac{2}{s+10}\right]\left[\dfrac{s+5}{s+10}\right]}{1 + \dfrac{s+5}{s+10}\left[3s + \dfrac{1}{s}\left[\dfrac{2}{s+10}\right]\right]} = \frac{2s(s+5)}{3s^4 + 46s^3 + 170s^2 + 102s + 10}.$$

Example 6.15 Comparison of proportional, derivative, and integral feedback
Figure 6.9 diagrams a system for controlling the angular position of a robotic arm. The controller comprises a voltage-controlled motor and its associated electronic control hardware. Given an input voltage, $r(t)$, the motor rotates a shaft by an angle proportional to $r(t)$. The angle of the arm, $p(t)$, is controlled by this shaft. In response to a sudden change in $r(t)$, the arm does not move instantaneously to its final position because of inertia. *When the feedback connection is disabled,* the transfer function of the controller and arm together is given by

$$G(s) = C(s)A(s) = \frac{10}{s+5}.$$

That is, the open-loop system has a time constant, τ_{ol}, of 0.20 s.

One may modify the response of the robotic arm by introducing negative feedback, as shown in the figure. Consider first the use of *proportional feedback*. In this case, $F(s) = k$, where k is a constant. Given that $G(s)$ may be expressed as $G(s) = b/(s + a)$, the closed-loop transfer function in the presence of proportional feedback will be

$$H_p(s) = \frac{G(s)}{1 + F(s)G(s)} = \frac{b}{s + (a + kb)}.$$

Therefore, the system is still first-order, but the pole location has changed. If the parameters (a, b) are positive, then the new time constant, $\tau_p = 1/(a + kb)$, is smaller

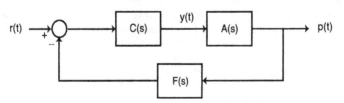

FIGURE 6.9. Model of a control system for a robotic arm.

than the original time constant, $\tau_{ol} = 1/a$, and the step response achieves its new steady state more quickly than without feedback. If, for example, $k = 1$ for the robotic arm system, then the new time constant is 0.067 s, or about one-third of τ_{ol}. This improvement in response, however, has a price. Using the Final Value theorem, the final value of $p(t)$ when $r(t)$ is a unit-step function is

$$\lim_{t \to \infty}[p(t)] = \lim_{s \to 0}[sP(s)] = \lim_{s \to 0}\left[\frac{sb}{s + a + kb}\frac{1}{s}\right] = \frac{b}{a + kb}.$$

Note that without feedback the final value of $p(t)$ would be b/a. Therefore, the steady-state gain has been decreased in the presence of proportional feedback. Defining an error, $e(t) = r(t) - p(t)$, the final value of $e(t)$ is larger with proportional feedback.

For *derivative feedback*, the feedback transfer function has the form $F(s) = ks$. Now the closed-loop transfer function is

$$H(s) = \frac{\dfrac{b}{s + a}}{1 + ks\dfrac{b}{s + a}} = \frac{b}{s(1 + kb) + a}.$$

Again, the system is still first-order, but the time constant now is $\tau_d = (1 + kb)/a$, or, assuming $k > 0$, $\tau_d = (1 + kb)\tau_{ol}$, and thus the closed-loop system responds more slowly than the system without feedback. One may easily show that the steady-state response to a step input is the same for both the open-loop and closed-loop systems.

Finally we may consider the effect of *integral feedback*, in which case $F(s) = k/s$. The closed-loop response becomes

$$H(s) = \frac{\dfrac{b}{s + a}}{1 + \dfrac{k}{s}\dfrac{b}{s + a}} = \frac{bs}{s^2 + as + kb}.$$

Integral feedback adds another pole to the closed-loop transfer function, converting the first-order system into a second-order system. Notice that adjusting the feedback gain will alter the natural frequency of the system but not its damping factor. Perhaps the most profound effect of adding integral feedback occurs in the steady-state response to a step input. This response is

$$\lim_{s \to 0}(sP(s)) = \lim_{s \to 0}\left[s\frac{bs}{s^2 + as + kb}\frac{1}{s}\right] = 0.$$

The system has become a highpass filter with a gain of zero at zero frequency.

6.8   BIOMEDICAL APPLICATIONS OF LAPLACE TRANSFORMS

Example 6.16   Simple postural control model   A mass on an inverted pendulum (Fig. 6.10) provides a simple model of the use of functional neuromuscular stimulation (FNS) to maintain an upright posture. This model assumes that the entire body mass, M, is placed at the center of gravity of the body and is balanced atop a straight "stick" leg. T is the effective torque resulting from artificial activation of the leg muscles using electrical stimuli generated by a pulse generator and applied to intramuscular wire electrodes or skin surface electrodes. This applied torque acts against three torque components of the system: (1) a torque due to circular acceleration of the body mass, (2) an equivalent torque due to rotational resistance at the ankle joint, and (3) the torque due to the component of the gravitational force that is perpendicular to the "leg" at a distance L from the rotation point. The equation of motion of this system is

$$T(t) = ML^2\ddot{\theta}(t) + R\dot{\theta}(t) + MgL\theta(t). \qquad (6.36)$$

Here the approximation $\sin(\theta) \approx \theta$ for small angles has been used in the third term on the r.h.s. of Eq. (6.36). Taking Laplace transforms and solving for the transfer function between the applied torque and the resulting angle of rotation (assuming zero initial conditions) yields

$$H(s) = \frac{\theta(s)}{T(s)} = \frac{1/ML}{Ls^2 + (R/LM)s + g}.$$

The poles of $H(s)$ are located at $s_{1,2} = -(R/2ML^2) \pm (\sqrt{R^2/M^2L^2 - 4gL}/2L)$.

Consequently, if there is no rotational resistance, the poles are at $s_{1,2} = \pm j\ g/L$, implying that $h(t)$ comprises an undamped oscillation. For the more likely case that

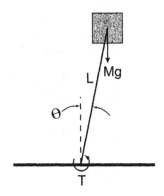

FIGURE 6.10. Modeling postural control as the balancing of an inverted pendulum. Body mass, M, is assumed to be atop a rigid "stick" leg of length L. T is the net torque developed by applying electrical stimuli from a pulse generator to appropriate leg muscles.

$R > 0$, there are two possibilities. If $R^2 > 4gL^3M^2$, then both poles of $H(s)$ are on the negative real axis and the response of this system to a step change in applied torque, T, is dominated by the time constant associated with the pole nearest to $s = 0$. If $0 < R^2 < 4gL^3M^2$, then $H(s)$ has complex conjugate poles and $h(t)$ comprises a damped oscillation. This last case (small but nonzero R) seems most likely physiologically.

Example 6.17   Iron kinetics in blood   Iron is an important element that is necessary for the production of hemoglobin, myglobin, and cytochromes. Dietary iron is transported in the blood (plasma) to erythropoietic and nonerythropoietic tissues (Fig. 6.11). There is a faster reflux of iron from the latter and a slower reflux from the former due to ineffective erythropoiesis. Some plasma iron is lost to other tissues and some is taken up directly by red cells and not recirculated. One may develop a compartmental model of iron kinetics in order to predict the plasma iron concentration in response to a bolus injection of a tracer such as ^{59}Fe. The following mass balance equations may be written for the model of Fig. 6.11:

$$V_1 \frac{dC_1}{dt} = -k_1 C_1(t) + a_{12} C_2(t) + a_{13} C_3(t) + M\delta(t);$$

$$V_2 \frac{dC_2}{dt} = a_{21} C_1(t) - a_{12} C_2(t);$$

$$V_3 \frac{dC_3}{dt} = a_{31} C_1(t) - a_{13} C_3(t);$$

$$V_4 \frac{dC_4}{dt} = a_{41} C_1(t);$$

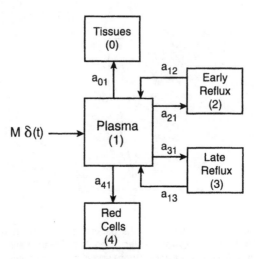

FIGURE 6.11. Compartmental model of iron tracer kinetics. See text for explanation.

where V = volume of a compartment, C = concentration, M is the mass of the bolus of tracer, $k_1 = a_{01} + a_{21} + a_{31} + a_{41}$, and all concentrations are zero at $t = 0$. Let us express these relationships in terms of the following dimensionless variables:

$$\theta_i = C_i/[M/V_i], \ \tau = t/[V_1/k_1], \ \delta(\tau) = \delta(k_1 t/V_1) = (V_1/k_1)\delta(t).$$

After some algebra we find

$$\frac{d\theta_1}{dt} = -\theta_1 + \alpha_{12}\theta_2 + \alpha_{13}\theta_3 + \delta(\tau);$$

$$\frac{d\theta_2}{dt} = \alpha_{21}\theta_1 - \beta_{12}\theta_2;$$

$$\frac{d\theta_3}{dt} = \alpha_{31}\theta_1 - \beta_{13}\theta_3;$$

$$\frac{d\theta_4}{dt} = \alpha_{41}\theta_1;$$

where

$$\alpha_{12} = (V_1/k_1)a_{12}, \ \alpha_{13} = (V_1/k_1)a_{13}, \ \alpha_{21} = (V_1/V_2)(a_{21}/k_1), \ \alpha_{41} = (V_1/V_4)(a_{41}/k_1);$$

$$\alpha_{31} = (V_1/V_3)(a_{31}/k_1), \ \beta_{12} = (V_1/V_2)(a_{12}/k_1), \ \beta_{13} = (V_1/V_3)(a_{13}/k_1).$$

Taking the Laplace transforms of these equations, we obtain:

$$(s + 1)\Theta_1(s) = \alpha_{12}\Theta_2(s) + \alpha_{13}\Theta_3(s) + 1;$$

$$(s + \beta_{12})\Theta_2(s) = \alpha_{21}\Theta_1(s);$$

$$(s + \beta_{13})\Theta_3(s) = \alpha_{31}\Theta_1(s);$$

$$s\Theta_4(s) = \alpha_{41}\Theta_1(s).$$

Combining and solving for the normalized plasma iron concentration yields

$$(s + 1)\Theta_1(s) = \frac{\alpha_{12}\alpha_{21}}{s + \beta_{12}}\Theta_1(s) + \frac{\alpha_{13}\alpha_{31}}{s + \beta_{13}}\Theta_1(s) + 1,$$

from which we determine

$$\Theta_1(s) = \frac{(s + \beta_{12})(s + \beta_{13})}{(s + 1)(s + \beta_{12})(s + \beta_{13}) - \alpha_{12}\alpha_{21}(s + \beta_{13}) - \alpha_{13}\alpha_{31}(s + \beta_{12})}.$$

Thus we see that the kinetics of plasma iron concentration following a bolus injection are those of a third-order system having also two zeros. Given specific parameter values, one could use partial fractions to expand the above expression and then solve for $\theta_1(t)$.

Introductory Example Revisited

Consider again the introductory example for this chapter, in which it was desired to determine the effect of filtering on the measured response of a two-compartment model of drug distribution dynamics following a bolus infusion of the drug into the first compartment. This response before filtering may be calculated by solving the differential equation using Laplace transforms. Taking the Laplace transform of both sides of Eq. (6.1), allowing for nonzero initial conditions, one obtains

$$s^2 Y(s) - sy(0) - \dot{y}(0) + a[sY(s) - y(0)] + bY(s) = k_1 X(s).$$

The general solution for $Y(s)$ is

$$Y(s) = \frac{sy(0) + \dot{y}(0) + ay(0)}{s^2 + as + b} + \frac{k_1}{s^2 + as + b} X(s).$$

Using the previously specified values for a, b, $y(0)$, $\dot{y}(0)$ and letting $x(t) = 0$, we calculate $Y^*(s)$ as

$$Y^*(s) = \frac{0.5}{s^2 + 0.255s + 0.0025}.$$

Note that the denominator may be factored into $(s + 0.0117)(s + 0.2133)$. As specified previously, the filtered signal, $y_f(t)$, is the output of a first-order lowpass filter with unity gain and cutoff frequency of 0.2513 rad/s and input $y^*(t)$. Thus, if $H(s)$ is the transfer function of the filter,

$$Y_f(s) = Y^*(s)H(s) = \frac{0.5}{(s + 0.0117)(s + 0.2133)} \frac{0.2513}{s + 0.2513}. \qquad (6.37)$$

Expanding Eq. (6.37) using partial fractions,

$$Y_f(s) = \frac{c_1}{s + 0.0117} + \frac{c_2}{s + 0.2133} + \frac{c_3}{s + 0.2513}. \qquad (6.38)$$

It is readily shown that $c_1 = 2.60$, $c_2 = -16.60$, $c_3 = 14.00$. Therefore, the inverse Laplace transform of Eq. (6.38) is identical to Eq. (6.2).

6.9 SUMMARY

The Laplace transform models a signal as a sum of exponentially decaying sinusoids having various frequencies and rates of decay. Similar to the case with the Fourier transform, an infinite number of such signals are required to reconstruct an arbitrary function, $x(t)$. The bilateral Laplace transform requires specification of the region of the s-plane over which it converges. By considering only time functions that are zero for all $t < 0$, one may define a unilateral Laplace transform for which each transform can be assumed to correspond to a unique function of time. Its region of convergence includes all of the s-plane to the right of its rightmost pole. The unilateral Laplace transform exists for most such functions (if they are integrable over all finite time intervals), unless one cannot find any constant, c, such that the magnitude of the function grows in time more slowly than e^{ct}.

If a Laplace transform, $X(s)$, of a function $x(t)$ has the form of a proper rational polynomial, and if $X(s)$ only has poles in the left-half plane and/or on the imaginary axis, then the Fourier transform, $X(\Omega)$, may be determined from $X(s)$. If there are only poles in the left-half plane, then $X(\Omega) = X(s)|_{s=j\Omega}$.

The usual approach for calculating the inverse Laplace transform is the method of partial fractions. Typically, one factors the denominator polynomial into a product of factors of the form $(s - p_i)$ where p_i may be real or complex. Real poles give rise to time functions of the form $c_i e^{p_i t} u(t)$, whereas pairs of complex conjugate poles correspond to time functions having the form $c_i e^{\text{Re}\{p_i\}t} \cos(\text{Im}\{p_i\}t + \measuredangle p_i) u(t)$.

For an LTIC system having the impulse response $h(t)$, the Laplace transform $H(s)$ is the transfer function of the system. If its input signal has a Laplace transform, then the output signal has a Laplace transform given by the product of $H(s)$ with the Laplace transform of the input. Because the Laplace transform of a derivative accounts for the initial value of a signal at $t = 0$, from the Laplace transforms of the terms in a system equation one may solve for the complete output response of a system including both the transient response due to the initial conditions (zero-input response) and the response to the specific input signal (zero-state response). This capability is important in biomedical applications involving transient signals, for which the steady-state analyses based on frequency response functions are not readily applicable. The method of Laplace transforms also provides a convenient approach for analyzing the transfer function, frequency response, and transient behaviors of systems comprising multiple subsystems connected in series, parallel, or feedback configurations.

Because in many cases the frequency response of a filter is derivable from its transfer function, it will be possible to define discrete-time filters whose frequency responses are related to those of specific continuous-time filters. To do so we must first develop an understanding of Fourier transforms applied to discrete-time signals and their relationship to a discrete-time counterpart of the Laplace transform known as the Z-transform. These topics are discussed in the next two chapters.

EXERCISES

6.1 Calculate the Laplace transforms of the following signals:
a. $x(t) = t[u(t) - u(t-2)]$;
b. $w(t) = \cos(5\pi t)\cos(7\pi t)$;
c. $z(t) = t\cos^2(\Omega t)u(t)$;
d. $v(t) = e^{-3t} * e^{-5t}, t \geq 0$;
e. $g(t) = P_1(t-0.5) - P_2(t-2) + P_1(t-3)$.

6.2 Given that $X(s) = 2/(s+4)$, find the Laplace transforms of the following signals:
a. $y(t) = x(t) * x(t)$;
b. $r(t) = e^{-t/4}x(t)$;
c. $v(t) = x(2t-5)u(2t-5)$;
d. $z(t) = x(t)\sin(\Omega_0 t)$.

6.3 Determine the inverse Laplace transforms of the following functions:

a. $X(s) = \dfrac{2}{s^2+7s+12}$;

b. $Y(s) = \dfrac{s}{s+1}$;

c. $Z(s) = e^{-2s}\dfrac{5}{s^2}$;

d. $H(s) = \dfrac{4}{s^2(s^2+2s+6)}$;

e. $G(s) = \dfrac{3s+2}{s^3+11s^2+36s+36}$.

6.4 Determine the impulse response of the LTIC system having the transfer function

$$H(s) = \frac{Y(s)}{X(s)} = \frac{4(s+10)}{s^2+2s+4}.$$

6.5 Second-order, lowpass, Butterworth filters have a transfer function given by $H(s) = \Omega_1^2/(s^2+\sqrt{2}\Omega_1 s+\Omega_1^2)$, where Ω_1 is the cutoff frequency. Determine the step response of these filters.

6.6 A filter has the impulse response $h(t) = (t-1)^2 e^{-2(t-1)}u(t-1)$. Calculate its transfer function and frequency response.

6.7 The impulse response of an LTIC system is $h(t) = e^{-t}\cos(628t)$. What is its response to the unit ramp signal, $x(t) = tu(t)$?

6.8 A negative feedback system comprises a plant having the transfer function

$$G(s) = \frac{s+0.1}{(s+1)(s+0.05)}$$

and a feedback loop having the transfer function $F(s) = 2$. What is the overall transfer function of this system?

6.9 Let the system depicted in Fig. 6.7 be configured as a negative feedback system described by the two coupled differential equations:

$$\dot{y}(t) = y(t) + v(t), \qquad \dot{y}_f(t) = 2y_f(t) + y(t).$$

Use Laplace transforms to determine the transfer function relating $y(t)$ to $x(t)$, and calculate the step response of this system.

6.10 Arterial blood pressure, $p(t)$, is regulated by a negative feedback loop involving baroreceptors. A simple model of this regulatory system is shown in Fig. 6.12. Arterial blood pressure is monitored by baroreceptors whose output is apparently compared to a reference pressure, $r(t)$. Brainstem neural mechanisms modulate blood pressure through control of sympathetic and parasympathetic activities to the heart and vasculature. The parameters given in the figure provide rough approximations to the temporal responses of these systems.

  a. Determine the transfer function relating $p(t)$ to the reference input, $r(t)$.

  b. What is the differential equation relating $r(t)$ and $p(t)$?

  c. The brainstem controller includes an unspecified gain, K. If K is such that the transfer function of the closed-loop system has poles on the imaginary axis, then $p(t)$ will oscillate even if $r(t)$ is constant. Find the value of K that will lead to an oscillation in blood pressure.

6.11 Exercise 3.18 described an electrophysiology recording table that was subjected to floor vibrations. Consider this same table and let $R = 100$ N/m/s, $K = 0.1$ N/mm, $M = 5$ kg.

  a. Recall, or solve for, the system equation for which vertical floor displacement is the input and vertical table displacement is the output.

  b. Determine the transfer function and the frequency response of this system.

  c. A motor running at 1500 rpm causes sinusoidal displacements of the floor

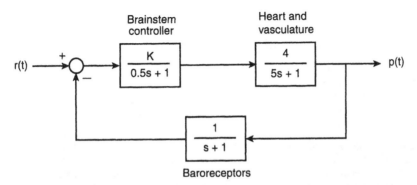

FIGURE 6.12. A linearized model of the control of arterial blood pressure, $p(t)$, by baroreceptor feedback. $r(t)$—reference input. K is the gain of the brainstem neural control mechanisms at $\Omega = 0$.

with amplitude 2 mm. What will be the amplitude of the steady-state sinusoidal displacements of the table?

d. Using MATLAB, determine the response of the (initially quiescent) table to a sudden floor displacement of 1 mm that lasts for 100 msec.

6.12   Fig. 6.13 depicts a simple model of an excised frog muscle pulling a weight along a surface. The muscle model includes a series elastance, K, a parallel viscous resistance, R, and an active state generator, F. Movement of the weight, M, is opposed by frictional resistance, R_f.

a. Derive the differential equation relating the displacement of the weight, $x(t)$, to the force, $f(t)$, generated by the active state generator.

b. In response to a brief electrical stimulus applied via intramuscular electrodes, the active state generates the transient force shown in Fig. 6.13(c). If the force (grams) is given by $f(t) = 2[e^{-0.05t} - e^{-0.2t}]$, calculate the resulting movement of the weight. Let $K = 1$ mm/gram, $R = 0.1$ grams/(mm/s), $R_f = 0.2$ grams/(mm/s), $M = 5$ grams.

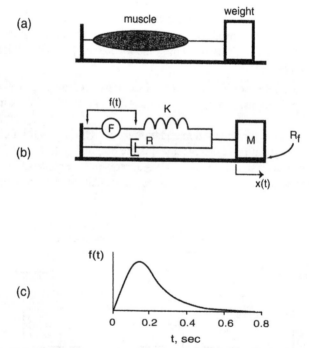

FIGURE 6.13. (a) Schematic of an excised frog leg muscle that is attached to a rigid pole at one end and to a weight of mass M at the other. (b) Mechanical diagram of the muscle-weight system. F: active state generator; K: series elastance; R: parallel viscous resistance; R_f: frictional resistance. (c) Theoretical active state force generated as a result of applying a brief electrical stimulus to the muscle via intramuscular wire electrodes.

6.13 One method for measuring changes in the intracellular concentration of a variety of chemical species is to attach a fluorescent tag to the species of interest. This chemical tag fluoresces after exposure to light of a certain frequency (that is different for different tagging chemicals). After a flash of light of appropriate frequency is applied to the tissue, the intensity of fluorescent light is proportional to the concentration of the tagged species. Assume that in response to a light flash a particular tagging compound fluoresces with an intensity that is proportional to the species concentration, k_i, and that it decays exponentially with a time constant of 2 msec. The fluorescent light is detected by a photomultiplier tube and amplified by an electronic system having a lowpass cutoff frequency of 1000 Hz. Compare the electronic output from the photomultiplier tube and amplifier with the specified time course of intensity of fluorescent light. Why is it possible to measure changes in concentration even though the electronic signal is distorted relative to the known light intensity?

6.14 Some biomedical transducers operate on the principle that a physical action (e.g., a force, pressure, or movement) causes a change in the inductance of a coil of wire. (Such transducers are called "variable reluctance" transducers.) Inductance L_2 in Fig. 6.14 is a variable inductance. It is connected into a "transducer coupler" circuit, with which it forms a bridge circuit as shown.

    a. Let $H(s) = E_{out}(s)/E_{in}(s)$. Determine $H(s)$ for a fixed value of L_2.

    b. Find the steady-state relationship between the magnitudes of e_{in} and e_{out} when the former is a sinusoidal signal. If e_{in} has a constant magnitude, determine the dependence of the magnitude of e_{out} on L_2.

6.15 Oxygen sensors for use in blood often have a membrane over the sensor through which oxygen must diffuse in order to be detected. This process limits the rapidity of response of the sensor to a change in oxygen level of the blood. A Clark-style electrode may have a time constant of several seconds. Assume that such an electrode has a time constant of 15 seconds. The electrode is situated in a small

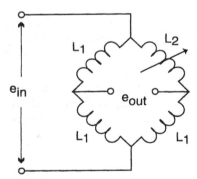

FIGURE 6.14. A bridge circuit containing a variable reluctance. Typically the bridge circuit, less L_2, resides within an electronic amplifier known as a transducer coupler, while L_2 is contained within a biomedical transducer that is connected to the coupler via cabling.

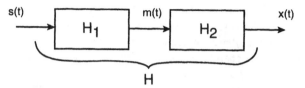

FIGURE 6.15. Conceptual framework for the cell culture experiment of Exercise 6.16. $s(t)$ = inflow rate of growth factor; $m(t)$ = concentration of growth factor; $x(t)$ = normalized density of cells.

chamber of a blood gas analyzer and has been exposed to air for a long time, then a laboratory technician abruptly fills the chamber with blood sampled from a patient. If the partial pressure of oxygen, P_{O2}, in the blood is 60 Torr, determine the time course of the *error* in the reading from the oxygen electrode. How long should the technician wait after injecting the sample before taking the reading?

6.16   A small number of cells is placed in a cell culture dish and the cells are allowed to grow in the presence of a growth factor added to the culture medium. Let $x(t)$ be the normalized density of cells (i.e., fraction of the surface of the culture dish that is covered by cells), and assume that $x(0) \approx 0$. Growth factor is added to the dish at a constant rate, $s(t)$, of 2.5 ng/min for 60 minutes. Since the growth factor is consumed by the cells, its concentration changes with time. By periodic sampling and chemical analysis, the investigator ascertains that the concentration of growth factor is $m(t) = 150[e^{-t/300} - e^{-0.03t}]$ ng. The investigator assesses $x(t)$ by measuring the transmission of light through the culture dish using appropriate lens systems and a photomultiplier tube. By curve-fitting this data she determines that the fractional covering of the surface varies according to the relation $x(t) = 1 - 0.5[e^{-t/1200} - e^{-t/300}]$. Note that the unit of time is "minutes". This system can be modeled as shown in Fig. 6.15.

a. Calculate the apparent transfer function of each subsystem in Fig. 6.15, and of the overall system that relates $x(t)$ to $s(t)$.

b. The investigator wishes to determine the effects of adding a possible inhibitor of cellular metabolism to the culture dish. One could argue that the inhibition will be most apparent if the substance is delivered at the time of maximum rate of growth of the cells. Determine this time.

c. Is it possible to modify $s(t)$ to increase the maximum rate of growth while keeping the total mass of delivered growth factor constant? Test your suggestion(s) by solving explicitly for $x(t)$ using the modified $s(t)$ as the input.

7

MODELING SIGNALS AS SUMS OF DISCRETE-TIME SINE WAVES

7.1 INTRODUCTION

Chapter 5 developed the Fourier series and Fourier transform as tools for modeling continuous-time signals as sums of sine waves. This chapter will extend both of these methods to discrete-time signals. Discrete-time signals arise naturally in biomedical situations in which measurements are acquired in association with discrete events. For example, the amount of blood ejected from the heart on each beat (stroke volume) and the volume of air inspired on each breath (tidal volume) are measurements from discrete events. In both of these cases the "time" variable is the index of an ordered sequence of events (i.e., heart beats or breaths). As these examples show, the ordered sequence of events may be indexed relative to variables other than time. A common example is the indexing of spatial sampling in imaging. Viewing the intensity values of each spatial pixel in an MR image as a "discrete-time" signal, one could ask, for example, whether there is a spatial (either linear or two-dimensional) sinusoidal variation in pixel intensity.

This concept is not restricted to MR images but may be applied to any array of physical sensors, such as a CCD array sensing intensity of fluorescent light over a tissue culture dish or an array of scintillation detectors exposed to an x-ray field. (It is important to recognize that measurements that are associated with discrete events often are continuously-graded variables. On the other hand, some truly continuous-time variables can assume only discrete levels. An example of the latter is the number of cells in a culture dish as a function of time.)

Discrete-time variables also may originate from measurements that intentionally count (or average) signals over a pre-selected time period. In Chapter 1, Fig. 1.4 depicts an example of this type of signal based on counting the number of action potentials generated by a neuron in one-second intervals. Another example is the count of the number of offspring per generation (the "time" variable) in studies of population dynamics. A histogram plot may be considered a "discrete-time" variable because it counts the number of events occurring within preselected intervals of an intensity variable and presents the results in an ordered sequence of these pre-selected intervals. Fig. 1.2(f) shows such an example.

Of course, discrete-time variables are generated most commonly by repetitive sampling of continuous-time variables, as discussed in Chapter 1. (We shall assume that this sampling occurs with a uniform inter-sample time.) The discrete-time variables so created are also ordered (i.e., indexed) sequences of numbers and, in this sense, are not different from the examples of indexed sequences given above. Since one often is concerned with associating the characteristics of a continuous-time signal with properties of the biomedical system from which it originated, a relevant issue is the relationship between models of discrete-time signals that were obtained by sampling and the Fourier series or Fourier transform representations of the continuous-time signals that were sampled. This important issue will be a major focus of this chapter.

7.2 INTERACTIVE EXAMPLE: PERIODIC OSCILLATIONS IN THE AMPLITUDE OF BREATHING

Some discrete-time signals are periodic, or nearly so, and can be modeled as a sum of sinusoids, as this example demonstrates. The amount of air inhaled on each breath (tidal volume) is not constant even when the environmental and other conditions of the subject are as constant as one can achieve. This variability in tidal volume is especially evident in certain disease situations, such as congestive heart failure, and in neonates. Figure 7.1 shows a record of tidal volume from a normal-term human infant during rapid-eye-movement (REM) sleep plotted as a discrete-time variable versus "breath number" (solid symbols connected by dotted lines). It appears that this signal might be modeled as a noisy sine wave. The fitting process is

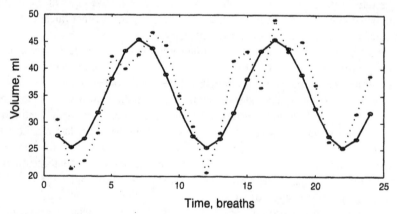

FIGURE 7.1. Breath by breath tidal volumes of a human infant in REM sleep (Hathorn, 1978). Dotted line: tidal volumes. Solid line: "best fit" sinusoid of period 10 breaths.

identical to that utilized for CT signals, and the file `fitsine.m` was used to fit a discrete-time sine wave with a period of 10 breaths to these data. The approximating sinusoid has an amplitude of 10.1 ml and a phase of –2.242 rad and is shown with open symbols connected by solid lines. The sinusoid is a reasonable approximation to the data but it should be noted that such "periodic" behaviors typically last only for a few cycles before the characteristics, such as amplitude or general shape, change. The reason may be that this process is not being driven by a periodic stimulus (as was the case for circadian variations studied in Chapter 5). Therefore, a sinusoidal approximation may be used to represent the data plotted in the figure but it does not provide insights into the physiological mechanisms generating the variations in tidal volume, which are not strictly periodic.

One could proceed as was done for the data of Fig. 5.1 and fit another sinusoid to the error between the two signals of Fig. 7.1. The interested reader will find these data in the file `vtdata.mat`. Because longer records of such data reveal that the signal is not truly periodic, strictly speaking this approach is not appropriate. Rather, one should apply a DT analog of the CT Fourier transform. Later we will re-analyze these data from that perspective. First, however, we will develop the concept of the discrete-time Fourier series as a foundation for discussing the discrete-time Fourier transform.

7.3 THE DISCRETE-TIME FOURIER SERIES

A discrete-time sequence, $x[n]$, is periodic with period N if $x[n + N] = x[n]$ for all n. To emphasize the periodicity of the signal, it will be symbolized as $\tilde{x}[n]$. Of course the functions $\cos[\omega_0 n]$, and $\sin[\omega_0 n]$, *may* fail to be periodic if there is no integer value of N that satisfies the condition for periodicity. However, given any N there is always a frequency ω_0 having the period N. That frequency is $\omega_0 = 2\pi/N$.

Two real-valued periodic sequences, $\tilde{x}_1[n]$ and $\tilde{x}_2[n]$, both having period N, are said to be orthogonal if

$$\sum_{k=0}^{N-1} \tilde{x}_1[k]\tilde{x}_2[k] = 0. \tag{7.1}$$

Note that either or both functions may have fundamental periods that are submultiples of N—for example, $N/2$, $N/3$. It is only necessary that there be some period N that is common to both. In particular, if $\omega_0 = 2\pi/N$, then the functions $e^{jk\omega_0 n}$, $k = 1$, $2, \ldots$, are mutually orthogonal and form a set of basis functions for expanding discrete-time periodic signals. That is, the periodic signal $\tilde{x}[n]$ can be decomposed into a summation of complex exponential functions of the form

$$\tilde{x}[n] = \sum_k \tilde{X}[k]e^{jk\omega_0 n}, \qquad \omega_0 = \frac{2\pi}{N},$$

where $\tilde{X}[k]$ are the weighting coefficients.

But since there are only N unique values of $\tilde{x}[n]$, one needs at most N terms in the summation. This result may be seen by noting that $e^{j\frac{2\pi}{N}(k+N)n} = e^{j\frac{2\pi k}{N}n}e^{j2\pi m} = e^{j\frac{2\pi k}{N}n}$. That is, the $(k+N)$-th harmonic of the fundamental complex exponential function is identically equal to the k-th harmonic. Thus the summation comprises only N unique functions. Consequently, the expansion of a periodic DT signal may be written as

$$\tilde{x}[n] = \frac{1}{N}\sum_{k=0}^{N-1}\tilde{X}[k]e^{j\frac{2\pi k}{N}n}, \qquad 0 \le n \le N-1. \tag{7.2}$$

Eq. 7.2 is the *Discrete Fourier series* for $\tilde{x}[n]$ and the $\tilde{X}[k]$'s are the Discrete Fourier series (DFS) coefficients. (The $1/N$ term is equivalent to the $1/K_0$ term in the CT orthogonal function representation. It is customary to include it in the DFS synthesis equation rather than in the analysis equation for the coefficients as was done for the CT Fourier series.)

To evaluate the DFS coefficients we first multiply both sides of Eq. (7.2) by $e^{-j\frac{2\pi}{N}nr}$ and then sum over one cycle. Thus

$$\sum_{n=0}^{N-1}\tilde{x}[n]e^{-j\frac{2\pi}{N}nr} = \frac{1}{N}\sum_{n=0}^{N-1}\sum_{k=0}^{N-1}\tilde{X}[k]e^{j\frac{2\pi}{N}kn}e^{-j\frac{2\pi}{N}nr}.$$

$$= \frac{1}{N}\sum_{k=0}^{N-1}\tilde{X}[k]\sum_{n=0}^{N-1}e^{j\frac{2\pi}{N}(k-r)n}. \tag{7.3}$$

Recalling the previous result that $\sum_{n=0}^{N-1}a^n = (1-a^N)/(1-a)$, one may show that the rightmost summation in Eq. (7.3) evaluates to N when $k = r$ and to zero otherwise. Therefore, letting r equal k leaves only one term on the r.h.s. of Eq. (7.3)—that is, $\tilde{X}[k]$—and we conclude that

$$\tilde{X}[k] = \sum_{n=0}^{N-1}\tilde{x}[n]e^{-j\frac{2\pi}{N}nk}, \qquad 0 \le k \le N-1. \tag{7.4}$$

Finally, note that one could evaluate Eq. 7.4 for $k > N-1$ but due to the equality of the k-th and $(k+N)$-th harmonics (as noted above), $\tilde{X}[k+N] = \tilde{X}[k]$. In other words, the DFS coefficients also have period N.

Example 7.1 The signal in Fig. 7.2(a) has period $N = 10$ and may be represented as a Discrete Fourier series with $\omega_0 = 2\pi/10$ and DFS coefficients

$$\tilde{X}[k] = \sum_{n=0}^{9}\tilde{x}[n]e^{-j\frac{2\pi}{10}nk} = \sum_{n=0}^{4}e^{-j\frac{2\pi}{10}nk} = \frac{1-\left(e^{-j\frac{\pi}{5}k}\right)^5}{1-e^{-j\frac{\pi}{5}k}}. \tag{7.5}$$

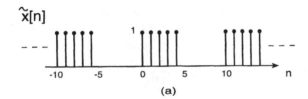

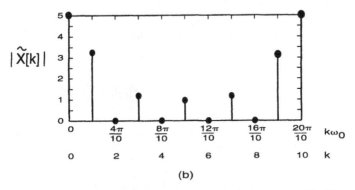

(b)

FIGURE 7.2. (a) A periodic discrete-time signal, and (b) the magnitudes of its Discrete Fourier series coefficients graphed versus frequency and harmonic number.

Using Euler's formulas, Eq. (7.5) may be rewritten as

$$\tilde{X}[k] = \frac{e^{-j\frac{\pi k}{2}}\left[e^{j\frac{\pi k}{2}} - e^{-j\frac{\pi k}{2}}\right]}{e^{-j\frac{\pi k}{10}}\left[e^{j\frac{\pi k}{10}} - e^{-j\frac{\pi k}{10}}\right]} = e^{-j\frac{4\pi k}{10}}\left[\frac{\sin\left[\dfrac{\pi k}{2}\right]}{\sin\left[\dfrac{\pi k}{10}\right]}\right]. \tag{7.6}$$

The DFS coefficients are plotted as a function of both k and $k\omega_0$ in Fig. 7.2(b)

Some Properties of Discrete Fourier series

The DFS operation is linear in the sense that if $\tilde{x}[n]$, $\tilde{y}[n]$ both have period N and one defines $\tilde{z}[n] = a\tilde{x}[n] + b\tilde{y}[n]$, then the DFS coefficients of the latter can be found from those of the component signals as $\tilde{Z}[k] = a\tilde{X}[k] + b\tilde{Y}[k]$.

Time-shifting of a periodic function adds a phase shift to its DFS coefficients. Thus, if $\tilde{x}[n]$ has DFS coefficients $\tilde{X}[k]$, then the DFS coefficients of $\tilde{x}[n - m]$ are $e^{-j\frac{2\pi}{N}km}\tilde{X}[k]$. If m is greater than N, then the m in the exponent of the complex exponential term must be expressed modulo N. This result is easily proven using Eq. (7.4).

If $\tilde{x}[n]$ has DFS coefficients $\tilde{X}[k]$, then $\tilde{x}^*[n]$ has DFS coefficients $\tilde{X}^*[-k]$. Again, this result is easily proven from Eq. (7.4). If the DT function is real, so that $\tilde{x}^*[n] = \tilde{x}[n]$, then $\tilde{X}^*[-k] = \tilde{X}[k]$, implying that $|\tilde{X}[-k]| = |\tilde{X}[k]|$ and $\angle \tilde{X}[-k] = \angle \tilde{X}[k]$.

Note that $\tilde{X}[0] = \sum_{n=0}^{N-1} \tilde{x}[n]e^{-j0} = \sum_{n=0}^{N-1} \tilde{x}[n]$.

Periodic Convolution

Given two period-N signals $\tilde{x}_1[n]$, $\tilde{x}_2[n]$, having DFS coefficients $\tilde{X}_1[k]$, $\tilde{X}_2[k]$, respectively, one can define a third Discrete Fourier series whose coefficients are $\tilde{X}_3[k] = \tilde{X}_2[k] \cdot \tilde{X}_1[k]$. To find the function $\tilde{x}_3[n]$ that has the specified DFS coefficients, $\tilde{X}_3[n]$, first note from Eq. (7.4) that

$$\tilde{X}_3[k] = \tilde{X}_2[k] \cdot \tilde{X}_1[k] = \sum_{m=0}^{N-1} \sum_{r=0}^{N-1} \tilde{x}_1[m]\tilde{x}_2[r]e^{-j\frac{2\pi}{N}k(m+r)}. \tag{7.7}$$

Substituting Eq. (7.7) into the definition of the Discrete Fourier series (Eq. (7.2)), and utilizing the common shorthand notation $W_N = e^{-j\frac{2\pi}{N}}$,

$$\tilde{x}_3[n] = \frac{1}{N}\sum_{k=0}^{N-1} \tilde{X}_3[k]W_N^{-nk} = \sum_{m=0}^{N-1} \tilde{x}_1[m]\sum_{r=0}^{N-1} \tilde{x}_2[r]\left[\frac{1}{N}\sum_{k=0}^{N-1} W_N^{-k(n-m-r)}\right] \tag{7.8}$$

Now since

$$\frac{1}{N}\sum_{k=0}^{N-1} W_N^{-k(n-m-r)} = \begin{cases} 1, & r = (n-m) + l \cdot N \\ \\ 0, & otherwise \end{cases} \quad , \quad l \text{ an integer,}$$

Eq. (7.8) becomes

$$\tilde{x}_3[n] = \sum_{m=0}^{N-1} \tilde{x}_1[m]\tilde{x}_2[n-m]. \tag{7.9}$$

Eq. 7.9 is the *periodic convolution* of the two period-N sequences. Unlike regular convolution, the summation is evaluated over one cycle only.

Example 7.2 Discrete Fourier series   The signal shown in Fig. 7.3 has a period of $N = 4$. Letting the cycle start from $n = 0$, its DFS coefficients are

$$\tilde{X}[k] = \sum_{n=0}^{3} \tilde{x}[n]e^{-j\frac{2\pi}{4}kn} = 2 + (1)\,e^{-j\frac{\pi}{2}k} + (0)e^{-j\frac{\pi}{2}(2k)} + (1)e^{-j\frac{\pi}{2}(3k)}$$

$$= 2 + e^{-j\frac{\pi}{2}k} + e^{j\frac{\pi}{2}k} = 2\left(1 + \cos\left[\frac{\pi}{2}k\right]\right), \qquad 0 \le k \le 3.$$

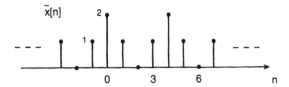

FIGURE 7.3. A periodic DT signal with a period of four.

Evaluating this expression we find $\tilde{X}[0:3] = [4, 2, 0, 2]$, , where we have used the $i{:}j$ notation to indicate the range of indices of a sequence (as in MATLAB, where a sequence is called a vector). Therefore, by Eq. (7.2) this periodic function may be represented in closed-form as

$$\tilde{x}[n] = \frac{1}{N} \sum_{k=0}^{N-1} \tilde{X}[k] e^{-j\frac{2\pi}{N}nk} = 1 + 0.5e^{j\frac{\pi}{2}n} + 0.5e^{j\frac{3\pi}{2}n}.$$

Consider the difference if the cycle of the signal of Fig. 7.3 is considered to start at $n = -2$. Now

$$\tilde{X}[k] = \sum_{n=-2}^{1} \tilde{x}_1[n] e^{-j\frac{\pi}{2}kn} = 0 + (1)e^{j\frac{\pi}{2}k} + 2 + (1)e^{-j\frac{\pi}{2}k} = 2\left(1 + \cos\left[\frac{\pi}{2}k\right]\right).$$

That is, so long as the time origin itself is not changed, the DFS coefficients are independent of the point in the cycle at which one starts the summation. Consider, however, the effect of shifting the time origin. In Fig. 7.3 let the point $n = 0$ be shifted to the left by two units of time to create a new periodic function $\tilde{x}_1[n] = \tilde{x}[n-2]$. Now

$$\tilde{X}_1[k] = \sum_{n=0}^{3} \tilde{x}[n] e^{-j\frac{\pi}{2}kn} = 0 + e^{-j\frac{\pi}{2}k} + 2e^{-j\frac{\pi}{2}(2k)} + e^{-j\frac{\pi}{2}(3k)}$$

$$= e^{-j\frac{\pi}{2}k} + 2(-1)^k + e^{j\frac{\pi}{2}k} = 2\left((-1)^k + \cos\left[\frac{\pi}{2}k\right]\right).$$

Evaluating this last equation, $\tilde{X}_1[0:3] = [4, -2, 0, -2]$ and $\tilde{x}_1[n] = 1 - 0.5e^{j\frac{\pi}{2}n} - 0.5e^{j\frac{3\pi}{2}n}$. Thus, when the time origin is changed, the magnitudes of the DFS coefficients, $\tilde{X}[k]$, are not altered but their angles are changed. The reader can confirm this result by applying the relationship discussed previously for the effect on DFS coefficients of time-shifting a periodic function.

Example 7.3 ***Noisy data*** For actual data that one expects should be periodic (e.g., the body temperature data of Fig. 5.1), additive noise may distort the true periodic signal if the available data samples only one cycle of the periodicity. If one can acquire data covering M cycles, then it is valid to average the M data points at corresponding time points in each cycle to generate an "average" cycle in which noise should be reduced. One may then find a Fourier series for this data. Another approach also is possible when M cycles of data are available. To illustrate, assume that one has acquired two cycles of noisy data (i.e., $M = 2$ and the total number of data points is $2N$). First determine the DFS coefficients for the signal with period $2N$:

$$\tilde{X}^{(2)}[k] = \sum_{n=0}^{2N-1} \tilde{x}[n]W_{2N}^{kn} = \sum_{n=0}^{N-1} \tilde{x}[n]W_{2N}^{kn} + \sum_{n=0}^{N-1} \tilde{x}[n+N]W_{2N}^{k(n+N)}. \qquad (7.10)$$

Note that $W_{2N}^{N} = -1$. Therefore,

$$\tilde{X}^{(2)}[k] = \sum_{n=0}^{N-1} \tilde{x}[n]W_{2N}^{kn}[1 + (-1)^k].$$

For odd k, the term in the brackets is zero. For even k, since $W_{2N}^{kn} = W_{N}^{kn/2}$, then

$$\tilde{X}^{(2)}[k] = 2\sum_{n=0}^{N-1} \tilde{x}[n]W_{N}^{kn/2} = 2\tilde{X}[k/2], \qquad (7.11)$$

where $\tilde{X}[k]$ is the DFS coefficient of the signal with period N assuming that the two cycles of sampled data are identical. If the two cycles are not identical, then the result from Eq. (7.11) for even k is twice the value of the DFS coefficient of the "average" cycle. This result generalizes to M cycles in the obvious manner.

Often discrete-time data will be samples of a continuous-time signal, such as body temperature, and one might wonder if there is a specific relationship between the Discrete Fourier series of the sampled signal and the Fourier series of the continuous-time signal. On the one hand, reconstruction of the CT signal from the Discrete Fourier series would be straightforward because the mapping between the discrete-time index, n, and continuous time would be obvious: $t = n \cdot \Delta t$, where Δt is the interval between consecutive data samples. On the other hand, the CT and DT Fourier series coefficients are not equal. The general relationship between the frequency components of a CT signal and those of a DT signal obtained by uniform sampling of the CT signal will be derived later in this chapter based on the Discrete-time Fourier transform.

7.4 FOURIER TRANSFORM OF DISCRETE-TIME SIGNALS

In Chapter 4, Eq. (4.25), the frequency response of an LSI system was defined as

$$H(e^{j\omega}) = \sum_{k=-\infty}^{\infty} h[k]e^{-j\omega k}$$

and it was noted that this function is continuous and periodic in ω with period $T = 2\pi$. Because it is periodic, it has a Fourier series representation (as a function of ω, not t) of the form

$$H(e^{j\omega}) = \sum_{k=-\infty}^{\infty} c_k e^{-j(1)k\omega}, \tag{7.12}$$

$$\text{with } c_k = \frac{1}{2\pi}\int_{-\pi}^{\pi} H(e^{j\omega})e^{j(1)\omega k}d\omega, \tag{7.13}$$

in which ω replaces t as the "time" variable, the fundamental period is 2π, and the fundamental frequency is $(2\pi/T) = (2\pi/2\pi) = 1$. Comparing Eqs. (4.25) and (7.12) we conclude that $c_k = h[k]$ because of the orthogonality of the complex exponential functions. That is, the impulse response of a DT system is also the set of coefficients in the Fourier series expansion (relative to ω) of the frequency response of the system. Conversely, from Eq. (7.13) one may express $h[k]$ as

$$h[k] = \frac{1}{2\pi}\int_{-\pi}^{\pi} H(e^{j\omega})e^{j\omega k}d\omega. \tag{7.14}$$

This latter equation describes $h[k]$ as the summation (integration) of an infinite number of discrete complex exponential functions where the weighting of frequencies in the range $(\omega, \omega + d\omega)$ is $(1/2\pi)H(e^{j\omega})d\omega$. This interpretation of $H(e^{j\omega})$ is analogous to that presented previously for the CT Fourier transform, $H(\Omega)$, and motivates defining $H(e^{j\omega})$ as the Discrete-*Time Fourier Transform (DTFT)* of $h[n]$. For any sequence, $x[n]$, that is absolutely summable, its DTFT is defined on the basis of Eq. (4.25) as

$$DTFT\{x[n]\} = X(e^{j\omega}) = \sum_{n=-\infty}^{\infty} x[n]e^{-j\omega n} \tag{7.15}$$

and the *inverse DTFT* (on the basis of Eq. (7.14)) as

$$x[n] = \frac{1}{2\pi}\int_{-\pi}^{\pi} X(e^{j\omega})e^{j\omega n}d\omega. \tag{7.16}$$

Like the frequency response function, the Discrete-time Fourier transform is a continuous function of frequency and is periodic in frequency with period 2π. As for the CTFT, the graph of $|X(e^{j\omega})|$ versus ω is the *magnitude spectrum* of $x[n]$ and the graph of $\angle X(e^{j\omega})$ versus ω is the *phase spectrum* of $x[n]$. Note that sequences that are not absolutely summable may have a DTFT if they are square summable.

Properties of the DTFT

The DTFT is a linear operator. Furthermore it is easy to show that if $x[n]$ is real, then $X^*(e^{j\omega}) = X(e^{-j\omega})$. Consequently, $|X(e^{j\omega})|$ will be an even function of frequency, whereas $\angle X(e^{j\omega})$ will be an odd function of frequency. Likewise, the $DTFT\{x[-n]\}$ $= X^*(e^{j\omega})$. Using this result one may show that the DTFT of the signal $\frac{1}{2}(x[n] +$

$x[-n]$), which is an even function of n, equals $\text{Re}\{X(e^{j\omega})\}$, whereas the DTFT of the odd function $\frac{1}{2}(x[n] - x[-n])$ equals $\text{Im}\{X(e^{j\omega})\}$.

It is trivial to show that

$$DTFT\{\delta[n]\} = 1. \tag{7.17}$$

If $y[n] = x[n - n_0]$, then

$$Y(e^{j\omega}) = \sum_{n=-\infty}^{\infty} x[n - n_0]e^{-j\omega(n-n_0)}e^{-j\omega n_0} = e^{-j\omega n_0}X(e^{j\omega}). \tag{7.18}$$

From Eq. (7.15) it is apparent that $DTFT\{e^{j\omega_0 n}x[n]\} = X(e^{j(\omega-\omega_0)})$. Consequently, using this result and Euler's relationship, $DTFT\{x[n]\cos[\omega_0 n]\} = \frac{1}{2}\{X(e^{j(\omega-\omega_0)}) + X(e^{j(\omega+\omega_0)})\}$. The corresponding result for multiplication by $\sin[\omega_0 n]$ is presented in Table 7.1. Some other properties of the DTFT, and transforms of some common signals, are given in Tables 7.1 and 7.2, respectively.

Example 7.4   Calculating the DTFT   The signal $x[n]$ is defined as

$$x[n] = \begin{cases} 2, & |n| \leq 2 \\ 0, & otherwise \end{cases}.$$

Its DTFT is

$$X(e^{j\omega}) = \sum_{n=-\infty}^{\infty} x[n]e^{-j\omega n} = \sum_{n=-2}^{2} 2e^{-j\omega n} = 2[e^{j2\omega} + e^{j\omega} + 1 + e^{-j\omega} + e^{-j2\omega}]$$

$$= 4[\cos(\omega) + \cos(2\omega) + \tfrac{1}{2}].$$

TABLE 7.1   Properties of the Discrete-Time Fourier Transform

Property	Time Domain	Frequency Domain
Linearity	$a\, x_1[n] + b\, x_2[n]$	$aX_1(e^{j\omega}) + bX_2(e^{j\omega})$
Time shift	$x[n - n_0]$	$e^{-j\omega n_0}X(e^{j\omega})$
Time reversal	$x[-n]$	$X(e^{-j\omega})$
Modulation	$x[n]e^{-j\omega_0 n}$	$X(e^{j(\omega-\omega_0)})$
Conjugation	$x^*[n]$	$X^*(e^{-j\omega})$
Convolution	$x[n] * y[n]$	$X(e^{j\omega})Y(e^{j\omega})$
Multiplication	$x[n]\, y[n]$	$\dfrac{1}{2\pi}X(e^{j\omega}) * Y(e^{j\omega})$
Frequency differentiation	$nx[n]$	$j\dfrac{dX(e^{j\omega})}{d\omega}$

TABLE 7.2 Discrete-Time Fourier Transforms of Common Signals

Signal	DTFT
$\delta[n]$	1
1	$2\pi\delta(\omega), \qquad |\omega| \le \pi$
$u[n]$	$\dfrac{1}{1-e^{-j\omega}} + \pi\delta(\omega), \qquad |\omega| \le \pi$
$-u[-n-1]$	$\dfrac{1}{1-e^{-j\omega}} - \pi\delta(\omega), \qquad |\omega| \le \pi$
$a^n u[n]$	$\dfrac{1}{1-ae^{-j\omega}}, \qquad |a| < 1$
$(n+1)a^n u[n]$	$\dfrac{1}{(1-ae^{-j\omega})^2}, \qquad |a| < 1$
$\sin[\omega_0 n]$	$-j\pi[\delta(\omega-\omega_0) - \delta(\omega-\omega_0)], \qquad |\omega_0|, |\omega| \le \pi$
$\cos[\omega_0 n]$	$\pi[\delta(\omega-\omega_0) + \delta(\omega-\omega_0)], \qquad |\omega_0|, |\omega| \le \pi$
$x[n]\sin[\omega_0 n]$	$\frac{j}{2}[X(e^{j(\omega+\omega_0)}) - X(e^{j(\omega-\omega_0)})]$
$x[n]\cos[\omega_0 n]$	$\frac{1}{2}[X(e^{j(\omega+\omega_0)}) + X(e^{j(\omega-\omega_0)})]$

Example 7.5 *DTFT of a pulse*   Define a pulse function, $P_{2q}[n]$, as

$$P_{2q}[n] = \begin{cases} 1, & |n| \le q \\ 0, & otherwise \end{cases}.$$

Its DTFT may be found by considering the function $x[n] = P_{2q}[n-q]$, from which $P_{2q}(e^{j\omega}) = e^{j\omega q}X(e^{j\omega})$. Thus

$$X(e^{j\omega}) = \sum_{n=0}^{2q} e^{-j\omega n} = \left[\frac{1-e^{-j\omega(2q+1)}}{1-e^{-j\omega}}\right] = \frac{e^{-j\omega\left(\frac{2q+1}{2}\right)}}{e^{-j\frac{\omega}{2}}}\left[\frac{e^{j\omega\left(\frac{2q+1}{2}\right)} - e^{-j\omega\left(\frac{2q+1}{2}\right)}}{e^{j\frac{\omega}{2}} - e^{-j\frac{\omega}{2}}}\right]$$

$$= e^{-j\omega q}\left[\frac{\sin([q+1/2]\omega)}{\sin(\omega/2)}\right].$$

Consequently,

$$P_{2q}(e^{j\omega}) = \left[\frac{\sin([q+1/2]\omega)}{\sin(\omega/2)}\right]. \tag{7.19}$$

Example 7.6 *DTFT of a constant*   Because of its periodicity, the DTFT $X(e^{j\omega}) = 2\pi\delta(\omega)$, $-\pi \le \omega < \pi$, comprises a train of impulses in frequency occurring at the frequencies $\pm 2k\pi$. By Eq. (7.16), the corresponding sequence is

$$x[n] = \frac{1}{2\pi}\int_{-\pi}^{\pi} 2\pi\delta(\omega)e^{j\omega n}d\omega = 1, \qquad \forall n.$$

Therefore, if $x[n] = A_0$, where A_0 is a constant, then $DTFT\{A_0\} = 2\pi A_0\delta(\omega)$.

Example 7.7 DTFT of a periodic signal   The periodic signal of Fig. 7.3 was analyzed previously and shown to have the Discrete Fourier series representation $\tilde{x}[n]$ $= 4 + 2e^{j\frac{\pi}{2}n} + 2e^{j\frac{3\pi}{2}n}$. Although the complex exponential function is not absolutely summable, we may define a generalized DTFT for this signal by utilizing the DTFT of a constant and the frequency shift associated with multiplication of a time signal by a complex exponential function. Thus, $DTFT\{Ae^{jk\omega_0 n}\} = 2\pi A\delta(\omega - k\omega_0)$, $|\omega| \le \pi$, and therefore, $DTFT\{\tilde{x}[n]\} = \sum_{k=0}^{N-1} 2\pi \tilde{X}[n]\delta(\omega - k\omega_0) = 8\pi\delta(\omega) + 4\pi\delta[\omega - (\pi/2)]$ $+ 4\pi\delta[\omega - (3\pi/2)]$.

7.5 PARSEVAL'S RELATION FOR DT NONPERIODIC SIGNALS

The derivation of Parseval's Relation for this case parallels that for CT nonperiodic signals (Eq. (5.38)) and only the result is presented here. For a nonperiodic, deterministic, discrete-time signal, $x[n]$, having a DTFT,

$$\sum_{k=-\infty}^{\infty} |x[k]|^2 = \frac{1}{2\pi}\int_{-\pi}^{\pi} |X(e^{j\omega})|^2 d\omega. \qquad (7.20)$$

As in the CT case, $|X(e^{j\omega})|^2$ is interpreted as a *deterministic energy spectral density* function for the signal, $x[n]$.

Since power is energy per unit time, for a bounded finite-length signal, $x[n]$, $0 \le n \le N - 1$, one may define its *deterministic power spectral density (PSD)* function as

$$P_X(e^{j\omega}) = \frac{|X(e^{j\omega})|^2}{N}. \qquad (7.21)$$

Example 7.8 PSD of DT biomedical data   Because the data of Fig. 7.1 are not truly periodic, it is more appropriate to analyze these data using the DTFT instead of the DFS. Table 7.3 presents the power spectral density function determined from direct calculation of Eq. (7.15) at 12 frequencies. Note that the mean level was removed from the data before the calculations were performed; therefore, $X(e^{j0}) = 0$. There is a large peak at 0.5236 rad/sample, which is equivalent to 0.0833 cycles/sample, or a period of 12 samples (i.e., breaths). This spectral peak reflects the major oscillatory component of the data. Note also the presence of smaller peaks at other frequencies (especially at the fourth harmonic of 0.5236), reflecting the fact that the data are not a pure sine wave.

TABLE 7.3 Power Spectral Density Function of the Data of Fig. 7.1

Frequency	Power
0	0
0.2618	3.222
0.5236	14.34
0.7854	7.921
1.0472	1.043
1.3090	5.625
1.5708	0.1229
1.8326	0.3210
2.0944	10.17
2.3562	0.2919
2.6180	0.4850
2.8798	0.00447

There is an important relationship between the PSD and the autocorrelation function, the latter defined in Chapter 2, Eq. (2.35), (for $x[n]$ a real-valued signal) as

$$R_x[m] = \frac{1}{N} \sum_{k=0}^{N-1} x[k + m]x[k].$$

In general, $x[n]$ may also be complex-valued and Eq. (2.35) must be revised by replacing one of the terms in the summation, say $x[k]$, by its complex conjugate. Therefore, the DTFT of $R_x[m]$ is

$$DTFT\{R_x[m]\} = \sum_{m=-\infty}^{\infty} \left[\frac{1}{N} \sum_{k=0}^{N-1} x^*[k]x[m + k] \right] e^{-j\omega m}$$

$$= \frac{1}{N} \sum_{k=0}^{N-1} x^*[k]e^{j\omega k} \sum_{m=-\infty}^{\infty} x[m + k]e^{-j\omega(m+k)}.$$

The two summations may be recognized as the DTFT of $x[k]$ and its complex conjugate. Thus

$$DTFT\{R_x[m]\} = \frac{1}{N}|X(e^{j\omega})|^2 = P_x(e^{j\omega}). \tag{7.22}$$

Equation (7.22) expresses a fundamental and very important relationship between the memory properties of a signal, which are reflected in $R_x[m]$, and its deterministic power spectral density function. Essentially this equation states that the power at each frequency is a reflection of the magnitude of the sinusoidal component at that frequency in its autocorrelation function.

The autocorrelation function of Eq. (2.35) defines the average behavior of $x[k + m]$ relative to $x[k]$ for a given time difference (or lag), m. Given two functions, $x[n]$, $y[n]$, both defined on $0 \le n \le N - 1$, one may define the average behavior of $y[k + m]$ relative to $x[k]$ in an analogous manner. That is, the *cross-correlation function*,

$$R_X[m] = \frac{1}{N} \sum_{k=0}^{N-1} y[k + m]x^*[k]$$

reveals the extent to which temporal variations in $x[n]$ are reflected in $y[n]$ after some time lag, m. Note that $R_{XY}[m]$ is not necessarily an even function; therefore, it must be calculated for negative values of m. One may also define a *cross-covariance function* between $x[n]$ and $y[n]$ as

$$C_{XY}[m] = \sum_{k=0}^{N-1} (y[k + m] - \bar{y})(x[k] - \bar{x})^* = R_{XY}[m] - \bar{y}\bar{x}^*.$$

The DTFT of $R_{XY}[m]$ is known as the *cross-spectral density function* (or simply, cross-spectrum), $P_{XY}(e^{j\omega})$.

7.6 OUTPUT OF AN LSI SYSTEM

Given an LSI system with input $x[n]$, impulse response $h[n]$, and output $y[n]$, let $x[n] = Ae^{j\omega_0 n}$. Then, since $e^{j\omega_0 n}$ is an eigenfunction of any LSI system, $y[n] = H(e^{j\omega_0})Ae^{j\omega_0 n}$. Now if $x[n]$ is the summation of complex exponential functions given by $x[n] = (1/2\pi)\int_{-\pi}^{\pi} X(e^{j\omega})e^{j\omega n}d\omega$, then the output must be $y[n] = (1/2\pi)\int_{-\pi}^{\pi} H(e^{j\omega}) X(e^{j\omega})e^{j\omega n}d\omega$. Since by the definition of the DTFT, $y[n] = (1/2\pi)\int_{-\pi}^{\pi} Y(e^{j\omega})e^{j\omega n}d\omega$, equating these two results yields

$$Y(e^{j\omega}) = H(e^{j\omega})X(e^{j\omega}). \tag{7.23}$$

Note that since we know that $y[n] = h[n] * x[n]$, this result establishes that

$$DTFT\{x[n] * h[n]\} = H(e^{j\omega})X(e^{j\omega}),$$

which is true for any $x[n]$ and $h[n]$ that have DTFTs.

Based on Eq. (7.23), we may derive an expression for the power spectrum of the output of the LSI system. Thus

$$P_y(e^{j\omega}) = \frac{|Y(e^{j\omega})|^2}{N} = \frac{|H(e^{j\omega})X(e^{j\omega})|^2}{N} = |H(e^{j\omega})|^2 P_x(e^{j\omega}) \tag{7.24}$$

Example 7.9 Ideal DT filter The frequency response of an ideal DT lowpass filter is shown in Fig. 7.4. The impulse response of this filter may be determined from Eq. (7.16) to be

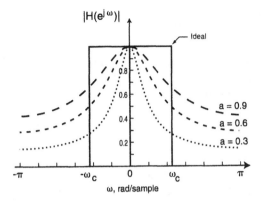

FIGURE 7.4. Frequency response of an ideal DT lowpass filter with cutoff frequency ω_c (solid line) and of a first-order lowpass filter for various values of the parameter a (dashed lines), normalized to unity gain at $\omega = 0$.

$$h[n] = \frac{1}{2\pi}\int_{-\pi}^{\pi} H(e^{j\omega})e^{j\omega n}d\omega = \frac{1}{2\pi}\int_{-\omega_c}^{\omega_c} e^{j\omega n}d\omega = \frac{\omega_c}{\pi}\frac{\sin[\omega_c n]}{\omega_c n}, \quad \forall n.$$

Clearly this (and every ideal) filter is noncausal.

Example 7.10 Causal DT filter A first-order DT filter has the impulse response $h[n] = a^n u[n]$, $|a| < 1$. Its frequency response is

$$H(e^{j\omega}) = \sum_{n=-\infty}^{\infty} h[n]e^{-j\omega n} = \sum_{n=0}^{\infty} a^n e^{-j\omega n} = \sum_{n=0}^{\infty}(ae^{-j\omega})^n = \frac{1}{1 - ae^{-j\omega}}. \quad (7.25)$$

This lowpass frequency response is also shown in Fig. 7.4 for three values of the parameter a. The reader can demonstrate that the frequency response of a system having the impulse response $h'[n] = (-a)^n u[n]$, $|a| < 1$, is $H'(e^{j\omega}) = 1/(1 + ae^{-j\omega})$. This latter frequency response has maxima at $\omega = \pm\pi$ and a minimum at zero; therefore, it represents a highpass filter.

Example 7.11 Digital filter design by impulse invariance One approach to designing a digital filter is to let the unit-pulse response of the digital filter equal the values of the impulse response of a CT (analog) filter at times $t = nt_s$, where n is an integer and t_s is a basic sampling interval. This approach is known as impulse invariance design and it will be studied in detail in Chapter 8. Here we shall examine the design of a simple, first-order, digital filter having the unit-pulse response, $h_D[n]$, based on sampling the CT impulse response $h_A(t) = a_1 e^{-a_2 t}u(t)$. Thus, let $h_D[n] = bh_A(nt_s)$, where b is a constant to be determined. Now $H_A(\Omega) = a_1/(a_2 + j\Omega)$, which is the frequency response of a CT lowpass filter with a cutoff frequency equal to a_2 rad/s. For the DT system, $h_D[n] = bh_A(nt_s) = ba_1(e^{-a_2 t_s})^n u[n]$. By Example 7.10, its frequency response is

$$H_D(e^{j\omega}) = \frac{ba_1}{1 - e^{-a_2 t_s} e^{-j\omega}}.$$

Let us choose $b = (1 - e^{-a_2 t_s})/a_1$ so that the magnitude of the DT frequency response is unity at zero frequency. The frequency responses of both filters are shown in Fig. 7.5. Notice that the resulting digital filter is also a lowpass filter, but there is not a simple relationship between the DT frequency response and the CT frequency response. Furthermore, the magnitude of the former never approaches zero unless $t_s \ll 1$. This result is a consequence of the periodicity of the DT Fourier transform. In effect, it is impossible (by this method) to map all of the CT frequencies, $0 \le \Omega \le \infty$, into the DT frequency range $0 \le \omega \le \pi$. (The specific relationship between Ω and ω will be explored later in this chapter.) Although the impulse invariance design method has this limitation, Chapter 8 will demonstrate that under the right conditions it can be a viable method for designing digital filters.

Example 7.12 Frequency response via the DTFT   Savitsky-Golay smoothing of data is used commonly in analytical chemistry. The five-point version of this filter is given by the relationship

$$y[n] = -0.0857\, x[n] + 0.3429\, x[n-1] + 0.4856\, x[n-2] + 0.3429\, x[n-3]$$
$$- 0.0857\, x[n-4], \tag{7.26}$$

where the $x[n]$ may represent any sequence of data values from an analyzer and n may represent, for example, a series of wavelengths at which light absorption is tested. One may use the DTFT to characterize the frequency response of this filter. By Eq. (7.23), the frequency response may be determined by testing the filter with a known input and dividing the DTFT of the resulting output by that of the input. For example, let $x[n] = P_5[n-2]$. The resulting output, $y[n]$, can be calculated either from Eq. (7.26) or by convolution (Fig. 7.6(a)). Thus

$$Y(e^{j\omega}) = \sum_{n=0}^{\infty} y[n] e^{-j\omega n} = \sum_{n=0}^{8} y[n] e^{-j\omega n}$$

$$= -0.0857\, e^{-j\omega(0)} + 0.2572\, e^{-j\omega} + 0.7428\, e^{-j2\omega} + 1.0857\, e^{-j3\omega} + 1.0000\, e^{-j4\omega}$$

$$+ 1.0857\, e^{-j5\omega} + 0.7428\, e^{-j6\omega} + 0.2572\, e^{-j7\omega} - 0.0857\, e^{-j8\omega}. \tag{7.27}$$

The DTFT of the input is

$$X(e^{j\omega}) = 1 + e^{-j\omega} + e^{-j2\omega} + e^{-j3\omega} + e^{-j4\omega}. \tag{7.28}$$

The frequency response of the five-point Savitsky-Golay smoothing filter is

$$H(e^{j\omega}) = \frac{Y(e^{j\omega})}{X(e^{j\omega})},$$

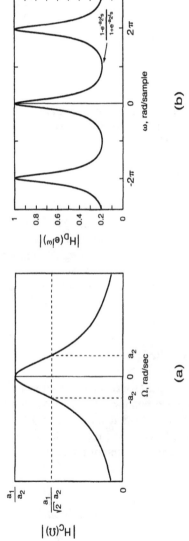

FIGURE 7.5. (a) Frequency response of a CT first-order lowpass filter with cutoff frequency a_2 and impulse response $h_C(t)$. (b) Normalized frequency response of a digital filter whose unit-pulse response is defined to be $h_D[n]$ $= h_C(nt_s)$, where t_s is the reciprocal of the sampling frequency.

247

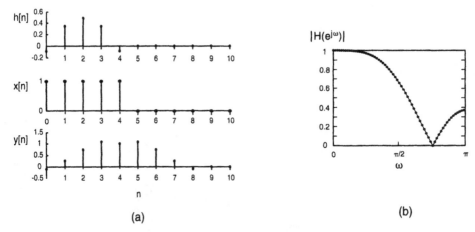

FIGURE 7.6. (a) Impulse response (h[n]) for a five-point Savitsky-Golay filter (determined by letting $x[n] = \delta[n]$ in Eq. (7.26)), test input (x[n]), and resulting output (y[n]) signals. (b) Actual frequency response of the filter (solid line) and estimate of the frequency response (circles) obtained by dividing the DTFT of y[n] by the DTFT of x[n]

which may be calculated at various frequencies (circles, Fig. 7.6(b)). Also shown in the figure is the actual frequency response calculated directly from the DTFT of $h[n]$ (solid line). Obviously the two calculations are superimposable. (A note of caution is appropriate here: This method of estimating the frequency response deteriorates if the output signal contains noise (or if the *measured* input signal contains noise that does not pass through the system). In such cases the method may still provide a starting point for estimating the frequency response.)

7.7   RELATION OF DFS AND DTFT

There is a relationship between the Discrete Fourier series and the DTFT that is the equivalent of the relationship between the CT Fourier series and the CTFT that was discussed in Chapter 5. As before, it is a relationship between the Fourier transform of a finite-length, nonperiodic function and the Fourier Series of a periodic function that is constructed by replicating the non-periodic function, as shown in Fig. 7.7. Note that

$$x[n] = \begin{cases} \tilde{x}[n], & 0 \le n \le N-1 \\ 0, & otherwise, \end{cases}$$

where $N = 6$ in the example of Fig. 7.7 and ω_0 for the periodic function is $2\pi/N$. The DTFT of $x[n]$ is $X(e^{j\omega}) = \sum_{n=0}^{N-1} x[n]e^{-j\omega n}$, whereas the DFS coefficients for the periodic extension of $x[n]$ are

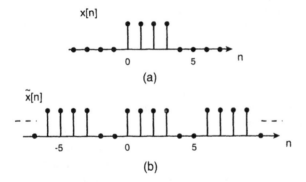

FIGURE 7.7. (a) A finite-duration DT signal. (b) The periodic signal constructed by replicating x[n] modulo 6.

$$\tilde{X}[k] = \sum_{n=0}^{N-1} \tilde{x}[n]e^{-j\omega n}\Bigg|_{\omega=2\pi k/N} = X(e^{j\omega})\big|_{\omega=2\pi k/N}, \qquad 0 \le k \le N-1. \quad (7.29)$$

That is, the DFS coefficients equal the DTFT of the nonperiodic function evaluated at integer multiples of ω_0, the fundamental frequency of the periodic signal constructed by replicating the nonperiodic signal.

This relationship has a graphical interpretation. Consider a complex number, z, and the two-dimensional complex plane defined by the orthogonal axes, Re$\{z\}$ and Im$\{z\}$ (Fig. 7.8). Let $z = e^{j\omega}$, so that $|z| = 1$, $\measuredangle z = \omega$. Then letting the frequency vary over the range $-\pi \le \omega \le \pi$ is equivalent to letting z traverse the unit circle in the

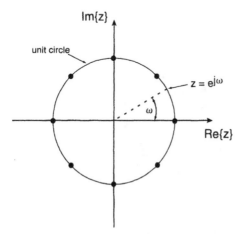

FIGURE7.8. The unit circle in the z-plane, showing (dots) the locations $z = e^{j\omega}$ at which the Discrete Fourier series coefficients for $\tilde{x}[n]$ are equal to the DTFT of x[n] for N = 8.

complex plane (also called the z-plane). *Consequently, one may consider that the DTFT, $X(e^{j\omega})$, is a function that is evaluated at points around the unit circle of the z-plane.* The DFS coefficients are equal to $X(e^{j\omega})$ evaluated at specific values of frequency—that is, integer multiples of $2\pi/N$—and therefore at specific points on the unit circle that are specified by $z = e^{j\frac{2\pi}{N}k}$. Thus there are N points, equally spaced around the unit circle, at which the value of the DTFT specifies the DFS coefficients (and vice versa). These points are indicated in Fig. 7.8 for $N = 8$. Note that $\tilde{X}[0]$ is always evaluated at $\omega = 0$ (i.e., $z = 1$).

Example 7.13   Figure 7.2(a) presented a periodic signal whose Discrete Fourier series was analyzed earlier in this chapter. The DFS coefficients are given by Eq. (7.6) and are plotted in Fig. 7.2(b). Consider the nonperiodic signal formed by taking one cycle of the periodic signal over the range $-5 \le n \le 4$. It is easy to show that

$$X(e^{j\omega}) = \sum_{n=0}^{4} e^{-j\omega n} = e^{-j2\omega}\frac{\sin(5\omega/2)}{\sin(\omega/2)}.$$

The magnitude of this DTFT is graphed in Fig. 7.9(a). The solid circles indicate the values $|\tilde{X}[k]|$ versus $k(2\pi/N)$ from Eq. (7.6). Figure 7.9(b) indicates the locations on the unit circle at which $X(e^{j\omega})$ equals the values of the DFS coefficients.

7.8   WINDOWING

Windowing is a type of amplitude modulation in which a signal, $x(t)$ or $x[n]$, is multiplied by a finite-duration signal, $w(t)$ or $w[n]$, for the purpose of: (1) forcing the resulting signal to be zero outside of a chosen range of n, and; (2) controlling the

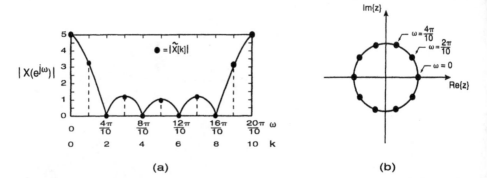

FIGURE 7.9. (a) Magnitude of the DTFT of the pulse signal of Example 7.13 (solid line) and magnitudes of the DFS coefficients (dots) of the pulse train with period 10 that is a periodic extension of the pulse (Fig. 7.2(a)). (b) Locations on the unit circle corresponding to the frequencies at which $\tilde{X}[k] = X(e^{j\omega})$.

abruptness of the transition to zero. A continuous-time example was presented in Fig. 5.12, in which an infinite-duration sine wave was multiplied by a unit-pulse function to create a finite-duration sinusoidal signal. When used in a windowing sense, the unit-pulse function is known also as a *rectangular window*. Because multiplication in the time domain is equivalent to convolution in the frequency domain, the effect of truncating an infinite-duration sine wave by a rectangular window is to transform its Fourier transform from one having only impulses (Fig. 5.12(c)) to one in which the Fourier transform of the window function (Fig. 5.12(b)) is replicated and centered at the locations of the original impulses (Fig. 5.12(d)). If the signal being windowed has a more complex Fourier transform, then the Fourier transform after windowing will be more complex *but it always will extend over a wider frequency range than that of the unwindowed signal*. Note from Fig. 5.12(b) that the width of the main lobe of the transform of the pulse function decreases as the pulse width, a, increases. Consequently, in the limit as pulse width increases, the transform of the windowed function approaches that of the unwindowed function.

A typical application of windowing is to represent a real-world signal that has been observed for a finite time as the product of a signal of infinite duration and a windowing function. Since one cannot observe an infinite amount of data for a physical signal, a real-world finite-duration signal has implicitly been multiplied by a rectangular window that is as long as the finite-duration signal itself. Although the underlying infinite-duration signal will be unknown, it is important to recognize that the Fourier transform of the finite-duration signal differs from that of the infinite-duration signal due to windowing. This effect may be important when estimating the energy spectral density (i.e., the squared magnitude of the Fourier transform) of a signal, as discussed below.

The abrupt onset and termination of the rectangular window in the time domain have undesirable consequences in the frequency domain. Recall that abrupt changes in a signal imply greater content of high-frequencies in its Fourier transform. Therefore, because the Fourier transform of the windowed signal equals the convolution of the Fourier transforms of the unwindowed signal and the window function, the frequency content of the windowed signal will extend to higher frequencies than that of the unwindowed signal. Furthermore, a large peak in the Fourier transform of the unwindowed signal will cause large artifactual signal components at neighboring frequencies in the Fourier transform of the windowed signal because of the breadth (in frequency) of the Fourier transform of the rectangular window. In some applications, such as estimating the energy spectral density of the unwindowed signal, these distortions of the magnitude spectrum of the Fourier transform are undesirable. They may be reduced (but not eliminated) by multiplying the finite-duration signal by another window function that has less high-frequency content. Many alternative window functions have been defined that have less high-frequency content than the rectangular window and all achieve this objective by imposing a smooth (rather than abrupt) transition from unity gain to zero gain. Typically one chooses a window length equal to the data length, then multiplies the signal, $x[n]$, by the chosen window, $w[n]$, before performing the desired data analysis. Some of the more common DT window functions are described in Table 7.4 and plotted in Fig. 7.10.

TABLE 7.4 Window functions

Type	Time Domain	Frequency Domain
Rectangular	$w_R[n] = \begin{cases} 1, & 0 \le n \le N-1 \\ 0, & \text{otherwise} \end{cases}$	$W_R(e^{j\omega}) = e^{-j\omega(N-1)/2}\dfrac{\sin(\omega N/2)}{\sin(\omega/2)}$
Bartlett	$w_B[n] = \begin{cases} n/M^*, & 0 \le n \le M \\ 2 - n/M, & M+1 \le n \le N-1 \\ 0, & \text{otherwise} \end{cases}$	$W_B(e^{j\omega}) = W_R^2(e^{j\omega})/M$
Hamming	$w_M[n] = \begin{cases} 0.54 - 0.46\cos\left(\dfrac{2\pi n}{N-1}\right), & 0 \le n \le N-1 \\ 0, & \text{otherwise} \end{cases}$	$W_M(e^{j\omega}) = \frac{1}{4}W_R(e^{j(\omega-\pi/M)}) + \frac{1}{2}W_R(e^{j\omega}) + \frac{1}{4}W_R(e^{j(\omega+\pi/4)})$
Hann	$w_N[n] = \begin{cases} \dfrac{1}{2}\left[1 - \cos\left(\dfrac{2\pi N}{N-1}\right)\right], & 0 \le n \le N-1 \\ 0, & \text{otherwise} \end{cases}$	$W_M(e^{j\omega}) = \frac{1}{2}W_R(e^{j\omega}) + \frac{1}{4}[W_R(e^{j(\omega-2\pi/N)}) + W_R(e^{j(\omega+2\pi/N)})]$
Truncated Gaussian[⊕]	$w_G[n] = \begin{cases} \exp(-\frac{1}{2}[2\alpha(n-M)/(N-1)])^2, & 0 \le n \le N-1 \\ 0, & \text{otherwise.} \end{cases}$	$W_G(e^{j\omega}) = \dfrac{\sqrt{2\pi}}{2\alpha}\exp\left(-\dfrac{1}{2}[\omega/\alpha]^2\right) * W_R(e^{j\omega})$

*$M = (N-1)/2$

#This result may be written as $W_R(e^{j\omega}) = e^{-j\omega(N-1)/2}D_N(e^{j\omega})$, where D_N is the Dirichlet kernel.

⊕ α is a parameter controlling the rate of decay of $w_G[n]$.

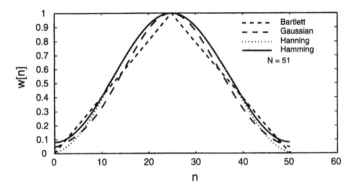

FIGURE 7.10. Time-domain graphs of various 51-point windows.

A window function is characterized in the frequency domain by three parameters (Table 7.5, Fig. 7.11) determined from the magnitude of its Fourier transform: the width of its main lobe, the peak side lobe level, and the average rate of change of magnitude at frequencies well away from the main lobe (rolloff). To discuss the importance of these properties, let $y[n] = x[n]w[n]$, so that $Y(e^{j\omega}) = X(e^{j\omega}) * W(e^{j\omega})/(2\pi)$. If $|X(e^{j\omega})|$ contains two peaks at frequencies that are separated by less than the main lobe width of $W(e^{j\omega})$, then these two peaks will be merged together in $Y(e^{j\omega})$ and may not be reliably discernible. Therefore, to be able to detect signal components that are close in frequency, one must utilize a window with a main lobe whose width is less than the difference in frequencies of the two peaks. However, even if the main lobe is narrow, if the side lobes are large or if their amplitudes do not fall quickly enough, then a signal of large amplitude at one frequency may obscure a small signal at neighboring frequencies due to the sidelobes in $W(e^{j\omega})$, a

TABLE 7.5 Properties of Some Window Functions

Type	Main Lobe Width*	Highest Sidelobe (dB)	Rolloff Rate (dB/octave)
Rectangular	$0.89\dfrac{2\pi}{N}$	−13.3	−6
Bartlett	$1.28\dfrac{2\pi}{N}$	−26.5	−12
Hann	$1.44\dfrac{2\pi}{N}$	−31.5	−18
Hamming	$1.30\dfrac{2\pi}{N}$	−43	−6
Gaussian#	$1.33\dfrac{2\pi}{N}$	−42	−6

*Half-power width.
#$\alpha = 2.5$.

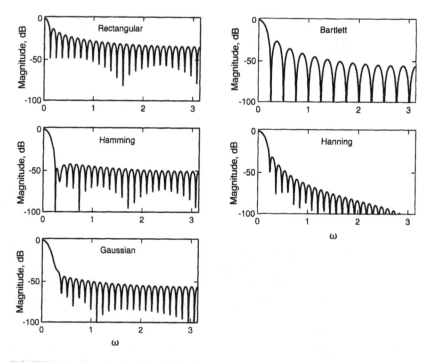

FIGURE 7.11. Magnitude of the DTFT of various window functions, all of length $N = 51$.

phenomenon known as *leakage*. (For this reason the rectangular window is almost never used for spectral estimation, even though it has the narrowest main lobe.) Reducing side lobe amplitude always requires increasing the main lobe width. Consequently, a large variety of window functions have been proposed, each displaying a particular trade-off between main lobe width and side lobe amplitude. Figures 7.10 and 7.11 show just a few of these window functions. The Hann(ing) window is a good compromise because it has a high rolloff and its main lobe width is comparable to that of other common windows. The Hamming window is also popular for spectral estimation. (See Exercises 7.20 and 7.22.)

A second application of windowing is for design of FIR filters. The method, which is based on truncating the impulse response of an IIR filter, will be discussed in Chapter 8 along with other methods of digital filter design.

Example 7.14   Windowing   Windowing must be utilized with care, as this example shows. Consider the analysis of the energy spectral density of a finite-length observation of the signal $x[n] = \sin[(2\pi/25)n]$, which has a period of $N_0 = 25$, when one has either 64 or 640 samples of the signal. In both cases a Bartlett window is applied (Fig. 7.12(a)). For $N = 64$, there are only two complete cycles of the sine wave and windowing severely distorts the sine wave, whereas the distortion is relatively less, at least for a few cycles, when $N = 640$. The Fourier transforms of the

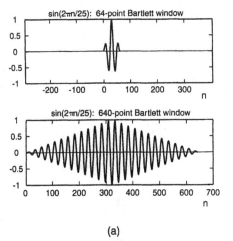

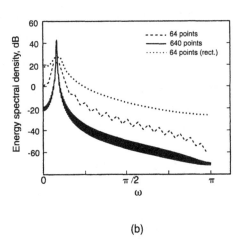

(a)   (b)

FIGURE 7.12. Effect of windowing on the energy spectral density function of a sinusoid. (a) A sine wave multiplied by either a 64-point (top) or a 640-point (bottom) Bartlett window. (b) Energy spectral density function for the two windowed sine waves (dashed, solid) and for the same sine wave windowed by a 64-point rectangular window (dotted).

windowed signals were calculated and their squared magnitudes are plotted in Fig. 7.12(b). Because the width of the main lobe of the Fourier transform of the window is inversely proportional to the window length (in time), the peak in the spectrum of the 64-point record is broader and the energy is less well localized in frequency. The energy spectrum of the 64-point signal truncated by a rectangular window is also given in the figure. Its peak is even less well defined than the peak associated with the 64-point Bartlett window.

7.9 SAMPLING

In many applications it is desired to determine the magnitude and phase spectra of a CT signal, $x_C(t)$, $0 \le t \le T_0$, when one has acquired the sequence of values of this signal, $x[n]$, $0 \le n \le N - 1$, by sampling at some constant rate $f_S = 1/T$ such that $t = nT$ and $T_0 = NT$. We will now derive an explicit relationship between the CTFT of $x_C(t)$ and the DTFT of $x[n]$, from which one may calculate the Fourier transform of the CT signal from that of the DT signal.

The CTFT and its inverse are defined as:

$$X_C(\Omega) = \int_{-\infty}^{\infty} x_C(t) e^{-j\Omega t}\, dt; \tag{7.30}$$

$$x_C(t) = \frac{1}{2\pi} \int_{-\infty}^{\infty} X_C(\Omega) e^{j\Omega t}\, d\Omega. \tag{7.31}$$

$x[n]$ has a DTFT, $X(e^{j\omega})$, from which one can evaluate $x[n]$ as

$$x[n] = \frac{1}{2\pi} \int_{-\pi}^{\pi} X(e^{j\omega}) e^{j\omega n} d\omega. \tag{7.32}$$

Now from the sampling process one can say that

$$x[n] = x_C(nT) = \frac{1}{2\pi} \int_{-\infty}^{\infty} X_C(\Omega) e^{j\Omega nT} d\Omega. \tag{7.33}$$

Following Oppenheim and Schafer (1975), Eq. (7.32) may be rewritten as

$$x[n] = x_C(nT) = \frac{1}{2\pi} \sum_{r=-\infty}^{\infty} \int_{(2r-1)\pi/T}^{(2r+1)\pi/T} X_C(\Omega) e^{j\Omega nT} d\Omega.$$

After a change of variables, in which we let $\beta = \Omega - (2\pi r/T)$, the previous equation becomes

$$x[n] = \frac{1}{2\pi} \sum_{r=-\infty}^{\infty} \int_{-\pi/T}^{\pi/T} X_C\left(\beta + \frac{2\pi r}{T}\right) e^{j\beta nT} e^{j2\pi rn} d\beta$$

$$= \frac{1}{2\pi} \int_{-\pi/T}^{\pi/T} \left[\sum_{r=-\infty}^{\infty} X_C\left(\beta + \frac{2\pi r}{T}\right) \right] e^{j\beta nT} d\beta.$$

The preceding equation is almost an expression of the inverse DTFT (Eq. (7.32)) except that the integration is not calculated over the full range $(-\pi, \pi)$. By another change of variables, $\omega = \beta T$, this equation may be written as

$$x[n] = \frac{1}{2\pi} \int_{-\pi}^{\pi} \frac{1}{T} \left[\sum_{r=-\infty}^{\infty} X_C\left(\frac{\omega}{T} + \frac{2\pi r}{T}\right) \right] e^{j\omega n} d\omega. \tag{7.34}$$

Comparing Eq. (7.34) to Eq. (7.32), one concludes that

$$X(e^{j\omega}) = \frac{1}{T} \sum_{r=-\infty}^{\infty} X_C\left(\frac{\omega}{T} + \frac{2\pi r}{T}\right), \tag{7.35}$$

Finally, since β is just a CT frequency variable, it may be replaced by a more familiar symbol so that the equation relating the CT and DT frequency variables may be written as $\omega = \Omega T$.

Eq. (7.35) is known as the *sampling theorem*. It expresses the DTFT as a summation of an infinite number of replications of the CTFT in which: (1) each replication is scaled by $1/T$ and shifted along the frequency axis by some integer multiple, r, of the sampling frequency, Ω_s, where $\Omega_s = 2\pi f_s = (2\pi/T)$, and; (2) the CT frequency variable, Ω, is set equal to the DT frequency variable, ω, scaled by the sampling frequency $f_s = (1/T)$ This process is diagrammed in Fig. 7.13. Note that the relationship between the CT and DT frequency variables implies that the frequency $\Omega = \Omega_s$ maps to the DT frequency $\omega = \Omega_s T = 2\pi$. (In terms of Hz, the sampling frequency f_s

maps to the DT frequency 1.0.) *An observation of fundamental importance is apparent from Fig. 7.13:* if the CTFT is such that $X_c(\Omega) = 0$, for $|\Omega| \geq (\pi/T) = \Omega_S/2$, then there is no overlapping of the replicates of $X_c(\Omega)$ in $X(e^{j\omega})$ and

$$X(e^{j\omega}) = \frac{1}{T}X_c\left(\frac{\omega}{T}\right), \quad -\pi \leq \omega \leq \pi. \tag{7.36}$$

In this case, $X_c(\Omega)$ may be recovered completely from $X(e^{j\omega})$ through the reverse relationship

$$X_c(\Omega) = T X(e^{j\Omega T}), \quad -\frac{\Omega_S}{2} \leq \Omega \leq \frac{\Omega_S}{2}. \tag{7.37}$$

That is, when the CTFT of a signal $x(t)$ is zero for all frequencies above some $f = f_1$ (or $\Omega = \Omega_1$), then the CTFT of $x(t)$ is recoverable from the DTFT of its samples, $x[n]$, as long as the sampling frequency is higher than $2f_1$ (or $2\Omega_1$). This condition is

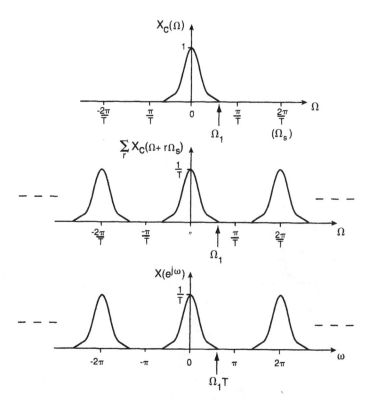

FIGURE 7.13. The effect of sampling on the Fourier spectrum. Top: Fourier transform of a band-limited CT signal, $x(t)$. Middle: Replication of the CTFT, with replicates centered at integer multiples of the sampling frequency, that is implied by the sampling theorem. Bottom: DTFT of the sampled signal, $x[n]$.

called the *Nyquist criterion* and the minimum sampling frequency that permits recovery of the CTFT (i.e., $2f_1$ or $2\Omega_1$) is known as the *Nyquist frequency*.

Figure 7.14 indicates the overlapping of replicates of $X_C(\Omega)$ that occurs when the CTFT has nonzero magnitude at frequencies greater than one-half of the sampling frequency. This phenomenon, known as *aliasing*, produces a DTFT that is distorted relative to the CTFT. The CTFT cannot be uniquely determined from the DTFT in the presence of aliasing unless the range of nonzero magnitudes of the CTFT is finite and is known a priori (which is not usual).

Since one can recover the CTFT from the DTFT in the absence of aliasing, in theory one also should be able to recover the signal $x_C(t)$. Furthermore, since the DTFT is derived from the sampled signal, $x[n]$, then $x_C(t)$ should be derivable from $x[n]$. Considering that $X_C(\Omega) = 0$, $|\Omega| \geq \pi/T$, then $x_C(t)$ may be expressed as

$$x_C(t) = \frac{1}{2\pi} \int_{-\pi/T}^{\pi/T} X_C(\Omega)e^{j\Omega t}d\Omega = \frac{1}{2\pi} \int_{-\pi/T}^{\pi/T} T X(e^{j\Omega T})e^{j\Omega t}d\Omega$$

$$= \frac{T}{2\pi} \int_{-\pi/T}^{\pi/T} \left[\sum_{k=-\infty}^{\infty} x_C(kT)e^{-j\omega kT} \right] e^{j\Omega t}d\Omega$$

$$= \sum_{k=-\infty}^{\infty} x_C(kT) \left[\frac{T}{2\pi} \int_{-\pi/T}^{\pi/T} e^{-j\Omega(t-kT)}d\Omega \right].$$

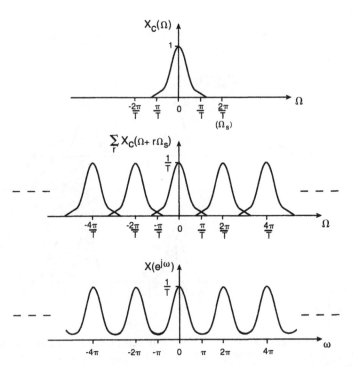

FIGURE 7.14. Same as Fig. 7.13 except that the sampling frequency is less than the Nyquist frequency.

Recognizing that the term in the brackets is a sinc function, we find that

$$x_c(t) = \sum_{k=-\infty}^{\infty} x_c(kT) \frac{\sin\left[\frac{\pi}{T}(t - kT)\right]}{\frac{\pi}{T}(t - kT)}, \qquad \forall t. \qquad (7.38)$$

Figure 7.15 diagrams the reconstruction process implied by Eq. (7.38). Each sinc function is displaced so that it is centered at one of the original sampling times, $t = kT$. Furthermore, *each sinc function evaluates to zero at every other sample time.* This result implies, however, that at all times between sampling times every DT sample contributes to the reconstruction of the CT signal. Consequently this process may be computationally intensive to implement.

Example 7.15 Sampling The CT signal $x_c(t) = \cos(50t) + \cos(70t)$ is sampled every 31.416 msec to produce the DT signal $x[n]$. Assume that both $x_c(t)$ and $x[n]$ are infinitely long. The DTFT of $x[n]$ may be determined by applying the sampling theorem to the CTFT of $x_c(t)$. Since $T = 0.031416$ sec, the sampling frequency is

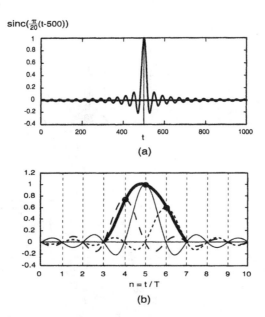

FIGURE 7.15. Reconstruction of a CT signal, $x(t)$, from its unaliased samples, $x[n]$. (a) The sinc function of the reconstruction formula (Eq. (7.38)), centered at $t = 500$ ($T = 1$). (b) Partial reconstruction of $x(t)$ based on three samples of $x[n]$ and the corresponding sinc functions. $x[4]$, $x[5]$, $x[6]$ are shown as heavy dots. Each corresponding sinc function is zero at every other sampling time. Between sampling times the CT function (heavy line) is the sum of the three sinc functions.

$\Omega_S = 2\pi(1/.031416) = 200$ rad/s. Since $X_C(\Omega) = \pi[\delta(\Omega - 50) + \delta(\Omega + 50) + \delta(\Omega - 70) + \delta(\Omega + 70)]$, and $\Omega T = 2\pi(\Omega/\Omega_S)$, then

$$X(e^{j\Omega T}) = \frac{1}{T} \sum_{r=-\infty}^{\infty} X_C(\Omega + r\Omega_S)$$

$$= \frac{1}{.031416} \sum_{r=-\infty}^{\infty} \pi[\delta(\Omega - 50 + 200r) + \delta(\Omega + 50 + 200r) + \delta(\Omega - 70 + 200r)$$

$$+ \delta(\Omega + 70 + 200r)].$$

These functions are plotted in Fig. 7.16. The solid and dashed impulses indicate the locations along the ω axis of the replicates of the primary features of $X_C(\Omega)$.

Example 7.16   Nyquist frequency The signal $x(t) = \sin(1256t)$ is to be sampled in order to calculate its energy spectral density function. At what sampling frequency (f_S), and for what finite length of time (T_0), should it be sampled (assuming that sampling starts at $t = 0$)?

Solution: The theoretical CTFT of $x(t)$ is $X(\Omega) = j\pi[\delta(\Omega + 1256) - \delta(\Omega - 1256)]$. Blindly applying the Nyquist criterion, we would say that the sampling frequency

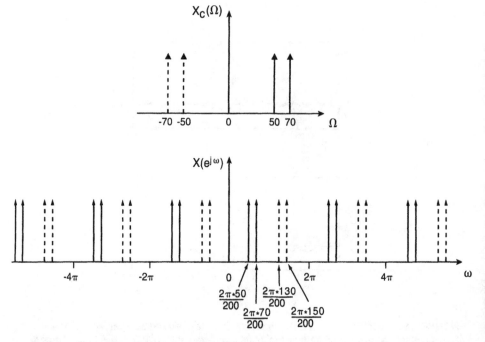

FIGURE 7.16. The CTFT (top) of a signal comprising two sine waves with frequencies 50 and 70 rad/s, and the DTFT (bottom) of the signal obtained by sampling at 200 rad/s. Solid and dashed impulse functions indicate where the replicates of the original impulses of $X_C(\Omega)$ appear in $X(e^{j\omega})$.

should be greater than $2 * 1256 = 2512$ rad/s. Since T_0 is to be finite, however, the signal that will be sampled is $y(t) = x(t) w_R(t)$, where $w_R(t)$ is a rectangular window of length T_0. Therefore, we expect that the CTFT of $y(t)$ will extend to frequencies above 1256 rad/s. Since the half-width of the main lobe of $W_R(\Omega)$ is $2\pi/T_0$, to a first approximation the CTFT will be nonzero for $1256 - 2\pi/T_0 \le \omega \le 1256 + 2\pi/T_0$. If sampling occurs at 400 Hz (equivalent to 2512 rad/s), we cannot prevent aliasing. If sampling occurs at 500 Hz, then one needs to ensure that $1256 + (2\pi/T_0) < 250(2\pi)$, or $T_0 > 0.020$ s. Another criterion related to T_0 is the amount of artifactual "spread" in frequency we find acceptable in the energy spectrum. This smearing of the true spectrum is due to the finite width of the main lobe of $W_R(\Omega)$. (See Fig. 5.12(b)). If we wish to localize the frequency of the peak energy in $x(t)$ to within some uncertainty, say 5 Hz, then the half-power width of the window should be no more than $2\pi \cdot 5 = 31.41$ rad/s. Using Table 7.5, this condition implies that $0.89(2\pi/T_0) \le 31.41$, or $T_0 \ge 0.1782$ sec. If we let $T_0 = 0.200$ sec and sample at 500 Hz (i.e., $T = 0.002$ sec), then we will acquire 100 samples of $y(t)$. The DTFT of $y[n]$ will exhibit negligible aliasing and the frequency resolution relative to ω will be approximately $\Delta\omega = (\Delta\Omega)T = 0.89(2\pi/T_0)T = 0.89(2\pi/0.2)(0.002) \approx 0.0178\pi$. Note that because of windowing the energy of the signal will be "spread out" across frequencies near $\Omega_0 = 1256$, and the total energy may be estimated as the integral of the energy spectrum over $\Omega_0 - (2\pi/T_0) \le \Omega \le \Omega_0 + (2\pi/T_0)$, rather than as the height of the peak at Ω_0.

7.10   THE DISCRETE FOURIER TRANSFORM (DFT)

Just as the DT signal $x[n]$ can result from sampling a CT signal $x(t)$, the Discrete Fourier transform (DFT) of a signal $x[n]$ is a sampled version of the Discrete-Time Fourier transform (DTFT) of $x[n]$. In this case the sampling occurs at equal increments of frequency, not time. Actually the DFT is derived from the Discrete Fourier series and it is the relationship between the DFS coefficients and the DTFT, expressed in Eq. (7.29), that permits interpreting the DFT as a sampling of the DTFT.

Consider the Fourier analysis of a nonperiodic, finite-length DT signal, $x[n]$, $0 \le n \le N-1$. This signal may be inherently finite in length or may be the consequence of applying a time window to a signal of infinite duration. Its DTFT is

$$X(e^{j\omega}) = \sum_{n=-\infty}^{\infty} x[n]e^{-j\omega n} = \sum_{n=0}^{N-1} x[n]e^{-j\omega n}. \tag{7.39}$$

In general, this function cannot be expressed in closed form and one must calculate it explicitly for any frequency of interest. One approach is to evaluate $X(e^{j\omega})$ on a sufficiently dense set of frequencies that there are unlikely to be "important" variations in this function between sampled frequencies. A fairly conservative sampling is to choose the set of frequencies that are integer multiples of the lowest frequency for which one complete cycle can be represented in the available data, $x[n]$—that is, the frequency with period N, or $1/N$ cycles/sample (or $2\pi/N$ rad/sample). Note that the fastest periodic change possible in $x[n]$ has a period of two, or a frequency of 0.5 cycles/sample (i.e., π rad/sample). As long as N is not trivially small, then some integer multiple of $1/N$ will be close to 0.5. In fact, if N is small, then the effect of the implicit rectangular window in $x[n]$ will spread out any frequency content at 0.5 to nearby frequencies so the frequency sampling does not have to be exactly at 0.5.

The DTFT is periodic in ω with period 2π, and over one cycle of the DTFT there will be N frequencies that are multiples of $2\pi/N$. These frequencies are exactly the frequencies that occur in a Discrete Fourier series when a periodic function has a fundamental period of N. Consider constructing a periodic extension of the nonperiodic, finite-length signal $x[n]$ by defining

$$\tilde{x}[n] = \sum_{r=-\infty}^{\infty} x[n + rN] = x[n]_{\text{mod}(N)}, \qquad \forall n.$$

Its DFS coefficients are given by

$$\tilde{X}[k] = \sum_{n=0}^{N-1} \tilde{x}[n]e^{-j\frac{2\pi}{N}kn} = \sum_{n=0}^{N-1} x[n]e^{-j\frac{2\pi}{N}kn} = X(e^{j\omega})|_{\omega=2\pi k/N} = X\left(e^{j\frac{2\pi}{Nk}}\right), \quad (7.40)$$

where $X(e^{j\omega})$ is the DTFT of $x[n]$.

The DFS coefficients are defined for all k but are periodic with period N. Define the *Discrete Fourier transform*, $X[k]$, of the nonperiodic, finite-length signal, $x[n]$, as one cycle of the DFS coefficients—that is,

$$X[k] = \begin{cases} \tilde{X}[k] & 0 \le k \le N-1 \\ 0, & \text{otherwise.} \end{cases} \qquad (7.41)$$

By Eq. (7.40),

$$X[k] = X\left(e^{j\frac{2\pi k}{N}}\right), \qquad 0 \le k \le N-1. \qquad (7.42)$$

In other words, the DFT of $x[n]$ equals the DTFT of $x[n]$ evaluated at the N frequencies

$$\omega_k = \frac{2\pi k}{N}, \qquad 0 \le k \le N-1.$$

Because it is closely related to the DFS coefficients, the DFT has properties similar to those of the DFS coefficients; however, because the latter are periodic whereas the DFT is not, some of the properties of the DFS do not transfer directly to the DFT. Rather, they must be specified relative to modulus N (Table 7.6).

An important advantage of the Discrete Fourier transform is that its computation can be implemented very efficiently compared to calculating the summation of Eq. (7.40) directly. This efficiency is primarily a result of the uniform spacing of the frequencies ω_k and symmetries involved in evaluating the complex exponential terms.

TABLE 7.6   Properties of the Discrete Fourier Transform

Property	Time Domain	DFT
Time shift	$x[n + n_0]$	$W^{-kn_0}X[k]$
Modulation	$W^{k_0 n}x[n]$	$X[(k + k_0)_{\text{mod}\ (N)}]\ w_R[n]$

For example, consider the evaluation of

$$X[k] = X\left(e^{j\frac{2\pi k}{N}}\right) = \sum_{n=0}^{N-1} x[n]e^{-j\frac{2\pi}{N}kn},$$

which may be written as

$$X[k] = \sum_{n=0}^{N/2-1} x[2n]e^{-j\frac{2\pi}{N}k(2n)} + e^{j\frac{2\pi k}{N}} \sum_{n=0}^{N/2-1} x[2n+1]e^{-j\frac{2\pi}{N}k(2n)}. \qquad (7.43)$$

Now each of the summations in Eq. (7.43) may be recognized as an $(N/2)$-point DFT of half of the original data signal. Each of these summations in turn may be decomposed into two $(N/4)$-point DFT calculations using one-fourth of $x[n]$. This process is repeated until one achieves N/2 two-point DFT calculations. Thus the calculation of the original DFT becomes N/2 calculations of simple DFTs plus a large number of additions and multiplications by $e^{j2\pi k/N}$. On the order of $N \log(N)$ mathematical operations are required using this decomposition method compared to N^2 operations to evaluate Eq. (7.40) directly. For $N = 1024$, the decomposition method requires about 5000 mathematical operations and the direct method about 10^6! This decomposition method is an example of a *decimation-in-time Fast Fourier transform (FFT) algorithm*. There are numerous variations on this scheme, as well as decimation-in-frequency FFT methods, and other approaches based on factoring N. *Most, but not all, of these FFT algorithms require that N be an integer power of 2!* In any case, the algorithms compute much faster if this latter condition is met.

The Inverse DFT

Calculating $x[n]$ from $X[k]$ is straightforward since the DFT, $X[k]$, of the nonperiodic signal, $x[n]$, equals the DFS coefficients of the periodic extension, $\tilde{x}[n]$. First one calculates the periodic signal as

$$\tilde{x}[n] = \frac{1}{N}\sum_{k=0}^{N-1} \tilde{X}[k]e^{j\frac{2\pi}{N}kn} = \frac{1}{N}\sum_{k=0}^{N-1} X[k]e^{j\frac{2\pi}{N}kn}, \qquad (7.44),$$

then

$$x[n] = \tilde{x}[n]w_R[n], \qquad w_R[n] = \begin{cases} 1, & 0 \le n \le N-1 \\ 0, & \text{otherwise.} \end{cases} \qquad (7.45)$$

Note that Eq. (7.44) is structurally similar to Eq. (7.4). Consequently, with only slight modifications the calculation of $\tilde{x}[n]$ may be accomplished using a Fast Fourier transform algorithm also. Practical implementation of Fourier analysis for DT signals usually involves calculation of the direct and inverse Discrete Fourier transform of size N (Eqs. (7.40) and (7.44)–(7.45)) via an FFT algorithm.

Circular Convolution

Let $x_1[n]$ and $x_2[n]$ both be nonzero only for $0 \le n \le N-1$. If we form the periodic extensions of these signals, $\tilde{x}_1[n]$, $\tilde{x}_2[n]$, in the usual manner, and define a new set of DFS coefficients as the product of the DFS coefficients of these periodic functions—that is, $\tilde{X}_3[k] = \tilde{X}_1[k]\tilde{X}_2[k]$—then from Eq. (7.9) we know that

$$\tilde{x}_3[n] = \sum_{m=0}^{N-1} \tilde{x}_1[m]\tilde{x}_2[n-m]. \qquad (7.9)$$

Equation (7.9) describes the periodic convolution of two periodic sequences. Now define the nonperiodic function $x_3[n]$ by taking one cycle of $\tilde{x}_3[n]$, so that

$$x_3[n] = \tilde{x}_3[n]w_R[n] = \begin{cases} \tilde{x}_3[n], & 0 \le n \le N-1 \\ 0, & \textit{otherwise.} \end{cases} \qquad (7.46)$$

From Eq. (7.9), $x_3[n]$ may be expressed as

$$x_3[n] = \left[\sum_{m=0}^{N-1} \tilde{x}_1[m]\tilde{x}_2[n-m] \right] w_R[n] = \left[\sum_{m=0}^{N-1} x_1[m]x_2[n-m]_{\mathrm{mod}(N)} \right] w_R[n]. \qquad (7.47)$$

Equation (7.47) defines the *circular convolution* of two finite-length sequences that are both defined over the same indices, $0 \le n \le N-1$. Circular convolution differs from normal convolution because expressing $x_2[n-m]$ relative to mod(N) means that $x_2[n]$ "wraps around" and repeats itself *as if it were periodic*, as in the example below.

Example 7.17   Circular convolution   Let

$$x_1[n] = \delta[n-2], \; x_2[n] = \begin{cases} N-n, & 0 \le n \le 5 \\ 0, & \textit{otherwise.} \end{cases}$$

(Fig. 7.17). Let $x_3[n]$ be the circular convolution of these two signals. Then $x_3[1] = \sum_{m=0}^{5} \delta[m-2]x_2[1-m]_{\mathrm{mod}(6)}$.

$x_2[1-m]_{\mathrm{mod}(6)}$ is plotted in Fig. 7.17(b). $x_3[1]$ equals the value of this function when $m = 2$. Thus $x_3[1] = x_2[5] = 1$. Similarly, $x_3[2]$ is the value of $x_2[2-m]_{\mathrm{mod}(6)}$ when $m = 2$, or $x_3[2] = x_2[0] = 6$ (Fig. 7.17(c)). Finally, $x_3[n]$, which is nonperiodic, is shown in Fig. 7.17(d).

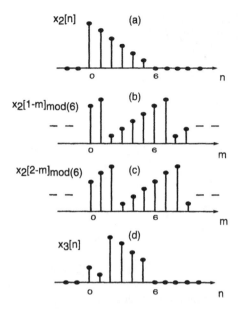

FIGURE 7.17. Demonstration of circular convolution of $\delta[n - 2]$ with the function $x_2[n]$. (a) $x_2[n]$. (b), (c) $x_2[n]$ after time reversal, replication modulo 6, and delayed one or two samples, respectively. (d) The result of the circular convolution.

From a computational viewpoint, calculation of the circular convolution of two sequences can be fast because it may be accomplished in the frequency domain by multiplying the DFTs of the two sequences using an FFT method. Circular convolution may be utilized to compute the standard linear convolution between two finite-length sequences, $x[n]$ and $y[n]$, if both signals are modified by adding N zeros to each first. That is, define

$$x_1[n] = \begin{cases} x[n], & 0 \leq n \leq N-1 \\ 0, & N \leq n \leq 2N-1 \end{cases}$$

and similarly for $y_1[n]$. Because of the zero-padding, the circular convolution of these length $2N$ sequences is equal to the standard linear convolution of the length N sequences. Therefore, we may use the DFT to calculate $Z[k] = X_1[k]\, Y_1[k]$, where $z[n] = x[n] * y[n]$.

Use of the FFT to Calculate the CTFT of a Signal x(t)

This application of the DTFT is so pervasive that the process will be summarized here. Given that it is desired to calculate the CTFT of a signal $x(t)$ that is observed on the interval $0 \leq t \leq T_0$, we may determine the CTFT at N individual frequencies after sampling $x(t)$ to obtain $x[n]$, $0 \leq n \leq N-1$. Let the sampling frequency be

$\Omega_S = 2\pi f_S = 2\pi(1/T)$ and assume that the CTFT of $x(t)$ is $X_C(\Omega)$ and the DTFT of $x[n]$ is $X(e^{j\omega})$. In general, by Eq. (7.35) we know that

$$X(e^{j\omega}) = \frac{1}{T}\sum_{r=-\infty}^{\infty}X_C\left(\frac{\omega}{T} + \frac{2\pi r}{T}\right).$$

If, however, there is no aliasing (i.e., if $|X_C(\Omega)| = 0$, $|\Omega| \geq \pi f_S = \frac{1}{2}\Omega_S$), then

$$X(e^{j\omega}) = \frac{1}{T}X_C\left(\frac{\omega}{T}\right), \qquad -\pi \leq \omega \leq \pi; \qquad (7.48a)$$

or,

$$X_C(\Omega) = TX(e^{j\Omega T}), \qquad -\pi f_S \leq \Omega \leq \pi f_S; \qquad (7.48b)$$

or,

$$X_C(2\pi f) = TX(e^{j2\pi fT}), \qquad -\tfrac{1}{2}f_S \leq f \leq \tfrac{1}{2}f_S. \qquad (7.48c)$$

Now make a periodic extension of $x[n]$ such that

$$\tilde{x}[n] = x[n]_{\text{mod}(N)} = \sum_{r=-\infty}^{\infty} x[n + rN].$$

The DFS coefficients of this periodic function may be calculated using an FFT method as

$$\tilde{X}[k] = \sum_{n=0}^{N-1} \tilde{x}[n]e^{-j\frac{2\pi}{N}kn} = X\left(e^{j\frac{2\pi k}{N}}\right),$$

and the DFT of $x[n]$ is given by one cycle of the DFS coefficients:

$$X[k] = \begin{cases} \tilde{X}[k], & 0 \leq k \leq N-1 \\ 0, & \text{otherwise.} \end{cases}$$

Note that the frequency of the k-th component of the DFS is $\omega_k = (2\pi k/N)$. Then from Eq. (7.48a),

$$X_C\left(\frac{\omega}{T}\right)\Big|_{\omega=2\pi k/N} = TX(e^{j\omega})\big|_{\omega=2\pi k/N} = T\tilde{X}[k] = TX[k].$$

Let $f_k = (k/N)f_S$ and note that $2\pi f_k = 2\pi k/NT$. Substituting this result into the previous equation yields

$$X_C(2\pi f_k) = TX(e^{j2\pi f_k T}) = TX[k], \qquad 0 \leq k \leq N-1. \qquad (7.49)$$

That is, the DFT, $X[k]$, scaled by T, equals the CTFT at the frequency $\Omega_k = (2\pi k/N)f_S$, or $f_k = (k/N)f_S$.

The deterministic power spectral density function of $x(t)$ is

$$P_X(\Omega) = \frac{|X(\Omega)|^2}{T_0} = \frac{T^2|X(e^{j\Omega T})|^2}{NT} = TP_X(e^{j\Omega T}), \tag{7.50}$$

where

$$P_X(e^{j\omega}) = \frac{|X(e^{j\omega})|^2}{N}.$$

At the specific frequencies corresponding to the DFT, combining Eqs. (7.49) and (7.50) yields

$$P_X(2\pi f_k) = \frac{|X_c(2\pi f_k)|^2}{NT} = \frac{T|X[k]|^2}{N}. \tag{7.51}$$

In some cases one may wish to identify the frequency response of a CT system by exciting it with a known input, $x(t)$, measuring the resulting output, $y(t)$, and then estimating the frequency response as $H(\Omega) = Y(\Omega)/X(\Omega)$. This estimate may be evaluated at the DFT frequencies if $x(t)$ and $y(t)$ are sampled simultaneously (without aliasing) to produce $x[n]$ and $y[n]$, $0 \le n \le N - 1$. Then

$$H(2\pi f_k) = \frac{TY(e^{j2\pi f_k T})}{TX(e^{j2\pi f_k T})} = \frac{Y[k]}{X[k]}. \tag{7.52}$$

Two final comments about calculating the CTFT via the DFT for practical signals should be noted. First, usually one removes the mean level from $x[n]$ before calculating the DFT because the effect of the implicit windowing of $x[n]$ is to spread the component at zero frequency (represented by the mean level of $x[n]$) to adjacent frequencies. This effect may hide signal components if the mean level is even modestly large compared to variations about the mean level. Second, if the available FFT algorithm requires that $N = 2^p$, where p is an integer, then it may be necessary to augment the sampled data by adding enough zeros to the end of $x[n]$ to meet this requirement. This procedure does not alter the DTFT (or the DFT) of the signal but it does change the frequencies ω_k and f_k at which the transforms are evaluated. Of course, if one calculates the power spectral density, then it is necessary to divide the squared magnitude of the DFT by the original value of N rather than the augmented value.

7.11 BIOMEDICAL APPLICATIONS

Frequency Response of a Transducer

Measuring the unknown frequency response of a transducer is difficult unless one can compare its response to another "gold standard" transducer having a wideband response (with respect to the signals to be measured). To determine the frequency response of a Celesco® model LCVR pressure transducer with model CD10 carrier demodulator, its pressure input port was connected in parallel with the pressure input port of an Endevco® model 4428 transducer and demodulator having a known constant frequency response for frequencies up to 10 KHz. This common connection was itself connected to a balloon. The output voltages from the two demodulators were sampled by a laboratory computer at 2000 Hz. A rapid pulse of pressure was applied to the balloon, causing a high-frequency transient pressure wave simultaneously at both transducers.

The two signals are shown in Fig. 7.18(a). The signal from the "standard" is $p_S(t)$, and the signal from the transducer under test is $p_T(t)$. The CTFT, $P_S(\Omega)$, was determined via Eq. (7.51) from the DFT of the sampled signal, $p_S[n]$. It confirmed that the frequency content of the signal was well within the range of uniform gain of the frequency response of the transducer used as the standard. $p_S(t)$ then was considered as the input to the test transducer and $p_T(t)$ as its output. The graphs of $p_S(t)$ and $p_T(t)$ indicate that the two signals were similar but not identical. The DFT of each sampled signal was determined, and the CT frequency response of the transducer under test was calculated at the corresponding frequencies via Eq. (7.52) (Fig. 7.18(b)). (The dashed part of the plot represents the frequency range for which the input signal contained components of very small amplitude; therefore, the calculated gain is unreliable.) The frequency response of the test transducer has a (3 dB) cutoff frequency of approximately 225 Hz.

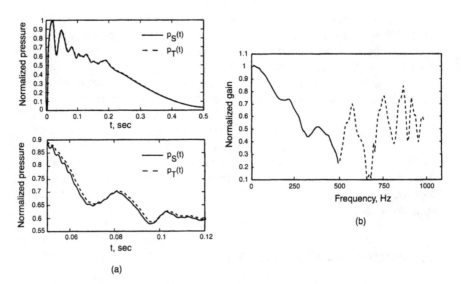

FIGURE 7.18. (a) The responses of two pressure transducers at two amplifications. $p_S(t)$—from the standard transducer, $p_T(t)$—from the transducer under test. (b) Relative gain of the transducer under test determined as the magnitude of the ratio of the DTFT of $p_T[n]$ to the DTFT of $p_S[n]$. Dashed line indicates unreliable data (see text).

Other Applications from the Literature

The DTFT has been utilized to characterize the tendency of neurons to exhibit oscillatory discharge patterns. For example, Gdowski and Voigt (1998) studied auditory neurons. The authors recorded the responses of these neurons to short tone bursts repeated at 1-second intervals, from which the autocorrelogram of the neuron was calculated. The autocorrelogram is a measure of the probability of a neuron discharging in a time interval $(t_0 + k \cdot \Delta t, t_0 + (k + 1) \cdot \Delta t)$ given that it discharged in some preceding interval $(t_0, t_0 + \Delta t)$. The time is discretized as $t_0 = n \cdot \Delta t$ and the probability is averaged over all values of t_0. The calculation is repeated for a sufficiently large range of k to capture the range of nonzero autocorrelation in the discharge pattern. The top two tracings of Fig. 7.19 present two examples (although the discrete points corresponding to each $k \cdot \Delta t$ are connected by lines in the graphs). The power spectral density function (Fig. 7.19, bottom traces) is determined by calculating the DTFT of the autocorrelogram. Each neuron was then characterized by the magnitude and frequency of the largest peak in its power spectrum, as indicated for these two examples.

An example of discrete-time filtering is found in flow cytometry. In Fig. 7.20 lung fibroblast cells were stained with a fluorescent dye (Nile Red) to reveal the presence of lipids. The intensity of fluorescence from each cell is an indicator of

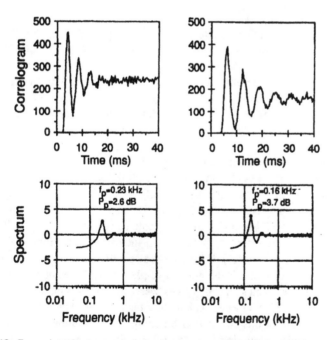

FIGURE 7.19. Examples of autocorrelograms (top) and spectral density functions (bottom) for two auditory neurons. See text for explanation of the autocorrelogram. (Godowski and Voight, 1998. Reprinted with permission from ANNALS OF BIOMEDICAL ENGINEERING. 1998, Biomedical Engineering Society.)

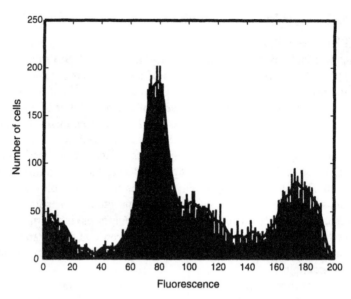

FIGURE 7.20. Histogram: flow cytometry data. Solid curve: same data after filtering with a Remez filter. See text for details. (Data courtesy of Margaret Bruce.)

the amount of lipid in the cell. The cells were then passed through a flow cytometer that counted the number of cells having intensities, I, within specific bins $(\bar{I}_k, \bar{I}_k + \Delta I)$ where $\bar{I}_k = k \cdot \Delta I$. The figure plots the number of cells vs. intensity. Again these data are discrete because the intensity axis is broken into discrete intervals and the intensity axis could be specified by the integer k, where $k = \bar{I}_k/\Delta I$. These data were smoothed by applying a DT filter designed using the Remez method (see Chapter 8) with an order of 21, a passband of $0 \le \omega \le 0.2\pi$, and a stopband of $0.3\pi \le \omega \le \pi$.

In another application Mao and Wong (1998) describe an optical method for measuring the average beat frequency of cultured cilia that entailed shining a 632.8-nm laser light on the cells and collecting both backscattered photons from the beating cilia that underwent a Doppler frequency shift and photons from the stationary background. Mixing of the two photon streams at a photomultiplier tube (PMT) produced an interference pattern with intensity fluctuations at the frequency of ciliary beating. Figure 7.21(a) shows a record of 2048 photon counts from the PMT, where each count is the number of photons received in a 4-msec window (or channel). The 100 most recent values were used to calculate the energy spectral density (Fig. 7.21(b)) by first calculating the autocorrelation function and then computing the FFT of this correlation function. The variations of the spectra with time were revealed by plotting them on a three-dimensional graph (Fig. 7.21(b)). The peak in the energy spectral density was interpeted as the average ciliary beat frequency.

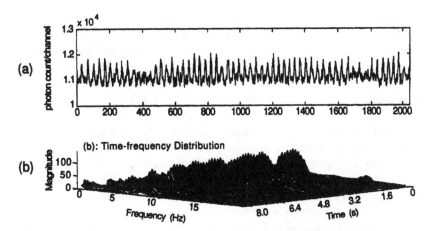

FIGURE 7.21. (a) Count of photons arriving at a photomultiplier tube in 4 msec time intervals. (b) PSD of photon count signal, taking 100 samples at a time, with succeeding spectra displaced correspondingly along the time axis. (Mao and Wong, 1998. Reprinted with permission from ANNALS OF BIOMEDICAL ENGINEERING. 1998, Biomedical Engineering Society.)

7.12 SUMMARY

Discrete-time signals may originate from measurements related to discrete events (such as the heartbeat), from processes that count or average data over fixed time intervals, or from sampling of continuous-time signals. The Discrete Fourier series (DFS) decomposes a periodic DT signal into a sum of DT complex exponential functions. In contrast to the CT Fourier series, if the period of a DT periodic function is N, then there are only N unique terms in the DFS since the sequence of DFS coefficients is also periodic with period N. The DFS satisfies many properties that are analogous to those of the CT Fourier series.

Nonperiodic DT signals can have a Discrete-time Fourier transform (DTFT) if they are absolutely (or square) summable. The DTFT is a continuous function of frequency, ω, and is periodic with period 2π. A version of Parseval's Relation establishes that the sum of the squared magnitudes of the values of a DT signal is proportional to the integral of the squared magnitude of its DTFT over one cycle. This result motivates the definition of the deterministic power spectral density (PSD) of a bounded, finite-length DT signal as the squared magnitude of its DTFT divided by its length, N. The PSD equals the DTFT of the autocorrelation function of the signal. For an LSIC system with unit-pulse response, $h[n]$, its frequency response is the DTFT of $h[n]$. Consequently, the DTFT of its output equals the frequency response multiplied by the DTFT of its input.

For a periodic DT signal, $\tilde{x}[n]$, with period N, each DFS coefficient, $\tilde{X}[k]$, multiplies a term $e^{j\frac{2\pi}{N}kn}$ in the DFS summation. If a nonperiodic signal, $x[n]$, is generated by taking one cycle of the periodic signal, then its DTFT is related to the DFS coefficients as $X(e^{j\frac{2\pi}{N}k}) = \tilde{X}[k]$. This relationship motivates defining a Discrete Fourier transform (DFT) to be one cycle of the DFS coefficients. The DFT then specifies the DTFT at N frequencies corresponding to N points uniformly spaced around the unit circle of the complex plane. The corresponding frequencies are $\omega_k = 2\pi k/N$, $0 \le k \le N - 1$. Because the DFT may be calculated very efficiently using various Fast Fourier transform (FFT) algorithms, this approach is the most common computational implementation for calculating values of the DTFT.

Windowing is an amplitude-modulation process in which one signal is multiplied by a second, finite-duration signal in order to force the resulting signal to be zero outside of a specified time interval. The shape of the windowing signal determines the rate at which the windowed signal approaches zero. The DTFT of the windowed signal is proportional to the convolution of the DTFTs of the two signals that are multiplied. CT signals that are sampled for a finite length of time can be visualized as infinite-duration signals that have been windowed.

Sampling of CT signals at a uniform sampling rate preserves the Fourier components of the CT signal if the sampling rate is greater than twice the highest frequency for which a Fourier component of the CT signal is nonzero (the Nyquist frequency). In such a case the DTFT of the sampled signal is equal to the Fourier transform of the CT signal, scaled by the sampling rate, $1/T$, with $\omega = \Omega T$ and $-\pi \le \omega \le \pi$. If the sampling frequency is too low, the DT signal is said to be aliased and there is no longer a simple relationship between the DTFT and the CTFT. The PSD of the CT signal equals the PSD of the sampled signal multiplied by T if there is no aliasing.

Biomedical applications of the DFS, and especially of the DFT, are extensive. They include analysis of the CTFT based on analysis of a sampled signal, investigation of periodic or nearly periodic behavior using the PSD, and analyses of frequency responses for smoothing of DT signals and other types of filtering. The next chapter will investigate further the many varieties of DT filters.

EXERCISES

7.1 Each of the signals below is periodic over the interval shown but its fundamental period may be shorter. Sketch and identify one period of each periodic signal, and calculate its Discrete Fourier series.

a. $\tilde{x}[n] = \begin{cases} 1, & -5 \le n \le 5, n \ne +2, -2 \\ -1, & n = +2, -2; \end{cases}$

b. $\begin{Bmatrix} n \\ \tilde{y}[n] \end{Bmatrix} = \begin{Bmatrix} -6 -5 -4 -3 -2 -1\ 0\ 1\ 2 \\ 0\ \ 1\ \ 2\ \ 1\ \ 0\ \ 1\ 2\ 1\ 0; \end{Bmatrix}$

c. $\tilde{z}[n] = \sum_{k=-\infty}^{\infty} \delta[n - 10k]$.

7.2 Consider the problem of calculating the DTFT, $F(e^{j\omega})$, of the signal $f[k]$ by making a periodic extension of $f[k]$ and calculating its DFS coefficients, where

$$f[k] = \begin{cases} 1, & k = 0, 1, 2 \\ 0, & k > 2 \end{cases}$$

a. $f[k]$ is a sequence of length 3. Calculate the DFS coefficients for the periodic extension which has period equal to three. Remembering that $F(e^{j\omega})$ is a continuous function of ω, what problem do you encounter in this case when trying to determine the DTFT from the DFS coefficients?

b. Now define another sequence, $f_1[k]$, which has length eight, where

$$f_1[k] = \begin{cases} 1, & k = 0, 1, 2 \\ 0, & 3 \le k \le 7. \end{cases}$$

Make a periodic extension of this sequence and find its DFS coefficients. Can you use these DFS coefficients to evaluate the DTFT $F(e^{j\omega})$?

7.3   A continuous-time signal $x(t)$ has the CTFT given by

$$X(\Omega) = \begin{cases} 2, & |\Omega| \le 200 \\ 0, & |\Omega| > 200. \end{cases}$$

Sketch the DTFT, $X(e^{j\omega})$ of the signal $x[n]$ which results from sampling $x(t)$ with the sampling interval:

a. $T = 2\pi/300$.

b. $T = 2\pi/600$.

7.4   Consider a continuous-time LTI system having the impulse response $h(t) = 5e^{-2t}u(t)$. If this system is given the input $x(t) = u(t)$, its output will be $y(t) = 5[1 - e^{-2t}]u(t)$.

a. Sample $x(t)$ and $y(t)$ using the sampling interval $T = 0.10$ sec. to obtain the discrete-time variables $x[n] = u[n]$, and $y[n] = 5(1 - e^{-0.2n})u[n]$ Now assume that $x[n]$ was the input to a discrete-time LTI system and $y[n]$ was the resulting output. Calculate the frequency response, $H(e^{j\omega})$ of this DT system.

b. Assuming there is no aliasing, we have calculated a specific relationship between the Fourier transform of a CT signal and the DTFT of a DT signal obtained by sampling the CT signal. That is, if the signal being sampled is $h(t)$, then $H(\Omega) = T \cdot H(e^{j\Omega T})$ Using your answer from part a, calculate the CT frequency response $H(\Omega)$ based on this relationship. What is the actual frequency response of the CT system described above? Why doesn't the relationship between the CTFT and the DTFT give the right answer for $H(\Omega)$?

7.5   Graph each of the following DT signals and determine its DTFT:

a. $x[n] = \frac{1}{2}\delta[n + 1] + \delta[n] + \frac{1}{2}\delta[n - 1]$;

b. $x[n] = u[n + 3] - u[n - 4]$;

c. $x[k] = (0.5)^k \cos[6k]u[k]$.

7.6   The signal $x[n]$ has the DTFT $X(e^{j\omega}) = 2/(1 + 0.3e^{-j\omega})$. Determine the DTFT of the following signals:

a. $w[n] = x[n - 3]$;

b. $y[n] = e^{j2n}x[n]$;

c. $z[n] = x[n] * x[n]$.

7.7   In normal young subjects heart rate varies with respiration, often following a nearly-sinusoidal variation. Since heart rate can be measured only once per cardiac cycle, any measurement of heart rate is necessarily a sequence of numbers. Call this sequence $r(n)$. For the present purposes I will approximate heart rate variations by the following:

$$r[n] = \bar{r} + 6 \sin\left[\frac{2\pi}{12}n\right],$$

where $\bar{r}$ is the mean heart rate, and generate a sequence of 32 simulated heart rate values from this equation. Assuming that I do not "know" the frequency of the heart rate variations, I subtract the mean level from the values of $r[n]$, make a periodic extension of this (zero-mean) finite-length sequence and then calculate the Discrete Fourier series of the resulting periodic sequence. The magnitudes of the coefficients I get are shown in Table 7.7.

TABLE 7.7   DFS Coefficients for Simulated Heart Rate Data

1	1.9392305e+001
2	2.2639238e+001
3	4.4907360e+001
4	7.5184972e+001
5	1.6348469e+001
6	8.3686034e+000
7	5.3752340e+000
8	3.8723213e+000
9	3.0000000e+000
10	2.4484381e+000
11	2.0806101e+000
12	1.8275775e+000
13	1.6515308e+000
14	1.5305354e+000
15	1.4515163e+000
16	1.4067908e+000
17	1.3923048e+000
18	1.4067908e+000
19	1.4515163e+000
20	1.5305354e+000
21	1.6515308e+000
22	1.8275775e+000
23	2.0806101e+000
24	2.4484381e+000
25	3.0000000e+000
26	3.8723213e+000
27	5.3752340e+000
28	8.3686034e+000
29	1.6348469e+001
30	7.5184972e+001
31	4.4907360e+001
32	2.2639238e+001

a. Specify the frequency that corresponds to each DFS coefficient.

b. Explain why the maximum DFS coefficient is not at the known sinusoidal frequency in the data.

c. Explain why the DFS coefficient at the frequency nearest to the known sinusoidal component in the data is not the only nonzero coefficient. Include diagrams to complement your explanation.

7.8 Use MATLAB to generate a sequence of 32 data points from the equation for $r[n]$ in the previous exercise, then calculate the corresponding DFT coefficients using the fft command. You should get the same answers given in Table 7.7. Now generate 256 data points (instead of 32) and calculate the new DFT coefficients. Make plots of the magnitudes of the old DFT coefficients versus frequency and of the new DFT coefficients versus frequency. Compare these two graphs.

7.9 Generate 128 data points using the equation $r(n) = \bar{r} + 6 \sin[(2\pi/16)n]$ Evaluate the corresponding DFT coefficients and plot their magnitudes vs. frequency. Why are there now nonzero values only at the known sinusoidal frequencies in the data?

7.10 The response of a linear, discrete-time, shift-invariant system to the input

$$x[n] = \begin{cases} 1, & n \geq 0 \\ 0, & n < 0 \end{cases}$$

is $y[n] = (n + 1)\{u[n] - u[n - 2]\} + 3u[n - 3]$.

a. Determine the unit-pulse response of this system.

b. Determine its frequency response.

7.11 A periodic discrete-time signal is given by $\tilde{x}[k] = -2\{1 - \cos[\frac{1}{2}\pi k]\}$ Specify its period, N, and calculate all of its unique Discrete Fourier series coefficients. If the finite-length signal $x[k]$ consists of one cyle of $\tilde{x}[k]$, specify the frequencies at which $X(e^{j\omega})$ equals $\tilde{X}[k]$.

7.12 One method for decimation of a discrete-time signal, $x[n]$, involves replacing every other value of the signal by zero. Mathematically this process can be expressed as:

$$y[n] = \begin{cases} x[n], & n \text{ even} \\ 0, & n \text{ odd} \end{cases}.$$

Equivalently, we could say that $y[n] = x[n]p[n]$, where $p[n] = \Sigma_k \delta[n - 2k] = \frac{1}{2}(1 + (-1)^n)$. Express the DTFT $Y(e^{j\omega})$ in terms of $X(e^{j\omega})$.

Hint: Think of other ways to express $(-1)^n$.

7.13 Conceptually, one could design a digital FIR filter that approximates the frequency response of an IIR filter by truncating the impulse response of the IIR filter. For a simple example, consider two filters whose impulse responses are given by: $h_1[k] = a^k u[k]$, $|a| < 1$, and $h_2[k] = a^k u[k]$, $0 \leq k \leq 1$, with $h_2[k] = 0$, $k > 1$.

a. Find the frequency response of both filters and compare them. How well does the FIR filter approximate the frequency response of the IIR filter?

b. The difference equation for the first filter is $y[k] + ay[k-1] = x[k]$. Determine the difference equation for the second filter.

7.14 The signal $x(t) = [10\sin(40\pi t)]P_1(t - 0.5)$ is sampled with $T = 5$ msec for $0 \le t < 1$ to form the sequence $x[n]$, $0 \le n \le 199$. Sketch the magnitude of the DTFT of $x[n]$. If $X(e^{j\omega})$ is evaluated using a DFT method, at what specific frequencies will one be able to calculate the values of $X(\Omega)$?

7.15 Let $x[n]$ be the input to an LSI system having the frequency response $H(e^{j\omega})$ and $y[n]$ be its output.

a. Let $x[n] = 2\cos[100n]$, $h[n] = (-0.8)^n u[n]$, and calculate the power spectral density of $y[n]$.

b. If $x[n]$ was acquired by sampling the signal $x(t) = 2\cos(1000t)$ with $T = 0.10$ sec, calculate the power spectral density of $x(t)$, $P_X(\Omega)$.

c. If one uses a DFT with $N = 1024$ points for calculating the spectral density of $x(t)$ from part b, at what DT and CT frequencies can one evaluate the spectral density function?

7.16 Since in CT an ideal differentiator has the frequency response $H_d(\Omega) = j\Omega$, one might suppose that a DT approximation of this CT system would have the frequency response $H_d(e^{j\omega}) = j\omega$. If the input to the DT system, $x[n]$, is acquired by sampling the signal $x(t)$ without aliasing, determine whether the output of the DT system will be the corresponding samples of the output of the CT system—that is, of dx/dt.

7.17 The delay time or center of gravity of a DT signal, $x[n]$, is defined as

$$N_D = \frac{\displaystyle\sum_{n=-\infty}^{\infty} nx[n]}{\displaystyle\sum_{n=-\infty}^{\infty} x[n]}.$$

Express N_D in terms of the DTFT of $x[n]$.

7.18 A simple form of model for a natural population is an age structured model. Assume some population exists in only two age classes (juveniles or adults) and that juveniles produce f_0 offspring per year and adults produce f_1 offspring per year. Further assume that juveniles survive to adulthood per year with probability p_0 and adults do not survive beyond one year. Then if $j[k]$ is the number of juveniles, $a[k]$ is the number of adults, and $x[k]$ is the number of juveniles added per year from an external source (such as a neighboring population), the population may be described by the relationship

$$\begin{pmatrix} j[k+1] \\ a[k+1] \end{pmatrix} = \begin{pmatrix} f_0 & f_1 \\ p_0 & 0 \end{pmatrix}\begin{pmatrix} j[k] \\ a[k] \end{pmatrix} + \begin{pmatrix} x[k] \\ 0 \end{pmatrix}.$$

a. This system has two outputs, $j[k]$ and $a[k]$. Find the impulse response of both outputs when $x[k]$ is the input.

b. Calculate the DTFT of both impulse responses and compare their frequency responses. Let $f_0 = 0.25$, $f_1 = 0.50$, $p_0 = 0.90$.

7.19   The periodontal membrane connects the root of a tooth to the underlying bone and acts as a shock absorber. To test the biomechanical properties of this membrane, Okazaki et al. (1996) applied an impulsive force to the tip of a maxillary incisor tooth and measured the resulting acceleration of the tooth using a miniature accelerometer. The force was applied using a computer-controlled "hammer" and was quantified also as the acceleration of the hammer. Figure 7.22 shows data from one subject that is available in digitized form in the file tooth.mat. The two variables in this file are the accelerations of the hammer and tooth, $a_H[n]$ and $a_T[n]$, named acch and acct, respectively. They were sampled with an intersample interval of 0.1453 msec.

a. Using the MATLAB function fft calculate the DTFT of the acceleration of the tooth, $a_T[n]$, and plot the magnitude of this transform verus frequency. Similarly determine the DTFT of $a_H[n]$ and plot its magnitude versus frequency.

b. Calculate and graph the frequency response of this system.

c. If one assumes that the input is "sufficiently impulse-like", one may determine the frequency response simply from the DTFT of $a_T[n]$. Is this assumption valid? Explain.

7.20   The distributed force and torque exerted by the ground on the foot of a standing subject can be represented by a single net force vector (called the ground reaction force) located at a ground position called the center of force. The file sway.mat comprises samples of the antero-posterior location of the center of force, $c(t)$, when a normal young subject attempts to stand still on top of a force plate with his eyes closed. (These data were obtained using a modified Equitest[R] system.) $c(t)$ was sampled at 50 Hz for 20 seconds.

a. Use the MATLAB function fft to calculate $C(e^{j\omega})$, the DTFT of the sampled signal $c[n]$, and then calculate and plot the power spectral density function of $c[n]$.

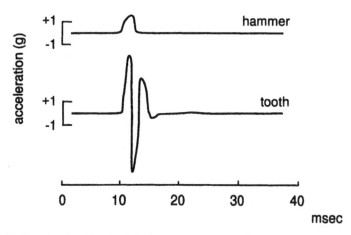

FIGURE 7.22. Acceleration signals of the hammer and tooth from the file tooth.mat. (Data from Okazaki et al., 1996. Reprinted with permission from ANNALS OF BIOMEDICAL ENGINEERING. 1996, Biomedical Engineering Society.)

b. Determine the power spectral density function of $c(t)$ and identify the frequencies of any periodic oscillations present in $c(t)$.

7.21 The file hrs.mat contains beat by beat heart rate data from a human subject. Consider this signal as a DT variable with the basic unit of time being one heartbeat. Remove the mean value from this signal—call the new signal $hr[n]$—and calculate the autocorrelation function of $hr[n]$. (*Suggestion:* Use the MATLAB function xcov.) Can you detect any oscillatory behavior in $hr[n]$ based on examination of its autocorrelation function? If so, what are the periods? Now use fft to calculate the DTFT of the autocorrelation function—that is, the power spectral density function—and determine whether there are any peaks in this latter function that are indicative of oscillatory behavior in $hr[n]$.

7.22 Calculate the PSD of the signal $c[n]$ from sway.mat after windowing this signal with a Hamming window, and compare this result to the PSD calculated in Exercise 7.20(a).

8

NOISE REMOVAL AND
SIGNAL COMPENSATION

8.1 INTRODUCTION

Removing noise from a signal probably is the most frequent application for signal processing. As discussed in earlier chapters, the distinction between noise and desired signal is an heuristic judgment that must be determined by the user. Often the distinction is obvious—for example, 60-Hz interference in an EEG recording or temperature-related drift in the reading of a pH electrode. In other situations the user depends on experience and indirect knowledge of the signal and its source to decide whether or not certain features of a signal should be considered noise. For example, since the frequency components of the EEG are principally below 25 Hz, one would suspect that prominent signals in the EEG at much higher frequencies have another source such as the electromyogram activity of neck, jaw, or extraocular muscles. Once the distinction between the desired signal and noise has been resolved, then removal of the noise with the least possible distortion of the desired signal is the goal.

In concept, one may specify the frequency response of a linear filter that can effect the separation of the frequency components of the noise from those of the desired signal. In many cases, however, the necessary frequency response may not be similar to that of the simple first- and second-order filters that have been discussed as examples in earlier chapters. In other cases the separation between the passband and the stopband of a simple filter may be inadequate either because the stopband gain is not small enough or because the frequency components of signal and noise are too close together relative to the width (in frequency) of the transition between the passband and stopband. In these situations it is necessary to design a better filter.

Because biomedical applications often involve the acquisition of continuous-time signals by digital sampling, one has the choice of filtering either before sampling using a continuous-time filter (*analog filter*), or after sampling using a discrete-time filter (*digital filter*). Digital filters are more versatile and more convenient to modify for different applications. Analog filters, however, also have certain advantages and there are pertinent reasons to employ them. First, to prevent aliasing one *always* should filter a CT signal with an analog filter before sampling it. Second, even if it is incomplete, any noise removal that one can effect with an analog filter before sampling reduces the requirements on any subsequent digital filter. Finally, the properties of some classes of analog filters are well-studied and these filters provide a foundation for designing "equivalent" digital filters.

Because much filtering of biomedical signals is accomplished digitally, this chapter focusses on digital filters. (Furthermore, the reader will recognize that many newer biomedical instruments that acquire analog signals process these signals by digitizing them, and often include built-in digital filters.) The chapter begins with an example of analog filtering that will be discussed later from the viewpoint of digital filtering. Because digital filters are designed using Z-transforms, this topic occupies the first half of the chapter.

8.2   INTRODUCTORY EXAMPLE: REDUCING THE ECG ARTIFACT IN AN EMG RECORDING

When an electromyogram (EMG) is recorded from muscles on the torso, the electrocardiogram (ECG) is commonly superimposed on the EMG signal because of the physical proximity of the heart and the large electrical amplitude of the ECG compared to many EMG signals. An example is presented in the bipolar recording of the EMG from the lower ventrolateral rib cage of a human subject shown in Fig. 8.1(a). The four bursts of EMG activity correspond to the contraction of muscles of the chest during the inspiration phase of four breaths. To better visualize the EMG signal due to breathing, or for subsequent quantitative analysis of the signal, it is desirable to filter out the ECG artifacts. Let us proceed by estimating the frequency range of the components of the EMG and ECG signals individually.

Using Eq. (7.21) we may calculate separately the deterministic power spectral density functions for short segments of data that comprise mainly EMG and mainly ECG components. First we examine a segment of 256 samples (acquired by sampling at 1000/s) encompassing the ECG signal near $t = 2.30$ s. Its power spectrum is plotted in Fig. 8.1(b) as the dotted line. Next we calculate the power spectrum for each of two 384-sample test segments of the EMG signal acquired from the middle portions of the second and third bursts, between the neighboring ECG signals. The average of these two spectra is also depicted in Fig. 8.1(b). Because it is concentrated in frequency and large in amplitude, the ECG has much greater peak power than the EMG. Furthermore, the frequency contents of the two signals seem to overlap below about 30 Hz. Therefore, removal of the ECG by filtering the EMG recording with a highpass filter will necessarily remove some EMG signal components also. Nevertheless, let us ascertain the degree to which the ECG can be reduced through application of a second-order filter.

A second-order Butterworth highpass filter was designed and the EMG recording of Fig. 8.1(a) was filtered using the MATLAB function lsim. Through trial and error it was found that a cutoff frequency of 70 Hz produced (visually) the largest

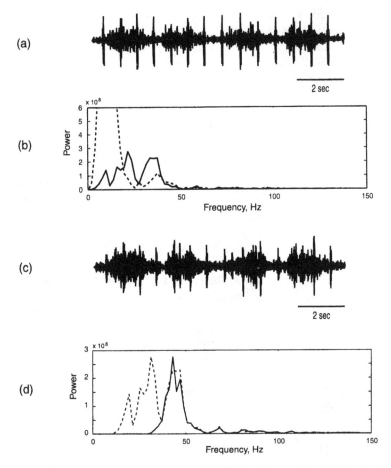

FIGURE 8.1. (a) Recording of electromyogram (EMG) from lower chest wall of a human subject, showing respiratory muscle EMG and electrocardiogram (ECG). (b) Power spectra of EMG (solid) and ECG (dotted), the latter truncated because of its large amplitude. See text. (c) Recording from 'a' after being passed through "optimal" second-order Butterworth filter. (d) Power spectrum of EMG before (dotted) and after (solid) filtering.

reduction of the ECG without severe degradation of the amplitude of the EMG signal. The filtered signal is presented in Fig. 8.1(c), and Fig. 8.1(d) compares the average power spectrum of the two EMG test segments before and after filtering. There is a definite visual improvement in the EMG recording but the power spectrum shows that filtering has essentially removed the EMG frequency components below about 30 Hz. Although the ECG signal components below 30 Hz are also greatly attenuated, the large original amplitude of these components offsets the small gain of the filter, the result being that the product of these two factors yields signal components in the filter output that are comparable in magnitude to the EMG signal. Thus the ECG is diminished but still very visible in the filtered signal!

Later we will address this same problem using digital filtering, with moderately better success because it is easy to implement high-order digital filters that separate passband and stopband more precisely. A markedly more successful approach to the difficult problem of ECG removal from EMG signals involves the application of a topic beyond the current text, namely, adaptive filtering.

8.3 EIGENFUNCTIONS OF LSI SYSTEMS AND THE Z-TRANSFORM

Eigenfunctions

Consider an LSI system with unit-pulse response $h[n]$, having input $x[n]$ and output $y[n]$. Assume that the input may be expressed as a linear combination of some set of basis functions, so that $x[n] = \Sigma_k a_k \phi_k[n]$. The output of this system then will be $y[n] = x[n] * h[n] = \Sigma_k a_k \phi_k[n] * h[n]$. In other words the output may be expressed as a weighted sum of another set of basis functions, $\theta_k[n]$, defined by $\theta_k[n] = \phi_k[n] * h[n]$. Indeed the weighting factors, a_k, are the same for both $x[n]$ and $y[n]$! This result is general as long as $x[n]$ is definable in terms of a set of basis functions. If the basis of $x[n]$ is such that the basis functions for $y[n]$ are scaled versions of those for $x[n]$—that is, if $\theta_k[n] = b_k \phi_k[n]$, then the members of the basis set $\phi_k[n]$ are known as *eigenfunctions* of the LSI system and the coefficients b_k are *eigenvalues*. If an eigenfunction is the input to an LSI system, the output is the input signal scaled by the eigenvalue. If the eigenfunctions form a basis set for the input signal, $x[n]$, then they also form a basis set for $y[n]$.

Consider now the input signal $x[n] = z^n$, where z is any complex number. By convolution, the resulting output of the system with unit-pulse response $h[n]$ is

$$y[n] = \sum_{m=-\infty}^{\infty} h[m]z^{n-m} = z^n \sum_{m=-\infty}^{\infty} h[m]z^{-m} = H(z)z^n,$$

where

$$H(z) = \sum_{m=-\infty}^{\infty} h[m]z^{-m}. \tag{8.1}$$

From the above definitions, z^n must be an eigenfunction of any LSI system and $H(z)$ must be the corresponding eigenvalue. $H(z)$ is known as the *Z-transform* of $h[n]$. That $H(z)$ is a generalized frequency response of the LSI system may be seen by expressing the complex number, z, in terms of its magnitude and angle, $z = re^{j\theta}$, and letting $r = 1$. Noting that on the unit circle $z = e^{j\omega}$, Eq. (8.1) implies that

$$H(z)|_{z=e^{j\omega}} = \sum_{m=-\infty}^{\infty} h[m]e^{-j\omega m} = H(e^{j\omega}). \tag{8.2}$$

That is, the Z-transform, $H(z)$, evaluated on the unit circle in the complex plane equals the frequency response of the LSI system.

Note that z^{cn} is also an eigenfunction of LSI systems for any finite value of c. Therefore, any function $x[n]$ that may be expressed as a summation of terms in powers of z is describable as a sum of eigenfunctions of LSI systems. This class of functions is very large, especially when one considers power series expansions of functions like sines and exponentials.

Consequently, as with previous transforms, we will extend the Z-transform to functions other than those that are unit-pulse responses of LSI systems. In general, we define the Z-transform of the sequence $x[n]$ as

$$X(z) = \sum_{m=-\infty}^{\infty} x[m]z^{-m} \triangleq Z\{x[n]\}. \tag{8.3}$$

Note that Eq. (8.2) implies that $X(z)|_{z=e^{j\omega}} = X(e^{j\omega})$. That is, the DTFT of $x[n]$ equals the Z-transform of $x[n]$ evaluated on the unit circle, assuming that the defining summation for $X(z)$ converges on the unit circle. That $X(z)$ may converge for some values of z but fail to converge on the unit circle may be appreciated by writing $X(z)$ in the form

$$X(z = re^{j\theta}) = \sum_{m=-\infty}^{\infty} x[m]r^{-m}e^{-jm\theta}.$$

Clearly this summation may fail to converge for $r = 1$ and still converge for $r \neq 1$. Consequently, the Z-transform of a DT signal may exist even though its DTFT does not.

Region of Convergence of the Z-Transform

In general, the Z-transform, Eq. (8.3), will converge for some values of z but not for others; therefore, it is necessary to specify the region of convergence of $X(z)$ in the z-plane. As with Laplace transforms, two different functions may have the same expression for their Z-transforms but the regions of convergence will be different. A common example is the pair of real-valued functions $x[n] = a^n u[n]$, $y[n] = -a^n u[-n-1]$. Evaluating their respective Z-transforms,

$$X(z) = \sum_{m=0}^{\infty} a^m u[m]z^{-m} = \sum_{m=0}^{\infty} (a^{-1}z)^{-m} = \sum_{m=0}^{\infty} (az^{-1})^m = \frac{1}{1-az^{-1}} = \frac{z}{z-a}, \qquad |z| > |a|.$$

$$Y(z) = -\sum_{m=-\infty}^{-1} a^m z^{-m} = -a^{-1}z \sum_{m=0}^{\infty} (a^{-1}z)^m = \frac{-a^{-1}z}{1-a^{-1}z} = \frac{z}{z-a}, \qquad |z| < |a|.$$

Note that the ROC for each transform is specified based on the criterion for convergence of the corresponding summation. Figure 8.2 indicates the ROC for each of these Z-transforms.

Convergence of the Z-transform of an arbitrary DT function $x[n]$ requires that

$$|X(z)| = \left| \sum_{m=-\infty}^{\infty} x[m]z^{-m} \right| < \infty.$$

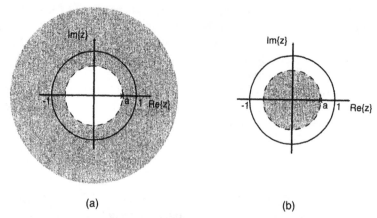

(a)   (b)

FIGURE 8.2. Regions of convergence of: (a) $x[n] = a^n u[n]$; (b) $x[n] = -a^n u[-n - 1]$.

A sufficient condition for convergence is that the terms in this summation be absolutely summable since $|\sum_{m=-\infty}^{\infty} x[m]z^{-m}| \leq \sum_{m=-\infty}^{\infty} |x[m]z^{-m}|$. Therefore, $X(z)$ converges if

$$\sum_{m=-\infty}^{\infty} |x[m]z^{-m}| = \sum_{m=-\infty}^{\infty} |x[m]r^{-m}| |e^{j\theta m}| = \sum_{m=-\infty}^{\infty} |x[m]r^{-m}| < \infty. \qquad (8.4)$$

Note also that the (finite) boundaries of the ROC of a Z-transform are circular. This result may be demonstrated by assuming that an arbitrary Z-transform $X(z)$ converges for some value of z, corresponding to $z_1 = r_1 e^{j\theta_1}$. But by Eq. (8.4), $X(z)$ also will converge for any $z = r_1 e^{j\theta}$—that is, everywhere on a circle of radius r_1.

Certain general properties of the Z-transform often permit an easy determination of the ROC. Consider first a bounded, finite-duration signal, $x[n]$, $n_1 \leq n \leq n_2$. Its Z-transform may be written as

$$X(z) = \sum_{n=n_1}^{n_2} x[n]z^{-n}$$

$$= x[n_1]z^{-n_1} + x[n_1 + 1]z^{-(n_1+1)} + \ldots + x[n_2 - 1]z^{-(n_2-1)} + x[n_2]z^{-n_2}.$$

So long as $x[n]$ is bounded, $X(z)$ will be finite for any value of z, except possibly for $z = 0$, or $z = \infty$. If $n_1 < 0$, then $X(z)$ will not converge for $z = \infty$ since $z^{-n_1}|_{z \to \infty} \to \infty$ for negative n_1. In fact, if the ROC of $X(z)$ includes $z = \infty$, then $x[n]$ must be causal (although it may be infinite in duration). If $n_2 > 0$, then $X(z)$ will not converge for $z = 0$ since $z^{-n_2}|_{z \to 0} \to \infty$. Therefore, the ROC of a finite-duration sequence includes all of the z-plane, with the possible exceptions of $z = 0$ or ∞.

A *right-sided sequence* is a sequence $x[n]$ for which $x[n] = 0$, $n < n_1$. Its Z-transform is $X(z) = \sum_{n=n_1}^{\infty} x[n]z^{-n}$. From the preceding discussion, $X(z)$ will not converge for $z = 0$ nor, if $n_1 < 0$, for $z = \infty$. If $n_1 \geq 0$, then $z = \infty$ is in the ROC. To determine other properties of the ROC, assume that $X(z)$ converges for some $z = z_1$. Then the condition of Eq. (8.4) will be met for any z such that $|z| > |z_1|$. But since $X(z)$ cannot converge at any pole, then z_1 cannot be inside the circle defined by the magnitude of the largest pole of $X(z)$. Consequently, the ROC for a right-sided sequence includes all of the z-plane outside of its largest pole, except for $z = \infty$ if $n_1 < 0$ (e.g., Fig. 8.2(a)). Conversely, if it is known that the ROC of $X(z)$ is $|z_1| < |z| < \infty$, one may conclude that $x[n]$ is right-sided, and if $z = \infty$ is included in the ROC, then $x[n]$ is causal. Note that for the DTFT of a right-sided sequence $x[n]$ to exist (which requires that $X(z)$ converges on the unit circle), all of the poles of its Z-transform must lie *inside* the unit circle.

A *left-sided sequence* is one for which $x[n] = 0$, $n > n_2$. The convergence properties of the Z-transform of a left-sided sequence mirror those of a right-sided sequence. The ROC includes all of the z-plane inside the circle defined by the magnitude of the smallest pole of $X(z)$, but will not include $z = 0$ if $n_2 > 0$. If a DT signal is the sum of a right-sided sequence and a left-sided sequence, then its ROC will be the region of the z-plane that is common to the regions of convergence of the individual sequences, if such exists. It will have the form $|z_1| < |z| < |z_2|$ which describes an annulus centered at $z = 0$.

Example 8.1 Common Z-transforms Z-transforms of many simple DT signals are easily computable directly from the definition. Thus

$$\text{for } x[n] = \delta[n], \qquad X(z) = \sum_{n=-\infty}^{\infty} \delta[n]z^{-n} = 1, \qquad \forall z;$$

$$\text{for } x[n] = \delta[n - N], \qquad X(z) = z^{-N}, \qquad |z| > 0;$$

$$\text{for } x[n] = u[n], \qquad X(z) = U(z) = \sum_{n=0}^{\infty} z^{-n} = \frac{1}{1 - z^{-1}}, \qquad |z| > 1.$$

Note that $U(z)$ also may be written as

$$U(z) = \frac{z}{z - 1}.$$

For $x[n] = a^n u[n]$, as shown previously,

$$X(z) = \frac{1}{1 - az^{-1}}, \qquad |z| > a.$$

Using the last result, if $x[n]$ is generated by sampling the CT function $x(t) = e^{bt}u(t)$ at a sampling frequency $1/T$, so that $x[n] = e^{-nbT}u[n]$, then $X(z) = 1/(1 - e^{-bT}z^{-1})$, $|z| > e^{-bT}$.

Before discussing properties of the Z-transform we should note that the definition of Eq. (8.3) is properly designated the *bilateral Z-transform*. An alternative characterization, for which the lower limit of the summation in Eq. (8.3) is $m = 0$, defines the *unilateral Z-transform*. Because the unilateral Z-transform assumes that the sequence is zero for $n < 0$ (i.e., a right-sided sequence), there is a unique sequence corresponding to each Z-transform and it is not necessary to specify the ROC. This form of the Z-transform is especially useful for dealing with causal systems with right-sided sequences as inputs and we shall consider it later. For now we address only the bilateral Z-transform.

8.4 PROPERTIES OF THE BILATERAL Z-TRANSFORM

Linearity: The Z-transform is a linear operator. Thus, if $x[n]$ and $y[n]$ have the transforms $X(z)$ and $Y(z)$ defined on regions of convergence R_x and R_y, respectively, then

$$Z\{ax[n] + by[n]\} = aX(z) + bY(z),$$

with ROC at least the intersection of R_x and R_y. This property is easily proven using the basic definition, Eq. (8.3).

Example 8.2   Linearity   Let $x[n] = \delta[n] + 2(0.6)^n u[n]$. Then $X(z) = 1 + 2[1/(1 - 0.6z^{-1})]$. The ROC of $X(z)$ is the region $|z| > 0.6$.

Example 8.3   Z-transforms of trigonometric functions   The Z-transforms of trigonometric functions may be calculated using Euler's formulas and the Z-transform of $a^n u[n]$. Thus, if $x[n] = \cos[\omega_0 n]u[n]$,

$$X(z) = Z\left\{\frac{1}{2}e^{j\omega_0 n}u[n] + \frac{1}{2}e^{-j\omega_0 n}u[n]\right\} = \frac{1}{2}\frac{1}{1 - e^{j\omega_0}z^{-1}} + \frac{1}{2}\frac{1}{1 - e^{-j\omega_0}z^{-1}},$$

with the ROC of both terms being $|z| > 1$. The terms in $X(z)$ may be combined to yield

$$X(z) = \frac{1 - z^{-1}\cos(\omega_0)}{1 - 2z^{-1}\cos(\omega_0) + z^{-2}}.$$

Similarly,

$$Z\{\sin[\omega_0 n]u[n]\} = \frac{z^{-1}\sin(\omega_0)}{1 - 2z^{-1}\cos(\omega_0) + z^{-2}}.$$

Time shift: If $Z\{x[n]\} = X(z)$ then $Z\{x[n - d]\} = z^{-d}X(z)$. Again, this result is easily demonstrable from Eq. (8.3). Since delaying a signal by one unit of discrete time corresponds to multiplying its Z-transform by z^{-1}, often z^{-1} is called a *unit-delay operator*.

Scaling: If $Z\{x[n]\} = X(z)$, $Z\{a^n x[n]\} = X(a^{-1}z)$. That is, scaling of the amplitude of $x[n]$ by the exponential term, a^n, corresponds to dilation ($a > 1$) or compression ($a < 1$) of the z-plane. If the ROC of $X(z)$ is $|z_1| < |z| < |z_2|$, the ROC for $X(a^{-1}z)$ is $|az_1| < |z| < |az_2|$. These results follow directly from the definition.

Example 8.4 Scaling Let

$$x[n] = \begin{cases} 1, & 0 \le n \le N-1 \\ 0, & \textit{otherwise.} \end{cases}$$

$X(z)$ is found using Eq. (8.3) as $X(z) = \sum_{n=0}^{N-1} z^{-n} = (1 - z^{-N})/(1 - z^{-1})$ unless $z = 1$, in which case $X(z) = N$. Since $x[n]$ has finite duration and is causal, its ROC is the entire z-plane except $z = 0$. Now let $y[n] = 0.5^n x[n]$. By the scaling property, $Y(z) = X(z/0.5) = (1 - 0.5^N z^{-N})/(1 - 0.5z^{-1})$.

Multiplication by n: $Z\{nx[n]\} = -z(dX(z)/dz)$. This property is demonstrable by differentiating both sides of Eq. (8.3). Thus $dX(z)/dz = \sum_{n=-\infty}^{\infty} x[n](-n)z^{-n-1} = -z^{-1}\sum_{n=-\infty}^{\infty}(nx[n])z^{-n}$. Dividing by z^{-1} proves the desired result.

Example 8.5 Let $v[n] = nu[n]$. Then $V(z) = -z(dU(z)/dz)$ where $U(z)$ is the Z-transform of the DT unit-step function. Thus $V(z) = -z^{-1}(-1/(1 - z^{-1})^2) = (z^{-1}/(1 - z^{-1})^2)$.

Time reversal: If $x[n]$ has the transform $X(z)$, then $x[-n]$ has the transform $X(z^{-1})$. Furthermore the boundaries of the ROC for $x[-n]$ are the reciprocals of those for $x[n]$. This property follows from Eq. (8.3) after a change of variables—for example, let $k = -m$.

Time-domain convolution: Given $Z\{x[n]\} = X(z)$, $Z\{v[n]\} = V(z)$. If $y[n] = x[n] * v[n]$, then

$$Y(z) = X(z)V(z). \tag{8.5}$$

To prove this result, compute the Z-transform of $y[n]$:

$$Y(z) = \sum_{n=-\infty}^{\infty} y[n]z^{-n} = \sum_{n=-\infty}^{\infty} \left[\sum_{k=-\infty}^{\infty} x[k]v[n-k] \right] z^{-n}$$

$$= \sum_{n=-\infty}^{\infty} v[n-k]z^{-(n-k)} \sum_{k=-\infty}^{\infty} x[k]z^{-k}.$$

This equation may be written as

$$Y(z) = \sum_{m=-\infty}^{\infty} v[m]z^{-m} \sum_{k=-\infty}^{\infty} x[k]z^{-k} = V(z)X(z).$$

The immediate application for this property is the determination of the Z-transform of the zero-state output response of an LSI system given the Z-transforms of its input signal and its unit-pulse response. This application will be explored in depth later in the chapter. In order to determine the time response of the output, it will be necessary to determine the inverse Z-transform of $Y(z)$. Methods to calculate inverse Z-transforms will be addressed shortly.

A useful, related property permits the indirect calculation of the cross-correlation function of two sequences. Recall that $R_{vx}[m] = \sum_{n=-\infty}^{\infty} v[n]x[n-m] = x[n] * v[-n]$. Taking the Z-transform of both sides of this equation and using the time reversal property,

$$R_{xv}(z) = X(z)Z\{v[-n]\} = X(z)V(z^{-1}). \tag{8.6}$$

Similar to the case for convolving two sequences, the cross-correlation of two sequences may be calculated by multiplying in the Z-domain, then evaluating the inverse Z-transform.

Initial value theorem: If $x[n]$ is *causal*, then $x[0] = \lim_{z\to\infty} X(z)$. To prove this result, write $X(z)$ as $X(z) = x[0]z^0 + \sum_{n=1}^{\infty} x[n]z^{-n}$. Taking the indicated limit, clearly all the terms after the summation sign approach zero as $z \to \infty$ (since $X(z)$ exists for a causal sequence) and the theorem is proven. These and other properties are summarized in Table 8.1.

TABLE 8.1 Properties of the Z-Transform

Property	Time Domain	Z-Domain
Linearity	$ax[n] + by[n]$	$aX(z) + bY(z)$
Scaling by a^n	$a^n x[n]$	$X\left(\dfrac{z}{a}\right)$
Multiply by n	$nx[n]$	$-z\dfrac{d}{dz}X(z)$
Time convolution	$x[n] * y[n]$	$X(z)Y(z)$ (bilateral Z-transform)
	$x[n]u[n] * y[n]u[n]$	$X(z)Y(z)$ (unilateral Z-transform)
Right shift (delay)	$x[n-k]$	$z^{-k}X(z)$ (bilateral Z-transform)
	$x[n-k]$	$z^{-k}X(z) + \sum_{i=1}^{k} x[-i]z^{-k+i}$ (unilateral Z-transform)
	$x[n-k]u[n-k]$	$z^{-k}X(z)$ (unilateral Z-transform)
Left shift	$x[n+k]$	$z^{k}X(z)$ (bilateral Z-transform)
	$x[n+k]u[n]$	$z^{k}X(z) - \sum_{i=0}^{k-1} x[i]z^{k-i}$ (unilateral Z-transform)
Amplitude modulation	$\cos[\omega_0 n]x[n]$	$\frac{1}{2}[X(ze^{j\omega}) + X(ze^{-j\omega})]$
	$\sin[\omega_0 n]x[n]$	$\frac{j}{2}[X(ze^{j\omega}) + X(ze^{-j\omega})]$
Initial value	$x[0]$	$= \lim_{z\to\infty} X(z)$ ($x[n]$ causal)
Final value	$\lim_{n\to\infty} x[n]$	$= \lim_{z\to 1}[(1 - z^{-1})X(z)]$ ($X(z)$ rational; magnitudes of poles of $(1 - z^{-1})X(z) < 1$)

8.5 POLES AND ZEROS OF Z-TRANSFORMS

As seen from the examples presented in Table 8.2, many common Z-transforms have the form of rational functions—that is, a ratio of two polynomials in z^{-1} (or in z). Therefore, as was the case for Laplace transforms, the numerator and denominator polynomials may be factored so that the transform has the form

$$X(z) = \frac{N(z)}{D(z)} = \frac{b_0 + b_1 z^{-1} + \ldots + b_M z^{-M}}{a_0 + a_1 z^{-1} + \ldots + a_N z^{-N}}$$

$$= \frac{b_0}{a_0} \frac{\Pi_{k=1}^{M}(1 - z_k z^{-1})}{\Pi_{k=1}^{N}(1 - p_k z^{-1})} = \frac{b_0}{a_0} z^{-M+N} \frac{\Pi_{k=1}^{M}(z - z_k)}{\Pi_{k=1}^{N}(z - p_k)}, \tag{8.7}$$

TABLE 8.2 Some Common Z-Transforms

Signal	Z-Transform	ROC
$\delta[n]$	1	all z
$u[n]$	$\dfrac{z}{z-1}$	$\|z\| > 1$
$nu[n]$	$\dfrac{z}{(z-1)^2}$	
$n^2 u[n]$	$\dfrac{z(z+1)}{(z-1)^3}$	
$\dfrac{n(n-1)}{2!} u[n]$	$\dfrac{z}{(z-1)^3}$	
$a^n u[n]$	$\dfrac{z}{z-a}$	$\|z\| > \|a\|$
$na^n u[n]$	$\dfrac{az}{(z-a)^2}$	$\|z\| > \|a\|$
$(n+1)a^n u[n]$	$\dfrac{z^2}{(z-a)^2}$	
$(n+1)(n+2)a^n u[n]$	$\dfrac{z^3}{(z-a)^3}$	
$\cos[\omega_0 n] u[n]$	$\dfrac{z^2 - z\cos(\omega_0)}{z^2 - 2z\cos(\omega_0) + 1}$	$\|z\| > 1$
$\sin[\omega_0 n] u[n]$	$\dfrac{z\sin(\omega_0)}{z^2 - 2z\cos(\omega_0) + 1}$	$\|z\| > 1$
$a^n \cos[\omega_0 n] u[n]$	$\dfrac{z^2 - az\cos(\omega_0)}{z^2 - 2az\cos(\omega_0) + a^2}$	$\|z\| > \|a\|$
$a^n \sin[\omega_0 n] u[n]$	$\dfrac{az\sin(\omega_0)}{z^2 - 2az\cos(\omega_0) + a^2}$	$\|z\| > \|a\|$

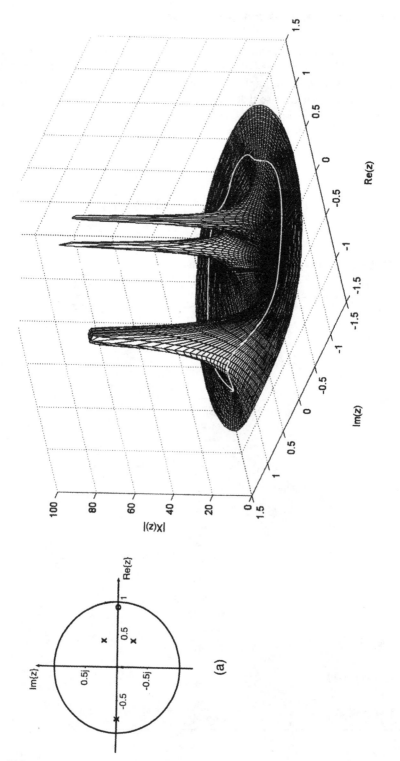

FIGURE 8.3. (a) Locations in the z-plane of poles and zeros of $X(z) = 10(z - 0.9)/(z^3 - 0.44z + 0.16)$. (b) Magnitude of $X(z)$. White line indicates projection of unit circle onto the magnitude function.

where the a_i and b_i coefficients are constants, and the z_k and p_k are the zeros and poles of $X(z)$, respectively. Assuming $N > M$, then $X(z)$ also has $N - M$ zeros at $z = 0$. Note from Eq. (8.7) that if one knows the poles and zeros of a Z-transform, then it is possible to construct $X(z)$ to within a gain constant.

The representation of Eq. (8.7) is advantageous when one wishes to obtain insights into the DTFT, $X(e^{j\omega})$. To see this, consider the graph (Fig. 8.3(a)) of the locations in the z-plane of the poles and zeros of the Z-transform of a right-sided sequence,

$$X(z) = \frac{10(z - 0.9)}{z^3 - 0.44z + 0.16}.$$

$X(z)$ has a zero at $z = 0.90$, a simple pole at $z = -0.80$, and complex conjugate poles at $z = 0.40 \pm j0.20$. Its ROC will include the unit circle. Recall that $X(e^{j\omega})$ equals the Z-transform, $X(z)$, evaluated on the unit circle. Knowledge of the locations of the zeros and poles of $X(z)$ permits one to estimate the shape of the magnitude of the DTFT. Thus, when a zero of $X(z)$ lies on the unit circle, then $|X(e^{j\omega})| = 0$ at the corresponding frequency. Furthermore, because a Z-transform has continuous derivatives in its ROC, if a zero lies near, but not on, the unit circle, the magnitude of the DTFT must be close to zero for frequencies corresponding to points on the unit circle that are proximate to the zero of $X(z)$. Similarly, when a pole is near the unit circle, the magnitude of $X(z)$ (and of the DTFT) at nearby points on the unit circle should be very large. By such arguments one may obtain qualitative insights into the variation of $|X(e^{j\omega})|$ with frequency. The magnitude of $X(z)$ in the present example is shown in Figure 8.3(b), from which one can conclude that $|X(e^{j\omega})|$ has the following properties: (1) it is nearly zero around $\omega = 0$ (i.e., near $z = 1$) because of the zero at $z = 0.90$ (just visible behind the twin, complex conjugate peaks in the figure); (2) for $0 < \omega < \pi/2$, it will exhibit a small peak due to the "skirt" of the complex conjugate peak; (3) for $3\pi/4 < \omega < 5\pi/4$, there will be a larger peak because of the pole at $z = -0.80$. Of course, $|X(e^{j\omega})|$ in the range $0 > \omega > -\pi$ is the mirror image of itself for $0 < \omega < \pi$ due to the symmetry properties of the DTFT.

8.6 THE INVERSE Z-TRANSFORM

The inverse Z-transform specifies the calculation of $x[n]$ from $X(z)$. The most general method utilizes the Cauchy integral theorem that may be given in one form as

$$\frac{1}{2\pi j} \oint_C z^{k-1} dz = \begin{cases} 1, & k = 0 \\ 0, & k \neq 0, \end{cases} \tag{8.8}$$

where C is a contour in the z-plane that encircles the origin. We apply this theorem to the definition of the Z-transform, Eq. (8.3), by multiplying both sides of this latter equation by z^{k-1} and integrating over a contour C that lies entirely within the ROC of $X(z)$. Thus we obtain

$$\frac{1}{2\pi j} \oint_C X(z)z^{k-1}dz = \frac{1}{2\pi j} \oint_C \sum_{n=-\infty}^{\infty} x[n]z^{-n+k-1}dz$$

$$= \sum_{n=-\infty}^{\infty} x[n]\frac{1}{2\pi j} \oint_C z^{-n+k-1}dz. \qquad (8.9)$$

By Eq. (8.8), the last term of Eq. (8.9) is nonzero only when $k = n$. Therefore, only the $x[n]$ term from the summation remains, and

$$x[n] = \frac{1}{2\pi j} \oint_C X(z)z^{n-1}dz. \qquad (8.10)$$

Equation (8.10) is the general definition of the *inverse Z-transform*. Eq. (8.10) may be evaluated using a residue theorem that states that the integral around the contour, C, is equal to the summation of the residues of all poles of $X(z)z^{n-1}$ contained within C, where the residue of a pole at $z = z_j$ is defined as $\text{Res}_j = (z - z_j)X(z)z^{n-1}|_{z=z_j}$. This method is always valid for any Z-transform and any value of N. Since in many cases $X(z)$ has the form of a rational function, however, one may avoid contour integration and evaluate the inverse Z-transform using a partial fraction approach similar to that used with Laplace transforms.

Before describing the partial fraction method, it should be noted that if $X(z)$ can be put into the form of the summation in Eq. (8.3), then $x[n]$ may be inferred directly. If $X(z)$ is a rational function, this form may be achieved by dividing the numerator polynomial by the denominator polynomial. There are complications if $x[n]$ contains both causal and non-causal signals, but assuming that $x[n]$ is causal, then the described division will produce the result

$$X(z) = x[0] + x[1]z^{-1} + x[2]z^{-2} + x[3]z^{-3} + \dots,$$

from which $x[n]$ is available by inspection. Of course, if one requires more than just the first few values of $x[n]$, this method may become extremely tedious.

Inverse Z-Transforms by Partial Fractions

As was the case for inverse Laplace transforms, the goal is to express $X(z)$ as a sum of Z-transforms, each of which has an identifiable inverse. One may start with $X(z)$ expressed as either a function of z^{-1} or a function of z but the latter is more common. Thus, given a valid Z-transform, $X(z)$, in the form

$$X(z) = \frac{N(z)}{D(z)}, \qquad (8.11)$$

first one must ensure that $X(z)$ is proper by dividing $N(z)$ by $D(z)$ until the order of the numerator, M, is strictly less than the order, N, of $D(z)$. Thus $X(z)$ will have the form

$$X(z) = \sum_{k=0}^{K} c_k z^k + \frac{N_1(z)}{D(z)}. \tag{8.12}$$

Since the inverse transform of the summation term in Eq. (8.12) is $\sum_{k=0}^{K} c_k \delta[n + k]$, it remains only to determine the inverse transform of the second term, a proper rational polynomial.

For simplicity, assume we begin with $X(z)$ of Eq. (8.11) already a proper rational function with $M < N$. If $X(z)$ is expressed as a function of z^{-1}, then multiply both $N(z)$ and $D(z)$ by z^N so that now

$$X(z) = \frac{b_0 z^N + b_1 z^{N-1} + \ldots + b_M z^{N-M}}{z^N + a_1 z^{N-1} + \ldots + a_{N-1} z + a_N}. \tag{8.13}$$

Here we have assumed that $a_0 = 1$ without any loss of generality since one may always divide both $N(z)$ and $D(z)$ by a_0 if it does not equal one. As will become apparent, in order to produce terms having identifiable inverse transforms it is necessary to expand $P(z) = (X(z)/z)$ using partial fractions rather than $X(z)$ itself. Assuming that $D(z)$ comprises only simple, nonrepeated factors of the form $(z - p_i)$, then the expansion will have the form

$$P(z) = \frac{X(z)}{z} = \frac{N(z)}{z \prod_{i=1}^{N}(z - p_i)} = \frac{c_0}{z} + \frac{c_1}{z - p_1} + \ldots + \frac{c_N}{z - p_N}. \tag{8.14}$$

We proceed exactly as for finding inverse Laplace transforms. Defining $p_0 = 0$, the c_i's may be evaluated as

$$c_i = (z - p_i)\frac{X(z)}{z}\bigg|_{z=p_i}, \qquad i = 0, 1, \ldots, N. \tag{8.15}$$

Multiplying Eq. (8.14) by z,

$$zP(z) = X(z) = c_0 + \frac{c_1 z}{z - p_1} + \ldots + \frac{c_N z}{z - p_N}. \tag{8.16}$$

Either by inspection or by using Table 8.2, the inverse Z-transform of Eq. (8.16) is

$$x[n] = c_0 \delta[n] + c_1(p_1)^n u[n] + \ldots + c_N(p_N)^n u[n], \tag{8.17}$$

assuming that $x[n]$ is causal.

Assuming that all of the coefficients in $X(z)$ are real, then if $X(z)$ has a complex root, p_i, it also will have another root that is the complex conjugate of p_i. By convention we assume that this latter root is indexed as p_{i+1}. Consequently, $p_{i+1} = p_i^*$ and the expansion coefficients c_i and c_{i+1} will be related by $c_{i+1} = c_i^*$. Therefore, if c_i

and p_i are expressed as $c_i = |c_i|e^{j\alpha_i}$, $p_i = |p_i|e^{j\beta_i}$, this complex pole pair will have a causal inverse transform given by

$$Z^{-1}\left\{\frac{c_i}{z - p_i} + \frac{c_i^*}{z - p_i^*}\right\} = |c_i|e^{j\alpha_i}|p_i|^n e^{jn\beta_i}u[n] + |c_i|e^{-j\alpha_i}|p_i|^n e^{-jn\beta_i}u[n]$$

$$= 2|c_i||p_i|^n \cos[\beta_i n + \alpha_i]u[n]. \qquad (8.18)$$

Example 8.6 Inverse Z-transform by partial fractions   To find the (causal) inverse Z-transform of

$$X(z) = \frac{z^{-1} + z^{-3}}{1 - z^{-1} - z^{-2} - 2z^{-3}} = \frac{z^2 + 1}{z^3 - z^2 - z - 2}$$

we first find a partial fraction expansion (using, for example, the MATLAB function residue) as shown below:

$$\frac{X(z)}{z} = \frac{c_0}{z} + \frac{c_1}{z - p_1} + \frac{c_2}{z - p_2} + \frac{c_3}{z - p_3},$$

where $p_1 = -\frac{1}{2} - j(\sqrt{3}/2)$, $p_2 = -\frac{1}{2} + j(\sqrt{3}/2)$, $p_3 = 2$.
Evaluating the coefficients in the usual manner,

$$c_0 = -\tfrac{1}{2}, \qquad c_1 = 0.0714 + j0.2062, \qquad c_2 = c_1^*, \qquad c_3 = 0.3571.$$

One may express p_1 and c_1 as $p_1 = (1)e^{j\frac{4\pi}{3}}$, $c_1 = 0.2182e^{j1.237}$. Then by Eqs. (8.17) and (8.18),

$$x[n] = \frac{1}{2}\delta[n] + 0.3571(2)^n u[n] + 2(0.2182)(1)^n \cos\left[\frac{4\pi}{3}n + 1.237\right]u[n].$$

Repeated Poles

Repeated poles in Z-transforms are handled in the same way as in Laplace Transforms—by taking derivatives of the partial fraction expansion and, after each derivative, evaluating the c_i coefficient that is not multiplied by some power of $(z - p_j)$. Thus, in the case of a simple pole repeated r times, $X(z)/z$ may be expanded as

$$\frac{X(z)}{z} = \frac{c_0}{z} + \frac{c_1}{z - p_1} + \frac{c_2}{(z - p_1)^2} + \ldots + \frac{c_r}{(z - p_1)^r} + \frac{c_{r+1}}{z - p_2} + \ldots + \frac{c_N}{z - p_{N-r}}.$$

The coefficients for the repeated poles are evaluated from the relationship

$$c_{r-1} = \frac{1}{i!}\left[\frac{d^i}{dz^i}\left\{(z-p_1)^r\frac{X(z)}{z}\right\}\right]\Bigg|_{z=p_1}. \qquad (8.19)$$

Table 8.2 indicates some causal time responses corresponding to poles of the form $(z-p_1)^k$.

8.7 POLE LOCATIONS AND TIME RESPONSES

Given an input signal, it is possible to calculate the exact output of an LSI system having a known impulse response, $h[n]$, (assuming both the impulse response and the input signal are Z-transformable); however, there are many situations in which graphical convolution may be useful to approximate the output. In other cases $H(z)$ may be known but not $h[n]$. In these cases it is useful to be able to approximate $h[n]$ from knowledge of the poles of $H(z)$. Using Eqs. (8.17) and (8.18) one may sketch the shape of the function $h[n]$ for various possible simple or complex poles. Because of linearity, if $H(z)$ comprises a sum of such poles, then $h[n]$ will be the equivalent summation of the corresponding time responses. Fig. 8.4(a–d) depicts the time responses corresponding to simple poles. Note that the unit circle is the boundary between responses that decay with time and those that increase in an unbounded fashion. One can conclude that a BIBO stable system must have its poles inside (or possibly on) the unit circle. Note that a pole to the left of the imaginary axis corresponds to a time function that alternates in sign. From Eq. (8.18) we expect that a pair of complex conjugate poles will correspond to an amplitude-modulated DT cosine waveform. We see from Fig. 8.4(e–f) that the unit circle also divides the z-plane in this case, so that complex conjugate poles inside the unit circle correspond to sinusoids with decaying amplitudes whereas time responses related to poles outside the unit circle increase in amplitude without bound. Complex conjugate poles on the unit circle correspond to constant-amplitude cosine signals. Finally, Fig. 8.4(g–h) indicates the responses for a repeated simple pole. Note the biphasic nature of the signal when the poles are inside the unit circle. Note also that a repeated pole at $z = 1$ does not correspond to a bounded signal, as it does when there is a single pole at this location.

8.8 THE UNILATERAL Z-TRANSFORM

Recall that the output of an n-th order linear system (i.e., the solution of its difference equation) for $n > n_0$, where n_0 is an arbitrary reference time, can be determined from knowledge of: (1) the system variables at time $n = n_0$, and; (2) the input for $n \geq n_0$. To obtain the same information about the output using the bilateral Z-transform, however, requires knowledge of the input for all time. The unilateral Z-trans-

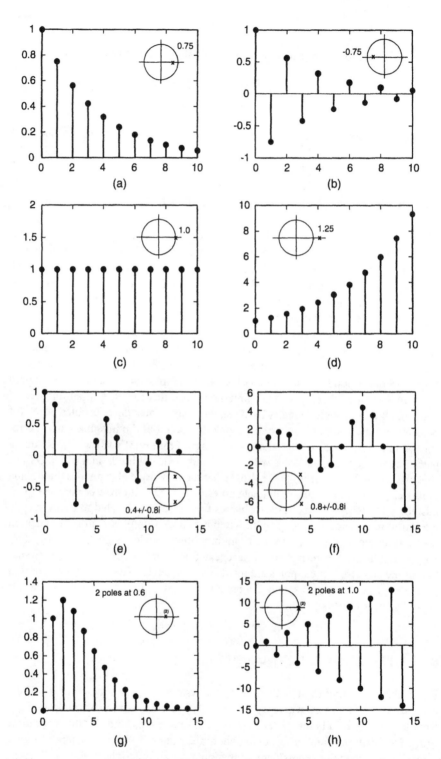

FIGURE 8.4. Time responses corresponding to various pole locations in the z-plane. Insets indicate pole locations relative to the unit circle.

form circumvents this problem. The unilateral Z-transform of a signal $x[n]$ is defined as

$$X^+(z) \triangleq \sum_{n=0}^{\infty} x[n]z^{-n}. \tag{8.20}$$

Obviously, for any $x[n]$ the unilateral Z-transform equals the bilateral Z-transform of $x[n]u[n]$. Furthermore, if $x[n]$ is causal, then $x[n] = x[n]u[n]$ and the unilateral and bilateral Z-transforms are equal. In this case the ROC of the unilateral transform is the same as that of the bilateral transform—that is, the exterior of a circle defined by the largest pole. Likewise, in this case there is a unique relationship between $X^+(z)$ and $x[n]$ and it is unnecessary to specify the ROC except as it relates to boundedness of $x[n]$ (or stability of an impulse response, $h[n]$).

The unilateral Z-transform manifests most of the same properties as the bilateral Z-transform; the important exceptions relate to time shifts of $x[n]$. If $x[n]$ has the transform $X^+(z)$, then the delayed signal $y[n] = x[n - k]$, $k > 0$, has the unilateral Z-transform $Y^+(z) = z^{-k}[X^+(z) + \sum_{n=1}^{k} x[-n]z^n]$, $k > 0$. This result is proven easily from the definition, Eq. (8.20). One may interpret this result as follows: If $y[n]$ equals $x[n]$ delayed by k samples, then for $n \geq 0$, $y[n]$ contains k samples that are absent from $x[n]$. Therefore, in addition to multiplying $X^+(z)$ by z^{-k}, it is necessary to add the Z-transform of the new samples, as reflected by the summation term in the above result. For example, if $y[n] = x[n - 1]$, then $Y^+(z) = z^{-1}X^+(z) + x[-1]$. If $y[n] = x[n - 2]$, then $Y^+(z) = z^{-2}X^+(z) + x[-1]z^{-1} + x[-2]$.

A similar consequence applies when a signal is advanced in time. If $y[n] = x[n + k]$, where $k > 0$, then $Y^+(z) = z^k[X^+(z) - \sum_{n=0}^{k-1} x[n]z^{-n}]$. The interpretation parallels that for time delaying, except that now the first k samples of $x[n]$ (i.e., $x[0]$ through $x[k - 1]$) are "lost" due to the time advance.

Use of the unilateral Z-transform is principally for calculating the output of a linear filter when it is not initially at rest or when its input is causal. Note, however, that the bilateral Z-transform also is applicable when a filter is initially at rest and the input is causal. Here we first solve a specific example, then in the next section we will consider the general problem of this type and finally the explicit case of a filter initially at rest having a causal input.

Example 8.7   Unilateral Z-transform   Consider a causal filter described by the difference equation

$$y[n] - ay[n - 1] = x[n], \qquad x[n] = b^n u[n], \qquad y[-1] = c_1.$$

Since in this case $X^+(z) = X(z)$, one may apply the unilateral Z-transform to both sides of this equation, yielding

$$Y^+(z) - az^{-1}Y^+(z) - ay[-1] = \frac{1}{1 - bz^{-1}}.$$

Therefore,

$$Y^+(z) = \frac{ac_1}{1 - az^{-1}} + \frac{1}{1 - az^{-1}} \frac{1}{1 - bz^{-1}}. \tag{8.21}$$

The r.h.s. of Eq. (8.21) comprises a sum of two signals, the response to the initial condition (zero-input response) and the specific response to the input signal (zero-state response). Since the input is causal, we may assume that the response will be causal and therefore the unilateral and bilateral Z-transforms are equal. Thus the inverse transform of Eq. (8.21) may be determined by a partial fractions method to be

$$y[n] = c_1 a^{n+1} u[n] + \frac{b^{n+1} - a^{n+1}}{b - a} u[n].$$

8.9 ANALYZING DIGITAL FILTERS USING Z-TRANSFORMS (DT TRANSFER FUNCTIONS)

In practice, one is often concerned with LSIC systems at rest having causal inputs. Such cases may be analyzed using either the bilateral or unilateral Z-transform and from here on we shall drop the "+" superscript in the unilateral Z-transform unless it is necessary to emphasize the explicit use of this form of the transform.

We now consider the generic LSI system of Fig. 8.5. Just as the frequency response of an LSI system is defined as the DTFT, $H(e^{j\omega})$ of its unit-pulse response, $h[n]$, the *transfer function* of an LSI system is the Z-transform, $H(z)$, of $h[n]$. From the convolution property we already know that one may determine the Z-transform of the output from the relationship $Y(z) = H(z)X(z)$. As the previous example indicated, $Y(z)$ may be determined also by way of the difference equation of the system. The general difference equation of an LSI system is

$$y[n + N] = -\sum_{k=0}^{N-1} a_{N-k} y[n + k] + \sum_{k=0}^{M} b_{M-k} x[n + k]. \tag{8.22}$$

Assuming that the system is initially quiescent, the Z-transform may be applied to this equation to yield

$$z^N Y(z) = -\sum_{k=0}^{N-1} a_{N-k} z^k Y(z) + \sum_{k=0}^{M} b_{M-k} z^k X(z).$$

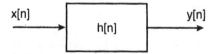

FIGURE 8.5. A generic LSI system.

Solving for $Y(z)$:

$$Y(z) = \frac{\sum\limits_{k=0}^{M} b_{M-k} z^k}{z^N + \sum\limits_{k=0}^{N-1} a_{N-k} z^k} X(z). \tag{8.23}$$

Since $Y(z) = H(z) X(z)$, we conclude that

$$H(z) = \frac{\sum\limits_{k=0}^{M} b_{M-k} z^k}{z^N + \sum\limits_{k=0}^{N-1} a_{N-k} z^k}. \tag{8.24}$$

Equation (8.24) may be rewritten as

$$H(z) = \frac{\sum\limits_{k=0}^{M} b_{M-k} z^k}{\sum\limits_{k=0}^{N} a_{N-k} z^k} = \frac{B(z)}{A(z)}, \qquad \text{where } a_0 = 1, \tag{8.25}$$

and

$$A(z) = z^N + a_1 z^{N-1} + \ldots + a_{N-1} z + a_N, \quad B(z) = b_0 z^M + b_1 z^{M-1} + \ldots + b_{M-1} z + b_M. \tag{8.26}$$

That is, if $H(z)$ has the form of a rational polynomial function, then the coefficients of the difference equation are obtainable by inspection of $H(z)$, and vice versa. If the filter is also causal (i.e., LSIC), then either the unilateral or the bilateral Z-transform may be applied to determine $H(z)$.

Example 8.8   DT transfer function   An LSIC filter has the transfer function $H(z)$ $= (z^{-1} - 2z^{-2})/(1 - 3z^{-1} + 3z^{-2})$. To determine its difference equation we may multiply by (z^2/z^2) to obtain the form of Eq. (8.25). This step, however, is unnecessary as we may simply set $H(z) = Y(z)/X(z)$ and cross-multiply, obtaining

$$Y(z)[1 - 3z^{-1} + 3z^{-2}] = X(z)[z^{-1} - 2z^{-2}].$$

Utilizing bilateral Z-transforms the inverse Z-transform operation yields

$$y[n] - 3y[n-1] + 3y[n-2] = x[n-1] - 2x[n-2].$$

This same equation would result from applying the unilateral Z-transform and assuming that initial conditions are zero.

Example 8.9   Output response of an LSIC filter   To calculate the zero-state step response of a filter having the transfer function $H(z) = (z - 1)/(z^2 + z)$, first we set $x[n] = u[n]$. The Z-transform of the output is given by

$$Y(z) = H(z)X(z) = \frac{z-1}{z(z+1)} \frac{z}{z-1} = \frac{1}{z+1}.$$

Now

$$\frac{1}{z+1} = 1 - \frac{z}{z+1}; \therefore y[n] = \delta[n] - (-1)^n u[n].$$

Example 8.10   Frequency response   For calculating the power spectrum many biomedical signals can be modeled as the output signal of a linear filter whose input is random noise (see Chapter 9). If the signal has been sampled, then the modelling filter is an LSIC system. For the electromyogram (EMG) signal recorded by large surface electrodes overlying a skeletal muscle, a reasonable first approximation is obtained by using a filter of the type $H(z) = b_0/(1 + a_1 z^{-1} + a_2 z^{-2})$. Some realistic values for the parameters (taken from the report by Bower et al. (1984), assuming a sampling frequency normalized to one) are: $b_0 = 2.0$, $a_1 = -1.35$, $a_2 = 1.7$. To resolve the type of filter represented by this transfer function, we evaluate the frequency response as

$$H(e^{j\omega}) = H(z)|_{z=e^{j\omega}} = \frac{b_0}{1 + a_1 e^{-j\omega} + a_2 e^{-j2\omega}}$$

$$= \frac{2.0}{1 - 1.35(\cos(\omega) - j\sin(\omega)) + 1.7(\cos(2\omega) - j\sin(\omega))}.$$

Thus

$$|H(e^{j\omega})| = 2[(1 - 1.35\cos(\omega) + 1.7\cos(2\omega))^2 + (1.35\sin(\omega) - 1.7\sin(2\omega))^2]^{-0.5}.$$

This frequency response function is plotted in Fig. 8.6. It resembles the frequency response of an underdamped, second-order, lowpass filter. If uncorrelated random noise is the input to this filter, then its output will have a power spectrum that approximates the power spectrum of the EMG signal.

8.10   BIOMEDICAL APPLICATIONS OF DT FILTERS

Biomedical applications for LSIC filters are abundant. Almost every sampled signal requires some sort of filtering for noise removal and we shall present several examples later. The first example, however, demonstrates that digital filters may be ap-

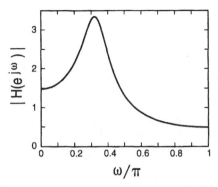

FIGURE 8.6. Frequency response of a two-pole model of an EMG signal.

plied to any signal that is sampled at uniform increments of another variable. Most commonly the "other" variable is time but it need not be.

Example 8.11   Measurement of tissue pH using an optical dye   Neutral Red is a dye that absorbs light in differering amounts at different frequencies. Furthermore, the wavelength of light at which maximum absorption occurs is altered in the presence of hydrogen ions which attach to the dye in proportion to their concentration. At very high pH (i.e., low H^+ concentration) the wavelength of maximum absorption is around 465 nm, whereas at very low pH it is around 560 nm. To measure the pH of a thin strip of tissue, the tissue first is bathed in Neutral Red. Then a variable-wavelength light source is directed at the tissue so that light passes through the tissue to a photomultiplier tube. If percent transmission of light is measured at many discrete frequencies, the pH of the tissue may be derived from the ratio of the percent transmission at 560 nm and 465 nm.

Percent transmission is measured at many frequencies because if the data are noisy, one might need to smooth the data in the vicinity of the two critical wavelengths in order to estimate the true percent transmission better. To illustrate how a digital filter could be applied in this situation, we create an artificial example that can be solved analytically. Assume that the percent transmission, p, was measured at 1 nm increments in wavelength from 450 to 650 nm, so that we may create a signal, $p[n]$, where $p[0]$ is the transmission at 450 nm, $p[1]$ is the transmission at 451 nm, and so forth, and $p[n]$ comprises 201 sample points. Thus $n = 15$ corresponds to 465 nm and $n = 110$ corresponds to 560 nm. For illustration, assume that $p[n]$ is a sine wave of the form

$$p[n] = 100 - 50 \sin\left[\frac{2\pi}{400}n\right] + 10 \sin\left[\frac{2\pi}{5}n\right]$$

(Fig. 8.7(a)), where the first two terms on the right describe the "actual" absorption curve and the third term represents an added, high-frequency, noise. To reduce the

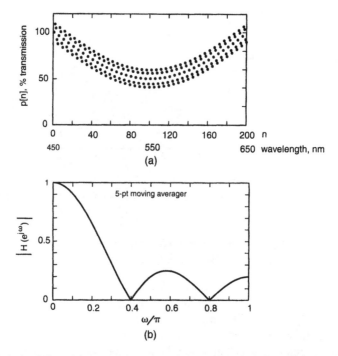

FIGURE 8.7. A simulation of the percent transmission, as a function of wavelength, of a tissue stained with Neutral Red dye. (b) Frequency response of a five-point moving average filter.

noise effect, we will use a filter having the unit-pulse response, $h[n] = 0.2$, $0 \le n \le 4$, and $h[n] = 0$, otherwise. This FIR digital filter is a five-point moving averager that will be used to average across samples in wavelength instead of samples in time. Its frequency response can be determined from $H(z)$. Thus

$$H(z) = \sum_{n=0}^{4} 0.2z^{-n} = 0.2\frac{1 - z^{-5}}{1 - z^{-1}}, \qquad \text{and}$$

$$H(e^{j\omega}) = 0.2\frac{1 - e^{-j5\omega}}{1 - e^{-j\omega}} = 0.2e^{-j2\omega}\frac{e^{j2.5\omega} - e^{-j2.5\omega}}{e^{j0.5\omega} - e^{-j0.5\omega}}.$$

The magnitude of the frequency response is

$$|H(e^{j\omega})| = 0.2\left|\frac{\sin(2.5\omega)}{\sin(0.5\omega)}\right|,$$

where "frequency" is relative to phenomena that are periodic along the wavelength axis instead of along the time axis. This latter function is plotted in Fig. 8.7(b). The filter has a lowpass characteristic and, at the frequency of the "noise" (i.e., 0.4π) its gain is zero. If we filter $p[n]$ using this $H(e^{j\omega})$, the noise will be removed entirely. Obviously in a realistic situation one could not design a filter to remove the noise completely, but the essential point is that digital filters are not restricted to functions of time.

Another common biomedical application for Z-transforms is in the design of systems for closed-loop control. For example, a digital controller might be designed to control the position of an arm prosthesis using stepper motors. In such systems a sensor provides feedback to a control system from the biomedical system being controlled. The control system combines this information with a model of the dynamic properties of the controlled system to predict its output at some future time. If the predicted output differs from the desired output, then the controller modifies the input to the controlled system so that the predicted output response is correct. In many such applications it is either not possible to develop detailed realistic models of the controlled system, or too computationally intensive to compute predicted outputs for such models. A feasible alternative is to implement a linear approximation to the controlled system and use it to predict the needed change in output from the current level. Typically the parameters of this linear model are reevaluated periodically to ensure that the model is still sufficiently accurate. For further development of this topic the reader is referred to the many textbooks on closed-loop control systems.

8.11 OVERVIEW: DESIGN OF DIGITAL FILTERS

The design of digital filters is an extensive topic whose practical implementation is eased considerably by the availability of modern computer software. Often we will refer to commands in the MATLAB Signal Processing Toolbox that may be utilized for designing specific types of digital filters. The intent here is to familiarize the reader with the principles and algorithms for digital filter design and with the advantages and limitations of various design methods for different biomedical applications so that design software may be utilized effectively.

Although there are situations in which it is desirable to filter inherently discrete signals (as some previous examples have demonstrated), a very common problem is the need to remove noise from a sampled CT signal as shown in Fig. 8.8. Because it

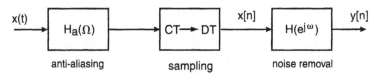

FIGURE 8.8. Scheme for removing noise from a sampled CT signal, $x(t)$. Some noise removal is accomplished by the anti-aliasing filter but usually it is necessary to design a digital filter to remove noise from $x[n]$.

is important to prevent aliasing by lowpass filtering of the signal $x(t)$ before sampling, it seems reasonable to accomplish noise removal with this filter as well. This goal is not always possible. For example, the noise might occur at frequencies near the important frequency content of $x(t)$ and, since anti-aliasing filters are analog filters, usually it is not possible to achieve the narrow transition bands that are possible with digital filters except in very costly analog filters. Furthermore, low-frequency noise cannot be removed by the anti-aliasing filter. Also, one may need to preserve all of the signal content of $x(t)$ for one application but, at the same time, require only a subset of this frequency content for another application. A typical example is the analysis of saccadic eye movements: To determine the exact durations and amplitudes of saccades it is necessary to preserve high-frequency content of the movement signal (at the cost of having to distinguish a noisy "blip" from a saccade by some other criteria), whereas if one only needs to count the number of large-amplitude saccades per unit time, then the signal may be lowpass filtered with a lower cutoff frequency.

Previously it was established that an LSI system (i.e., a digital filter) may be classified as an IIR (infinite impulse response) or FIR (finite impulse response) filter depending on the duration of its impulse response. The design methods for IIR and FIR filters differ and we will consider these two classes of filters separately. In addition, a digital filter may be designed to be (in some sense) "similar to" a specific analog filter, or it may be designed by considering only its DT filtering properties. Because practical (i.e., causal) analog filters possess infinitely-long impulse responses, their natural complements are IIR digital filters. One may truncate an infinitely-long impulse response and develop an approximating FIR digital filter, of course, but it is likely that the filter order will need to be quite high in order to preserve the "similarity" to the analog filter. We will find that some simple digital filter design methods that begin from an analog filter specification are useful only for lowpass and some restricted bandpass filters. These filters have the advantage, however, that their transfer functions can be represented by a few coefficients. FIR filters may require many coefficients but contemporary design methods can produce elegant, highly selective filters. Here we initially focus on the design of lowpass filters. Subsequently we will develop the means for transforming a lowpass digital filter into a highpass or bandpass filter.

An important question is whether a digital filter must be causal. If the entire signal to be filtered is available in digital form, then this requirement is unnecessary. Noncausal filters have the advantage that they can be designed to produce zero phase shift. Thus, one does not need to compensate when determining the *timing* of events in sampled signals relative to external (i.e., unsampled) events such as the occurrence of a triggering event. (It is imperative, however, that *signals* being *compared* should be filtered identically!) On the other hand, for real-time applications such as control of a prosthesis causal filters usually are mandatory and, as mentioned above, designs based on analog filters necessarily consider causal filters only. Because causal filters are always applicable, we assume causality unless specified otherwise. This assumption has nontrivial implications for the resulting filter design. First, by the Paley-Weiner Theorem (Proakis and Manolakis, 1996; *p.* 616) the frequency response of a causal system cannot be zero except at a finite number of non-adjacent frequencies. Consequently, as we established previously, an ideal filter cannot be causal. Second, the frequency response cannot be constant over a finite frequency range and the transition from passband to stopband can not be infinitely sharp. Third, although it has not been shown here, the real and imaginary parts of the frequency response of a causal filter are interdependent and the designer cannot specify both arbitrarily.

Often IIR filters are specified (in the transform domain) as the ratio of two polynomials as in Eq. (8.25). Such filters also are known as recursive filters since the current output is used to determine future outputs (Eq. (8.22)). The transfer function of FIR filters is a single, finite-length polynomial—for example, $B(z)$ in Eq. (8.25), with $A(z) = 1$. FIR filters are also called *nonrecursive filters*.

8.12 IIR FILTER DESIGN BY APPROXIMATING A CT FILTER

Because many biomedical signals are continuous-time, one often has knowledge of the desired signal and the noise components of a signal relative to CT frequency. Thus it is a direct step to specify the parameters of an analog filter for noise removal. If noise removal is to be accomplished after sampling the data signal, one must design a digital filter that effects the same separation of frequency components that the analog filter would have achieved. This situation is diagrammed in Fig. 8.9. The data signal, $x(t)$, could be filtered effectively by the CT filter, $H(s)$, but it has been sampled at the frequency $f_s = 1/T$ to produce the signal $x[n] \equiv x(nT)$. Can one determine a digital filter, $H(z)$, so that $y[n] = y(nT)$? To address this question we need to establish a relationship between the CT transform variable 's' and the DT transform variable 'z'.

Given a CT signal $x(t)$ such that $x(t) = 0$, $t < 0$, define the DT signal $x[k] = x(kT)$. We will use the artifice of representing the "sampled" signal as a train of CT impulse functions. Thus

$$x_s(t) = \sum_{k=0}^{\infty} x(kT)\delta(t - kT). \qquad (8.27)$$

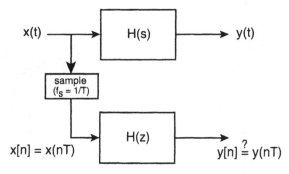

FIGURE 8.9. Parallels between filtering of a CT signal, $x(t)$, and filtering of the sampled signal, $x[n]$.

The Laplace transform of Eq. (8.27) is

$$X_s(s) = \sum_{k=0}^{\infty} x(kT) L\{\delta(t - kT)\} = \sum_{k=0}^{\infty} x(kT) e^{-sTk}.$$

On the other hand, we may calculate the Z-transform of $x[k]$ and evaluate it at $z = e^{sT}$, yielding

$$X(z)|_{z=e^{sT}} = \sum_{k=0}^{\infty} x[k] z^{-k}|_{z=e^{sT}} = \sum_{k=0}^{\infty} x(kT) e^{-sTk} = X_s(s). \qquad (8.28)$$

Equation (8.28) establishes a relationship between the Laplace Transform of the sampled signal considered as a CT impulse train and the Z-transform of the corresponding DT signal through the relationship $z = e^{sT}$. Appropriately the relationship between the CT and DT transform variables depends on the sampling interval, T!

To better understand this result, we consider the mapping of s-plane regions onto z-plane regions via the relationship $z = e^{sT}$, as depicted in Fig. 8.10. It is easy to show that $s = 0$ maps to $z = 1$, and that the points A-E in the s-plane map to the indicated points in the z-plane. For instance, C is the point $s = j(\pi/T)$, which maps to $z = e^{j\frac{\pi}{T}T} = -1$. Consequently each strip of the imaginary axis in the s-plane of width $2\pi/T$ maps to one complete traversing of the unit-circle in the z-plane. Furthermore each corresponding portion of the l.h.s. of the s-plane (such as the shaded section) fills the interior of the unit-circle. Therefore the mapping $z = e^{sT}$ is a many-to-one mapping. It is apparent that T must be small enough (and f_s high enough) that $H(\Omega) \approx 0$, $\Omega \geq 2\pi/T$, in order for there to be a simple, direct relationship between the CT frequency response, $H(s)|_{s=j\Omega}$ and the DT frequency re-

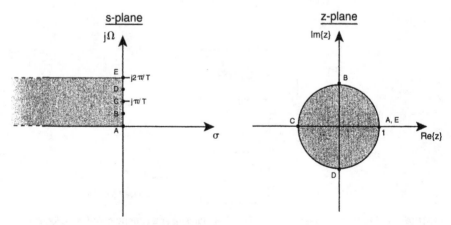

FIGURE 8.10. The mapping from the s-plane to the z-plane implicit in the relationship $z = e^{sT}$.

sponse, $H(z)|_{z=e^{j\omega}}$. In practice, this criterion means that all of the poles of $H(s)$ must lie between the lines $s = -j\pi/T$ and $s = +j\pi/T$. This criterion also implies that there will be negligible or no aliasing, in which case we have previously established that

$$H(\Omega) = TH(e^{j\omega})|_{\omega=\Omega T} = TH(z)|_{z=e^{j\Omega T}}. \tag{8.29}$$

8.13 IIR FILTER DESIGN BY IMPULSE INVARIANCE

An obvious approach to approximating a CT filter by a DT filter is to sample the CT impulse response, $h[n]$, and set the DT impulse response, $h[k]$, equal to this sampled signal—that is, $h[k] = h(kT)$. The method is exemplified assuming that $H(s)$ has only simple poles but it is readily extended to other situations. Assume that $H(s)$ is given by

$$H(s) = \sum_{i=1}^{N} \frac{c_i}{s - p_i}.$$

Then the corresponding impulse response has the form

$$h(t) = \sum_{i=1}^{N} c_i e^{p_i t} u(t).$$

Sampling this signal at intervals of T produces $h[k]$,

$$h[k] = h(kT) = \sum_{i=1}^{N} c_i e^{p_i kT} u[k]$$

and

$$H(z) = \sum_{k=0}^{\infty} \left(\sum_{i=1}^{N} c_i e^{p_i kT} \right) z^{-k} = \sum_{i=1}^{N} c_i \sum_{k=0}^{\infty} (e^{p_i kT} z^{-1})^k = \sum_{i=1}^{N} \frac{c_i}{1 - e^{p_i T} z^{-1}}. \tag{8.30}$$

That is, for each pole, p_i, in $H(s)$ one finds a pole in $H(z)$ at $z = e^{p_i T}$. The residue at this pole is c_i, the residue of the corresponding pole in $H(s)$. To calculate $H(z)$, one first determines the poles of $H(s)$, then writes out the terms of the summation of Eq. (8.30), and finally combines these terms into a single ratio of polynomials of the form of Eq. (8.25). This approach is known as the method of *impulse invariance design*. Note that the sampling of $h(t)$ must occur at a high enough frequency to prevent aliasing. For this reason the impulse invariance method cannot be used to design highpass filters. Recall also that the CT and DT frequency variables are related by $\omega = \Omega T$.

An advantage of impulse invariance design is that the digital filter will be BIBO stable if the analog filter is BIBO stable. To demonstrate this result, assume that

$H(s)$ is BIBO stable, having all of its poles in the left-half s-plane. All of these poles will map to poles in $H(z)$ that are inside the unit circle. But this placement of poles is the criterion for BIBO stability of an LSI system.

A disadvantage of the direct impulse invariance method is that the zero-frequency gains of the CT and DT filters are (usually) greatly different, that of the DT filter usually being larger. Because of the scale factor, T, in Eq. (8.29), often this method is implemented by multiplying the calculated $H(z)$ by T so that $x[k]$ and $x(t)$ have comparable amplitudes. As a consequence, the scale factor, T, would have to be removed when applying Eq. (8.29).

Aside: One should not assume that matching of the impulse responses of the CT and DT systems ensures that the outputs of the two systems will be matched for other inputs. In general, even the step responses of the two filters will not be equal. However, this approach of response matching is not limited to impulse responses. One could also design the digital filter by matching the step responses, or any other responses, of the CT and DT filters if it were important to achieve that specific matching.

Example 8.12 Impulse invariance design   Let $H(s) = a/(s + a)$, with $a = 0.10$. Thus $p_1 = -0.10$. Set $T = 0.01$ s. Using Eq. (8.30) and including the scaling by T, the transfer function of the digital filter is

$$H(z) = T\frac{c_1}{1 - e^{p_i T}z^{-1}} = T\frac{a}{1 - e^{-aT}z^{-1}} = \frac{aTz}{z - e^{-aT}}. \qquad (8.31)$$

Substituting the given values, $H(z) = 10^{-3}z/(z - e^{-0.001})$. The frequency response of the DT filter is shown in Fig. 8.11. The cutoff frequency of this filter is

$$\omega_c = \Omega_c T = 0.10(0.01) = 0.001 \text{ rad/sample}.$$

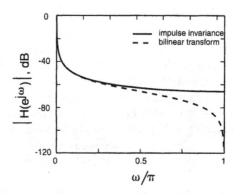

FIGURE 8.11. Frequency response of a digital filter based on a first-order analog filter using impulse invariance design (solid) and bilinear transformation design (dashed).

8.14 IIR FILTER DESIGN BY BILINEAR TRANSFORMATION

Another approach to designing a digital filter based on the transfer function of an analog filter is to explicitly utilize the relationship $z = e^{sT}$. The problem, of course, is that it is difficult to manipulate this transcendental equation into an exact, closed-form, polynomial equation. One may utilize polynomial division to derive the relationship

$$e^{sT} = 1 + sT + \frac{1}{2}(sT)^2 + \ldots \approx \frac{1 + \dfrac{T}{2}s}{1 - \dfrac{T}{2}s}$$

and obtain an approximation given by $z = (1 + (T/2)s)/(1 - (T/2)s)$, or, solving for s,

$$s = \frac{2}{T}\frac{z-1}{z+1}. \tag{8.32}$$

(This result is also obtainable by numerical integration of the differential equation of a CT system using the trapezoidal rule.) To convert a given CT transfer function, $H(s)$, to a corresponding DT transfer function, $H(z)$, one substitutes the r.h.s. of Eq. (8.32) for each occurrence of 's' in $H(s)$. This method is known as *bilinear transformation design* because the equations relating 's' and 'z' are bilinear mappings of the complex plane onto itself.

In order to determine the relationship between the frequency variables ω and Ω it is necessary to solve Eq. (8.32) for the mapping from the $j\Omega$-axis of the s-plane to the unit circle of the z-plane. Thus, setting $s = j\Omega$ and $z = e^{j\omega}$ we obtain $j\Omega = (2/T)(e^{j\omega} - 1)/(e^{j\omega} + 1)$. Solving for the relationships between the two frequency variables, we have

$$\omega = 2\tan^{-1}\left[\frac{\Omega T}{2}\right], \quad \text{or} \quad \Omega = \frac{2}{T}\tan\left[\frac{\omega}{2}\right]. \tag{8.33}$$

That is, the simple relationship $\omega = \Omega T$ no longer applies when comparing the frequency variables of the CT and DT filters if bilinear transformation design is used. This distortion of the linear relationship between the frequency variables (Fig. 8.12) is known as *frequency warping*. The effect of frequency warping on the frequency response of the digital filter will be demonstrated in the following example. The filter designer must be particularly aware of the effect of frequency warping on the cutoff frequency of the digital filter. As the cutoff frequency of the CT filter is changed, that of the DT filter changes nonlinearly. In addition, frequency warping distorts the shape of the frequency response, as shown in Fig. 8.13 and some of the later examples. Note that all types of filters may be designed using this method so long as one corrects for the effects of frequency warping.

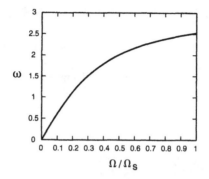

FIGURE 8.12. Relationship between the DT frequency variable and the CT frequency variable when bilinear transform design is utilized (frequency warping).

Example 8.13   Bilinear transformation design   Let us design the digital filter of the previous example using bilinear transformation. Substituting Eq. (8.32) for '*s*' in $H(s)$, we obtain

$$H(z) = \frac{a}{\dfrac{2}{T}\dfrac{z-1}{z+1} + a} = \frac{aT(z+1)}{(aT+2)z + (aT-2)}.$$

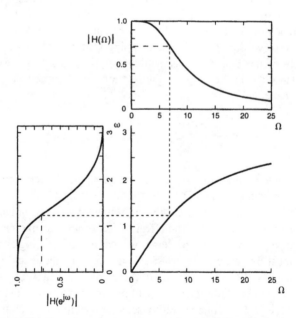

FIGURE 8.13. Distortion of the frequency response of a digital filter due to frequency warping. The CT filter is a second-order Butterworth filter with cutoff frequency of 7.56 rad/s. The digital filter, designed by bilinear transformation, has a cutoff frequency of 0.4π rad/sample.

Using the given parameter values,

$$H(z) \approx \frac{5 \cdot 10^{-4}(z + 1)}{z - 1}.$$

The cutoff frequency of the digital filter is

$$\omega_c = 2 \tan^{-1}\left[\frac{\Omega_c T}{2}\right] = 2 \tan^{-1}[0.0005] \approx 0.0573 \text{ rad/sample.}$$

Thus impulse invariance and bilinear transformation designs have produced DT filters having different cutoff frequencies. This result reflects the fact that frequency warping causes the frequency response of the filter designed by bilinear transformation to differ from that of the filter designed by impulse invariance (Fig. 8.11).

If one wished to achieve the same DT cutoff frequency using bilinear transformation as that resulting from impulse invariance design, one could "pre-warp" the cutoff frequency of the analog filter. Pre-warping is achieved by solving the second part of Eq. (8.33) for Ω_c given the desired ω_c. Then the analog filter is re-specified using this new cutoff frequency. When bilinear transformation design is applied to the new $H(s)$, the digital filter will have the desired cutoff frequency.

The user must recognize that two different relationships between Ω and ω are involved when one considers filtering of a sampled analog signal using bilinear transformation design. The first is the relationship $\omega = \Omega T$ that describes the effect of sampling on the frequency content of the CT signal. The second is given by Eq. (8.33) and is the relationship between the frequency axes of the CT and DT filter. The former is used when the frequency ranges of desired signal and noise are known for the CT signal. From this relationship one determines the corresponding frequency ranges of desired signal and noise after sampling, from which the cutoff frequency of the digital filter may be specified. Knowing this cutoff frequency, one utilizes the frequency warping equation (Eq. (8.33)) to calculate the equivalent cutoff frequency for the analog filter and specifies $H(s)$ on this basis. Application of the bilinear transformation equation (Eq. (8.32)) to this $H(s)$ then will produce a digital filter having the specified DT cutoff frequency.

8.15 BIOMEDICAL EXAMPLES OF IIR DIGITAL FILTER DESIGN

It is straightforward to design digital versions of the standard analog filters, such as Butterworth, Chebychev, and elliptic filters, using impulse invariance or bilinear transformation designs. Although software such as MATLAB usually contains commands to accomplish digital filter designs by these methods (see butter, cheb1, cheb2, ellip, for example) the user must specify the necessary order and cutoff frequency of the filter. The following examples demonstrate this process for Butterworth and Chebychev filters.

Example 8.14   Digital Butterworth filter   Useful components in the EEG signal may extend up to 25 Hz. A conservative approach to analysis of such signals is to utilize a lowpass anti-aliasing filter having a cutoff frequency somewhat above 25 Hz—say, 35 Hz—and sample at ten times the highest desired frequency—for example, 250 samples per second. After sampling, the digitized EEG signal can be digitally filtered with a cutoff frequency near 25 Hz to improve the signal to noise ratio. To design a digital Butterworth filter for this purpose it is necessary to specify the order, N, and cutoff frequency, Ω_c, of the analog Butterworth filter that will achieve the cutoff frequency and transition band desired for the digital filter. For illustration we use impulse invariance design.

Although one could use a trial-and-error approach to selecting the filter parameters, an acceptable filter can be achieved in one step via a formal design procedure. Let the CT filter have unity gain at zero frequency. As depicted in Fig. 8.14(a), four design specifications for the digital filter can be determined from knowledge of the frequency content of the CT EEG signal: ω_p, ω_s specify the edge frequencies of the transition band, while δ_p, δ_s determine the minimum passband and maximum stopband gains. For the EEG signal let us choose the upper limit of the passband to be 25 Hz. Thus, since $f_s = 250$ Hz and $\omega = \Omega T$, $\omega_p = (25 * 2\pi)(250)^{-1} = 0.20\pi$. Let $\omega_s =$

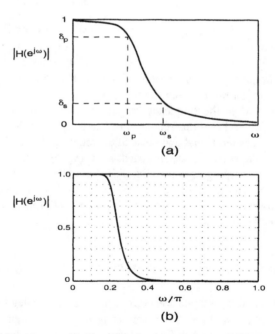

FIGURE 8.14. (a) Design specifications for a lowpass Butterworth filter. ω_p, δ_p: edge frequency of passband and minimum passband gain; ω_s, δ_s: edge frequency of stopband and maximum stopband gain. (b) Frequency response of a digital Butterworth filter designed using the MATLAB command butter after determining the order ($N = 6$) and cutoff frequency ($\omega_c = 0.22434\pi$) from impulse invariance design.

0.30π, which corresponds to 37.5 Hz. Choose $\delta_p \geq -1$ dB, $\delta_s \leq -15$ dB. These specifications may be converted to equivalent specifications on the design of an analog Butterworth filter. By impulse invariance design, the frequencies ω_p, ω_s correspond to CT frequencies Ω_p, Ω_s given by $\Omega_p = \omega_p/T = 0.20\pi/(250)^{-1} = 157.08$ rad/s, $\Omega_s = 235.62$ rad/s.

The general formula for gain of a unity-gain CT Butterworth filter is

$$|H(\Omega)|^2 = \cfrac{1}{1 + \left(\cfrac{\Omega}{\Omega_c}\right)^{2N}}. \tag{8.34}$$

Setting the gain at $\Omega = \Omega_p$ to be -1 dB implies that $20 \log_{10}|H(\Omega_p)| = -1$, or

$$\cfrac{1}{1 + \left(\cfrac{\Omega_p}{\Omega_c}\right)^{2N}} = 10^{-0.1}.$$

Likewise,

$$\cfrac{1}{1 + \left(\cfrac{\Omega_s}{\Omega_c}\right)^{2N}} = 10^{-1.5}.$$

These last two equations may be solved simultaneously for N and Ω_c either analytically or using the MATLAB command fmins. To solve analytically, subtract one from both sides then divide the equations, obtaining

$$\left[\frac{\Omega_p/\Omega_c}{\Omega_s/\Omega_c}\right]^{2N} = \frac{10^{0.1} - 1}{10^{1.5} - 1}.$$

Taking the logarithm of both sides and solving for N, we get $N = 5.88$. To assure meeting the design specifications, N is rounded up to the next larger integer. Thus $N = 6$ and the necessary cutoff frequency may be calculated by setting Eq. (8.34) equal to $(1/\sqrt{2})^2$. It is $\Omega_c = 176.2$ rad/s (28.0 Hz). From a table of design equations for analog Butterworth filters the general transfer function for a sixth order filter is

$$H(s) = \frac{\Omega_c^6}{(x^2 + 0.517\Omega_c s + \Omega_c^2)(s^2 + 1.4142\Omega_c s + \Omega_c^2)(s^2 + 1.9319\Omega_c s + \Omega_c^2)}.$$

Using the residues command in MATLAB, one may determine both the roots of the denominator polynomial and the residues of these poles after substituting the desired value for Ω_c. Then one creates $H(z)$ in partial fractions form as indicated in Eq. (8.30) and combines terms to obtain the digital filter, $H(z)$. These calculations may be performed using MATLAB; however, an alternative approach will be described after considering the use of bilinear transformation for this same design problem.

This filter design problem could have been solved using the bilinear transformation approach. The steps would vary from the above example in two regards. First, given ω_p, ω_s, it would be necessary to pre-warp to determine the corresponding frequencies for the analog filter. Second, after determining N and Ω_c one would evaluate the transfer function $H(s)$ as above and substitute for 's' using Eq. (8.32). Alternatively one could determine the value of ω_c that corresponds to Ω_c using the frequency warping equation (Eq. (8.33)) and specify this value and N as inputs to the MATLAB command butter. This command utilizes bilinear transformation to design the N-th order digital Butterworth filter having the specified parameters. Its outputs are the numerator and denominator coefficients of $H(z)$.

There is also an easier alternative than the partial fraction expansions required by the impulse invariance approach. After determining the necessary filter order we may specify a DT cutoff frequency $\omega_c = \Omega_c T$ and use the butter command to design the digital filter. Because of frequency warping the frequency response of the digital filter will not be exactly the same as if we had completed the impulse invariance design, but it is likely to meet our design specifications. For the present example this approach was taken and the final frequency response is shown in Fig. 8.14(b). For comparison with the specifications, the desired $\omega_p = 0.200\pi$ and that of the filter is 0.203π. Likewise, the desired $\omega_s = 0.300\pi$ and that achieved is 0.291π. The transfer function returned by butter is

$H(z) =$

$$\frac{6.108 * 10^{-4}z^6 - 3.665 * 10^{-3}z^5 + 0.0916z^4 + 0.0122z^3 + 0.00916z^2 + 3.665 * 10^{-3}z + 6.108 * 10^{-4}}{z^6 + 3.287z^5 + 4.880z^4 - 4.064z^3 + 1.980z^2 - 0.5312z + 0.06097}$$

Aside: Having designed a digital filter, one must consider how to implement this filter. In MATLAB it is perhaps easiest to specify the numerator and denominator polynomials of $H(z)$ as vectors which may be used in commands such as filter. If one is writing dedicated code in C or other programming language, one approach is to generate the difference equation corresponding to $H(z)$ (as was done earlier in this chapter) and implement a recursive solution to this equation.

Example 8.15   Digital Chebychev filter   Numerous investigators have demonstrated the presence of "high-frequency oscillations" in multi-unit (also called "mass discharge") recordings of the electrical activity of motor nerves to certain muscles of respiration, particularly in monopolar recordings from the phrenic nerves that innervate the diaphragm. Because neural activity has a broad bandwidth, differential amplifiers for phrenic nerve recordings may be set with a bandpass between 20 Hz and 5000 Hz, for example, and the signals may be sampled at a high frequency such as 10,000 Hz. Because the oscillations (known as HFOs) occur in the range of 50 Hz to 150 Hz, it is useful to lowpass filter the signals in order to better visualize the HFOs. Let us consider the design of a digital second-order Chebychev filter with a cutoff frequency corresponding to 200 Hz using bilinear transformation. The DT cutoff frequency will be

$$\omega_c = (200 * 2\pi)(10^{-4}) = 4\pi * 10^{-2} \text{ rad/sample.}$$

This frequency must be pre-warped to the corresponding cutoff frequency of the CT filter. Thus

$$\Omega_c = \frac{2}{T} \tan\left[\frac{\omega_c}{2}\right] = 2 * 10^4 \tan[0.02\pi] = 21.93 \text{ rad/s.}$$

The general transfer function of a second-order, unity-gain, Chebychev filter is

$$H(s) = \frac{0.50\Omega_c^2}{s^2 + 0.645\Omega_c s + 0.708\Omega_c^2}.$$

Substituting the pre-warped cutoff frequency,

$$H(s) = \frac{240.5}{s^2 + 14.14s + 340.5}.$$

Now after substituting Eq. (8.32), the DT transfer function is

$$H(z) = \frac{6 * 10^{-5}(z^2 + 2z + 1)}{z^2 - 2z + 1}.$$

8.16   IIR FILTER DESIGN BY MINIMIZATION OF AN ERROR FUNCTION

Another general approach to IIR filter design involves specifying an error function that is dependent on the parameters of the filter, then determining the parameter values that minimize the error function. This method may be applied in either the time domain or the frequency domain. Numerous methods of this type have been described and below we discuss one common time-domain method to exemplify the approach. It should be noted that these methods may be developed also from a standard ordinary least squares statistical basis as well as by the approach taken here.

Error Minimization in the Time Domain (Prony's Method)

Regarding the design methods discussed so far, the emphasis has been on achieving a desired frequency response. Time-domain methods are concerned with achieving a desired impulse response. The general problem may be stated thusly: Given a generic LSIC filter (Fig. 8.5) and a signal $x[n]$ (either measured or specified based on theoretical considerations), how does one choose the parameters of the filter so that its impulse response equals (or closely approximates) $x[n]$? A method due to Padé designs the filter so that its impulse response matches the first $M + N + 1$ elements of $x[n]$, where M and N are the orders of the numerator and denominator polynomials of $H(z)$. This method, however, requires high-order polynomials to match a significant portion of an IIR impulse response. Furthermore the method assumes that $x[n]$ is known exactly (i.e., is noise-free). Therefore, direct use of the Padé method with real data is problematic.

Prony's method has more general utility than Padé's method because it closely approximates $x[n]$ rather than exactly matching $M + N + 1$ elements of it. Assume that

$$H(z) = \frac{B(z)}{A(z)} = \frac{\displaystyle\sum_{l=0}^{M} b[l]z^{-1}}{\displaystyle\sum_{l=0}^{N} a[l]z^{-1}}, \qquad \text{where } a[0] = 1$$

and define an error $e'[n] = x[n] - h[n]$ where, as above, $x[n]$ is the desired impulse response of the filter, and $h[n]$ is the actual impulse response. Taking Z-transforms, $E'(z) = X(z) - B(z)/A(z)$. Now define another error term $e[n]$ such that

$$E(z) = A(z)E'(z) = A(z)X(z) - B(z). \tag{8.35}$$

By inverse Z-transforms, $e[n] = a[n] * x[n] - b[n]$. We may write the convolution term more explicitly, obtaining

$$e[n] = \begin{cases} x[n] + \displaystyle\sum_{l=1}^{N} a[l]x[n-l] - b[n], & 0 \le n \le M \\ x[n] + \displaystyle\sum_{l=1}^{N} a[l]x[n-l], & n \ge M+1. \end{cases} \tag{8.36}$$

The next step is to define an appropriate error measure to be minimized. By definition for Prony's method, the error to be minimized is

$$\varepsilon = \sum_{n=M+1}^{\infty} |e[n]|^2 = \sum_{n=M+1}^{\infty} e^*[n]e[n]. \tag{8.37}$$

This error is to be minimized by the appropriate choice of the coefficients $a[n]$ and $b[n]$. Although we typically deal only with real-valued parameters, it is simpler to develop the solution by allowing the $a[n]$'s and $b[n]$'s possibly to be complex. The error may be minimized by taking the partial derivative of the error with respect to each parameter, setting each partial derivative equal to zero, and solving the resulting N equations simultaneously. Thus, for any k in the range $1 \le k \le N$,

$$\frac{\partial \varepsilon}{\partial a^*[k]} = \sum_{n=M+1}^{\infty} \frac{\partial(e^*[n]e[n])}{\partial a^*[k]} = 0. \tag{8.38}$$

But by Eq. (8.36), for $n \geq M + 1$, $e^*[n] = x^*[n] + \sum_{l=1}^{N} a^*[l]x^*[n - l]$. Note that $a^*[k]$ only appears once in this expression—i.e., when $l = k$. Therefore, Eq. (8.38) may be written as

$$\sum_{n=M+1}^{\infty} \frac{\partial(e^*[n]e[n])}{\partial a^*[k]} = \sum_{n=M+1}^{\infty} e[n]x^*[n - k] = 0, \qquad 1 \leq k \leq N. \qquad (8.39)$$

Substituting for $e[n]$ from Eq. (8.36) for $n > M$ and rearranging terms, we find that

$$\sum_{l=1}^{N} a[l]\left[\sum_{n=M+1}^{\infty} x[n - l]x^*[n - k]\right] = -\sum_{n=M+1}^{\infty} x[n]x^*[n - k], \; 1 \leq k \leq N. \qquad (8.40)$$

Typically, one defines a type of correlation function

$$r_x[k, l] = \sum_{n=M+1}^{\infty} x[n - l]x^*[n - k] \qquad (8.41)$$

and then rewrites Eq. 8.40 as

$$\sum_{l=1}^{N} a[l]r_x[k, l] = -r_x[k, 0], \; 1 \leq k \leq N. \qquad (8.42)$$

To determine the values of the N coefficients $a[k]$, $1 \leq k \leq N$, one first calculates from $x[n]$ the values of $r_x[k, l]$ for $k = 1, 2, \ldots, N$, and $l = 0, 1, \ldots, N$. The N equations of Eq. (8.42) may then be written in the form

$$\begin{bmatrix} r_x[1, 1] & r_x[1, 2] & \cdots & r_x[1, N] \\ r_x[2, 1] & r_x[2, 2] & \cdots & r_x[2, N] \\ \vdots & \vdots & \vdots & \vdots \\ r_x[N, 1] & r_x[N, 2] & \cdots & r_x[N, N] \end{bmatrix} \begin{bmatrix} a[1] \\ a[2] \\ \vdots \\ a[N] \end{bmatrix} = -\begin{bmatrix} r_x[1, 0] \\ r_x[2, 0] \\ \vdots \\ r_x[N, 0] \end{bmatrix}. \qquad (8.43)$$

Equation (8.43) may be solved by any of the various algorithms for solving N linear algebraic equations in N unknowns.

Determining the values of the numerator coefficients is now straightforward using the upper part of Eq. (8.36). We simply set $e[k] = 0$ for $0 \leq k \leq M$ and solve for $b[k]$. Thus

$$b[k] = x[k] + \sum_{l=1}^{N} a[l]x[k - l], \; 0 \leq k \leq M. \qquad (8.44)$$

Filters designed by Prony's method are less sensitive to noise than those designed by Padé's method. One should note, however, that the error being minimized is *not* the difference between the desired and actual impulse responses. Furthermore, the values of M and N will not be known a priori and it is likely that a range of values for these parameters will need to be tested. If M and N are too small, the resulting $h[n]$ may not exhibit all of the dynamic features of $x[n]$. On the other hand, if M and N are too large, the model may reproduce some of the noise components of $x[n]$. In the end the user must decide what features of $x[n]$ need to be captured by the filter.

Another question that may arise is how to design an IIR filter when $x[n]$ is measured for a finite (often short) time—e.g., $0 \leq n \leq N_0$. In this situation one usually assumes that $x[n] = 0$ for $n > N_0$ and "hopes for the best". Especially in this situation these time-domain methods may be susceptible to producing an unstable filter.

If the signal, $x[n]$, that is measured is not actually the impulse response of a system, is there merit in designing a filter that models it as an impulse response? There are at least three potential applications for this type of modeling. First, the transfer function $H(z)$ typically will contain far fewer parameters than there are elements in $x[n]$. Thus $H(z)$ provides a compact representation of the signal for storage or transmission. Second, one may be able to utilize the parameters of $H(z)$ for classification of the signal—for example, as repesenting a normal or abnormal clinical measurement. Third, parametric modeling of a signal provides an alternative method for calculating the power spectrum of a signal, as the following example demonstrates.

Example 8.16   All-pole filter design by Prony's method   All-pole filters are filters for which $B(z) = b[0]$. That is, $M = 0$. Although it may require very large N to capture all of the behavior of an arbitrary $x[n]$ when $M = 0$, in many real situations the signal-to-noise ratio does not permit such detailed knowledge of $x[n]$ and consequently a lower order approximation is acceptable. Furthermore, it is not difficult to calculate all-pole models even for fairly large N—for example, $N > 50$. Additionally there are efficient recursive algorithms for modifying the coefficients of an all-pole filter adaptively as the properties of $x[n]$ change with time. Therefore, the design of all-pole filters has many important roles in signal processing.

In the all-pole case $H(z)$ may be written as

$$H(z) = \frac{b[0]}{1 + \displaystyle\sum_{k=1}^{N} a[k]z^{-k}}. \tag{8.45}$$

Prony's error function will be $\varepsilon = \sum_{n=1}^{\infty} |e[n]|^2$. However, since $e[0] = x[0]$ and is a constant that is independent of the $a[k]$ coefficients, one may define a new error function

$$\varepsilon_N = \varepsilon + |e[0]|^2 = \sum_{k=0}^{\infty} |e[n]|^2, \tag{8.46}$$

which has the same solutions for the $a[k]$'s as the original. Prony's equations are

$$\sum_{l=1}^{N} a[l]r_x[k, l] = -r_x[k, 0], \quad 1 \leq k \leq N. \tag{8.47}$$

Now according to Prony's method we define $r_x[k, l] = \sum_{n=n_0}^{\infty} x[n-l]x^*[n-k]$. In the original method $n_0 = M + 1 = 1$, but using the modified error function of Eq. (8.46) the lower limit of the summation for $r_x[k, l]$ is $n_0 = 0$. Also in this case $r_x[k, l] = r_x^*[l, k]$ and with some algebra we may show further that $r_x[k, l]$ depends on $k - l$ and not on the specific values of k and l. Consequently, assuming that the $x[n]$ are real-valued, Prony's equations in matrix form become

$$
\begin{bmatrix}
r_x[0] & r_x[1] & \cdots & r_x[N-1] \\
r_x[1] & r_x[0] & \cdots & r_x[N-2] \\
\vdots & \vdots & \vdots & \vdots \\
r_x[N-1] & \cdots & \cdots & r_x[0]
\end{bmatrix}
\begin{bmatrix}
a[1] \\
a[2] \\
\vdots \\
a[M]
\end{bmatrix}
= -
\begin{bmatrix}
r_x[1] \\
r_x[2] \\
\vdots \\
r_x[N]
\end{bmatrix}.
\tag{8.48}
$$

One uses any convenient linear equation technique to solve Eq. (8.48) for the vector $[a[1], a[2], \ldots, a[N]]'$. Because all of the entries on any diagonal of the matrix of autocorrelation values are equal, this matrix has a Toeplitz structure and it may be inverted using the Levinson-Durbin recursion discussed in Chapter 9.

Since the equation for the error function, Eq. (8.46), involves sums of products of values of $x[n]$ and also involves the $a[k]$ coefficients, it is not surprising that the minimum error may be determined as a function of the $r_x[k]$ values and the coefficients. After some algebra (see Exercise 8.23) one may show that

$$
\min \varepsilon_N = r_x[0] + \sum_{k=1}^{N} a[k]r_x[k],
\tag{8.49}
$$

where again it is assumed that the $x[n]$ are real-valued. By convention one lets $b[0] = \sqrt{\min \varepsilon_N}$ as this choice ensures that the impulse response of the filter and $x[n]$ have the same total energy.

Consider designing an all-pole model of the signal $x[n] = 0.5^n u[n] - 0.2^n u[n]$. Assume that samples of $x[n]$ are available for $0 \le n \le 25$ (Fig. 8.15). Since we know that the signal may be represented as the impulse response of a second-order system, choose $N = 2$. Using the MATLAB command xcorr one may calculate the values of $r_x[k]$, $0 \le k \le 2$. These are: $r_x(0:2) = [0.1528, 0.0972, 0.0528]$. Prony's equations (8.48) become

$$
\begin{bmatrix}
0.1528 & 0.0972 \\
0.0972 & 0.1528
\end{bmatrix}
\begin{bmatrix}
a[1] \\
a[2]
\end{bmatrix}
= -
\begin{bmatrix}
0.0972 \\
0.0528
\end{bmatrix}.
$$

The solution is $[a[1]\ a[2]]' = [-0.700, 0.100]'$. The minimum error is calculated from Eq. (8.49) to be 0.0900. Therefore we let $b[0] = 0.30$. Thus

$$
H(z) = \frac{b[0]}{A(z)} = \frac{0.30}{1 - 0.700z^{-1} + 0.100z^{-2}}.
$$

One may readily show that the roots of $A(z)$ are indeed 0.500 and 0.200. In this situation in which the data $x[n]$ are noise-free and $x[n]$ is very nearly zero for

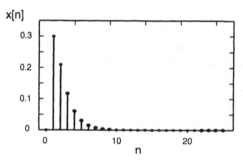

FIGURE 8.15. Samples of a data signal, x[n], used to design an all-pole filter.

$n > 25$, the impulse response of the all-pole filter is essentially an exact model of the signal.

If we had chosen $N > 2$, would the model be incorrect? We may solve the problem again for $N = 3$ by calculating one more value, $r_x[3]$. The reader may show that the solution now is $[a[1]\ a[2]\ a[3]]' = [-0.700, 0.100, 0.000]'$. In other words, the all-pole model correctly determines that there is no need for a non-zero $a[3]$ term. This encouraging result, however, is completely dependent on having noise-free data. Furthermore it is possible for one $a[k]$ coefficient to be nearly (or actually) zero even when $a[k + K]$ differs from zero for some positive value(s) of K.

Error Minimization in the Frequency Domain

For the sake of completeness an overview of methods involving error minimization in the frequency domain is given here. The basic approach is to assume a form for $H(z)$ as in Eq. (8.25), then substitute $z = e^{j\omega}$ to evaluate the frequency reponse as a function of the coefficients of $A(z)$ and $B(z)$. An error function is used that expresses the difference between the desired frequency response of the filter and $H(e^{j\omega})$. Typically this error function is

$$E = \sum_{j=1}^{J} [|H(e^{j\omega_j})| - |H_d(e^{j\omega_j})|]^2, \tag{8.50}$$

where $H_d(e^{j\omega})$ is the desired frequency response and the set of ω_j, $1 \leq j \leq J$, are frequencies at which the desired response is specified. Minimization of E with respect to the $a[k]$ and $b[k]$ coefficients in $A(z)$ and $B(z)$ is a complex, nonlinear optimization problem that must be solved using an iterative method. Once the solution is obtained it is important to compare the desired and actual frequency responses because most iterative optimization methods do not guarantee to converge to a global optimum. In addition, the actual frequency response may vary in a complex manner between the specified frequencies, ω_j.

8.17 FIR FILTER DESIGN

Basic Properties of FIR Filters

FIR filters have a singular advantage when compared to IIR filters: their phase responses can be made to be linear with frequency. In situations in which phase distortion needs to be minimized, FIR filters are probably the clear choice. FIR filters often can be designed that have narrow transition bands with low stopband gains although the filter order might be much higher than a comparable IIR filter. On the other hand, the output of an FIR filter is easy to calculate using direct convolution. Of course, the output of an IIR filter can be calculated easily by solving its difference equation recursively. Which approach is more useful may depend on the application.

A causal FIR filter has the impulse response

$$h[n] = \begin{cases} h_n, & 0 \le n \le M - 1 \\ 0, & n > M - 1 \end{cases}, \tag{8.51}$$

where the h_n's are constants.

The transfer function of this FIR filter has the general form

$$H(z) = \sum_{k=0}^{M-1} h[k]z^{-k} = \sum_{k=0}^{M-1} h_k z^{-k}$$

$$= z^{-(M-1)}(h[0]z^{M-1} + h[1]z^{M-2} + \ldots + h[M-2]z + h[M-1]), \tag{8.52}$$

where M is the order of the filter. Thus, this filter has $M - 1$ poles at $z = 0$ and $M - 1$ zeros.

As long as $h[n]$ is bounded, then the FIR filter is BIBO stable. The frequency response of the filter has the form

$$H(e^{j\omega}) = h[0] + h[1]e^{-j\omega} + h[2]e^{-j2\omega} + \ldots + h[M-1]e^{-j(M-1)\omega}. \tag{8.53}$$

If, in addition, $h[n]$ satisfies the symmetry condition, $h[M - 1 - n] = h[n]$, then (assuming M is odd)

$$H(e^{j\omega}) = e^{-j\omega(M-1)/2}\{(h[0]e^{j\omega(M-1)/2} + h[M-1]e^{-j\omega(M-1)/2})$$

$$+ (h[1]e^{j\omega(M+1)/2} + h[M-2])e^{-j\omega(M+1)/2} + \ldots$$

$$+ (h[(M-3)/2]e^{j\omega} + h[(M+1)/2]e^{-j\omega}) + h[(M-1)/2\}.$$

But since $h[0] = h[M - 1]$, and so on, each pair of terms in parentheses can be written as a cosine function. Therefore,

$$H(e^{j\omega}) = e^{-j\omega(M-1)/2}H_1(e^{j\omega}), \tag{8.54}$$

where $H_1(e^{j\omega})$ is a real-valued function. Consequently the phase of the frequency response is

$$\angle H(e^{j\omega}) = -\omega(M-1)/2. \tag{8.55}$$

That is, the phase is linear with frequency, corresponding to a time delay of $(M-1)/2$ samples. (The reader may show that the same conclusion applies if M is even and also if $h[M-1-n] = -h[n]$.)

There are two basic approaches to designing FIR filters. The first approach is based on windowing of the impulse response of an ideal digital filter. In this method there are many similarities to the previous discussion of windowing related to spectral estimation. The second approach involves algorithms to fit specifications on the frequency response of the desired filter. Note that the frequency response of an FIR filter is completely specified by M parameters (some of which may be redundant if the above symmetry condition on $h[n]$ is met). Conceptually one could determine the filter coefficients (i.e., $h[n]$) if given the frequency response at M frequencies. We shall see that implementation of this simple idea is not trivial!

Window Design of FIR Filters

The simplest method for designing an FIR filter is to truncate the impulse response of an ideal filter. For a lowpass design the frequency response of the ideal filter is specified by the desired cutoff frequency, ω_c (Fig. 8.16(a)). Being ideal, this filter has an impulse response, $h_i[n]$, that is infinitely long and noncausal (Fig. 8.16(b)). To create a causal signal that could be the impulse response of an FIR filter of desired order M, we shift $h_i[n]$ to the right by $(M-1)/2$ samples and then set all of its samples equal to zero for $n < 0$ and for $n > M-1$ (Fig. 8.16(c)). This new signal becomes $h[n]$, the impulse response of the FIR filter. Finally the frequency response of the FIR filter is calculated and compared to the desired ideal frequency response (Fig. 8.16(d)). The cutoff frequency will be very close to the desired one but the process of truncation introduces oscillatory variations in the frequency response that are the equivalent of the Gibbs' phenomenon observed with a truncated Fourier series. Usually oscillations having amplitudes as large as those in Fig. 8.16(d) are undesirable. By viewing this design process from a windowing perspective, we can identify means of altering these oscillations in the frequency response as well as adjusting the transition bandwidth. As in most situations, however, we will have to compromise between these two goals.

In the design process the step between Fig. 8.16(b) and 8.16(c) involved time-shifting the signal and then truncating it. The truncation step can be viewed as multiplying the shifted signal, $h_i[n-(M-1)/2]$, by a rectangular window function, $w_M[n]$, defined (as in Chapter 7) as

$$w_M[n] = \begin{cases} 1, & 0 \le n \le M-1 \\ 0, & otherwise. \end{cases} \tag{8.56}$$

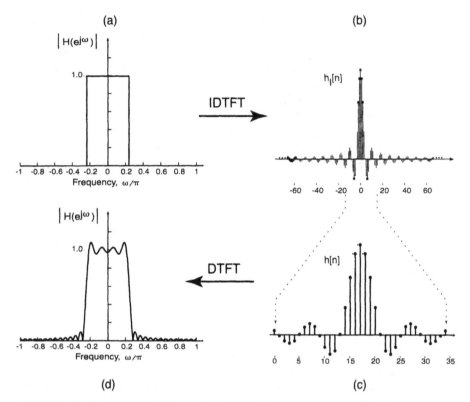

FIGURE 8.16. The process of FIR filter design by windowing. (a) First, the ideal filter template is specified. (b) Second, the unit-pulse response of the ideal filter is calculated using the inverse DTFT. (c) Third, to produce an FIR unit-pulse response, $h[n]$, the ideal unit-pulse response is truncated to the desired length and shifted to the right so that it is a causal signal. (d) Fourth, the frequency response of the filter having $h[n]$ as its unit-pulse response is calculated by DTFT.

Now $h[n]$ may be expressed as

$$h[n] = h_I[n - (M-1)/2]w_M[n]. \qquad (8.57)$$

Consequently, the frequency response of the FIR filter is

$$H(e^{j\omega}) = e^{-j\omega(M-1)/2}H_I(e^{j\omega}) * W_M(e^{j\omega}). \qquad (8.58)$$

The Fourier transform of the rectangular window function is shown in Chapter 7, Fig. 7.11. It is clear that the oscillations in $H(e^{j\omega})$ are generated by the convolution of the ideal filter response with the sidelobes of $W_M(e^{j\omega})$. Recalling that the widths of the lobes of $W_M(e^{j\omega})$ are inversely proportional to M, one might try increasing M

to reduce the sidelobe-generated oscillations. Furthermore, increasing M should reduce the main lobe width and decrease the width of the transition band of the FIR filter. Although one can achieve the second goal of reducing the transition width, increasing M only moderately reduces the sideband amplitudes relative to the zero-frequency gain (Fig. 8.17).

The reader will recall from Chapter 7 that for a given M the rectangular window has the narrowest transition band but the largest sidelobe amplitudes among all of the windows. Thus, by multiplying $h_I[n - (M - 1)/2]$ by a different window, we may achieve a reduction of the oscillations in the frequency response albeit (unless M is increased) at the expense of increasing the width of the transition band. Often this trade-off is acceptable. One then selects a window based mainly on the same considerations discussed in Chapter 7. Note that the transition bandwidth is directly proportional to the frequency resolution of the window. Note also that there are two disadvantages to increasing M. First, the effective time delay introduced through the linear phase shift will increase. Second, when filtering a finite-length data sequence, there will be an artifactual transient response at the beginning of the data having a duration of $M - 1$ samples. This transient occurs because the filter assumes that the input signal is zero for $n < 0$.

Example 8.17   Window design of an FIR filter   Consider the design of a 26th-order ($M = 26$) lowpass filter having a cutoff frequency of $\omega_c = 0.25\pi$. The frequency response of an ideal lowpass filter having an impulse response that is delayed by $(M - 1)/2$ samples is

$$H_I(e^{j\omega}) = \begin{cases} e^{-j\omega(M-1)/2}, & |\omega| \le \omega_c \\ 0, & |\omega| > \omega_c \end{cases}.$$

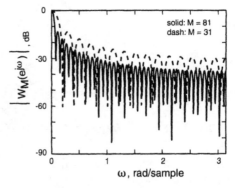

FIGURE 8.17. Magnitude of the DTFT of a rectangular window of length 31 (dashed) and 81 (solid).

Its impulse response is

$$h_I[n] = \frac{1}{2\pi} \int_{-\omega_c}^{\omega_c} e^{-j\omega(M-1)/2} d\omega = \pi \frac{\sin\left[\omega_c\left(n - \frac{M-1}{2}\right)\right]}{n - \frac{M-1}{2}} \qquad (8.59)$$

Using a rectangular window, $h[n]$ is just the 26 points of $h_I[n]$ from $n = 0$ through $n = 25$. This function and the magnitude of its DTFT are plotted in Fig. 8.18(a,c). For comparison, we recompute the filter using a Hamming window. In this case, $h[n] = h_I[n]w_{Ham}[n]$, where

$$w_{Ham}[n] = \begin{cases} 0.54 - 0.46 \cos\left[\dfrac{2\pi n}{M-1}\right], & 0 \le n \le M-1 \\ 0, & \text{otherwise} \end{cases}.$$

The corresponding impulse response and frequency response are shown in Fig. 8.18(b,c). Note the considerable reduction of the stopband gain achieved with the Hamming window at the expense of a wider transition band.

FIR Filter Design by Frequency Sampling

In the frequency sampling method we specify the desired frequency response at M discrete frequencies, then use these specifications to determine an M-point FIR impulse response, $h[n]$, $0 \le n \le M - 1$. Let the desired frequency response, $H_d(e^{j\omega})$, be specified at the frequencies

$$\omega_k = \frac{2\pi}{M}k, \quad 0 \le k \le \frac{M-1}{2} \text{ (M odd)}, \quad \text{or} \quad 0 \le k \le \frac{M}{2} - 1 \text{ (M even)}. \quad (8.60)$$

Recalling the symmetries of the DTFT of a real-valued $h[n]$, it is only necessary to specify the frequency response in the interval $[0, \pi)$. Now we set the actual frequency response, $H(e^{j\omega})$, equal to the desired frequency response at these frequencies. Thus

$$H(e^{j\omega_k}) = \sum_{n=0}^{M-1} h[n]e^{-jn(2\pi k/M)} \triangleq H_d(e^{j\omega_k}). \qquad (8.61)$$

By multiplying the second and third terms of Eq. 8.61 by $e^{j2\pi km/M}$ and summing from $k = 0$ to $k = M - 1$, the middle term of this equation reduces to $Mh[m]$. Therefore,

$$h[m] = \frac{1}{M} \sum_{k=0}^{M-1} H_d(e^{j\omega_k})e^{j2\pi km/M}, \qquad m = 0, 1, \ldots, M-1, \qquad (8.62)$$

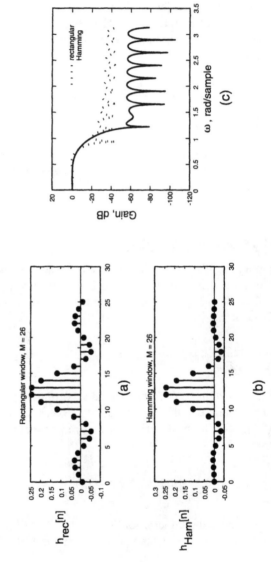

FIGURE 8.18. (a) Unit-pulse response of a rectangular window filter of length $M = 26$. (b) Unit-pulse response of a Hamming window filter of length $M = 26$. (c) Frequency responses of the filters from 'a' and 'b'.

where the symmetries of the frequency response are utilized for $k > (M - 1)/2$ (or $M/2 - 1$, if M is even). The reader may recognize Eq. (8.62) as an inverse Discrete Fourier transform representation of the M-point $h[n]$ in which $H[k] = H_d(e^{j\omega_k})$. Furthermore, by specifying a linear phase property for $H_d(e^{j\omega})$ as given by Eq. (8.55), $h[n]$ will assume the symmetry property $h[m] = h[M - 1 - m]$.

The major advantage of the frequency sampling method accrues when the desired gain at many of the frequency points is zero, thereby simplifying the computations in Eq. (8.62). One should expect, however, that the frequency response of the final filter may exhibit undesired behaviors between the specified frequency samples. For instance, specifying a desired frequency response that is an ideal filter will produce a final filter that exhibits large oscillations in gain as we saw above from the windowing method. Consequently, one should select one or more frequency points at which the specified gain lies between one and zero. These points will lie in the transition band of the filter. Rabiner has derived optimal solutions for the gains to be specified at these transition frequencies in order to minimize the amplitude of the oscillations (sidebands) of the final frequency response. Proakis and Manolakis (1996) summarize his findings and present tables of gains as a function of the number of transition frequencies, length of $h[n]$, and whether the frequencies ω_k are specified as in Eq. (8.60) or half-way between these frequencies. The major disadvantage of this method is that the frequency response must be specified at integer multiples of $2\pi/M$. Consequently, there is limited flexibility for selecting an arbitrary cutoff frequency, especially if one wishes to minimize the amplitude of sidebands.

Example 8.18   Frequency sampling design   To illustrate the method, we design a lowpass filter with $M = 33$ points having a desired gain of unity in the range $0 \le \omega \le 7(2\pi/33) = 0.4242\pi$. Note that the frequency sampling method will place 33 points in the interval $0 \le \omega < 2\pi$; therefore, the specified passband includes eight of these points corresponding to $k = 0, 1, \ldots, 7$, in Eq. (8.60). To reduce sidebands we place one frequency sample in the transition band. According to Appendix C of Proakis and Manolakis (1996), for $M = 33$ and a passband of eight points the gain at the transition sample should be 0.39039917. Therefore, we may specify the desired frequency response using MATLAB commands as: hdw = [ones(1,8),0.39039917,zeros(1,16),0.39039917,ones(1,7)]; Using the command hnc = ifft(hdw) a non-causal impulse response, hnc[n], then may be calculated. (It may be necessary to take the real part of hnc[n] in case roundoff errors lead to small nonzero imaginary parts.) Now the command h = fftshift(hnc) swaps the left and right halves of hnc and creates a causal FIR impulse response, $h[n]$ (Fig. 8.19(a)). Using freqz(h,1) we determine the actual frequency response shown in Fig. 8.19(b). Note that the maximum stopband gain is below –40 dB. The reader may repeat this procedure with no transition point and show that the maximum stopband gain now is above –20 dB. Finally, note that it is possible to reduce the maximum stopband gain further by including a second transition point; however, this benefit comes at the cost of widening the transition band.

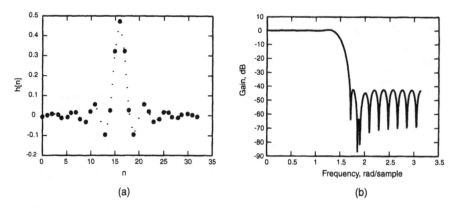

FIGURE 8.19. (a) Unit-pulse response and (b) frequency response of an FIR filter designed using the frequency sampling method. $M = 33$. There were eight sample points in the pass-band which extends to $0.4242\,\pi$ and one optimal transition point was used.

Equiripple FIR Filter Designs

One of the limitations of the above methods for FIR filter design is that they force the error to be small in the passband and permit the gain to vary greatly in the stop-band. An alternative approach would be to allow the error to be spread "uniformly" across both the passband and stopband. Methods based on this concept are known as equiripple designs. These methods require extensive computations and must be solved using a computer. Consequently, we will discuss the principles and utilize MATLAB functions for realizing actual designs. For simplicity our discussion will address the design of zero-phase FIR filters of order M (assuming M is odd) which are noncausal. A causal version of the same filter may be obtained by delaying the impulse response by $(M-1)/2$ samples.

Assume that $M = 2L + 1$ and that $h[-n] = h[n]$, $1 \le n \le L$, so that the frequency response of the filter may be expressed as

$$H(e^{j\omega}) = \sum_{n=-L}^{L} h[n]e^{-j\omega n} = h[0] + \sum_{n=1}^{L} 2h[n]\cos(\omega n). \qquad (8.63)$$

It is desired that this frequency response should meet the specifications shown in Fig. 8.20(a). That is, the gain in the passband should be constrained to lie between $1 - \delta_1$ and $1 + \delta_1$, whereas the stopband gain should lie between δ_2 and $-\delta_2$. The edge frequencies of the transition band are ω_p, ω_s. These frequencies, the gain specifiers (δ_1, δ_2), and the filter order M are to be determined. Clearly all five parameters cannot be specified independently.

The most common approaches to equiripple filter design begin by recognizing that $\cos(\omega n)$ may be expressed as a summation of powers of $\cos(\omega)$. Therefore, Eq. (8.63) may be written as

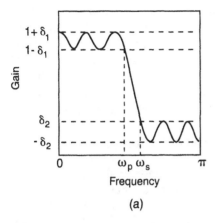 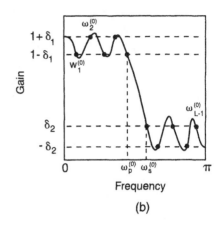

(a)   (b)

FIGURE 8.20. (a) Parameters for equiripple lowpass filter design. $1 + \delta_1$, $1 - \delta_1$: maximum and minimum gains in the passband; δ_2, $- \delta_2$: maximum and minimum gains in the stopband; ω_p, ω_s: edge frequencies of the passband and stopband. (b) The first step in the iteration. $\omega_i^{(0)}$ are the first guesses at the extremal frequencies. The extrema of the polynomial function through these points (solid curve) will be determined and utilized as the next estimate of the set of extremal frequencies.

$$H(e^{j\omega}) = \sum_{k=0}^{L} a_k(\cos(\omega))^k. \qquad (8.64)$$

A differentiable polynomial of order L, such as Eq. (8.64), has at most $L - 1$ local minima and maxima over one cycle of the cosine—i.e., over the open range $0 < \omega < \pi$. It is easy to show also that Eq. (8.64) has a maximum or minimum at both ends of this range and therefore there are at most $L + 1$ local extrema on the closed range. Let us now represent the desired frequency response using the subscript "d" and define an error function

$$E(\omega) = W(\omega)[H_d(e^{j\omega}) - H(e^{j\omega})], \qquad (8.65)$$

in which $W(\omega)$ is a weighting function and the error is evaluated over all frequencies in the passband and stopband. Using the approach of Parks and McClellan (Oppenheim and Schafer, 1975) we specify the parameters L, ω_p, ω_s and the ratio $K = \delta_1/\delta_2$. This ratio will determine the relative weighting to be given to errors in the passband and stopband, since we will define

$$W(\omega) = \begin{cases} \dfrac{\delta_2}{\delta_1}, & 0 \le \omega \le \omega_p \\ 1, & \omega_s \le \omega \le \pi \end{cases} \quad \text{and} \quad H_d(e^{j\omega}) = \begin{cases} 1, & 0 \le \omega \le \omega_p \\ 0, & \omega_s \le \omega \le \pi \end{cases}.$$

Now it is desired to minimize the maximum magnitude of $E(\omega)$ over the passband and the stopband.

Parks and McClellan presented a theorem from approximation theory (Alternation Theorem) which states that in order to obtain the optimum solution to the above minimization problem it is necessary and sufficient that the error function $E(\omega)$ exhibit $L + 2$ "alternations" on the closed interval $[0, \pi]$. That is, if a set of extremal frequencies are defined as $\omega_0 < \omega_1 \ldots < \omega_{L+1}$, then $E(\omega_i) = -E(\omega_{i-1}) = \pm \max|E(\omega)|$. The sufficiency condition of this theorem implies that if one can meet these conditions on the error function by any means, then one has obtained the optimum solution.

Parks and McClellan suggested an iterative procedure to determine the extremal frequencies: First, guess values for the $L - 1$ extremal frequencies in the open interval $(0, \pi)$. Then using a formula suggested by Rabiner (see Proakis and Manolakis, 1996; p. 645) δ_2 may be determined from knowledge of the extremal frequencies and the desired frequency response. Next, fit an interpolating polynomial to the points

$$H(e^{j\omega_1}) = 1 + \delta_1, \quad H(e^{j\omega_2}) = 1 - \delta_1, \quad \ldots, \quad H(e^{j\omega_p}) = 1 - \delta_1,$$

$$H(e^{j\omega_s}) = 1 + \delta_2, \quad \ldots, \quad H(e^{j\omega_{L-1}}) = \delta_2 \ (or -\delta_2).$$

(*Note:* The sign of the last term depends on how many frequency points are in the stopband.) In general, the extrema of this polynomial will not occur at the specified extremal frequencies (Fig. 8.20(b)). Therefore one determines the extremal frequencies of the interpolating polynomial function and utilizes them as the next guess of the optimal extremal frequencies. δ_2 then is recalculated as above and the process repeated. This procedure, often implemented using the Remez exchange algorithm (See the MATLAB command remez), is iterated until there is insignificant change in the extremal frequencies. To determine $h[n]$ one then samples the final interpolating polynomial at M equally-spaced frequencies and calculates the inverse DFT. Note that one is "stuck with" the final values of δ_1, δ_2. If these criteria are not acceptable, one must re-design the filter using a different value of M or of the edge frequencies of the transition band.

Example 8.19   Equiripple filter design   To demonstrate both the ease of designing such filters using software such as MATLAB and the trade-off between transition width and stopband gain of equiripple (and other) designs, we consider two different filters having $M = 51$ points. For the first filter, we choose $\omega_p = 0.225\pi$, $\omega_s = 0.35\pi$. The corresponding MATLAB command is b = remez(51, [0, omegap, omegas,1], [1, 1, 0, 0]), where omegap and omegas have been set equal to ω_p, ω_s. The frequency response of the final filter may be found, for example, from $[h, w] = freqz(b,1)$. It is shown in Fig. 8.21(a). Assume now that we wish to have a narrower transition band. Let $\omega_s = 0.25\pi$ and keep the other specifications the same. The frequency response of the new filter (Fig. 8.21(b)) has a narrower transition band but the maximum stopband gain is much higher than for the first filter. In addition there are larger oscillations of gain in the passband. The reader may demonstrate that this second filter can be improved by increasing M.

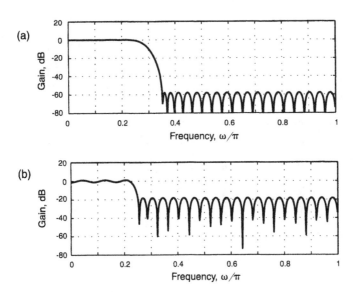

FIGURE 8.21. Frequency responses of 51-point equiripple lowpass filters designed using the MATLAB function remez. (a) $\omega_p = 0.255\pi$, $\omega_s = 0.35\pi$; (b) $\omega_p = 0.255\pi$, $\omega_s = 0.25\pi$.

8.18   FREQUENCY-BAND TRANSFORMATIONS

If one wishes to design a different digital filter than a lowpass filter, the design methods described in this chapter may still be used. It is possible to define transformations of the z-plane onto itself that will map a lowpass digital filter to another lowpass filter having a different cutoff frequency. The basis for such a transformation is derived in Oppenheim and Schafer (1975) and Proakis and Manolakis (1996). Furthermore, one may similarly define transformations from a lowpass filter to a highpass, bandpasss, or band reject filter. Here we state these results in Table 8.3 without deriving them. The procedure is straightforward. First a lowpass filter is designed using one of the above methods, then the appropriate transformation from the table is applied. To do so, first one uses the relationship from the rightmost column of the table to determine the parameter value(s) for the substitution formula given in the middle column. This parameter value depends on the cutoff frequencies of the lowpass and desired filters. Finally in the original $H(z)$ one substitutes the given formula for every occurrence of z^{-1}.

Although this approach may be utilized with any design method for lowpass filters, the resulting filters may not satisfy the same criteria as the original lowpass filter. Note also that it is possible to specify designs other than lowpass using the frequency sampling and equiripple methods so that no transformation is required. Finally, note that frequency transformations may be done for analog filters so that impulse invariance or bilinear transformation may be applied to other than lowpass filters. Be warned, however, that the impulse invariance method should not be used

TABLE 8.3 Equations for Transforming a Lowpass Digital Filter Having Cutoff Frequency ω_L into a Different Filter Type

Filter Type	Transformation Equation	Design Formula(s)	Comments
Lowpass	$z^{-1} \rightarrow \dfrac{z^{-1} - \alpha}{1 - \alpha z^{-1}}$	$\alpha = \dfrac{\sin\left(\dfrac{\omega_L - \omega_p}{2}\right)}{\sin\left(\dfrac{\omega_L + \omega_p}{2}\right)}$	ω_p = desired cutoff frequency
Highpass	$z^{-1} \rightarrow -\dfrac{z^{-1} + \alpha}{1 + \alpha z^{-1}}$	$\alpha = \dfrac{\cos\left(\dfrac{\omega_L + \omega_p}{2}\right)}{\cos\left(\dfrac{\omega_L - \omega_p}{2}\right)}$	ω_p = desired cutoff frequency
Bandpass	$z^{-1} \rightarrow -\dfrac{z^{-2} + \dfrac{2\alpha k}{k+1} z^{-1} + \dfrac{k-1}{k+1}}{\dfrac{k-1}{k+1} z^{-2} + \dfrac{2\alpha k}{k+1} z^{-1} + 1}$ ω_2, ω_1 = desired upper and lower cutoff frequencies	$\alpha = \dfrac{\cos\left(\dfrac{\omega_2 + \omega_1}{2}\right)}{\cos\left(\dfrac{\omega_2 - \omega_1}{2}\right)}$,	$k = \cot\left(\dfrac{\omega_2 - \omega_1}{2}\right)\tan\left(\dfrac{\omega_L}{2}\right)$
Bandstop	$z^{-1} \rightarrow -\dfrac{z^{-2} + \dfrac{2\alpha}{k+1} z^{-1} + \dfrac{1-k}{k+1}}{\dfrac{1-k}{k+1} z^{-2} + \dfrac{2\alpha}{k+1} z^{-1} + 1}$ ω_2, ω_1 = desired upper and lower cutoff frequencies	$\alpha = \dfrac{\cos\left(\dfrac{\omega_2 + \omega_1}{2}\right)}{\cos\left(\dfrac{\omega_2 - \omega_1}{2}\right)}$,	$k = \tan\left(\dfrac{\omega_2 - \omega_1}{2}\right)\tan\left(\dfrac{\omega_L}{2}\right)$

if the passband extends close to one-half of the sampling frequency because of the possibility of aliasing.

8.19 BIOMEDICAL APPLICATIONS OF DIGITAL FILTERING

It was mentioned previously that digital filtering is encountered often during the processing and analysis of biomedical data, sometimes in a framework that may not be recognized as digital filtering—for example, see the second differentiator example below. Here several examples are provided to illustrate the procedures for selecting filter parameters and designing basic digital filters using MATLAB software.

Example 8.20 Removing noise from an ultrasound signal The velocity of movement of red blood cells in a vessel (assumed to reflect the local fluid flow velocity) may be measured by passing an ultrasound beam through the vessel wall and determining the Doppler shift of frequency of the reflected beam. Since a 2-MHz ultrasound wave passes through the temporal region of the skull without excessive attenuation, it is possible to measure blood flow velocity in arteries of the brain of humans. Usually blood flow velocity in the middle cerebral artery (V_{MCA}) is measured. Since at any time there is a distribution of velocities, a transcranial Doppler ultrasound device often outputs a weighted average of the velocities as the instantaneous "mean" velocity. The upper tracing of Fig. 8.22(c) is an example of such a signal (data courtesy of Nicole Cleary).

This signal was sampled 100 times per second and it was desired to remove the high-frequency noise using a digital filter. It was estimated that the signal components above 10 Hz were due to noise; therefore the desired cutoff frequency was 10 Hz, corresponding to a DT cutoff frequency, 'omegac', of $\omega_c = \Omega_c T = 2\pi(10)(0.01) \approx 0.628$ rad/sample. Two 26-point FIR filters were compared: one based on windowing design using a Hamming window, the second designed using the Remez equiripple method. The first was designed using the MATLAB command firl with 'order' = 25: bham = firl(25,omegac). The second filter design used the remez command: brem = remez(25,[0, omegac, omegac + 0.1, 1],[1, 1, 0, 0]). Their unit-pulse responses are plotted in Fig. 8.22(a). These responses are similar but the equiripple filter exhibits somewhat larger oscillations away from the central peak. The frequency responses were calculated using freqz and they are graphed in Fig. 8.22(b). The differences between the two filters are quite clear. The Hamming window filter has a smaller stopband gain but its transition band is about twice the width of that of the equiripple filter.

Figure 8.22(c) demonstrates the real differences in output from these two filters when they are applied to the V_{MCA} signal. More high-frequency components are present in the output of the equiripple filter, but this filter also more faithfully reproduces the amplitude of the sharp increase in V_{MCA} at the beginning of each flow pulse. For both filters, however, the "spike-like" noises in the input signal cause small oscillations in the filtered signals that resemble other fluctuations reflecting actual

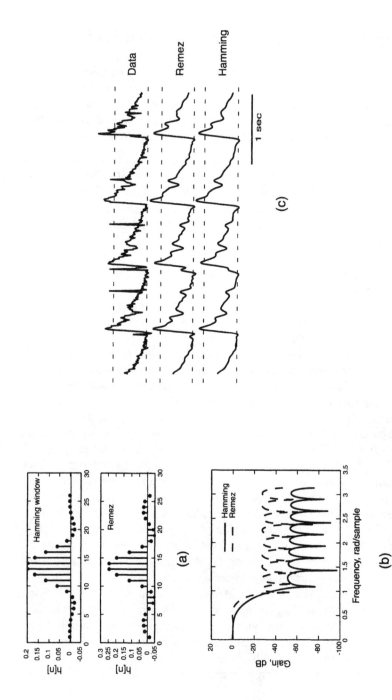

FIGURE 8.22. (a) Unit-pulse responses and (b) frequency responses of a 26th-order Hamming window lowpass filter and a 26th-order equiripple lowpass filter, both having a cutoff frequency ≈ 0.628 rad/sample. (c) Ultrasound measurement of MCA blood flow velocity (top), and the same signal after filtering by the equiripple filter (middle) and the Hamming window filter (bottom).

334

changes in blood flow velocity. Consequently, either of these filters may be adequate for analyses that average the flow velocity over each flow pulse, for example, but may be inadequate for analyses that address the details of the time-course of the signal within each flow pulse. (In the latter case, nonlinear filtering using a median filter has been found to provide much better rejection of the "spike-like" noise.) These signals are available in the file cbfxmpl.mat and the reader may wish to test other filter designs.

Example 8.21 Design of a digital differentiator using an equiripple method
Often it is desired to obtain the derivative of a sampled biomedical signal. One may be tempted to evaluate successive differences between samples of the signal, dividing by the sampling interval to calculate the derivative, but this method is very sensitive to noise. A better method is to design a digital filter having a frequency response that approximates that of a true differentiator. In CT the frequency response of a differentiator is $H(\Omega) = j\Omega$. That is, the magnitude of $H(\Omega)$ increases linearly with frequency. The Remez algorithm may be used to design a digital filter having a gain linearly proportional to frequency if three modifications are made. First, the linear gain must be restricted to a frequency range $(0, \omega_{max})$, where $\omega_{max} < \pi$. That is, we select the maximum frequency so that the range of linear gain encompasses all of the important components of the signal to be differentiated. Second, the function that weights the errors, $W(\omega)$, is defined so that it is proportional to $1/\omega$, $0 \leq \omega \leq \omega_{max}$, so that the relative weighting decreases as the gain increases. Finally, we note that the unit-pulse response of an ideal digital differentiator is

$$h[n] = \frac{1}{2\pi} \int_{-\pi}^{\pi} H(e^{j\omega}) e^{j\omega n} d\omega = \frac{1}{2\pi} \int_{-\pi}^{\pi} (j\omega) e^{j\omega n} d\omega = \frac{\cos[\pi n]}{n}, \qquad n \neq 0,$$

and $h[0] = 0$. Thus $h[n]$ satisifes the condition $h[n] = -h[-n]$, which is opposite to the usual symmetry condition we have imposed on FIR filters. Suffice it to say that one can rederive the iterative algorithm for equiripple design using this condition of anti-symmetry. This design is also implemented in the MATLAB command remez.

The signal to be addressed here is the pressure in the left ventricle of the heart (P_{LV}). The maximum derivative of this signal (during a given heart beat) is often considered an index of myocardial contractility. Fig. 8.23(a) presents a sample of P_{LV} obtained from the file dog_heart_data available from the Signal Processing Information Database at Rice University (http://spib.rice.edu/spib.html). (Note that the vertical scale represents analog-to-digital converter numbers.) After subsampling this data to an effective sampling rate of 200/s, a differentiator was designed to have a linear frequency response in the range $0 \leq \omega \leq 0.4\pi$, corresponding to a CT frequency range of $0 \leq 2\pi f \leq 2\pi(40)$. The filter was obtained using remez as bder = remez(21, [0, 0.4 * pi, 0.6 * pi,1], [0, 1, 0, 0,], 'differentiator'). Its unit-pulse and frequency responses are plotted in Fig. 8.23(b,c). Note the anti-symmetry of $h[n]$ and the linear gain of the frequency response. Fig. 8.23(d) shows a portion of the original signal and its derivative calculat-

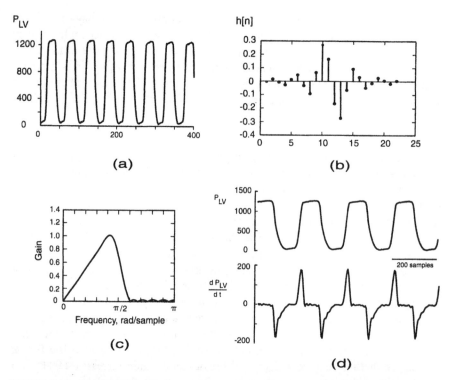

FIGURE 8.23. (a) Left ventricular pressure signal. See text for description. (b) Unit-pulse response and (c) frequency response of a digital differentiator. (d) A portion of the signal from 'a' (top) and its derivative (bottom)—that is, output of the digital differentiator filter.

ed using bder with the filter function. The derivative signal was shifted left by 10 points to compensate for the phase shift of the FIR differentiator. Visually the derivative signal appears to reflect small fluctuations in P_{LV} without being unduly sensitive to noise. As expected, the maximum derivative occurs during the upstroke of the ventricular pressure.

Example 8.22 Numerical derivative as an FIR filtering problem Numerous time-domain algorithms for calculating the derivative of a sampled function have been developed based on numerical approximations to the signal or its derivative. A simple approach is to fit a polynomial function to several consecutive data points, then evaluate the derivative of the polynomial function at its middle point. This approach may be viewed as a type of FIR digital filter. Consider fitting a second-order polynomial of the form $y = at^2 + bt + c$ to three consecutive data points then evaluating the derivative, $dy/dt = 2at + b$, at the middle point.

The problem is simplified by assuming that the time points are $t = -1, 0, 1$ and later scaling the calculated derivative by the sampling interval. If y_1, y_2, and y_3 are the values of the data at the three time points, one may write three equations (one for each time point) in three unknowns using the polynomial function:

$$\begin{bmatrix} 1 & -1 & 1 \\ 0 & 0 & 1 \\ 1 & 1 & 1 \end{bmatrix} \begin{bmatrix} a \\ b \\ c \end{bmatrix} = \begin{bmatrix} y_1 \\ y_2 \\ y_3 \end{bmatrix}. \tag{8.66}$$

Solving by inverting the 3×3 matrix, one obtains

$$\begin{bmatrix} a \\ b \\ c \end{bmatrix} = -\frac{1}{2} \begin{bmatrix} -1 & 2 & -2 \\ 1 & 0 & -1 \\ 0 & -2 & 0 \end{bmatrix} \begin{bmatrix} y_1 \\ y_2 \\ y_3 \end{bmatrix}. \tag{8.67}$$

Now the derivative at the middle point is

$$\left. \frac{dy}{dt} \right|_{t=0} = [2at + b]|_{t=0} = b.$$

From Eq. (8.67), $b = \frac{1}{2}(y_3 - y_1)$. One may implement this calculation as a digital filter that enumerates one-half of the difference between points that are two samples apart. Such a filter would have the unit-pulse response

$$h[n] = \begin{cases} -\frac{1}{2}, 0, \frac{1}{2} & n = 0, 1, 2 \\ 0, & otherwise \end{cases}.$$

This FIR filter will introduce a delay of one sample. If the data are already stored digitally, one could implement a noncausal filter instead by making the nonzero portion of $h[n]$ begin at $n = -1$.

Example 8.23   Initial EMG filtering example revisited   In the example at the beginning of this chapter a second-order, Butterworth, analog, highpass filter was applied to an EMG recording to reduce the ECG contamination (Fig. 8.1). Improvement in the signal was moderate (Fig. 8.1(c)). After sampling the EMG signal at 500/sec, a digital highpass filter was designed for the same objective. Knowing that the cutoff frequency of the "best" analog filter was approximately 70 Hz., we attempted to design an equiripple filter having a cutoff frequency near $\omega = 2\pi(70)(0.02) = 0.880$ rad/sample. In MATLAB notation (as a fraction of half of the sampling frequency) this frequency is 0.28. By trial and error it was found that a filter order of 91 would produce a stopband gain less than -80 dB if the transition band width was at least 0.1π. Thus the filter was designed using the command bhp = remez(90, [0, 0.225, 0.33, 1], [0, 0, 1, 1]). Its frequency response is shown in Fig. 8.24(a). The cutoff frequency is approximately 0.29π, which corresponds to a CT frequency of 72.5 Hz. The sampled EMG signal was processed through this filter (Fig. 8.24(b)). There is a noticeable improvement over the results using the analog filter (Fig. 8.1(c)) due to the lower stopband gain and narrower transition band of the digital filter. The reader may wish to test various window design filters to see if further improvement is possible.

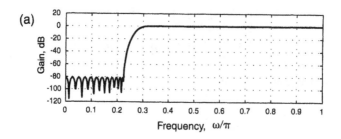

FIGURE 8.24. (a) Frequency response of a 91-point equiripple highpass filter with transition band $0.225\pi \leq \omega \leq 0.33\pi$. (b) EMG from Fig 8.1(a) after filtering by the highpass filter of 'a'.

8.20 SUMMARY

This chapter has introduced the related concepts of Z-transforms and digital filters. The Z-transform of a DT signal (or sequence) $x[n]$ is a representation of the signal in terms of a weighted sum of eigenfunctions of LSI systems (Eq. (8.10)) in which the eigenfunctions have the form $z^n = r^n e^{jn\theta}$. The region of convergence of the bilateral Z-transform must be specified in order to uniquely relate the transform to a sequence whereas the unilateral Z-transform assumes that $x[n]$ is zero for $n < 0$. The Z-transform is a linear operator and numerous basic properties simplify the computation of Z-transforms. If the Z-transform $X(z)$ converges on the unit circle, then the relationship $X(e^{j\omega}) = X(z)|_{z=e^{j\omega}}$ follows. That is, the DTFT of $x[n]$ equals the Z-transform of $x[n]$ evaluated on the unit circle. The most common approach for determining the inverse of a Z-transform is the method of partial fraction expansion.

Z-transforms are especially useful for describing the input-output properties of LSI systems. If $x[n]$ is the input, $h[n]$ is the unit-pulse response of the system, and $y[n]$ is the output, then $Y(z) = H(z) X(z)$. $H(z)$ is the transfer function of the LSI system. From this result one may calculate the output for a given input. If one uses the unilateral Z-transform, initial conditions may also be handled conveniently.

There are two basic approaches for designing digital filters. One approach begins by specifying an appropriate analog filter, then utilizes some specific relationship between CT and DT frequency responses. The most common approaches of this type are the impulse invariance and bilinear transformation methods, which produce IIR digital filters. The former maps the poles of the CT filter into poles of the DT filter through the relationship $p_i \rightarrow e^{p_i T}$. That is, only the poles are mapped via the relationship $z = e^{sT}$. The bilinear transformation method maps the entire s-plane into the z-plane by an approximation to this latter relationship. Because of frequency warping with this method, one must pre-warp the cutoff frequencies of the CT filter so that they map to the desired cutoff frequencies for the DT filter. IIR filters also may be designed using time-domain optimization methods, such as Prony's method. This method is particularly useful for designing all-pole filters.

A common method for designing FIR filters is the windowing method. By this approach the IIR unit-pulse response of an ideal filter is multiplied by a finite-length window function to create an FIR response. The choice of the window function determines the stopband gain and the width of the transition band. In the frequency sampling method, the desired frequency response is specified at a number of frequencies equal to the desired length of the unit-pulse response. By placing some of these frequencies in the transition band, one may control the stopband gain and the amplitude of ripples in the gain. A third design method for FIR filters allows the gain in both the stopband and passband to fluctuate. Based on properties of polynomials and approximation theory, the order of the filter can be related to the order of a polynomial function that passes through the points of the desired frequency response. Iterative techniques have been developed to solve the resulting nonlinear optimization problem to determine the unit-pulse response. These latter methods are known generally as equiripple design methods, and often are implemented using the Remez algorithm.

Digital filters are encountered frequently in the processing of biomedical signals that are sampled, and often such filters are embedded in biomedical instrumentation. Therefore, it is important to understand how these filters are modifying the measured signals even if one is not involved in the design of digital filters.

EXERCISES

8.1 Determine the Z-transforms of the following causal sequences:
  a. $f[n] = n + \sin[2n]$;
  b. $y[k] = e^{0.1k}$;
  c. $v[m] = 6\delta[m - 2] + 2^m$;
  d. $w[n] = \sin[4\pi n + 0.1\pi]$;
  e. $h[k] = k(1)^k + k(-1)^k$;
  f. $x[n] = 0.2^n \cos[0.1\pi n]$, $|z| > 0.2$;
  g. $p[m] = (m - 3)u[m - 3]$;
  h. $s[n] = (n - 3)u[n]$.

8.2 Determine the sequences having the given Z-transforms (assuming a right-hand sequence if necessary):
  a. $X(z) = 2z^2 + 5z - 1$, $|z| < \infty$;
  b. $Y(z) = 2z^2 + 5z^{-1} - 1$, $0 < |z|$;
  c. $V(z) = \dfrac{z + 1}{(z^2 + 0.9z + 0.2)(z - 0.8)}$;

d. $H(z) = \dfrac{0.5z^{-1}}{(1 - 0.5z^{-1})^2}$;

e. $W(z) = \dfrac{0.5z^{-1}}{1 - z^{-1} + 0.5z^{-2}}$;

f. $E(z) = \dfrac{1 - 0.8z^{-1}}{(1 - 0.5z^{-1})^2(1 - 0.2z^{-1} + 0.15z^{-2})}$;

g. $F(z) = \dfrac{(z - 1)^2}{z - \frac{1}{2}}$;

h. $Q(z) = \dfrac{4(2 - z^{-1})}{8 + 6z^{-1} + z^{-2}}$.

8.3   Consider the two sequences $x[n] = a^n u[n]$ and $y[n] = a^n(u[n] - u[n - N])$. Graph and compare the poles and zeros of $X(z)$ and $Y(z)$ for $N = 4$.

8.4   An LSIC system has the transfer function $H(z) = z/(2z^2 - 3z + 1)$.
  a. Determine the difference equation of this system.
  b. Evaluate its unit-pulse response.
  c. Calculate its frequency response.

8.5   Derive the inverse Z-transform of Eq. (8.21).

8.6   An LSIC system having the unit-pulse response $h[n] = b^n u[n]$ receives the input signal $x[n] = a^n u[n]$. Calculate its zero-state response. What is this response when $a = b$?

8.7   The unit-step response of an LSIC system is found to be $y[n] = (\frac{1}{3})^n(u[n] - u[n - 3])$. Determine the transfer function of this system.

8.8   An LSIC system is described by the difference equation

$$y[n] - 0.75y[n - 1] + 0.125y[n - 2] = 2x[n].$$

Find its transfer function. Is this system BIBO stable?

8.9   An LSIC system has the unit-pulse response $h[n] = [(\frac{1}{2})^n - (\frac{1}{3})^n]u[n]$. Determine its transfer function and frequency response.

8.10   In Example 8.16 (All-pole filter design by Prony's method) a second-order filter was designed. Repeat the design for a third-order filter and show that $a[3] = 0$.

8.11   Derive the equivalent of Eq. (8.55) when M is even and show that the phase is also linear with frequency.

8.12   A common way of smoothing a noisy sequence is to add together the current value of the signal, plus 1/2 of the preceding value, plus 1/4 of the value two samples previously. This procedure can be described as an LSI system which has an impulse response given by the equation $h[n] = \frac{4}{7}(\frac{1}{2})^n(u[n] - u[n - 3])$. Calculate the transfer function of this system. Determine the magnitude and phase of its frequency response.

8.13 Design a digital filter using the MATLAB command fir1. Let the unit-pulse response of the filter have a length of 81 points and let the cutoff frequency be $\omega_c = 0.25\pi$. Use a Hann window. Also design a filter having the same length and cutoff frequency using the remez command. Plot the magnitude of the frequency response versus frequency for both filters. Which one would be better for separating two frequency components in the input signal located at 0.20π and $0.30\ \pi$?

8.14 Design a digital Butterworth filter to meet the following specifications for removing noise from a sampled ECG signal:

$$\omega_p = 0.25\,\pi, \qquad \omega_s = 0.40\,\pi, \qquad \delta_1 = -3\ \text{dB}, \qquad \delta_2 = -20\ \text{dB}.$$

8.15 Repeat the preceding problem using an equiripple design.

8.16 A DT filter has the transfer function $H(z) = (z - 0.1)/[(z - 0.8)(z - 0.05)]$. This filter was designed using the impulse invariance method with a sampling interval of $T = 0.05$ s. Determine the CT filter $H(s)$ upon which this design was based.

8.17 The impulse response of an analog filter is given by $h(t) = e^{-0.9t}u(t)$. Let $h[n]$ denote the unit sample response and $H(z)$ denote the transfer function for the digital filter designed from this analog filter using impulse invariance. Determine $H(z)$ including T as a parameter. Is the digital filter BIBO stable for all positive values of T? If not, for what values of T is it stable?

8.18 The transfer function $H(s)$ of an analog filter is $H(s) = s/[(s + 5)(s + 2)]$. Determine $H(z)$ for a digital filter designed from this analog filter using impulse invariance.

8.19 Repeat Example 8.18 (Frequency sampling design) with no transition point and compare the resulting frequency response with the one obtained in this example.

8.20 In Example 8.19 (Equiripple filter design) it was found that making the transition band narrower caused more ripples in the gain and a higher stopband gain. Repeat the design having the narrower transition band and demonstrate that increasing M (e.g., let $M = 75$) reduces these problems.

8.21 Figure 8.25 shows the frequency response of a digital filter.
   a. Sketch the frequency response of a continuous-time filter which will map to the frequency response in the figure if the digital filter is derived from the CT filter using impulse invariance design.
   b. Repeat part a when the digital filter is derived from the CT filter using bilinear transformation.

8.22 An LTIC discrete-time system is described by the difference equation $y[n + 2] + \sqrt{2}\omega_0 y[n + 1] + \omega_0^2 y[n] = \omega_0^2 x[n]$.
   a. What is the frequency response, $H(e^{j\omega})$ of this system?
   b. What is the steady-state response of this system to the input $x[n] = \cos[\omega_0 n]$?
   c. This difference equation "looks like" a discrete form of the usual differential equation for a Butterworth filter. Would you say that this filter has the type of frequency response characteristic of Butterworth filters? Explain.

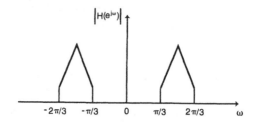

FIGURE 8.25. Frequency response of a digital filter

8.23 Derive Eq. (8.49), which expresses the minimum error in Prony's method in terms of the autocorrelation values and the coefficients of the filter.

8.24 A recording from an intravascular pH electrode is contaminated with a 20 Hz sinusoidal signal due to a preamplifier malfunction. The signal has been digitized by sampling at 50 Hz.

a. The user could have removed the 20 Hz signal before sampling by filtering with a system having a transfer function of the form $H(j\Omega) = \Omega_c/(j\Omega + \Omega_c)$. What value of Ω_c would give a gain of 0.1 (i.e., –20 dB) at 20 Hz ?

b. Since the user did not filter before sampling, he must filter the digitized signal. Use impulse invariance to design the digital equivalent of the analog filter above (use the value of Ω_c you found in part a) and sketch the magnitude of its frequency response versus the discrete-time frequency ω. Identify the value of ω which corresponds to 20 Hz and determine the gain of the digital filter at that frequency.

8.25 Design a digital filter by the windowing method using the MATLAB command `fir1`. The filter should have a cutoff frequency of 0.35π. (Note that `fir1` requires that you specify the cutoff frequency as a fraction of π.) First design a filter of length 15. Then design a filter of length 81. The output from `fir1` is the impulse response, $h[n]$, of your filter. For both cases use the `fft` command to find the DTFT of $h[n]$ (that is, the frequency response of the filter, $H(e^{j\omega})$) and plot the magnitude of the frequency response versus frequency for $0 \le \omega \le \pi$. Note that N is not a power of two for the FFT but MATLAB will calculate the FFT using a non-power-of-2 algorithm. Also calculate and plot the step responses and discuss the relative advantages and disadvantages of the filters from a time-domain perspective. Finally, repeat the steps above using the `remez` command to design a filter of length 81. Compare the frequency responses of the two 81-point filters. Which filter is better? Why?

8.26 Ion-selective electrodes are used to monitor the constituents of tissue baths. Although many respond slowly, having time constants of many seconds, some (such as solid-state pH electrodes) have time constants of response on the order of one second. If the output of such an electrode is sampled, say at 25/s, one would want to lowpass filter the sampled signal to remove high-frequency noise—for example, signal components above 2.5 Hz. Using the bilinear transformation method, design a second-order digital Butterworth filter for this application that meets the following specifications:

$$0.99 \le |H(e^{j\omega})| \le 1, \qquad 0 \le \omega \le 0.2\pi; \qquad |H(e^{j\omega})| \le 0.01, \qquad \omega > 0.4\pi.$$

Apply the specifications to determine the appropriate filter order and cutoff frequency, then use the MATLAB command butter to design the filter as discussed in the text.

8.27 Conceptually one could design an FIR filter by truncating the impulse response of an IIR filter. For a simple example consider two filters whose impulse responses are given by: $h_1[k] = a^k u[k]$, $|a| < 1$, $h_2[k] = a^k u[k]$, $0 \le k \le 1$, and $h_2[k] = 0$, $k > 1$.

a. Find $H(z)$ of both filters and compare the magnitudes of their discrete-time frequency responses.

b. The difference equation for the first filter is $y[k] = ay[k-1] = x[k]$. Determine the difference equation for the second filter.

c. If one wants to obtain an FIR lowpass digital filter, it is better to design a new filter using the windowing method than to truncate the impulse response of an actual IIR filter. Discuss at least three explicit reasons why the preceding statement is correct.

8.28 Develop a second-order, all-pole model of the EMG signal in the file c8emg.mat and use it to calculate the power spectrum of the EMG. Sample a 64-point segment of the EMG signal (emgf1), carefully avoiding the ECG waveforms, then use Prony's method to design the model. Use MATLAB to calculate the "auto-correlation" values that are needed and to solve the matrix equation for the coefficients. Determine the unit-pulse response, $h[n]$, of this model. Then calculate the normalized power spectral density function of $h[n]$, recalling from Chapt. 7 that $P_h(e^{j\omega}) = |H(e^{j\omega})|^2/N$. It is instructive to repeat this procedure utilizing other segments of the EMG signal in order to assess the variability in the spectrum.

8.29 In an earlier chapter the response of receptors in the airway of the lungs to stretching of the airway wall was described by a second-order differential equation in which $x(t)$ is the amount of applied stretch and $y(t)$ is the transmembrane voltage of the receptor. This equation was:

$$\frac{d^2y}{dt^2} + 2\frac{dy}{dt} + 4y = 10x + \frac{dx}{dt}.$$

This equation can be considered the input–output equation of a filter whose input is stretch and whose output is a neural signal. Using the bilinear transformation method, design a DT filter that models this CT system.

9

MODELING STOCHASTIC SIGNALS
AS FILTERED WHITE NOISE

9.1 INTRODUCTION

In previous chapters this text assumed that a biomedical signal could be modeled as the sum of a "true" signal plus added (usually white) noise. In many instances this model is too restrictive or even inappropriate because it implies that if a measurement were repeated under identical conditions, any differences between the two measurements are attributable to the added noise. The concept that each newly generated white-noise signal should differ from other white-noise signals may seem intuitively believable as an explanation for differences between two measurements, but this concept may apply to the "true" signal also. Each time a measurement is repeated, whether on the same subject or a different subject (where "subject" refers to any source of a signal, not just human subjects), some uncontrolled or unrecognized factors related to the subject are likely to differ from the time of the first measurement. Therefore, even after accounting for changes in the added noise, the underlying measurements are likely to differ (and to do so in ways that appear to be random). As a consequence, one is not able to discuss the exact temporal nature of the noise-free signal, because that signal differs among the data sets.

We assume that repeated measurements possess some properties that are consistent from one data set to the next and that one may obtain insight into these properties from each individual measurement. To obtain such insight, one must consider the probability that repeated measurements would resemble one another. If repeated measurements have been acquired from several subjects, then one may average these measurements at each point in time to generate an "average signal", but this average signal does not necessarily represent the measurement that one would obtain from any given subject in the absence of added noise. It seems clear that a different framework is needed for modeling such signals. This chapter first considers a specific example of these types of signals, then develops a general framework for addressing their properties, and finally formulates several methods for deriving the basic properties of these signals.

9.2 INTRODUCTORY EXERCISE: EEG ANALYSIS

The electroencephalogram (EEG) is a record of the electrical activity induced at the scalp by the flow of ionic currents across and along the membranes of neurons of the brain (Fig. 9.1). Because the cerebral cortex lies immediately beneath the scalp, the EEG recorded from the top of the head reflects activities of cortical neurons. For about 50 years it has been recognized that the cortical EEG changes its character as the state of consciousness changes. During relaxed wakefulness the cortical EEG (or just EEG) exhibits relatively low-amplitude, high-frequency voltages in the range of 8–12 Hz known as the α rhythm. Of course, this rhythm is not a pure oscillation like a sine wave and it is not the only signal component of the EEG. A trained neurologist, however, can recognize this dominant rhythm in a chart recording of the EEG (e.g., Fig. 9.1). Furthermore the intensity of α rhythm varies among recording sites and with other factors such as attentiveness. As a person enters the lighter stages of sleep (stages 1 and 2), the EEG develops more rhythms at lower frequencies (θ rhythm, 4–8 Hz) and exhibits some other features such as sleep spindles and k-complexes, both of which are irregularly occurring high-frequency bursts of activity. In deep sleep (stages 3 and 4) a relatively larger amplitude, low-frequency (<4 Hz) activity (δ rhythm) appears. In REM (rapid-eye movement) sleep, the state in which dreaming occurs, there is a predominant α-like rhythm. REM is distinguishable from wakefulness by the presence of eye movements which are detected by recording EMGs from extraocular muscles. During a typical night the sleep state of normal humans may change 15 times, encompassing several episodes of each of the above stages.

In many pathological situations patients have a disruption of their normal sleep patterns. In order to document this disruption it is necessary to record the EEG signals overnight and analyze them for sleep state (called "sleep scoring"). Usually a "score" is assigned every 30 seconds. This analysis is very labor intensive and many

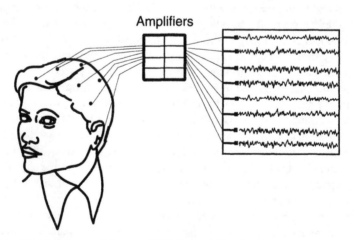

FIGURE 9.1 The electroencephalogram (EEG) is recorded simultaneously from many sites (e.g., 8, 16, or 32 sites) on the scalp and displayed on a monitor or chart recorder.

researchers have sought to develop automated analyses of digitized EEG signals. The example below explores the development of an automated system for EEG analysis.

Example 9.1   Analysis of EEG signals   The objective is to propose a method for analysis of an EEG signal that will discriminate its three basic rhythms (α, θ, δ). For true sleep staging one would need to incorporate information about sleep spindles, k-complexes, and extraocular and chin EMG signals, but for the present purpose these refinements will be ignored. The analysis will need to assign a sleep stage score to the EEG signal every 30 seconds, based on which of the three rhythms is dominant—for example, if δ rhythm is dominant, score the 30-second period as stage 3–4 sleep. Note that transitions in sleep state can occur in less than 30 seconds and that some patients with severe sleep disruption may leave a sleep stage less than 2 minutes after entering it.

Assume that the EEG signal is available in digitized form, having been sampled at 50 Hz. To help visualize this problem, the file eegsim.m can be used to create a different random sample of a simulated EEG signal each time it is invoked (Fig. 9.2).

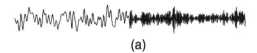

(a)

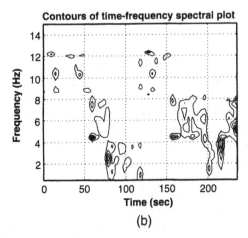

(b)

FIGURE 9.2   (a) 20 seconds of a simulated EEG signal from eegsim.m, showing an abrupt transition from a delta rhythm to an alpha rhythm. (b) Contours of a time-dependent spectral analysis, which indicate the presence of peaks in the power spectrum, for a simulated, 240-s, EEG signal. This analysis indicates that the simulated EEG signal begins as an alpha rhythm, then progresses to a theta then a delta rhythm. Around $t = 120$ s the rhythm changes abruptly to alpha again, then progresses to the slower rhythms. Note that these simulated EEG signals are idealized approximations of true EEG recordings.

This simulated signal is sampled at 50 Hz for approximately 4 minutes and contains examples of all three EEG rhythms. The signal will exhibit abrupt transitions among the three rhythms at random times and each rhythm will last for at least 15 seconds. Compared to a true EEG, this signal is simplified in several respects: (1) transitions in true EEGs occur over several seconds or longer; (2) real EEGs also contain spindles and k-complexes; (3) the simulated EEG ignores stage REM; (4) apparent sleep stages are shorter in the simulated signal than in normal subjects; (5) amplitude changes are less than occur naturally. Nonetheless, there is a considerable challenge in the analysis of the simulated EEG signal. One possible approach is to pass the EEG signal through three parallel bandpass filters having passbands of 0–4, 4–8, and 8–12 Hz, respectively. Then the relative amplitudes of the outputs would have to be quantified, perhaps by calculating their mean square values every 30 seconds. The reader is encouraged to implement such filters in MATLAB and test this method using the simulated EEG signal from eegsim.m. The simulated EEG is accessible in the MATLAB workspace as a vector name eeg. Note that eeg is updated each time the m-file is invoked.

To assist your interpretation of the simulated EEG signal, eegsim.m also analyzes the signal using a method derived from the material of the present chapter and presents graphs of those results (Fig. 9.2). A later exercise will dissect the analysis method implemented in eegsim.m and seek improvements to it.

9.3 RANDOM PROCESSES

Consider a situation in which a physician wishes to know the level of oxygen in the arterial blood of a critically ill patient. A nurse takes a blood sample and sends it to the Clinical Chemistry Lab for analysis and the lab reports a partial pressure of oxygen of 50 Torr. One hour later this process is repeated and the lab reports 55 Torr. Assuming that no treatment has been applied to the patient, could this result imply that the patient's clinical state has improved or is it possible that the difference in the two readings is due to some kind of random variability in blood oxygen level? Intuitively one might select the latter alternative. It is possible to formalize a description of the kind of random effect assumed to be occurring in this illustration.

Assume one wants to know the average blood pressure of normal humans. One approach to obtaining this information might involve measuring the blood pressure on many people and averaging the result. The population that you can most readily access is your group of peers, so let's assume that they agree to the placement of arterial pressure catheters for the sake of science! Now, as you examine the blood pressure signals (obtained simultaneously from all subjects) you are dismayed to observe that blood pressure is changing continuously in every subject. One may try to circumvent this new problem by averaging the blood pressure from each subject over some time interval. One problem, of course, is that one does not know whether this average is an adequate representation of the long-term average blood pressure of each subject.

There is another issue that is more fundamental. By assuming that it is possible to determine an average blood pressure that is a biologically-relevant measure of the population, one has assumed that the underlying biological mechanisms that determine blood pressure are structurally invariant from subject to subject except for possible random variations in their parameter values, the effects of which will be eliminated by averaging. Although this assumption seems reasonable for human biology, and indeed may be the only sensible way to approach an understanding of the human organism, it is essential to recognize that this conceptualization is being invoked by the simple act of averaging blood pressures across subjects.

The assumption discussed above addresses the differences in blood pressure between individuals but does not seem to account directly for the time variations of blood pressure in each subject. In fact, there are known mechanisms which explain some of this blood pressure variability, such as the coupling of heart rate with respiration, but even after accounting for these mechanisms blood pressure is not constant. To account for the remaining variations, recall the discussion of stochastic signals in Chapter 1. A stochastic (or random) signal is one whose future values cannot be predicted exactly (although one might be able to predict the probability of specified values occurring—for example, the probability that the future value will be greater than the present value). Random signals may arise from sources whose internal mechanisms truly vary randomly or they may represent mechanisms which we do not yet understand and so their activities appear unrelated to any explainable behaviors. In either case it is the continual influence of sources of random disturbances which is assumed to cause the remaining (unexplainable) variability of blood pressure in each subject. It is important to note that one does not know how these random effects actually influence blood pressure—we assume only that they do exert such an influence and are the cause of the variability in the measurement. This model of a biological process, in which there exist random parameter variations between subjects and random disturbances within a subject, is referred to as a random process model. It is also important to recognize that the true, undisturbed behavior of the random process could vary with time. Because of the randomness inherent in each measurement from an individual subject, however, the true function of time cannot be *exactly* determined from that measurement.

A *random process* is a physical process that is structurally similar in all instances in which it is implemented—for example, in each subject in the population—but its behavior (or output) is not identically the same in all implementations because: (1) there are random variations of its parameters among implementations, and (2) each implementation is "disturbed" by random noise. Figure 9.3 presents this concept. For whatever process is being considered—for example, the neural reflex process for control of blood pressure, the neuromuscular process regulating eye movements, the biochemical process regulating intracellular potassium concentration, etc.— each subject in the population is one implementation of the process (called a *realization* of the process). For the examples of blood pressure and eye movements, the "population" comprises a group of human subjects and each person represents one realization of the random process. For the intracellular potassium example, the "population" is a group of cells and each cell is one realization. A record of the time

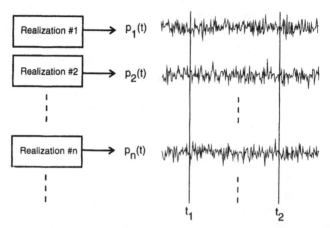

FIGURE 9.3 Conceptual scheme of a random process, as applied to a biomedical process. The process is assumed to be structurally identical in all realizations but parameter values differ randomly between realizations. In addition the output of the process is influenced by random disturbances within each realization. Therefore the output signals from the process, $p_i(t)$, do not exactly represent the true undisturbed behavior of the process. Each $p_i(t)$ is a sample function of the random process.

history of the signal of interest from a single realization (e.g., blood pressure, eye position, or potassium concentration) is referred to as a *sample function* of the random process (Fig. 9.3). Sample functions from different realizations of the same random process are different.

An important point is that the value of a single sample function at a specific time is formally a random variable. Although we will not probe this subject deeply, a *random variable* is a variable whose value depends on the outcome of a chance event such as the throw of a die or the spin of a roulette wheel. For example, a random variable, z, might be defined by assigning to z the number showing on the top face of a die each time it is thrown. In other words, the random variable assumes a new value each time the chance event—that is, a throw of the die—occurs. To extend this concept to random processes, consider a fixed point in time, t_1. Each sample function of the random process has a value at time t_1. Now a random variable, x, may be defined by randomly selecting a sample function from the ensemble of sample functions and assigning to the random variable the value of the selected sample function at time t_1. (The random selection of the sample function is the chance event.) Whenever another random selection of a sample function is made, a new value for x is determined. This example is completely analogous to the example based on throwing a die except in that case z could assume only six different values, whereas the sample functions of biomedical random processes are likely to assume many possible values.

If the values assigned to another random variable, y, are derived from the same set of sample functions but for a different time, t_2, then in general x and y will assume different values even if the same sample function is chosen. Likewise, if one selects a sequence of these sample functions and assigns values to x and y, as above, then the sequences of specific values of x and y will differ. Consequently a random variable, x, derived from random selection of sample functions of a random process is, in general, time-dependent. Usually this time-dependence is recognized by representing the random variable explicitly as $x(t)$.

In this book the focus will be on deriving representations (i.e., models) of sample functions from random processes in order to deduce the properties and parameter values of the random processes. The reader is *strongly cautioned* to keep in mind that each realization of a random process differs from every other one, and therefore the parameter values that are derived will differ among the various sample functions. Just as the mean blood pressures of individual subjects can be averaged to obtain a population mean blood pressure, it is necessary to average *any* parameter derived from sample functions of random processes in order to obtain the most reliable estimates of the mean parameter value within the population. The reliability of such averages as estimates of the population mean values, and whether an average from only one realization is useful, are important questions that will be addressed later in this chapter. These topics are considered in more depth in courses on statistical signal processing. Why then do we address random processes here? There are two reasons: First, it should be recognized that almost all biomedical signals can (and in many cases, should) be viewed as sample functions of random processes. Second, there are numerous situations in which determining population mean values is not the objective. Rather, one may want to predict the range of future values of an individual sample function, one may wish to determine if or when the features of an individual sample function change with time, or one may need to assay the signal for particular features such as an oscillation that may be obscured by stochastic disturbances. Such goals can be achieved with the introductory statistical understanding of random processes presented here.

It is customary to refer to a random process using a capital letter, such as $X(t)$ or $X[n]$, and to its *i-th* sample function using lower-case letters with a subscript, such as $x_i(t)$ or $x_i[n]$. Often it will be simpler to use unsubscripted function names, such as $x(t)$ or $x[n]$, to refer to any arbitrarily selected sample function of the random process and, by implication, to refer to the random process itself.

Example 9.2   White noise processes   The random number generator in a computer program like MATLAB is a simulation of a random process. Running the program one time generates one sample function of a "white noise" process. Recall from the discussion in Chapter 2 that a white noise signal is one for which memory is zero—that is, the autocorrelation between $x[n]$ and $x[n + m]$ is zero for all $m \neq 0$. A *white noise process* is simply a random process whose sample functions are white noise signals. Typically the computer program obtains a starting number—called a "seed"—from a source that is random relative to the operation of the program, often the real-time clock of the computer. The program then generates a sequence of numbers via rules which minimize the correlation between each new number and the ones that preceded it. (Note that this simulation is not perfect—for example, restarting with the same seed produces the same sequence of values.)

The present example uses the function randn to generate finite-length observations of three sample functions of the Gaussian random number generator in MAT-LAB and studies their properties. These sample functions are DT signals. The sample functions are plotted in Fig. 9.3 as $p_1(t)$, $p_2(t)$, and $p_n(t)$ with linear interpolation between consecutive points. Figure 9.4 demonstrates the application of time-averaging to estimate the mean value of the random process (which should be zero). For each time point N, an average has been calculated as $y_j[N] = (1/N)\Sigma_{k=1}^{N}x_j[k], j = 1, 2,$ 3, where the subscript "j" indexes the three original signals. Observe that the averages approach zero only for $\log(N) \approx 1.5$, or, $N \approx 32$. Also notice that the average does not necessarily approach zero monotonically as N increases (Fig. 9.4(b)).

One may generate other sample functions of this random process using a command like x1 = randn(2048,1); which will create a sequence (i.e., sample function) of 2048 Gaussian random numbers. Using the definition of power spectrum from Chapter 7, one may calculate the power spectrum of one of these sample functions and try to guess its theoretical properties.

Example 9.3 Poisson random processes   Poisson processes are very important in biology (as well as in the physical sciences) because they provide a framework for modeling events that occur randomly in time. The sample functions of a Poisson process are constructed as follows: Let each sample function be zero at $t = 0$. Now make a selection of a set of points, t_j, randomly in time. Then define $x_i(t)$ such that

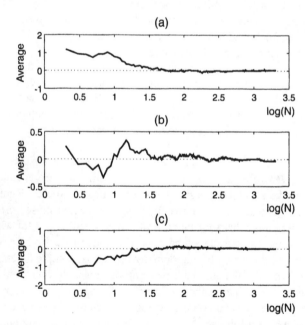

FIGURE 9.4 Estimating the mean of a zero-mean white noise process by averaging of data points of a single sample function. (a), (b), and (c) each represent one sample function. log(*N*) is the base-10 logarithm of the number of points from the sample function that are averaged.

$x_i(b) - x_i(a)$ equals the number of points from the set $\mathbf{t_j}$ in the interval (a, b). Thus, $x_i(t)$ is the number of points in the interval $(0, t)$. Each sample function is a "staircase" function (Fig. 9.5) that increments by one at every time point in $\mathbf{t_j}$. A new sample function is generated by selecting a new set of points, $\mathbf{t_j}$. The random process having these sample functions is a Poisson process.

One parameter characterizes a Poisson process—L, the average number of time points selected per second. Three sample functions of a Poisson process having $L = 0.2$ are drawn in Fig. 9.5. The probability of having k points occurring in an arbitrary time interval (a, b) has been shown to be

$$\text{Prob}\{k\, points\ in\ (a,\ b)\} = e^{-L(b-a)}\frac{[L(b-a)]^k}{k!}.$$

Letting $a = 0$ and $b = t$, and dropping the subscript "i" one concludes that

$$\text{Prob}\{k\, points\ in\ (0,\ t)\} = \text{Prob}\{x(t) = k\} = e^{-Lt}\frac{(Lt)^k}{k!}.$$

One may utilize this result, for example, to determine the probability that no event occurs in the interval $(0, 1)$. This probability is

$$\text{Prob}\{x(1) = 0\} = e^{-L}\frac{L^0}{0!} = e^{-L}.$$

Similarly, the probability that *at least* one event occurs in $(0, 1)$ is

$$\text{Prob}\{x(1) \geq 1\} = 1 - \text{Prob}\{x(1) = 0\} = 1 - e^{-L}.$$

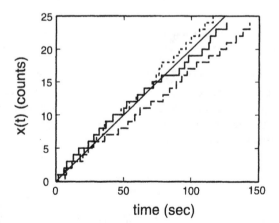

FIGURE 9.5   Three sample functions from a Poisson process with $L = 0.2$. Solid diagonal line is the time-dependent mean value of this random process.

Finally, one may define the average value of a random process at time t as the average of all of its sample functions at that time. Although we have not established a formal method of calculating the average value of the Poisson process at any time, t, one expects from Fig. 9.5 that the average value will increase with time.

Another type of Poisson process may be defined by taking the derivative of the sample functions of the Poisson process. Thus let

$$y_i(t) = \frac{dx_i(t)}{dt} \; \forall i.$$

These sample functions comprise trains of impulses occurring at the time points $\mathbf{t_j}$. This random process is known as a *Poisson impulse process*.

9.4 MEAN AND AUTOCORRELATION FUNCTION OF A RANDOM PROCESS

Because of the stochastic nature of sample functions of random processes, the properties of the collection of sample functions are describable only in a statistical sense. Given a random process, $X(t)$, the *mean value of the random process, $m_x(t)$, at any point in time, t,* is the average value of all of its sample functions at time t. That is, this average is computed *across the ensemble of sample functions* at a fixed time and therefore $m_x(t)$ is indeed a function of time. (Note that if the sample functions of a random process are DT functions, the mean value would be depicted as $m_x[n]$.) Recall that one may define a random variable, $\mathbf{x}(t)$, that assumes the values of the sample functions of a random process at any time t. Therefore, $m_x(t)$ is the mean value of $\mathbf{x}(t)$. The mean value of a random variable, $\mathbf{x}$, is determined by summing, over all possible values of $\mathbf{x}$, the product of the value of the random variable times the probability of that value occurring. In equation form,

$$E\{\mathbf{x}\} = \int_{-\infty}^{\infty} x \, \text{Prob}\{x \leq \mathbf{x} < x + dx\} dx = m_x(t), \tag{9.1a}$$

where $E\{\mathbf{x}\}$ is the mean or *expected value* of $\mathbf{x}$, and $\text{Prob}\{x \leq \mathbf{x} < x + dx\}$ is the probability that x lies between x and $x + dx$ and is known as the *probability density function (pdf)* of the random variable $\mathbf{x}$. If $\mathbf{x}$ may assume only discrete values, the above integral is replaced by a summation over all of the possible discrete values of x and the pdf expresses the probability that $\mathbf{x} = x$. Thus

$$m_x[n] = \sum_{x=-\infty}^{\infty} x \, \text{Prob}\{\mathbf{x} = x\}. \tag{9.1b}$$

For example, consider the random variable z defined above as the number on the top face of a fair die. The probability of any single number occurring on top is one-sixth. Thus one may calculate that

$$E\{z\} = \sum_{n=1}^{6} n \, \text{Prob}\{z = n\} = \frac{1}{6}[1 + 2 + 3 + 4 + 5 + 6] = 3.5.$$

Note that the "expected" value of z is a value that z cannot assume! However, if one averages enough values of z as the die is thrown repeatedly, this average will approach 3.5. Returning to a general random process, to determine $m_x(t)$ one integrates (or sums) over all possible values of the sample functions of the random process at time t. Necessarily, one must know the pdf of x(t) in order to compute $m_x(t)$. In the absence of other information, it must be assumed that $m_x(t)$ may change with time.

The *variance* of a random process, $X(t)$, is defined as the variance of the random variable $x(t)$. Thus,

$$V_x(t) = E\{|x(t) - m_x(t)|^2\} = \int_{-\infty}^{\infty} |x(t)|^2 \, \text{Prob}\{x(t) \le x(t) < x(t) + dx\} dx - [m_x(t)]^2 \quad (9.2a)$$

and also may be time-dependent. The variance of a DT random process, $X[n]$, is given by the summation

$$V_x[n] = \sum_{x=-\infty}^{\infty} x^2 \, \text{Prob}\{x = x\} - [m_x(n)]^2. \quad (9.2b)$$

Another measure of a random process is its autocorrelation function. In previous chapters auto- and cross-correlation functions played significant roles in the development of insights about physical systems from analyses of their signals. Autocorrelation functions have equal or greater roles in understanding random processes. The concept of the autocorrelation function of a random process, $X(t)$, having sample functions $x_i(t)$, derives from the understanding that at any two times, t_1 and t_2, its sample functions form the random variables $x(t_1)$ and $x(t_2)$. Therefore, one may assess the expected value of the product of the two random variables $x(t_1)$ and $x(t_2)$. This expected value is the autocorrelation function of the random process and is defined as

$$r_x(t_1, t_2) = E\{x(t_1)x(t_2)\}$$

$$= \int_{-\infty}^{\infty} \int_{-\infty}^{\infty} x(t_1)x(t_2) \, \text{Prob}\{x(t_1), x(t_2)\} dx(t_1)dx(t_2), \quad (9.3)$$

where $\text{Prob}\{x(t_1), x(t_2)\} \triangleq \text{Prob}\{x(t_1) \le x(t_1) < x(t_1) + dx_1 \text{ and } x(t_2) \le x(t_2) < x(t_2) + dx_2\}$. If the sample functions are DT signals,

$$r_x[n_1, n_2] = E\{x[n_1]x[n_2]\}$$

$$= \sum_{x[n_1]=-\infty}^{\infty} \sum_{x[n_2]=-\infty}^{\infty} x[n_1]x[n_2] \, \text{Prob}\{x[n_1] = x[n_1] \text{ and } x[n_2] = x[n_2]\}. \quad (9.4)$$

In Eqs. (9.3) and (9.4) the Prob terms are known as the *joint probability density functions* of the random variables. One may develop a visual interpretation of Eq.

(9.3) (or Eq. (9.4)) by referring to Fig. 9.3. The autocorrelation function is calculated for two specific times, t_1 and t_2. One first enumerates all of the possible paired values of $x(t_1)$ and $x(t_2)$, then one multiplies each paired value by its probability of occurrence and integrates (or sums) over all possible pairs of values. In general, the autocorrelation function of a random process must be evaluated for every possible combination of t_1 and t_2. Note that $r_x(t,t)$ (or $r_x[n,n]$) equals the variance of the random process (Eq. (9.2)).

If one defines probability as the relative frequency of occurrence of an event, then one may evaluate the Prob term in Eq. (9.4) as the ratio of the number of sample functions in which both $x[n_1] = x[n_1]$ *and* $x[n_2] = x[n_2]$ to the total number of sample functions. Then (assuming each sample function has the same probability of occurrence) for a subset of sample functions, $\{x_k[n], k = 1, 2, \ldots, K\}$, one may calculate an estimate of the autocorrelation function as

$$r_x^{(K)}[n_1, n_2] = \frac{1}{K} \sum_{k=1}^{K} x_k[n_1]x_k[n_2].\qquad(9.5)$$

Conceptually the autocorrelation function of Eq. (9.4) may be determined by taking the limit as $K \to \infty$ in Eq. (9.5). By this interpretation the autocorrelation function is the average across the sample functions of the product of two values from each sample function, in which the two values are always taken at times n_1 and n_2.

Recall from the discussion of correlation in Chapter 2 that the autocorrelation function of a deterministic DT signal $x[n]$, $0 \le n \le N - 1$, is defined as

$$r_x[m] = \frac{1}{N} \sum_{n=0}^{N-1} x[n]x[n + m], \qquad 0 \le m \le N - 1.\qquad(9.6)$$

The concept of autocorrelation was related to "memory" in the signal—that is, to what extent the fact that $x[n]$ differed from zero implied that $x[n + m]$ also is likely to differ from zero. The autocorrelation function of a DT random process is analogous to that for deterministic signals with one variation. For deterministic signals one increments n so that the calculation involves an average of products of two data points formed by moving along the time axis but for a random process the two times are fixed and one moves across the ensemble of sample functions. The difference in one's understanding of the autocorrelation functions in these two situations is subtle, but important. For deterministic signals the autocorrelation at lag M is an exact calculation of the average covariation of $x[n]$ and $x[n + m]$ *for that signal*. For random processes the autocorrelation function represents the average covariation of $x[n_1]$ and $x[n_2]$ over all of the sample functions.

9.5 STATIONARITY AND ERGODICITY

Stationarity

Although sample functions of random processes may be CT or DT functions, we shall focus on DT sample functions with the understanding that equivalent results apply to random processes having CT sample functions. In general, the mean and autocorrelation functions of a random process can be functions of time. In many practical cases, however, the mean value may not be time-dependent. For example, consider the measurement of muscle force during an isometric contraction. At least until fatigue ensues, it is reasonable to assume that the mean force would be constant. The mean value, $m_x[n]$, of a random process, $X[n]$, will be constant—that is, $m_x[n] = m_x$—if the probability density function Prob$\{x[n] = x\}$ is independent of n. A process meeting this condition is said to be *first-order stationary*. If one can model experimental data as sample functions from a random process for which there are well-understood mathematical models (such as white-noise processes, Poisson processes, or Brownian motion processes), then one may calculate $m_x[n]$ directly and test it for time-dependence. Many practical situations, however, may necessitate arguments for constancy of the mean based on knowledge of the physical process without complete knowledge of its density function.

In most cases (except for certain processes such as white-noise processes) the autocorrelation function, $r_x[n_1, n_2]$, is not independent of time. In many cases, however, it depends only on the time separation, $n_2 - n_1$, rather than on the absolute times, n_1 and n_2—a property that greatly enhances its usefulness. This property is guaranteed if the joint probability density function of $x[n_1]$ and $x[n_2]$ depends only on the time difference, $n_2 - n_1$, and a random process meeting this condition on the joint pdf is said to be *second-order stationary*. As above, it is possible to test this result for many mathematical models of specific random processes. If this condition on the joint density function is satisfied, then

$$r_x[n_1, n_2] = r_x[n_2 - n_1, 0] \equiv r_x[n_2 - n_1],$$

where $n_2 - n_1$ is known as the lag. Eq. (9.4) may then be written as

$$r_x[k] = E\{x[n]x[n + k]\}. \tag{9.7}$$

From the above discussion one may recognize the desirability of acquiring experimental data under conditions that restrict the time-dependence of the properties of the random process under study. In many cases the theoretical analyses to be developed below only require that the random process exhibit a type of restricted time-independence known as wide-sense stationarity. A random process, $X[n]$, is *wide-sense stationary* (w.s.s.) if the following conditions are met: (1) $m_x[n] = m_x$ (a constant independent of time); (2) $r_x[n_1, n_2]$ depends only on $n_2 - n_1$; (3) $r_x[0] - m_x^2$, which is the time-independent variance of the random process, σ_x^2, is finite. Wide-sense stationarity is a weaker condition than second-order stationarity discussed above (because condition (2) is a stipulation on the autocorrelation function, not on the joint pdf) but for processes for which the probability density function has a Gaussian form, the two are equivalent. For arbitrary data one must test all three criteria to assess wide-sense stationarity. These tests are difficult and beyond the present text, and usually we shall assume wide-sense stationarity when it cannot be proven.

The autocorrelation function of a w.s.s. process exhibits several important properties that will be stated but not derived here. Specifically: (1) the autocorrelation functions of such processes are even functions of lag; (2) $r_x[0] = E\{|x[n]|^2\} = \sigma_x^2 + m_x^2$ (a constant); (3) $|r_x[0]| \geq |r_x[k]|$, $\forall k$. In addition, one may define an *autocovariance function for a random process* as

$$c_x[n_1, n_2] = E\{[x[n_1] - m_x[n_1]][x[n_2] - m_x[n_2]]\}, \tag{9.8}$$

which, for a w.s.s. process, becomes (from Eq. (9.7))

$$c_x[k] = E\{x[n]x[n + k]\} - m_x^2 = r_x[k] - m_x^2. \tag{9.9}$$

Ergodicity

Often in practical situations it is only possible to observe one sample function of a random process, and to observe that sample function only for a limited time. It is possible to estimate the mean and autocorrelation functions of the random process even under these conditions, as long as certain criteria are met. Here only an outline of the rigorous derivation of these criteria will be presented. Consider that one has observed one sample function, $x[n]$, on the interval $0 \leq n \leq N-1$ from a w.s.s. random process, $X[n]$. One may calculate the mean of the observed data as a tentative estimate of the mean of the random process, as

$$\hat{m}_x[N] = \frac{1}{N} \sum_{n=0}^{N-1} x[n]. \tag{9.10}$$

To determine under what conditions this calculation provides an acceptable estimate of the true mean, m_x, requires an understanding that this estimate is a weighted summation of n random variables, $x[n]$, $n = 0, 1, \ldots, N-1$. Consequently, $\hat{m}_x[N]$ itself is a random variable and the appropriate measures of the properties of $\hat{m}_x[N]$ are its expected value and its variance (or standard deviation). Therefore, the question of whether $\hat{m}_x[N]$ is a "good" estimate of the true mean, m_x, only can be answered from a probabilistic viewpoint and then two issues must be considered. First, for sufficiently large n it must be true that the expected value of the random variable, $\hat{m}_x[N]$, must equal m_x. Second, at the same time the expected value of the variance of $\hat{m}_x[N]$ should go to zero—that is,

$$\lim_{N \to \infty} E\{|\hat{m}_x[N] - m_x|^2\} = 0. \tag{9.11}$$

One may evaluate the limits stated above using the definition from Eq. (9.10) and show that sufficient conditions for the time-average mean (Eq. (9.10)) of a sample function of a w.s.s. random process to converge to m_x in the sense of Eq. (9.11) are that $c_x[0] < \infty$ and

$$\lim_{k \to \infty} c_x[k] = 0, \tag{9.12}$$

where $c_x[k]$ is the autocovariance function of the random process. These conditions imply that the random process must have finite variance and be asymptotically uncorrelated. If the conditions are met, the w.s.s. random process is said to be *ergodic in the mean*.

If a random process is wide-sense stationary, then its autocorrelation function depends only on the lag and not the absolute times. Given only one sample function, $x[n]$, it is possible to calculate an estimate of the autocorrelation function in the same way that one calculates a deterministic autocorrelation function—that is, Eq. (9.6). That is, the estimated autocorrelation function is

$$\hat{r}_x[m] = \frac{1}{N} \sum_{n=0}^{N-1} x[n]x[n+m], \qquad 0 \le m \le N-1. \tag{9.13}$$

Consideration of the criteria under which this estimate is acceptable parallels that above for estimating the mean. The estimate of Eq. (9.13) converges in the mean-square sense to the true autocorrelation function of the random process if

$$\lim_{N \to \infty} \left[\frac{1}{N} \sum_{k=0}^{N-1} c_x^2[k] \right] = 0. \tag{9.14}$$

A w.s.s. random process whose autocovariance function satisfies this condition is said to be *autocorrelation ergodic*.

Testing for ergodicity is usually difficult when experimental data are acquired. Often the main check on the assumption of ergodicity is the reproducibility of the subsequent data analysis (although reproducibility does not guarantee ergodicity, of course). *In the following discussions we use the term ergodic to mean that a w.s.s. random process is both ergodic in the mean and autocorrelation ergodic.*

One may calculate $\hat{r}_x[m]$ from a sample function of a random process but, as with all calculations based on sample functions, the calculated autocorrelation function will be somewhat different for each sample function. Therefore, it is necessary to average these autocorrelation functions to obtain a measure of the average estimated autocorrelation function of the population of sample functions. Typically this averaging is performed as one might imagine – i.e., if $\hat{r}_x^{(k)}[m]$ is the autocorrelation function calculated from the k-th sample function and one has observed k sample functions, then the average autocorrelation function is

$$\bar{r}_x[m] = \frac{1}{K} \sum_{k=1}^{K} \hat{r}_x^{(k)}[m]. \tag{9.15}$$

Because K is usually small compared to the size of the population, one expects that $\bar{r}[m]$ will not exactly equal $r_x[m]$ except by chance, but it is likely to be closer to the true $r_x[m]$ than the autocorrelation $\hat{r}_x^{(k)}[m]$ calculated from any single sample function.

Example 9.4   Gaussian white-noise process   The autocorrelation function of a Gaussian white-noise process is easy to determine. The essence of white noise is that the value of the signal at any time is not influenced by its value at any other time. Therefore for white noise, $w[n]$, the autocorrelation function $r_w[m]$ is zero for all $m \neq 0$. Of course, as noted previously, the autocorrelation function at $m = 0$ is the variance, σ_w^2. The panels of Fig. 9.11 each display the autocorrelation functions of three individual sample functions of a Gaussian white-noise process calculated using Eq. (9.13), based on acquiring either 64 data points (Fig. 9.11(a)) or 2048 data points (Fig. 9.11(b)) per sample function. In both cases the mean level of the autocorrelation at non-zero lag tends to zero, but more reproducibly so with more data points.

Power spectrum of a random process

The power spectrum was derived previously as a measure of a single deterministic function, $x[n]$, as

$$P_x(e^{j\omega}) = \frac{1}{N}|X(e^{j\omega})|^2. \qquad (9.16)$$

The formal definition of the power spectrum of a w.s.s. random process is

$$S_x(e^{j\omega}) = \Im\{r_x[m]\}. \qquad (9.17)$$

That is, it is the Fourier transform of the autocorrelation function of the random process. Recall that $r_x[m]$ is defined formally by taking expected values across the ensemble of sample functions but one often estimates it by calculating $\hat{r}_x[m]$, which is evaluated from a single sample function, $x_i[n]$. When applied to $x_i[n]$, the power spectrum defined in Eq. (9.16) can be shown to be equivalent to calculating the Fourier transform of $\hat{r}_x[m]$. That is, the function

$$\hat{S}_x(e^{j\omega}) = \frac{1}{N}|X_i(e^{j\omega})|^2 \qquad (9.18)$$

is an estimate of the true power spectrum based on one sample function. The spectral estimate of Eq. (9.18) is known as the *periodogram*. The two definitions of the power spectrum (Eqs. (9.17) and (9.18)) are not equal, but for an ergodic, w.s.s., random process,

$$\lim_{N\to\infty}E\{\hat{S}_x(e^{j\omega})\} = S_x(e^{j\omega}).$$

Unfortunately, there is another serious deficiency in the power spectrum calculated from a single sample function. Although its expected value is the true power spectrum, $\hat{S}_x(e^{j\omega})$ does not converge smoothly to $S_x(e^{j\omega})$ as N increases. Consequently, power spectra based on finite-length observations of single sample functions exhibit a great deal of variability and it is necessary to average spectra from multiple sample functions or to impose some other smoothing procedure (see later).

Example 9.5   A biological random process   The file `ranproc2.mat` contains measurements of respiratory airflow (sampled at 75 Hz) from 8 different rats breathing at rest. Each column of `ranproc2` is a short-term observation of a sample function of the random process that generates breathing in rats, with each sample function representing one animal and comprising some 30–40 breaths. One may plot these data in one figure window—using, for example, the command `strips(ranproc2)`—and compare the breathing of the different rats. As seen in Fig. 9.6(a), each animal breathes differently although there are general similarities, suggesting that the underlying process might be the same in each animal but its parameters might vary between animals. Additionally, for each animal there are

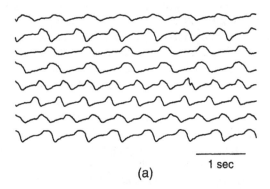

(a)   1 sec

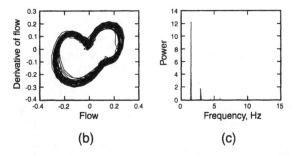

(b)   (c)

FIGURE 9.6 (a) Respiratory airflow patterns obtained from eight different anesthetized rats during resting breathing. Note general similarities across animals and cycle-to-cycle variations within each record, suggesting that the generation of the breathing pattern may be modeled as a random process. (b) Derivative of flow versus flow for the first tracing of Fig. 9.6(a). (c) Power spectrum (periodogram) of the first tracing of Fig. 9.6(a).

small differences among individual breaths. Therefore one might assume that breathing in each rat is generated by a random process which, in the absence of disturbances, would produce a breathing pattern that is the same for every breath. Because of parameter variations between animals and random disturbances, the airflow patterns (and breathing frequencies) differ among animals.

It is easiest to observe the interbreath variability of the breathing patterns by approximating the derivative of the airflow signal using the MATLAB command diff and graphing the derivative of airflow versus airflow (e.g., Fig. 9.6(b)). Also, using Eq. (9.18), one may calculate the estimated power spectrum of *each* of the eight signals (one of which is shown in Fig. 9.6(c)). Each has a sharp peak whose frequency represents the average respiratory frequency in that animal. Are these frequencies the same in every rat? From a random process viewpoint, does it make sense to average these frequencies across the eight subjects?

Example 9.6   Another biological random process   The file ranproc3.mat contains measurements of EMG signals (sampled at 500 Hz) from the tongue during a forceful contraction of the genioglossus muscle of the tongue. (See also Fig. 1.2(b).) Each column of ranproc3 contains 256 samples of EMG signal from one contraction and there are 8 columns of data, each from a different contraction but all from the same subject. Therefore each contraction (rather than each subject) is one realization of the random process that controls electrical activation of the genioglossus muscle in this subject and each column is a short-term observation of one sample function. One may graph these data in a common figure window as in the previous example. As seen for the two examples shown in Fig. 9.7(a), it is difficult to discern any distinguishing characteristics of these sample functions. We may calculate and compare the power spectra of these signals. As seen in the example of Fig. 9.7(b), there are no distinct peaks but (the reader may show that) all of the spectra have the same general shape, suggesting that they do represent the same underlying random process. It is not at all apparent, however, how to quantify these sample

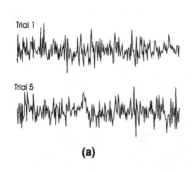

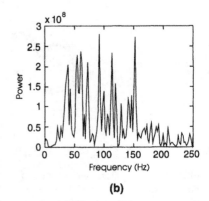

FIGURE 9.7   (a) Two trials of genioglossus EMG signal obtained during separate tongue muscle contractions. These tracings are the first and fifth records from the file ranproc3.mat. (b) Periodogram-based power spectrum of the fifth record from ranproc3.mat.

functions or what sort of average across the population would reveal fundamental information about the underlying random process. The remainder of this chapter will discuss methods for modeling of sample functions of random processes that are especially useful for situations like the present one.

9.6 GENERAL LINEAR PROCESSES

Properties of white-noise processes

Before proceeding it is necessary to formally define the properties of white noise processes because these processes will be fundamental to the methods for analyzing sample functions of arbitrary random processes. We will always assume that the true mean level of the sample functions of a white noise process is zero. The defining characteristic of a white noise process, $W[n]$, is that its sample functions, $w_k[n]$, are uncorrelated—that is, $r_w[m] = r_w[0]\delta[m] = V_w\delta[m]$, where V_w is the variance of the white-noise process, to which we assign the symbol σ_w^2. The true power spectrum of this process is

$$S_w(e^{j\omega}) = \Im\{\sigma_w^2\delta[m]\} = \sigma_w^2, \ \forall\omega. \tag{9.19}$$

That is, its power is the same at all frequencies and equals the variance of the process. (Depending on one's definition of the Fourier transform, power of a white-noise process can be defined as σ_w^2 or as $2\pi\,\sigma_w^2$. The former result is consistent with our definition of the DTFT.) Given a finite-length observation of a sample function, $w[n]$, $0 \le n \le N$, a white noise process, one can estimate the power level using a calculation of the statistical variance, s_w^2, as

$$s_w^2 = \frac{1}{N}\sum_{k=1}^{N}(w[k] - \hat{m}_w)^2, \tag{9.20}$$

where $\hat{m}_w$ is the mean level of $w[n]$ and is calculated according to Eq. (9.10).

Two white noise processes differ if their variances are unequal, but they also can differ in another fundamental way. Because of their lack of correlation, it is not possible to predict the next value of a white noise sample function. One can, however, keep measuring these values and derive the probabilities of occurrence of various values. For example, assume that some process may be able to take on only 50 possible values at its output. As long as the choice at each time point is un-correlated with preceding values, the resulting signal meets the criteria for white noise irrespective of the relative frequencies of occurrence of the 50 possible output values. But the temporal features of the sample function *will* depend on the frequencies of occurrence. That is, different white noise signals will result when the frequencies of occurrence of output values change. Unless stated otherwise, we will assume that the frequencies of occurrence—that is, the probability density function—of the output values of a white noise process can be described by a "normal" or Gaussian distribution (Fig. 9.8). A random variable, *x*, is Gaussian if its pdf is given by

$$f_x(a) = \frac{1}{\sigma_x \sqrt{2\pi}} e^{-\frac{(\alpha - m_x)^2}{2\sigma_x^2}},$$

where m_x and σ_x^2 are its mean and variance, respectively. To summarize, a white noise process has zero-mean sample functions, with frequencies of occurrence of their values describable by a normal distribution. By definition, the autocorrelation of a sample function is zero for $m \neq 0$. Thus for our case the only free parameter to describe a particular white-noise process will be its variance or power, σ_w^2.

Let us now consider the question of how to quantitatively describe biomedical signals that are viewed as sample functions of random processes. First, assume the sample functions are either inherently discrete-time or are sampled from CT functions. Because of the random disturbances inherent in a sample function, any description must include some characterization of these random disturbances. The difficulty is that these disturbances are not directly observable. Neither can one know the behavior of the random process in the absence of disturbances. Therefore, to simplify the problem let us consider only random processes whose mean values are time-invariant. If the mean is nonzero, its value can be *estimated* by calculating the mean value, $\bar{s}$, of an observed sample function, $s[n]$. Now the signal $x[n] = s[n] - \bar{s}$ represents the deviations of $s[n]$ from its mean level and if one can model $x[n]$, then it is trivial to add back the estimated mean value to obtain a model of $s[n]$. *Consequently, unless otherwise specified, it will be assumed that the mean value of a random process is zero.*

Given the assumption that $s[n]$ would approach a constant mean value in the absence of disturbances, it is obvious that $x[n]$ represents the effects of these random

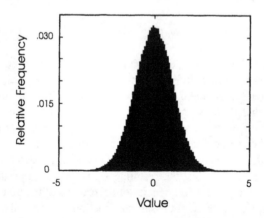

FIGURE 9.8 Histogram of 190,000 samples of zero-mean, unit-variance Gaussian white noise generated by the MATLAB command randn. Relative frequency at any value *x* is the proportion of samples having values between $x - 0.05$ and $x + 0.05$. The cumulative relative frequency between any two values, x_1 and x_2 (where $|x_2 - x_1| \gg 0.1$), is an estimate of the probability that a randomly selected sample of the white noise will have a value between x_1 and x_2.

disturbances on the output of the random process. Therefore, it is intuitively reasonable to develop a model of $x[n]$ in which $x[n]$ can be interpreted as the response of a system to a random input signal (Fig. 9.9). More specifically, a zero-mean random process $X[n]$ is a *general linear process* (*GLP*) if its sample functions can be represented as outputs of a linear, time-invariant, causal (LTIC) system whose inputs are sample functions of a zero-mean, (Gaussian) white-noise process. That is, each sample function, $x_k[n]$, of the random process is the result of a sample function, $w_k[n]$, of the white-noise process passing through the LTIC system. Using the convolution representation, the relationship between any input sample function, $w[n]$, and the resulting output sample function, $x[n]$, can be expressed as

$$x[n] = \sum_{i=1}^{\infty} h[i]w[n-i] + h[0]w[n], \qquad (9.21)$$

where we have assumed the most general case in which the upper limit of the summation is infinite. For an FIR filter whose impulse response has length N, the upper limit would be N − 1. In this form it is clear that $x[n]$ is expressed as a weighted summation of past and current values of the input, and this form is called the "moving average" (MA) form of a general linear process.

An equation similar to Eq. (9.21) can be written for any $x[n-i]$ and it will contain a term $h[0]w[n-i]$. Therefore one can solve for $w[n-i]$ and substitute into Eq. (9.21) to obtain:

$$x[n] = \frac{1}{h[0]} \sum_{i=1}^{\infty} h[i]\left\{ x[n-i] - \sum_{j=1}^{\infty} h[j]w[n-i-j] \right\} + h[0]w[n].$$

Now $x[n]$ is dependent on the present value of $w[n]$ and past values of both x and w. One may repeat this process by solving Eq. (9.21) for $w[n-i-j]$ in terms of $x[n-i-j]$ and, after repeated substitutions for $w[n-i-j]$,

$$x[n] = -\sum_{i=1}^{\infty} p[i]x[n-i] + h[0]w[n] \qquad (9.22)$$

where the values of $p[i]$ are functions of the $h[i]$, $0 \le i \le \infty$. The exact form of the $p[i]$ relationships may be difficult to determine, in general, but the essential point is that the dependence on *past* values of $w[n]$ can be eliminated. That this should be so can be recognized by considering that past values of $x[n]$ contain information about past

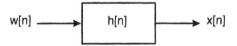

FIGURE 9.9 A General Linear Process model, representing a sample function, $x[n]$, of a random process as the output of a linear system that is excited by a sample function from a white noise process.

values of $w[n]$. Therefore it ought to be possible to express the current output value, $x[n]$, as a function of past values of $x[n]$, plus the current value of the input, $w[n]$. This form of the input–output relationship for a general linear process is termed the "autoregressive" (AR) form because $x[n]$ appears to be regressed on past values of itself.

There is a third form of the input–output relationship of a GLP which is derived by considering the LTIC system of Fig. 9.9 as a digital filter with transfer function $H(z)$. Let

$$H(z) = \frac{B(z)}{A(z)} = \frac{X(z)}{W(z)}, \tag{9.23}$$

$$\text{where} \qquad A(z) = 1 + \sum_{k=1}^{p} a(k)z^{-k} \tag{9.24}$$

$$\text{and} \qquad B(z) = b(0) + \sum_{k=1}^{q} b(k)z^{-k}. \tag{9.25}$$

Now cross-multiply the second and third terms of Eq. (9.23), then take inverse Z-transforms assuming all initial conditions are zero. Finally, solving for $x[n]$ one obtains

$$x[n] = -\sum_{k=1}^{p} a(k)x[n-k] + \sum_{k=0}^{q} b(k)w[n-k]. \tag{9.26}$$

This form of the input–output equation contains both "autoregressive" and "moving average" terms and is called the ARMA form. Note that this form involves only $p + q + 1$ coefficients even if the AR or MA form involves an infinite number of coefficients.

To complete the picture we should acknowledge that the first two input–output relationships also can be expressed in Z-transform form. It is easy to show that the equation for $H(z)$ based on the autoregressive form is

$$H(z) = \frac{X(z)}{W(z)} = \frac{h(0)}{1 + \sum\limits_{i=1}^{\infty} p(i)z^{-i}}. \tag{9.27}$$

Finally, if the moving average form (Eq. (9.21)) is expressed as

$$x[n] = \sum_{i=0}^{\infty} h[i]w[n-i],$$

then it is trivial to calculate that

$$H(z) = \sum_{k=0}^{\infty} h[k]z^{-k}, \tag{9.28}$$

which is just the definition of the Z-transform of $h[n]$. Note that all three forms of the equation for $H(z)$ must be mathematically equivalent since they describe the same relationship.

One might wonder whether there is any advantage to expressing an observed signal, $x[n]$, in the above forms when typically $w[n]$ cannot be observed. For the case of an FIR filter it is possible to determine the coefficients in these equations even without observing the input signal. Usually this form of modeling of a signal is not utilized to calculate the values of $x[n]$ that would result from a particular white noise sample function at the input. Instead, this form of model is used as a compact representation of the properties of the signal. Recall that the only free parameter for describing a stationary Gaussian white noise process is its variance. Therefore, the input white-noise variance, plus the gain, poles, and zeros of $H(z)$ contain much information about the random process, $x[n]$ (but of course do not provide a means to determine the values of its sample functions at each point in time). In fact, as will now be proven, knowledge of the above parameters permits one to calculate the power spectrum of $x[n]$.

First, recognizing that all three forms above can be represented in the ARMA form if one permits p or q to be infinite, consider the case when

$$H(z) = \frac{X(z)}{W(z)} = \frac{B(z)}{A(z)}. \tag{9.29}$$

Referring to Fig. 9.9, to calculate the power spectrum of $x[n]$ it is necessary first to determine $r_x[m]$. From its definition,

$$r_x[m] = E\{x[n]x[n+m]\} = E\left\{\sum_{l=-\infty}^{\infty} h[l]w[n-l]\sum_{k=-\infty}^{\infty} h[k]w[n+m-k]\right\}.$$

Interchanging summation and expectation, and using a shorthand symbol to represent a summation over all values of its index,

$$r_x[m] = \sum_l \sum_k h[l]h[k]r_w[m-k+l].$$

Letting $j = m - k$, and recalling that $r_w[i] = \sigma_w^2\delta[i]$, we obtain

$$r_x[m] = \sum_l \sum_j h[l]h[m-j]\sigma_w^2\delta[j+l] = \sigma_w^2 \sum_l h[l]h[m+l] = \sigma_w^2 h[k] * h[-k]. \tag{9.30}$$

Thus

$$S_x(e^{j\omega}) = \Im\{r_x[m]\} = \sigma_w^2 H(e^{j\omega})H^*(e^{j\omega}) = \sigma_w^2 |H(e^{j\omega})|^2. \tag{9.31}$$

Now since

$$H(e^{j\omega}) = \frac{B(e^{j\omega})}{A(e^{j\omega})}, \qquad \text{then}$$

$$S_x(e^{j\omega}) = \sigma_x^2 \left| \frac{B(e^{j\omega})}{A(e^{j\omega})} \right|^2 = \sigma_w^2 \left| \frac{b(0) + \sum_{k=1}^{q} b(k)e^{-j\omega k}}{1 + \sum_{k=1}^{p} a(k)e^{-j\omega k}} \right|^2. \tag{9.32}$$

Also recall from Chapter 7 that when a CT signal is sampled at a rate $f_s = 1/T$ for an observation time $T_0 = N \cdot T$ (i.e., $0 \le t \le (N-1)T$), then the power spectra of the CT and DT signals are related by $P_x(\Omega) = T \cdot P_x(e^{j\omega})|_{\omega=\Omega T}$. If $x(t)$ is a sample function of a random process that has been digitized as $x[n]$, then

$$\hat{S}_x(\Omega) = T\hat{S}_x(e^{j\Omega T}) = \hat{\sigma}_w^2 T \left| \frac{B(e^{j\Omega T})}{A(e^{j\Omega T})} \right|^2. \tag{9.33}$$

Example 9.7   Forms of the GLP model   Consider a signal that is modeled as the output of the simple linear system

$$H(z) = \frac{1}{1 - 0.5z^{-1}}.$$

To express this system in all three (i.e., AR, MA, and ARMA) forms, first note that it is given in the ARMA form, so by inspection

$$A(z) = 1 - 0.5z^{-1} \qquad \text{and} \qquad B(z) = 1.$$

Likewise, the given form can be considered an AR form with $h[0] = 1$ and

$$P(z) = 1 + \sum_{k=1}^{\infty} p(i)z^{-i} = A(z) = 1 - 0.5z^{-1}.$$

Finally, to obtain the MA form one can divide 1 by $A(z)$. To find the general solution let $A(z) = 1 - az^{-1}$. By long division one can show that

$$H(z) = \frac{1}{1 - az^{-1}} = 1 + \sum_{k=1}^{\infty} a^k z^{-k}.$$

Therefore, the MA form is $H(z) = 1 + \sum_{k=1}^{\infty} 0.5^k z^{-k}$.

Example 9.8   Power spectrum of an AR process   A signal $x[n]$ is generated by exciting the system from Example 9.7 with a sample function from a white-noise

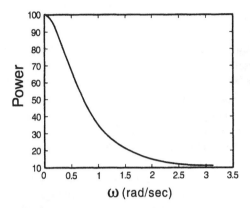

FIGURE 9.10 Power spectrum for Example 9.8.

process having a standard deviation (i.e., square root of the variance) of 5. What is the power spectrum of $x[n]$?

From Example 9.7 and Eq. (9.32), the power spectrum of $x[n]$ is

$$S_x(e^{j\omega}) = 25 \left| \frac{1}{1 - 0.5e^{-j\omega}} \right|^2 = \frac{25}{(1 - 0.5\cos(\omega))^2 + 0.25\sin^2(\omega)}.$$

This spectrum is plotted in Fig. 9.10.

Example 9.8 addressed the question of calculating the properties of a random process when the $a(k)$ and $b(k)$ coefficients are known. In typical biomedical applications, however, one is confronted with determining the $a(k)$ and $b(k)$ coefficients of a random process given a data set (i.e., an observation of one or a few of its sample functions). The usual approaches to solving this latter problem utilize the autocorrelation function extensively.

9.7 YULE-WALKER EQUATIONS

Now we are ready to address the central question of this chapter: How can one construct a model of a random process from a measured sample function? The procedure is very straightforward in concept. First one assumes that the measured signal originates from a linear system that is driven by a sample function from a white-noise process. Then one estimates the "best" linear system model which will transform white noise into a process that has characteristics of the measured signal. Since sample functions from a random process are all different, one does not attempt to reproduce the sample function exactly. Rather, one models basic properties of the random process that can be determined from the measured sample function. One candidate "basic property" is the mean of the random process, but the mean is not very informative because it conveys no information about the temporal variations of the sample functions. On the other hand, the autocorrelation function of a zero-mean random process quantifies the average correlation of the sample functions at two different times and therefore is a measure of the temporal variations of the sample functions. Furthermore, autocorrelation can be related to memory in physical processes (see Chapter 2). Thus the typical approach involves modeling the estimated autocorrelation function of the measured sample function because this function (e.g., Fig. 9.11) is an estimate of the true autocorrelation function of the random process. The issue then becomes: How does one determine the linear system that will transform uncorrelated white noise into an output having the same autocorrelation function as the data?

To begin, assume that $x[n]$ is a wide-sense stationary (w.s.s.), zero-mean random process. Therefore its autocorrelation and autocovariance functions are equal. Referring to Fig. 9.9, one can represent a sample function, $x_j[n]$, as the response of a linear system to a white-noise sample function, $w_j[n]$, using the General Linear Process (GLP) model in its ARMA form. Dropping the subscript "j", one can express (Eq. (9.26)) the ARMA(p,q) process as

$$x[n] = -\sum_{k=1}^{p} a(k)x[n-k] + \sum_{k=0}^{q} b(k)w[n-k].$$

Note that $r_x[m] = E\{x[n+m]x[n]\} = E\{x[n]x[n-m]\}$. Therefore, multiply Eq. (9.26) by $x[n-m]$ and apply the expectation operator to both sides. On the left side one obtains $r_x[m]$. Since the expectation of a constant is just the constant itself, the $a(k)$ and $b(k)$ terms can be moved outside of the expectation operator and we obtain

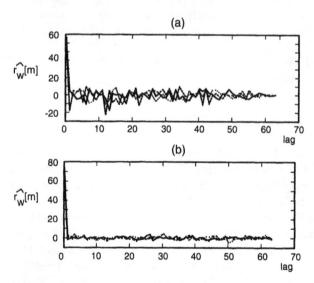

FIGURE 9.11 Estimated autocorrelation functions (i.e., $\hat{r}_x[m]$ vs. m) calculated (using Eq. (9.13)) from three different sample functions of a Gaussian white-noise process. (a) based on 64 data samples each; (b) based on 2048 data samples each.

$$r_x[m] = -\sum_{k=1}^{p} a(k)E\{x[n-k]x[n-m]\} + \sum_{k=0}^{q} b(k)E\{w[n-k]x[n-m]\}.$$

Since $x[n-k]x[n-m]$ can be written as $x[n]x[n-(m-k)]$, and $w[n-k]x[n-m]$ can be written as $w[n]x[n-(m-k)]$, this equation becomes

$$r_x[m] = -\sum_{k=1}^{p} a(k)r_x[m-k] + \sum_{k=0}^{q} b(k)r_{wx}[m-k]. \tag{9.34}$$

Here we have introduced an extension of the autocorrelation concept, the *cross-correlation function* between two variables. The definition of $r_{yx}[m]$ is an uncomplicated extension of that of $r_x[m]$—that is,

$$r_{yx}[m] = E\{x[n]y[n+m]\} = E\{y[n]x[n-m]\}. \tag{9.35}$$

To form a visual picture of cross-correlation, refer to Fig. 9.3. One must think of two random processes whose sample functions are observed simultaneously. Now instead of computing an expectation by taking values from one set of sample functions at two different times, one takes values from one set of sample functions at one time, n, and from the other set of sample functions at a second time, $n + m$. The cross-correlation function expresses the memory of signal x at time n that is present in signal y at time $n + m$.

To proceed we must evaluate $r_{wx}[m]$ for an arbitrary value of m (say $m = i$). By definition of the cross-correlation function and using the convolution representation for the output of an LSIC system

$$r_{wx}[i] = E\{x[n]w[n+i]\} = E\left\{w[n+i]\left[\sum_{k=0}^{\infty} h[k]w[n-k]\right]\right\}$$

$$= h[0]r_w[i] + \sum_{k=1}^{\infty} h[k]r_w[i+k].$$

Since $w[n]$ is white noise, $r_w[i] = \sigma_w^2\delta[i]$. Therefore,

$$r_{wx}[i] = \begin{cases} 0, & i > 0 \\ \sigma_w^2 h[0], & i = 0. \\ \sigma_w^2 h[-i], & i < 0 \end{cases} \tag{9.36}$$

Substituting this result into Eq. (9.34), one obtains the following general result for an ARMA(p,q) model of a random process known as the *Yule-Walker equation*:

$$r_x[m] + \sum_{k=1}^{p} a(k)r_x[m-k] = \begin{cases} \sigma_w^2\sum_{k=m}^{q} b(k)h[k-m], & 0 \le m \le q \\ \\ 0, & m > q \end{cases}. \tag{9.37}$$

For simplicity we can follow the convention of Hayes (1996) and define $c_q[m]$ to be

$$c_q[m] = \sum_{k=m}^{q} b(k)h[k-m] = \sum_{k=0}^{q-m} b(m+k)h[k]. \qquad (9.38)$$

Then for $0 \leq m \leq p + q$ Eq. (9.37) can be rewritten in matrix form as

$$
\begin{bmatrix}
r_x[0] & r_x[-1] & \cdots & r_x[-p] \\
r_x[1] & r_x[0] & r_x[-1] & \vdots \\
\vdots & r_x[1] & \cdots & \vdots \\
\vdots & \vdots & \vdots & \vdots \\
r_x[q] & r_x[q-1] & \cdots & r_x[q-p] \\
\hline
r_x[q+1] & r_x[q] & \cdots & r_x[q-p+1] \\
\vdots & \vdots & \vdots & \vdots \\
r_x[q+p] & \cdots & \cdots & r_x[q]
\end{bmatrix}
\begin{bmatrix}
1 \\
a(1) \\
a(2) \\
\vdots \\
a(p)
\end{bmatrix}
= \sigma_w^2
\begin{bmatrix}
c_q[0] \\
c_q[1] \\
\vdots \\
c_q[q] \\
\hline
0 \\
\vdots \\
0
\end{bmatrix}. \qquad (9.39)
$$

The horizontal dashed lines in the above equation serve only to indicate that the set of $p + q + 1$ equations implicit in this relationship can be separated into an upper set of $q + 1$ equations and a lower set of p equations. Using these equations, in concept one can develop a model of a random process from one of its sample functions by proceeding in four steps: First, calculate the estimated autocorrelation function values, $\hat{r}_x[m]$, from the measured sample function. Second, using $\hat{r}_x[m]$ in place of $r_x[m]$ in Eq. 9.39, solve the lower set of p equations for the $a(k)$ coefficients. Third, using these $a(k)$ values and the assumption that $b(0) = 1$, solve the upper set of $q + 1$ equations for the $b(k)$ coefficients and σ_w^2. Fourth, substitute the $a(k)$'s and $b(k)$'s into Eq. (9.26). When given an uncorrelated, unit-variance, white-noise process at its input, the resulting model (described by Eq. (9.26) or Eq. (9.23)) will produce an output random process whose autocorrelation function, $r_x[m]$, $0 \leq m \leq p + q$, will equal the autocorrelation values calculated from the data signal for the same range of m.

In practice, this approach and numerous variations of it are commonly used to develop such signal models. Below we will discuss some specific variations and their applications. Although easy to calculate, these approaches suffer from a potentially serious problem: they may be sensitive to inaccuracies in the values of the autocorrelation function. Unfortunately it may be difficult to obtain accurate estimates of autocorrelation values of a random process from the typically short, noisy data records common to biomedical applications. On the other hand, with adequate care and an awareness of potential problems, methods based on the Yule-Walker equation are quite viable for many biomedical signals and should be the user's first choice, at least as a general check on the solution since their properties have been studied extensively. There are, however, many variations and each has its advantages and pitfalls. One is well-advised to compare several of these variations when approaching a new application. MATLAB implements several of the common variants discussed below in its ar and armax commands.

9.8 AUTOREGRESSIVE (AR) PROCESSES

An autoregressive process of order p—symbolized as AR(p)—is a General Linear Process for which $q = 0$. In this case all but one of the $c_q[m]$ terms becomes zero. To evaluate the remaining term, $c_0(0)$, note that the GLP equation (Eq. (9.26)) evaluated at $n = 0$ yields $x[0] = b(0)w[0]$. But by convolution one also can write $x[0] = h[0]w[0]$. Therefore, $b[0] = h[0]$. (Note that this result did not assume an AR(p) process and is true for any GLP.) Now for an AR(p) process and for $m = 0$, Eq. (9.38) simplifies to $c_0[0] = b(0)h[0] = |b(0)|^2$. For all other values of m, $c_0[m]$ equals zero. Therefore the Yule-Walker equation becomes

$$r_x[k] = \sum_{j=1}^{p} a(j)r_x[k-j] = \sigma_w^2|b(0)|^2\delta[k], \qquad k \geq 0. \qquad (9.40)$$

Writing this equation in matrix form for the $p + 1$ equations indexed by $0 \leq k \leq p$, we have

$$\begin{bmatrix} r_x[0] & r_x[-1] & \cdots & \cdots & r_x[-p] \\ r_x[1] & r_x[0] & r_x[-1] & \cdots & : \\ : & : & : & : & : \\ : & : & : & : & : \\ r_x[p] & \cdots & \cdots & \cdots & r_x[0] \end{bmatrix} \begin{bmatrix} 1 \\ a(1) \\ a(2) \\ : \\ a(p) \end{bmatrix} = \sigma_w^2|b(0)|^2 \begin{bmatrix} 1 \\ 0 \\ 0 \\ : \\ 0 \end{bmatrix}. \qquad (9.41)$$

If the $p + 1$ autocorrelation values, $r_x[m]$, $0 \leq m \leq p$, are known, then the equation above produces $p + 1$ linear algebraic equations in the $p + 2$ unknowns $a(i)$, $1 \leq i \leq p$, $b(0)$, and σ_w^2. To solve this set of equations one chooses one of the latter two unknown values arbitrarily. That is, one can either assume that the input white noise has a variance of unity or that $h[0]$ equals 1. Consequently, developing a GLP model for an autoregressive process is simple if one knows $p + 1$ values of its autocorrelation function. Note that although Eq. (9.41) was written for values of k such that $0 \leq k \leq p$, Eq. (9.40) could be used to write $p + 1$ equations for any range of $p + 1$ contiguous values of k. Typically one uses the range shown in Eq. (9.41) because for many physical processes the autocorrelation function diminishes in magnitude as $|m|$ increases. Thus the calculation of $r_x[m]$ from a sample function is more susceptible to error due to noise for large $|m|$.

If the autocorrelation values of an AR(p) process are known exactly, then solving this set of equations will produce a GLP model of a random process whose autocorrelation function will exactly match that of the original AR(p) process *for all values of m!* That this should be so may be seen by writing Eq. (9.40) for some $k > p$. For an AR(p) process, one obtains a recursion through which the knowledge of $r_x[m]$, $0 \leq m \leq p$, permits one to calculate $r_x[k]$ for all other lags k:

$$r_x[k] = \sum_{j=1}^{p} a(j)r_x[k-j]. \qquad (9.42)$$

If the data originate from an actual AR(p) process, then the above equation applies both to the actual autocorrelation function and to the autocorrelation values calculated from the Yule-Walker equation.

Example 9.9 An AR(1) process   Consider a real AR(1) process, $x[n]$, driven by white noise having unity variance—that is, $\sigma_w^2 = 1$. Because $x[n]$ is real, $r_x[-k] = r_x[k]$ and the two Yule-Walker equations for this process are:

$$r_x[0] + r_x[1]a(1) = \sigma_w^2 b^2(0);$$

$$r_x[1] + r_x[0]a(1) = 0.$$

Solving these equations simultaneously we find

$$a(1) = -\frac{r_x[1]}{r_x[0]}, \qquad \sigma_w^2 b^2(0) = \frac{r_x^2[0] - r_x^2[1]}{r_x[0]}. \tag{9.43}$$

If $r_x[0]$ and $r_x[1]$ are unknown, they may be estimated from the data and these values substituted into the above equations to estimate $a(1)$ and $b(0)$. For example, Fig. 9.12 shows a 128-point Gaussian white-noise sequence, $w[n]$, having a variance of unity that was filtered by the system

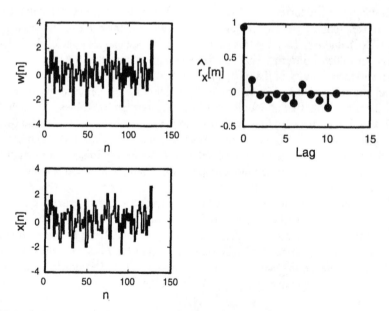

FIGURE 9.12 Data for Example 9.9. Left: Gaussian white-noise input to (top), and output from (bottom), the system $H(z) = 1/(1 - 0.275z^{-1})$. Right: First few lags of the autocorrelation function estimated from $x[n]$.

$$H(z) = \frac{1}{1 - 0.275z^{-1}}.$$

The resulting output time series, $x[n]$, and the first few values of its estimated autocorrelation function, calculated via Eq. (9.13), are also given in the figure. The estimated autocorrelation values are

$$\hat{r}_x[0] = 0.9156, \qquad \hat{r}_x[1] = 0.1736.$$

Thus, assuming that $b(0) = 1$, the estimated parameters for an AR(1) model of the time series $x[n]$ are found from Eq. (9.43) to be

$$\hat{a}(1) = -0.1896, \qquad \hat{\sigma}_w^2 = 0.8827.$$

For comparison, the correct value of $a(1)$ from $H(z)$ is -0.2750 and the white-noise variance was chosen to be 1.0000. This degree of error is a consequence of the finite data length and is typical for estimates based on one short data record (even though $x[n]$ contained no additive noise).

Similarly, if the parameters of the model are known, one can solve for the autocorrelation values as functions of these parameters. For the AR(1) process it is easy to show that

$$r_x[0] = \frac{b^2(0)}{1 - a^2(1)} \equiv \sigma_x^2 \qquad \text{and} \qquad r_x[k] = \{-a(1)\}^k r_x[0].$$

Power Spectrum of an AR(1) Process

Referring to Eq. (9.32) and recognizing that for an AR(1) process

$$B(z) = b(0) \qquad \text{and} \qquad A(z) = 1 - a(1)z^{-1},$$

the power spectrum of an AR(1) process is

$$P_x(e^{j\omega}) = \frac{\sigma_w^2 |b(0)|^2}{|1 + a(1) e^{-j\omega}|^2} = \frac{\sigma_w^2 |b(0)|^2}{1 + a^2(1) + 2a(1) \cos(\omega)}. \qquad (9.44)$$

The graph of Fig. 9.13 presents the power spectra for several AR(1) processes with various values for $a(1)$ and with $\sigma_w^2 |b(0)|^2 = 1$. Notice that when $a(1) > 0$, $x[n]$ exhibits a predominance of high-frequency power, the opposite being true when $a(1) < 0$.

Use of Estimated versus Actual Autocorrelation Function

Although we have noted that it is difficult to determine the exact autocorrelation values for a random process from a single sample function, so far little distinction

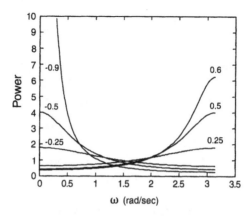

FIGURE 9.13 Theoretical power spectra for AR(1) processes having the indicated values for the parameter $a(1)$.

has been made between these exact values and estimates calculated from a single sample function using Eq. (9.13). Even for a w.s.s. ergodic random process the latter calculations have inherent errors when based on a finite set of data. For this reason it is customary to distinguish between actual and calculated (or estimated) autocorrelation function values, using the symbol $r_x[m]$ for the former and $\hat{r}_x[m]$ for the latter, as in the previous example. Likewise, when one calculates the $a(k)$ values based on $\hat{r}_x[m]$, and then evaluates the power spectrum using these estimated coefficients, $\hat{a}(k)$, the resulting power spectrum is only an estimate of the true power spectrum. As in Eq. (9.18), it is symbolized as $\hat{S}_x(e^{j\omega})$.

The inherent error in estimating $r_x[m]$ by calculating $\hat{r}_x[m]$ from real data has two components. First, even in the absence of noise the finite length of data causes a systematic error. Second, if noise contaminates the signal of interest, the resulting autocorrelation estimates are the sum of those for the signal and the noise processes. Because of the first type of error, even if the noise is uncorrelated white noise, the *estimated* autocorrelation values of the noise will not be zero (for $|m|>0$). Various techniques for dealing with these difficulties have been developed and, consequently, practical estimation of autoregressive models from data can be more complicated than shown here.

Another problem of considerable consequence arises from the question of whether an observed sample function from an unknown random process *should* be modeled as originating from an autoregressive process. A related issue is the identification of the types of problems that might develop if one applies an AR(p) model to a process that is not an AR(p) process. These questions can be addressed using the methods of model validation from the general topic of "system identification". These methods usually are discussed in a course on statistical signal processing or system identification and are beyond the present text. The reader is referred to the book by Ljung (1987) for extensive discussions of model validation.

If one is rather confident that the observed data can be represented as an AR(p)

process but is unsure of the value of p (i.e., the order of the model), there is a theoretical result that suggests a strategy for determining this value. Let the true order of an autoregressive process $x[n]$ be P and the AR coefficients of this process be $a_P[k]$, $1 \leq k \leq P$. Consider writing an AR(P + 1) model for this process in the form of Eq. (9.22). This equation will have AR coefficients $a_{P+1}[k]$, $1 \leq k \leq P +$ 1. Since $x[n]$ is actually an AR(P) process, there is no dependence of $x[n]$ at time n on $x[n - (P + 1)]$, although there is such a term in the AR(P + 1) form of Eq. (9.22). Therefore the coefficient of $x[n - (P + 1)]$, which is $a_{P+1}(P + 1)$, must be zero. Consequently, there are only P nonzero coefficients in the AR(P + 1) model and by necessity

$$a_{P+1}(k) = a_P(k), \qquad 1 \leq k \leq P$$

Based on this result, to find P for an unknown AR process one generates AR(p) models starting from $p = 1$, increasing p by one each time until $a_p(p) = 0$ and the coefficient values for $k < p$ do not change from the preceding (lower order) model. One then assumes that the true order is $P = p - 1$. Because of the errors noted above in the values of $\hat{r}_x[m]$, neither condition—that is, $a_p(p) = 0$, and $a_p(k) = a_{p-1}(k)$, $1 \leq k \leq p - 1$—is exactly satisfied and it is usually difficult to have confidence in choosing P. A possible alternative is to assume $b(0) = 1$ and increase p until $\hat{\sigma}_w^2$ is "sufficiently small". But since $\hat{\sigma}_w^2$ will always decrease as p becomes larger, one also needs to add a penalty related to the model order, p. The reader is referred to discussions of Akaike Information Criterion (AIC), Final Prediction Error (FPE), and Minimum Description Length (MDL) in books such as those by Ljung (1987) or Priestley (1981). Unfortunately none of these methods provides an absolute answer to the question of selecting the correct model order.

In very many applications the primary objective is to estimate the power spectrum of a signal (that is, the power spectrum of the random process from which the signal arose). In such situations it is often tolerable to choose a model order $p > P$. Because the high-order coefficients of such a model are expected to be zero, it is reasonable to assume that their contributions to the estimated power spectrum through $A(e^{j\omega})$ will be negligible. Based on this approach, various authors have recommended using $p = N/3$ or $p = N/2$, where N is the number of data points acquired from a single sample function. With a low signal-to-noise ratio, as often occurs with biomedical signals, these values of p may be too high, leading to poles whose amplitudes are determined by the noise components of the signal. That is, the errors in calculating autocorrelation function values cause the higher-order coefficients to be nonzero, producing additional poles in the transfer function (i.e., extra roots of $A(z)$). These unrealistic poles cause spurious peaks in the power spectrum. Therefore, it is useful to examine the power spectrum for several increasing orders of the AR model to assess which peaks are reproducible, recognizing of course that an AR(p) model cannot have more spectral peaks than $p/2$.

What if the data signal did not originate from a true AR(p) process? If it arose from a linear process, then it is theoretically possible to model it with an AR(p) structure although p might need to be very large. For many physical processes it is reasonable to assume that the signal at a particular time is independent of those parts of the input (and output) signals that are sufficiently far in the past. This assumption allows one to restrict p to a (possibly large) finite value. If the signal contains added uncorrelated noise, then one may expect to obtain reasonable estimates of the AR coefficients. On the contrary, if the added noise is not uncorrelated (or does not add linearly to the signal), then the noise autocorrelation will be included also in the $\hat{r}_x[m]$ values and the AR coefficients will be erroneous. This latter situation presents a difficult problem that can be addressed using advanced system identification methods.

Example 9.10   An AR(2) model   Assume the following estimated autocorrelation values have been calculated from some acquired data:

$$\hat{r}_x[0] = 10, \qquad \hat{r}_x[1] = 4, \qquad \hat{r}_x[2] = 2.$$

To construct an AR(2) model for this random process we need to solve the Yule-Walker equations for $0 \le k \le 2$. Using the simplified symbols r_k to represent $r_x[k]$, these equations are

$$r_0 = -a(1)r_{-1} - a(2)r_{-2} + \sigma_w^2 |b(0)|^2$$
$$r_1 = -a(1)r_0 - a(2)r_{-1}$$
$$r_2 = -a(1)r_1 - a(2)r_0.$$

Since $r_{-k} = r_k$, we can let $b(0) = 1$ and solve for $a(1)$ and $a(2)$ by multiplying the second equation by r_1 and the third equation by r_0, and then subtracting the resulting equations. Solving for these coefficients, we find

$$a(1) = \frac{r_1 r_2 - r_1 r_0}{r_0^2 - r_1^2}, \qquad a(2) = \frac{r_1^2 - r_0 r_2}{r_0^2 - r_1^2}. \tag{9.45}$$

Given the estimates $\hat{r}_x[m]$, $0 \le m \le 2$, from the data, one may use these estimates in Eq. (9.45) in place of the actual autocorrelation values to yield estimates of the AR coefficients:

$$\hat{a}(1) = -0.381, \qquad \hat{a}(2) = -0.0876.$$

Inserting these values into the first equation yields $\hat{\sigma}_w^2 = 8.381$.

The power spectrum of an arbitrary AR(2) process can be calculated from the power spectrum of a GLP, Eq. (9.32). The reader may derive the result that

$$S(e^{j\omega}) = \frac{\sigma_w^2 |b(0)|^2}{1 + a_1^2 + a_2^2 + 2a_1(1 + a_2)\cos(\omega) + 2a_2 \cos(2\omega)}. \tag{9.46}$$

Using the estimated AR coefficients, the estimated power spectrum for this process is

$$\hat{P}(e^{j\omega}) = \frac{8.381}{1.1528 - 0.6592 \cos(\omega) - 0.1752 \cos(2\omega)}.$$

This spectrum is plotted in Fig. 9.14.

Example 9.11   An AR model of heart rate variability   One can calculate a simple model of heart rate variability using an AR approach. For example, consider the time series of heart rate in the file hrv1.mat (which is similar to that shown in Fig. 1.2e). Assume that this signal originates from a random process represented by this one sample function, $c[n]$. The autocorrelation estimates for this data signal (after removing its mean level) were determined (using the MATLAB function xcov) to be

$$\hat{r}_c[0:3] = [6.5464, 5.7950, 4.8198, 3.4380] \cdot 10^3.$$

Using these values to solve for the coefficients of an AR(2) model yields

$$\hat{a}(1) = -1.078, \qquad \hat{a}(2) = 0.219.$$

Substituting these results into the Yule-Walker equation for $k = 0$, the estimated white-noise variance is $\hat{\sigma}_w^2 = 1354$. One may use these findings to construct a model of the data signal that will have the same autocorrelation values at lags 0, 1, and 2, as the data. By Eqs. (9.23)–(9.25) with $p = 2$ and $q = 0$,

$$H(z) = \frac{b(0)}{1 + a(1)z^{-1} + a(2)z^{-2}} = \frac{1}{1 - 1.078z^{-1} + 0.219z^{-2}}.$$

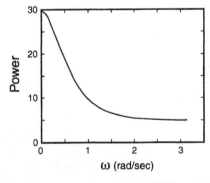

FIGURE 9.14   Estimated power spectrum for the AR(2) model of Example 9.10.

If one excites this model with a sample function from a white-noise process having a variance of 1354, then the output will be a sample function of a random process having the same autocorrelation values at lags 0, 1, and 2, as the data signal. Of course, one could also express this model in the time domain in the form of Eq (9.22) as

$$c[n] = -a(1)c[n-1] - a(2)c[n-2] + w[n] = 1.078c[n-1] - 0.219c[n-2] + w[n].$$

Example 9.12   *A better AR model for heart rate variability*   Not surprisingly, the second-order autoregressive model of the previous example is too simple to reveal all of the structure of the power spectrum of heart rate. Data in the file `hrv.mat` may be used to determine an autoregressive spectrum for `hrv`. The file contains 10 separate observations of beat-to-beat heart rate, organized in columns of 128 data points each.

The task is to determine the power spectrum for each of the 10 sample functions. Using the MATLAB function `ar`, one should determine the "best" AR spectrum for the first sample function by choosing an appropriate model order. One way to increase one's confidence in the estimated power spectrum is to determine what features of the spectrum are reproducible across several sample functions. Using the same model order chosen for the first sample function, one may calculate the estimated spectra from all 10 sample functions. If they seem consistent, then the ten $\hat{a}(k)$ values for each k may be averaged to obtain a final estimate of the power spectrum of *hrv* from the average AR coefficients. One striking advantage is evident when comparing the AR-based spectrum to the periodogram-based spectrum—the smoothness of the AR-based spectrum. This advantage is manifest especially when only one sample function is available (Fig. 9.15). But note the potential disadvantage of the smoothness imposed by using a low-order AR model: pertinent and important narrow-band features of the true spectrum might be lost.

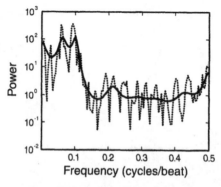

FIGURE 9.15   Two power spectral estimates for the data in column two of the file `hrv.mat`. Solid: from an AR(15) model. Dotted: from the periodogram (256 points, rectangular window).

9.9 MOVING AVERAGE (MA) PROCESSES

A moving average process of order q, symbolized as MA(q), is a General Linear Process for which $p = 0$. Referring to Eq. (9.26), a moving average process is described by the relationship

$$x[n] = \sum_{k=0}^{q} b(k)w[n - k]$$

Such a process represents $x[n]$ as the output of an FIR filter whose input is uncorrelated white noise, $w[n]$. Note that $q + 1$ coefficients are needed to specify an MA(q) process. Furthermore, it is trivial to prove that $h[k] = b(k)$. That is, the MA coefficients $b(k)$ form the impulse response of the filter. The power spectrum of a moving average process can be derived directly from Eq. (9.32) to be

$$S_x(e^{j\omega}) = \sigma_w^2 |B(e^{j\omega})|^2.$$

The Yule-Walker equations for these processes are found from Eq. (9.37) by letting $p = 0$. The terms in $a(k)$ drop out and one is left with

$$r_x[m] = \sigma_w^2 \sum_{j=0}^{q-|m|} b(j + m)b(j), \qquad |m| \le q \qquad (9.47a)$$

$$\text{and } r_x[m] = 0, \qquad |m| > q. \qquad (9.47b)$$

Two important conclusions follow from these results. First, for an MA(q) process the autocorrelation function is zero for all lags exceeding the magnitude of q. This result suggests one possible method for estimating the order of an unknown MA process from observation of one of its sample functions: Calculate $\hat{r}_x[m]$ for increasing values of m and determine that value of m for which the calculated autocorrelation function falls to zero. In practice this method provides only a rough guideline because of the errors inherent in $\hat{r}_x[m]$. The second conclusion is that the coefficients $b(k)$ are not linearly related to the autocorrelation values as was the case for AR processes. Therefore estimating the coefficients of an MA(q) process from knowledge of its autocorrelation function is more difficult.

Estimating the Coefficients of an MA(q) Process

To develop a moving average model of a random process from measurement of a single sample function, one utilizes the estimated autocorrelation values $\hat{r}_x[m]$. The procedure is more difficult than for AR processes because the combination of the nonlinear relationship noted above and the inherent errors in $\hat{r}_x[m]$ can produce large errors in the estimates of $b(k)$. One common approach is to model the process first as an AR(P) process, where $P \gg q$, then convert this AR model into an MA(q) model. Assume the AR(P) model has been estimated using methods discussed previously. The procedure for converting the AR(P) model into an MA(q) model can be derived by equating the transfer functions of the AR(P) and MA(q) processes, assuming that $b(0) = h(0) = 1$:

$$H(z) = \frac{1}{A_P(z)} = B_q(z),$$

from which one concludes that $B_q(z)A_P(z) = 1$. Note that the subscripts P and q are used to indicate the orders of the corresponding models. Taking inverse Z-transforms of both sides of the preceding equation, one obtains the following relationship between the $a(k)$ coefficients of $A_P(z)$ and the $b(k)$ coefficients of $B_q(z)$:

$$a(m) + \sum_{k=1}^{q} b(k)a(m-k) = \begin{cases} 1, & m = 0 \\ 0, & m \neq 0 \end{cases}. \tag{9.48(a)}$$

If the number of data points N is chosen such that $N > P \gg q$ in order to reduce the errors in $\hat{r}_x[m]$, then one can write Eq. (9.48(a)) for $1 < m \leq P$, (noting that by definition $a(0) = 1$):

$$\begin{bmatrix} a(0) & 0 & \cdots & 0 \\ a(1) & a(0) & \vdots & \vdots \\ \vdots & \vdots & \vdots & \vdots \\ a(P-1) & \cdots & \cdots & a(P-q) \end{bmatrix} \begin{bmatrix} b(1) \\ \vdots \\ \vdots \\ b(q) \end{bmatrix} = - \begin{bmatrix} a(1) \\ a(2) \\ \vdots \\ \vdots \\ a(P) \end{bmatrix}. \tag{9.48(b)}$$

This set of overdetermined linear equations can be solved using a standard least squares approach to find the estimated MA coefficients, $\hat{b}(k)$, $1 \leq k \leq q$. This approach to MA modeling is called *Durbin's method*.

Example 9.13   An MA(3) process   The signal in Fig. 9.16 was generated from a known MA(3) process having coefficients $[b(0), b(1), b(2), b(3)] = [1, .45, -.30, -.33]$. The estimated autocorrelation function, $\hat{r}_x[m]$, was calculated for $0 \leq m \leq 7$ and an AR(7) model was fit to these data. The AR coefficients were found to be:

$$[\hat{a}_0, \hat{a}_1, \ldots, \hat{a}_7] = [1, -.4032, .4966, .0065, .1200, .0848, -.0341, .0576].$$

These coefficients were then used in Eq. (9.48(b)) to solve for the coefficients of an MA(3) model using MATLAB matrix division for overdetermined equation sets (which provides the least squares solution). The estimated coefficients were

$$[\hat{b}(0), \hat{b}(1), \hat{b}(2), \hat{b}(3)] = [1, .3864, -.3541, -.3022].$$

The signal shown in Fig. 9.16 is available in the file ma3sig.m. The reader may repeat this procedure for other values of P (e.g., 10, 15, 30) and determine how

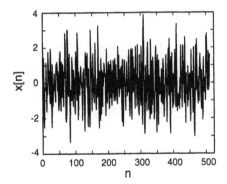

FIGURE 9.16 Sample function generated by passing a Gaussian white-noise signal through an MA(3) process with coefficients [1, 0.45, –0.30, –0.33].

large P must be in order to obtain good estimates of the known parameters. Note also whether those values of the autocorrelation function which theoretically should be zero are close to zero.

9.10 AUTOREGRESSIVE-MOVING AVERAGE (ARMA) PROCESSES

If both p and q are greater than zero, the General Linear Process is an ARMA(p,q) process. This form of the GLP is the most general because it allows one to directly express the dependence of the current output on its past values as well as on the current and past values of the driving white-noise input. The power spectrum of an ARMA(p,q) process, which is given in Eq. (9.22), has more flexibility than those of AR and MA processes, and therefore ARMA processes are desirable models for unknown signals. On the other hand, practical implementation of such models is frought with all of the difficulties associated with modeling of both AR and MA processes and, additionally, one must determine values for both p and q.

Estimating the Parameters of an ARMA(p,q) Model

The transfer function of an ARMA(p,q) model has the form

$$H(z) = \frac{B(z)}{A(z)} = \frac{\sum_{k=0}^{q} b(k)z^{-k}}{1 + \sum_{k=1}^{q} a(k)z^{-k}}. \tag{9.49}$$

Estimation of the $p + q + 1$ parameters of an ARMA(p,q) model is often accomplished in sequential steps in which the AR parameters are estimated first, followed by the MA parameters. Such a separation is made possible through use of the *high-*

er-order Yule-Walker equations, which are the set of equations formed by writing Eq. (9.37) for $q + 1 \leq m \leq q + K$, where K is to be determined. Note from Eq. (9.37) that the right-hand side of all of these equations is zero for this range of m. Conceptually one needs only to evaluate p of these equations (i.e., $K = p$) in order to uniquely determine the coefficients $a(k)$, $1 \leq k \leq p$, but the expectation that there will be errors in determining $\hat{r}_x[m]$ dictates that one should choose $K > p$ and solve for the coefficients using a least squares approach. Therefore these equations can be written in matrix form as

$$
\begin{bmatrix}
\hat{r}_x[q+1] & \hat{r}_x[q] & \cdots & \hat{r}_x[q-p+1] \\
\hat{r}_x[q+2] & \hat{r}_x[q+1] & \cdots & \vdots \\
\vdots & \vdots & \vdots & \vdots \\
\vdots & \vdots & \vdots & \vdots \\
\hat{r}_x[q+K] & \vdots & \vdots & \hat{r}_x[q+K-p+1]
\end{bmatrix}
\begin{bmatrix}
1 \\
a(1) \\
\vdots \\
a(p)
\end{bmatrix}
=
\begin{bmatrix}
0 \\
\vdots \\
\vdots \\
0
\end{bmatrix}.
\tag{9.50a}
$$

This equation set may be rearranged as follows to permit solving by an ordinary least squares method:

$$
\begin{bmatrix}
\hat{r}_x[q] & \cdots & \hat{r}_x[q-p+1] \\
\hat{r}_x[q+1] & \cdots & \vdots \\
\vdots & \vdots & \vdots \\
\vdots & \vdots & \vdots \\
\hat{r}_x[q+K-1] & \vdots & \hat{r}_x[q+K-p+1]
\end{bmatrix}
\begin{bmatrix}
a(1) \\
a(2) \\
\vdots \\
a(p)
\end{bmatrix}
= -
\begin{bmatrix}
\hat{r}_x[q+1] \\
\vdots \\
\vdots \\
\hat{r}_x[q+K]
\end{bmatrix}.
\tag{9.50b}
$$

Once one has obtained the $a(k)$ coefficients, the $b(k)$ coefficients are found by forming an MA model of the signal after "removing" the AR components. These components are removed by passing the original signal, $x[n]$, through a filter having the transfer function $A(z)$, as shown in Fig. 9.17. Since $A(z)$ is the reciprocal of the estimated AR component of the ARMA model, the output of this filter, $r[n]$, should be a pure moving average process whose parameters can be found using Durbin's method.

Example 9.14   An ARMA(1,1) model   The data, $x[n]$, shown in Fig. 9.17(b) were obtained by filtering unit-variance, Gaussian white noise by the system

$$
H(z) = \frac{1 - 0.50z^{-1}}{1 - 0.80z^{-1}}.
$$

To estimate the parameters of an ARMA(1,1) model for these data, first an AR(1) model was fit to the data using the higher-order Yule-Walker equations with $K = 2$. This process produced the estimated autoregressive fit

$$
[\hat{a}(0), \hat{a}(1)] = [1 \ -0.8013].
$$

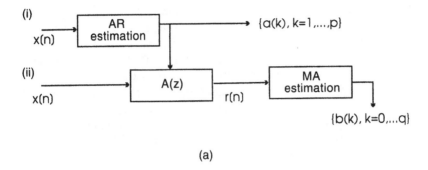

(a)

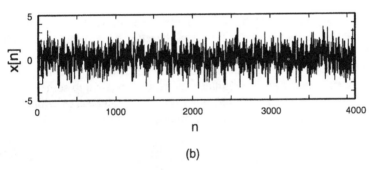

(b)

FIGURE 9.17 (a) ARMA model estimation procedure proceeds in two steps: (i) find AR coefficients; (ii) filter $x[n]$ to produce approximate MA time series, $r[n]$, and estimate its MA coefficients using Durbin's method. (b) Data used for Example 9.14.

A second time series, $r[n]$, was generated by filtering $x[n]$ by the system

$$H_1(z) = 1 - 0.8013\, z^{-1}.$$

An MA(1) model was fit to $r[n]$ using Durbin's method, yielding the following estimates:

$$[\hat{b}(0), \hat{b}(1)] = [1, -0.5026], \quad \hat{\sigma}_w^2 = 0.9765.$$

The fact that these estimated parameters are very close to the true parameters of $H(z)$ is due to the long data sequence (i.e., 4096 points) that was used. The signal $x[n]$ is available in the file armal2.mat. The reader may use the M-file quikarma.m to estimate ARMA(1,1) models for subsets of data taken from $x[n]$ and determine whether acceptable estimates result when short data sets are utilized.

The above example avoids the issue of determining values for p and q but clearly this is an essential step in the procedure of modeling a signal as the output of an ARMA(p,q) random process. As noted previously, if one is interested only in estimating the power spectrum of the process, then finding the best values for p and q is not necessary. One can generate the power spectra for various orders of the model (i.e., various combinations of p and q) and compare them, choosing values which seem to reproduce spectral features that are common to a range of orders but do not seem to generate spurious peaks and valleys. This procedure is more difficult than it may seem, but various indices have been proposed which provide guidelines (Ljung, 1987).

9.11 HARMONIC PROCESSES

Harmonic processes are an important class of random processes that generally cannot be modeled as General Linear Processes because their mean values often change with time. A simple form of *harmonic process* is a random process with sample functions that are sinusoidal signals whose frequencies are the same for each sample function but whose amplitudes and phases (relative to some time origin) vary randomly from one sample function to another. Assume each sample function, $x_j[n]$, can be represented by a cosine function of the form

$$x_j[n] = c_j \cos[\omega_0 n + \theta_j]. \tag{9.51}$$

Note that c_j and θ_j have definite, fixed values for a given sample function, $x_j[n]$, but vary randomly as a function of j. Therefore, c_j and θ_j are random variables. Typically it is assumed that the phase, θ_j, assumes values in $[0, 2\pi]$ with equal probability. In this case the harmonic process has a time-invariant mean of zero. If c_j and θ_j are uncorrelated and $E\{\theta_j\} = 0$, then the autocorrelation function of a such an harmonic process is

$$r_x[k, l] = E\{x_j[k]x_j[l]\} = E\{(c_j \cos[\omega_0 k + \theta_j])(c_j \cos[\omega_0 l + \theta_j])\}$$

$$= E\{\tfrac{1}{2}c_j^2[\cos[\omega_0 k + \theta_j - \omega_0 l - \theta_j] + \cos[\omega_0 k + \theta_j + \omega_0 l + \theta_j]]\}$$

$$= \tfrac{1}{2}E\{c_j^2[\cos[\omega_0[k - l]] + \cos[\omega_0[k + l] + 2\theta_j]]\}.$$

Note that the expectation is taken with respect to j. Therefore, the expected value of the cosine term having the random phase, $2\theta_j$, is zero and

$$r_x[k, l] = r_x[k - l] = \tfrac{1}{2}E\{c_j^2\} \cos[\omega_0(k - l)]. \tag{9.52}$$

That is, the autocorrelation function of this simple harmonic process is a cosine having the same frequency as the process itself.

This result can be generalized to the case of a process with periodic sample functions for which a Fourier series may be written. Thus, if each sample function may be expressed as

$$x_j[n] = \sum_{k=0}^{\infty} c_j(k) \cos[k\omega_0 n + \theta_j(k)],$$

then if $E\{c_j(k)c_j(l)\} = 0, \forall k, l,$

$$r_x[m] = \frac{1}{2} \sum_{k=0}^{\infty} E\{[c_j(k)]^2\} \cos[k\omega_0 m].$$

It is possible to extend this concept to situations in which the frequencies vary randomly across the sample functions as well, as in the example of airflow patterns (Fig. 9.6(a)). In such cases the constant ω_0 is replaced by a random variable, ω_j, in the above equations. Calculating the mean, variance, and autocorrelation function of these latter processes is more difficult and depends on the form of the distribution of ω_j.

Example 9.15   Biomedical data modeled as a harmonic process   The example shown in Fig. 9.18 is a test of a control algorithm for providing electrical stimulation of a leg muscle to produce rhythmic movement of the leg. The position of the upper leg of a seated subject was fixed and the position of the ankle joint in three-dimensional space was measured using an electromagnetic sensor (Flock of Birds™) and sampled at 50 Hz. From these data the instantaneous angle of the knee, $a[n]$, was determined during repetitive electrical stimulation of the quadriceps muscle. The stimulus (a train of electrical pulses) was delivered approximately once every 1.06 s and because of the regularity of the stimulus, the time series of knee angle is periodic and can be considered a sample function of an harmonic process. The Fourier series coefficients for this sample function can be found in the usual way. The first few terms of this series are

$a[n] = 11.04 + 18.90 \cos(5.928t - 1.986) + 13.54 \cos(11.855t + 2.4187)$

$\qquad + 4.99 \cos(17.783t + 0.2917) + 1.120 \cos(23.710t - 0.8413)$

$\qquad + 0.608 \cos(29.638t - 0.5287)$

The approximation to $a[n]$ based on these first five harmonics is shown as the dashed curve in Fig. 9.18. If another sample function were available, its Fourier series coefficients would be expected to be somewhat different from those above. A more precise model of this harmonic process would require the acquisition of more sample functions, with averaging of the Fourier series coefficients across the ensemble of sample functions.

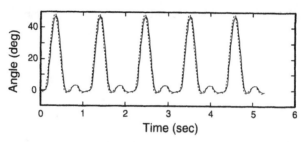

FIGURE 9.18 Solid: knee angle versus time during repetitive electrical stimulation of quadriceps muscle of a sitting subject. Dashed: reconstructed signal based on first five terms of Fourier series. (Courtesy of J. Reiss and J. Abbas.)

9.12 OTHER BIOMEDICAL EXAMPLES

Autoregressive modeling of time series has three particular advantages that render this approach especially useful for biomedical applications in which discrimination of pathophysiology from normal physiology is the goal. First, the parameters are easy to compute. Second, if one is interested mainly in the power spectrum of the time series, then it is not necessary to determine the optimal model order—only to find an order that is large enough to reveal the spectral features without introducing spurious peaks. Third, AR models provide high resolution spectra from single short data records (unlike Fourier spectra based on the periodogram). Consequently there exist numerous examples of AR models applied to biomedical problems. Akay (1994) presents several of these applications. Here two additional examples will be discussed.

AR Modeling of Electromyogram (EMG) Signals

Although EMG signals are very "noise-like," it is known that the spectral characteristics of an EMG signal can be altered both by the occurrence of fatigue of the muscles and by diseases of the muscle fibers. In both situations the biochemical processes responsible for generation and/or conduction of action potentials along muscle fibers are altered, thereby causing the shape of individual action potentials to be distorted from normal. Although neurologists diagnose specific muscle disorders by examining action potentials from individual muscle fibers, muscle fatigue often is detectable by examining the power spectrum of multi-unit EMG (MUEMG) recordings of the type obtained by placing large electrodes on the skin over a muscle or by implanting widely-separated wire electrodes into a muscle. During fatigue there is a general decrease of MUEMG power in high frequencies (roughly above 100 Hz) and an increase in low-frequency power (roughly below 60 Hz).

The diaphragm muscle is the primary muscle used for breathing and its ability to function continuously is critical. Fatigue of the diaphragm in certain diseases, such as chronic obstructive pulmonary disease (COPD) in adults and apnea of prematurity in newborn infants, has been suggested to play a role in the development of insufficient ventilation and much research has addressed the detection of fatigue of the diaphragm. An example of a diaphragm EMG recording obtained from surface electrodes on the lower rib cage of an adult human is shown in Fig. 9.19(a). Bower et al. (1984) have modeled the diaphragm MUEMG with an AR(2) model and attempted to detect fatigue by assessing changes in the autoregressive coefficients. Because the diaphragm EMG is nonstationary, they analyzed short data segments (i.e., 128 ms of EMG signal sampled at 1000 Hz), during which time they assumed that the signal was stationary. From the 128 data points they calculated $\hat{r}_x[0]$, $\hat{r}_x[1]$, $\hat{r}_x[2]$, and then evaluated $\hat{a}(1)$, $\hat{a}(2)$ as in Example 9.10. Then they calculated the power spectrum using Eq. (9.46), from which the mean frequency and the frequency of the maximum power were determined. The latter can be calculated from the expression for the power spectrum of an AR(2) process by taking the derivative with respect to frequency and setting it equal to zero. Thus

$$\frac{ds}{d\omega} = -\frac{\sigma_w^2 |b(0)| [-2a_1(1 + a_2) \sin(\omega) - 4a_2 \cos(2\omega)]\|^2}{(1 - a_1^2 + a_2^2 + 2a_1(1 + a_2) \cos(\omega) + 2a_2 \cos(2\omega))^2}.$$

Setting this derivative equal to zero at $\omega = \omega_{max}$ implies that

$$-2a_1(1 + a_2) \sin(\omega_{max}) - 4a_2 \cos(2\omega_{max}) = 8a_2 \sin(\omega_{max}) \cos(\omega_{max}).$$

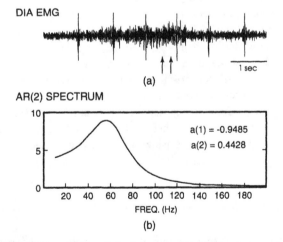

DIA EMG

1 sec

(a)

AR(2) SPECTRUM

a(1) = -0.9485

a(2) = 0.4428

FREQ. (Hz)

(b)

FIGURE 9.19   (a) Diaphragm EMG activity during one breath, from a human subject. Data for spectral analysis were sampled at 500 Hz during the 256-msec interval between the arrows. (b) AR(2) spectrum of the EMG signal, based on the indicated estimates of the autoregressive coefficients. The peak of the spectrum is at $F_{max} = 54.69$ Hz.

Solving for ω_{max},

$$\omega_{max} = \cos^{-1}\left[\frac{-a_1(1 + a_2)}{4a_2}\right].$$

The corresponding CT frequency is

$$F_{max} = \frac{1}{2\pi}\frac{\omega_{max}}{T} = \frac{1}{2\pi T}\cos^{-1}(-a_1[1 + a_2]/4a_2),$$

where T is the sampling interval and the coefficients are symbolized as a_1 and a_2 for convenience. (An example (not from the paper by Bower et al.) is presented in Fig. 9.19(b), in which F_{max} = 54.69 Hz.) In their example the subject was made to breathe through a flow resistance so that the diaphragm fatigued. The mean frequency and F_{max} both were observed to decrease with fatigue, but a more consistent response was a change in the value of $\hat{a}(1)$. For example, in one subject the average value of $\hat{a}(1)$ from 30 breaths before fatigue was −1.42+/−0.08, whereas during fatigue it was −1.58+/−0.07. The reader can demonstrate that this change is consistent with an increase in low-frequency power relative to high-frequency power in the spectrum of the EMG signal.

Modeling of Gastric Electrical Signals

During their contractions the muscles of the stomach generate electrical signals that can be recorded using electrodes on the surface of the abdomen (Fig. 9.20). The frequency of occurrence of these signals, and the frequencies contained in the signals when they occur, are both correlated with the presence or absence of gastric disease. Because the frequencies are low—typically 0.03–1.16 Hz—and the spectral peaks are fairly sharp, low-order AR models are useful for calculating the spectra of these signals because these models do not require the long data records and averaging needed for periodogram-based methods (Fig. 9.20(b)). Furthermore, by basing spectra on short data records it is possible to detect time-varying changes in spectral content of the signal, such as those which occur after a meal. Other investigators have applied ARMA models to such data (Chen et al., 1990). Since these spectra vary with time, more recent approaches to their characterization utilize advanced methods of time-frequency analysis which generate energy spectra that are inherently time-varying (Lin and Chen, 1994).

9.13 INTRODUCTORY EXAMPLE, CONTINUED

Return now to the exercise at the beginning of this chapter. The advantages of modeling the EEG signal using the methods of this chapter are the same advantages discussed above regarding gastric electrical signals. Both types of signals change with time, both are restricted in frequency content, and both have fairly discrete diagnos-

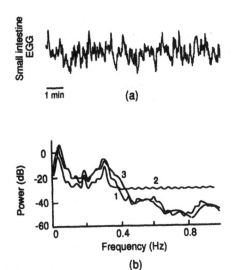

FIGURE 9.20 (a) A 10-minute sample of an EGG signal obtained from electrodes placed on the abdomen over the small intestine. Reproduced from Chen et al. (1993). (b) Power spectra of an EGG signal obtained by three different methods. Curve #1: from the periodogram; curve #2: from an AR(50) model, curve #3: from an ARMA(50,10) model. Reproduced from Chen et al. (1990).

tic frequency ranges. To finish that exercise you should now devise an analysis method based on AR, MA, or ARMA modeling of the EEG signal. To start, decode the m-file tfar.m, which is invoked by eegsim.m, and explain the method implemented in that program for estimating the spectrum of the EEG signal. Determine the parameters of the method such as the order(s) of the model, the number of data points used to evaluate one spectrum, and the time interval at which new spectra are calculated. Comment on the appropriateness of the values of these parameters. Can you suggest better values?

Finally, you should use the MATLAB command armax to estimate an ARMA model for the EEG signal generated from eegsim.m and determine the best orders p and q which allow you both to discriminate the three frequency bands without generating spurious peaks and to quickly detect when the state of consciousness (as reflected by the major frequency component in eeg) has changed. When you believe that you have optimized your algorithm, you may wish to test it on the actual EEG signal contained in the file montage.mat (Fig. 9.1(b)).

9.14 SUMMARY

This chapter has introduced the concept of a random process, discussed some properties of random processes, and presented some basic techniques for modeling a data time series as the output of a linear system driven by (sample functions from) a white-noise random process. A random process is a physical process which is structurally similar in every case in which it is physically realized, but its behavior for each realization differs due to random variations in its parameters and to random disturbances within each realization. A record of the temporal behavior of a single realization is termed a sample function of the random process. Many biomedical signals can be viewed as sample functions from random processes because they vary due to parameter variations among subjects and random disturbances within a subject.

Discussions were restricted to consideration of wide-sense stationary (w.s.s.), ergodic random processes. A random process is w.s.s. if its mean (looking across its sample functions) and its autocorrelation function are independent of the time reference at which they are evaluated. If a random process is autocorrelation ergodic, then its mean and autocorrelation function can be estimated from a sample function by methods which replace averaging across the ensemble of sample functions with averaging through time. One can state sufficient conditions for ergodicity, but in practice they are difficult to use. Often it is possible to detect a time-varying mean or a time-varying variance (which implies that the autocorrelation function is time-dependent) simply by calculating these measures over different segments of the data and comparing variations in their values with statistical confidence limits, as discussed in Chapter 12.

A General Linear Process (GLP) is a zero-mean random process whose sample functions can be modeled as outputs of a linear system that is excited by a white-noise process—that is, each sample function of the white-noise process presented at the input produces one sample function of the GLP random process at the output. This random process, however, is not a white-noise process since the white-noise sample functions have been filtered by the linear system. GLPs can be described by three equivalent representations: the AR, MA, and ARMA forms. The Yule-Walker equations describe the relationships between the parameters of a GLP and its autocorrelation function, and methods for developing a GLP model to represent observed data often begin with these equations. Because the autocorrelation function is unknown, it is estimated from the data and these values are used in the Yule-Walker equations to determine the parameters of the linear system which, when excited by white noise, will produce a random process that has the autocorrelation properties of the data. In fact the calculations are fairly easy and the more difficult problem is to determine the orders of the polynomials of the transfer function of the GLP. Choosing too small an order will cause one to lose some of the salient features of the data whereas selecting too high an order may introduce behaviors in the model that are not present in the data. It was suggested that an iterative approach be adopted wherein one evolves models of progressively higher orders until one obtains nonreproducible behaviors from the model. Reproducibility of these behaviors can be tested either by developing models for several different sample functions of the data or by exciting a model with several different simulated white-noise sample functions.

Finally we discussed harmonic processes, which have periodic sample functions. These processes often are not General Linear Processes because the mean is not

time-invariant but these models often are applicable to biomedical data. Their parameters can be determined through Fourier series modeling of their sample functions.

EXERCISES

9.1   Consider a manufacturing plant that produces automobile batteries with a
nominal voltage of 12 V. When examined closely, each battery produces a nominal
mean output voltage that approximates 12.00 V, but varies slightly with time.

a. Describe this manufacturing process as a random process and identify its realizations and its sample functions. Speculate on the factors that cause random variations between and within realizations. Do you think this random process could be
wide-sense stationary? Explain your answer.

b. The Quality Control department randomly removes batteries from the assem-
bly line just before packaging and measures their voltage levels continuously for
several hours to determine whether these levels meet manufacturing specifications.
Speculate on the estimated autocorrelation function and power spectrum of the random process that would be determined from each of these sample functions.

9.2   The generation of action potentials by a spontaneously firing neuron often
can be modeled as a sample function of a Poisson random process. Poisson random processes are models for processes which generate discrete events at some
mean rate, L, with the probability of k events occurring in any time interval $t_2 - t_1$
being

$$\text{Prob}(k) = e^{-L(t_2-t_1)} \frac{[L(t_2 - t_1)]^k}{k!}.$$

Its autocorrelation function is $r_P(\tau) = L\delta(\tau)$.

An experimenter measures the activity of the C1 neuron of the stomatogastric
ganglion of the lobster under resting conditions, in each of many ganglia that are
studied, in order to characterize the average resting behavior of this neuron. Formulate a random process description for this experimental study and identify the random process, its realizations, and its sample functions, and discuss its pdf and autocorrelation function.

9.3   In the above problem the average firing rate, L, of one neuron is found to be
2.187 events per second.

a. What is the estimated autocorrelation function of the random process based
on this one realization?

b. What is the probability that the neuron will *not* generate an action potential in
an arbitrary one-second interval?

9.4   It is desired to measure a time function $f(n)$ but we only have access to the
nonstationary random process $X(n) = f(n) + N(n)$, where $N(n)$ is a random process
that is uniformly distributed on the interval $(-A,A)$. However, $r_N[n]$ is nonzero only

on the interval $(-S, S)$ and its sum over this interval is "K". Use as an estimate of $f(n)$ the moving average given by

$$f_M(n) = \frac{1}{M} \sum_{i=n-M+1}^{n} X(i),$$

where $S \ll M$. Show that

$$E\{f_M(n)\} = \frac{1}{M} \sum_{i=n-M+1}^{n} f(n), \text{ and } \sigma_{f_M}^2 \approx \frac{K}{M}.$$

That is, M should be large enough that the variance is small, but small enough that the expected value is close to $f(n)$.

9.5   Consider a second-order, autoregressive random process, $x[n]$, which is derived by passing a white-noise process, $w[n]$, through a linear filter characterized by its impulse response, $h[n]$. Therefore, $x(n) = -a_1 x(n-1) + a_2 x(n-2) + w(n)$.

   a. If $H(z)$ is written as $B(z)/A(z)$, what are $A(z)$ and $B(z)$?

   b. Derive the equation for the power spectrum, $P_x(e^{j\omega})$, in terms of a_1, a_2, and σ_w^2.

   c. Write the Yule-Walker equations for this process.

9.6   Calculate and plot $r_x[n]$ and the power spectrum for an AR(1) process with $a(1) = -0.3$.

9.7   Calculate and plot $r_x[m]$ for an AR(2) process with $a(1) = 0.2$ and $a(2) = -0.5$. Let $0 \leq m \leq 5$.

9.8   Use MATLAB to generate a sample function of an AR(4) process having coefficients $[a(0), a(1), a(2), a(3), a(4)] = [1.00, -1.978, 2.853, -1.877, 0.904]$.

   a. Use MATLAB to find the power spectrum of your sample function using the periodogram method.

   b. What is the transfer function, $H(z)$, of the GLP model of this random process?

   c. Determine analytically the power spectrum of this random process and compare with your answer to part a above.

9.9   Generate 10 different sample functions, each of length 100 points, from the random process of Exercise 9.8. Use the function ar to estimate an AR(4) model from each sample function. Then use th2tf to find the corresponding transfer function for each AR(4) model. Finally, using freqz plot the magnitudes of the frequency responses of all 10 models on the same graph. Explain the variability in these responses.

9.10   The purpose of this MATLAB exercise is to examine the effects of added Gaussian noise on the ability to detect a sinusoidal signal by way of its power spectrum. Generate a 1 Hz sine wave having unit amplitude. Sample this signal every 0.0333 s to obtain 256 points. Call the sampled signal $y[n]$.

   a. Using psd, calculate the power spectrum of the noise-free sine wave, $y[n]$.

b. Let $z[n] = y[n] + V$*randn(256,1). Let V, the white-noise variance, be 1.00. Compare the power spectrum of $z[n]$ with that of $y[n]$.

c. Repeat part b for $V = 10, 30, 100, 300, 1000$. In each case compare the spectrum of the resulting $z[n]$ with that of $y[n]$.

9.11 Calculate an AR(2) model (i.e., the autoregressive coefficients and the white-noise variance) for a zero-mean random process driven by white noise which has autocorrelation values $r_X(0) = 0.90$, $r_X(1) = 0.81$, $r_X(2) = 0.70$.

9.12 A researcher obtains the power spectrum of Fig. 9.21 by analyzing her data using the periodogram method. (She has 1024 data points which she analyzed as one segment of 1024 points.) She thinks that the data exhibit two narrow peaks around 0.02 Hz and 0.08 Hz and a broad peak around 0.1 Hz. She asks for your help to find a "better" method of spectral analysis to see if these features actually are present in the spectrum. What do you recommend and why? *Be specific and concise.* Specify all data analysis parameters or options, including sampling frequency for data acquisition and the frequencies at which the spectral estimates will be calculated. *Note:* The researcher cannot obtain any more data from this experiment but she can do more experiments of the same type, if necessary. Due to limitations on the experimental design, however, the length of one data record is limited to that corresponding to the spectrum of Fig. 9.21.

9.13 The power spectrum of an electromyogram (EMG) signal typically spans the frequency range of 20–400 Hz. When the muscle fatigues, the spectrum of its EMG signal increases at low frequencies (e.g., below 100 Hz) and decreases at high frequencies. Several years ago some researchers proposed that they could detect muscle fatigue by calculating an AR(2) model of an EMG signal and looking for changes in the parameters of this model. In Table 9.1 are the calculated autocorrelation values for the diaphragm EMG from one subject before and during fatigue of the diaphragm muscle.

a. Calculate the AR(2) parameters for both conditions.

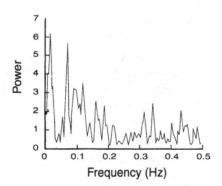

FIGURE 9.21 Power spectrum (periodogram) for Exercise 9.12.

TABLE 9.1 Autocorrelation Values for Exercise 9.13

Lag	0	1	2	3	4	5	6	7
Rest	1	0.83	0.47	0.08	-0.22	-0.37	-0.39	-0.26
Fatigue	1	0.9	0.66	0.36	0.07	-0.17	-0.32	-0.37

b. Do the changes in AR(2) parameters *qualitatively* reflect the expected changes in the power spectrum described above?

9.14 You have made the following observations from a random process, $x[n]$:

$$x[n] = [1.6371, 0.2468, \quad -1.1393, \quad -0.8585, \quad 0.4793, \quad 0.5608,$$
$$-1.1102, \quad -0.4374, \quad -0.7562, \quad -0.5905]$$

  a. Calculate an AR(2) model for this process.
  b. Calculate an MA(2) model for this process.
  c. Using `quikarma.m`, calculate an ARMA(1,1) model for this process.
  d. Of these three models, which one do you think is the "best" for these data? Why?

9.15 One popular method for determining the "best" order of an AR model is to find the model order, p, that minimizes the Akaike Information Criterion (AIC), which is calculated as

$$AIC(p) = N \log(\varepsilon_p) + 2p,$$

where ε_p is the mean-square modeling error and N is the number of data points. AIC penalizes both an increase in modeling error and an increase in number of parameters. For an AR(p) model an estimate of the modeling error is the estimated white-noise variance, $\hat{\sigma}_w^2$.

  a. For the first sample function (i.e., the first column) of the EMG signal in the file `ranproc3.mat`, estimate the "best" AR model using AIC as your criterion.
  b. Now apply this same model to each of the other columns in this data file. Calculate the mean and standard deviation for each autoregressive coefficient. How reproducible are the estimates of these coefficients across the eight sample functions?

9.16 It is known that AIC tends to overestimate the correct order for a true AR process when N is large but underestimates the correct order when an AR model is fit to a nonautoregressive process. Another measure of "goodness of fit" is Minimum Description Length proposed by Rissanen, which is defined as MDL(p) = $N \log(\varepsilon_p) + p(\log(N))$. Repeat Exercise 9.15 using MDL as your criterion.

9.17 In Example 9.12 you were asked to find the best AR model for the heart-rate signals in `hrv.mat` without any specific guide to selecting the best model. Repeat this exercise using AIC as your criterion.

  a. Find the best model (i.e., the one which minimizes AIC) for each of the 10 sample functions separately. Do you always obtain the same model order, p?

b. One approach to choosing a model order to apply to *all* of the sample functions is to select the highest order found for any individual sample function. Using this approach, evaluate the AR models for each sample function.

c. Average the 10 estimates of each autoregressive coefficient to determine an average model, then use MATLAB to determine the power spectrum of this model. You probably will want to write an m-file to calculate the spectrum of an arbitrary AR(p) model, or you can use the command freqz by specifying the appropriate transfer function polynomials.

d. Another common approach when one desires only the power spectrum (and not the exact model coefficients) is to let $p = N/4$, or even $p = N/3$. Use the first value and determine the power spectrum for each of the 10 sample functions. Do they exhibit any spurious peaks? If you find the average AR coefficients and then find the power spectrum, do you eliminate the spurious peaks?

9.18 An outlier in a time series is a data point for which one has evidence that either the measurement was erroneous or the behavior was influenced by a transient factor unrelated to the factors you wish to study. A single, isolated outlier can be replaced by an estimate of the true data value at that time point. That is, given the time series $x[n]$, with $x[j]$ an outlier, one can replace $x[j]$ with an estimate $\hat{x}[j]$ based on a random process model of the time series for $0 \leq n \leq j - 1$. First one determines an AR model of the usual form, as

$$x[m] = -\sum_{k=1}^{p} a(k)x[m-k] + w[m],$$

based on data in the time interval $0 \leq n < j$.

Using this model one then replaces $x[j]$ with its (one step-ahead) predicted value

$$\hat{x}[j] = -\sum_{k=1}^{p} a(k)x[j-k].$$

a. An electrical noise has contaminated the EEG signal of the file eegspike.m. Use a one step-ahead predictor based on an AR(15) model to determine a replacement value for the outlier.

b. If the data after the outlier are consistent with the data before the outlier, then one also can predict $x[j]$ based on the p data points *after* $x[j]$. Derive the (backwards) predictor equation for $\hat{x}[j]$ based on the AR(15) model from part "a" but using $x[n], j + 1 \leq n \leq j + p$ and calculate the predicted value.

c. Explain why the predicted values from parts "a" and "b" differ. Sometimes one averages the forward and backwards predictions to determine a replacement value. (*Note:* Usually one would also determine a backwards AR model—obtained by reversing the time series—for the backwards prediction rather than using the forward prediction model.)

9.19 In mathematical models of ligands attaching to receptors on a cell surface and initiating a series of biochemical reactions, the first step—the interaction of a ligand and a receptor—often is assumed to occur when random movement of the ligand causes it to bump into a receptor. Random movement may be modeled as Brownian motion which, in its simplest form, may be defined as the summation of white noise. Let $w[n]$ be zero-mean Gaussian white noise and $d[n]$ be discrete Brownian motion defined such that $d[n] = \sum_{k=0}^{n} w[k]$, $d[0] = 0$. Show that $E\{d[n]\} = 0$, but $d[n]$ is nonstationary because $E\{d^2[n]\} = \beta \cdot n$, where β is a constant.

10

SCALING AND
LONG-TERM MEMORY

10.1 INTRODUCTION

It is apparent that the methods from the previous chapter for modeling stochastic signals depend on approximating the autocorrelation properties of the signal. Although these approaches provide considerable flexibility if one selects a model of sufficiently high order, there are cases in which the combination of a relatively rapid decrease in correlation at short lags and a slow decline at longer lags is difficult to model using low-order ARMA models. This difficulty can be explained heuristically as follows: At sufficiently long lags the autocorrelation function of an ARMA model will be dominated by the pole closest to $z = 1$. At short lags the autocorrelation due to this pole will decay slowly. Thus, in order to retain significant autocorrelation values at long lags, its contribution at short lags must be large. The least-squares techniques used in ARMA estimation will weight errors in the autocorrelation function of the model at shorter lags more heavily (since the autocorrelation values are generally larger at short lags). Therefore an ARMA model is likely to underestimate the true autocorrelation function at large lags. If one needs to model long-term correlations that decay slowly, then another approach may be more parsimonious.

Signals possessing long-term correlation have been observed in many fields. Since the power spectrum is the Fourier transform of the autocorrelation function, one indicator of long-term correlation is the presence of broad-band, low-frequency power in the spectrum. In particular, if the low-frequency spectrum grows in amplitude as the length of the data window increases, then one can suppose that the autocorrelation function is going to zero slowly relative to the window length. This behavior has been reported for heart rate measured in human subjects (Fig. 10.1). The file `hrlong.mat` contains a sequence of 8100 heart beats from a human subject. For three different segments of the data the power spectrum was determined by fitting an AR(6) model using the Yule-Walker method. Notice (Fig.

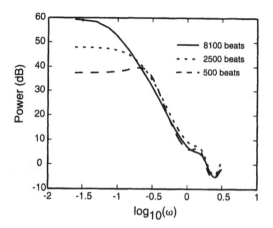

FIGURE 10.1. Spectral analyses of the heart rate data from file `hrlong.mat` using three different data lengths (500, 2500, and 8100 beats).

10.1) that the low-frequency power increases progressively as the data length increases, suggesting that the autocorrelation function is significantly non-zero at very long lags. Note also that the spectrum based on all 8100 beats can be well approximated by a straight line over most of its range (e.g., $-1.2 \leq \log(\omega) \leq 0.4$). This observation may be expressed as $20 \log(\hat{P}(e^{j\omega})) = A - B \log(\omega)$, or $\hat{P}(e^{j\omega}) = 10^{0.054}/\omega^{0.05B}$. Furthermore, this linear approximation likely fits the linear portions of the other two spectra also. Thus, there is the expectation that the linear spectrum (on the log-log plot) may represent the spectrum at even lower frequencies. Of course this approximation cannot be valid at zero frequency because the *log* term would become infinite and physical signals cannot possess infinite power. Such spectra are generically referred to as "$1/f$" spectra because of the reciprocal power-law dependence of power on frequency.

Numerous classes of signals exhibit long-term correlation and/or $1/f$ spectra. The fractal signals introduced in Chapter 1 belong to one such class. These signals have the property that they "look similar" at many levels of magnification. Fig. 10.2(a) shows an example of a fractal signal that was created using the spectral synthesis method (`fracssm.m`). The power spectra, based on fitting an AR(6) model to this signal, were determined for four different data lengths (i.e., 256, 500, 2048, and 5000 points) and are plotted in Fig. 10.2(b). Note that here the low-frequency power also increases with increasing data length and that the overall spectrum is approximately linear, especially for the two longer data lengths. This example suggests that a fractal signal might be a more appropriate model for the heart rate data discussed above (and numerous researchers have demonstrated that fractal models fit heart rate data well). This chapter will discuss the properties of fractal signals in detail, but first we will discuss the concept of self-similarity.

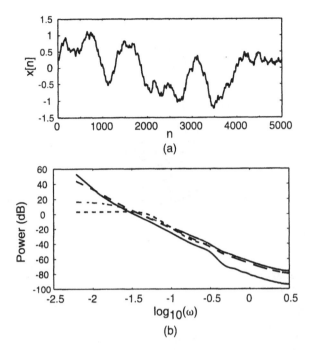

FIGURE 10.2. Analysis of a fractal signal ($H = 0.7$) generated using the spectral synthesis method. (a) The fractal signal. (b) Spectral analyses using four different data lengths: solid, 5000 points; long dashes, 2048 points; dot-dash, 500 points; short dashes, 256 points.

10.2 GEOMETRICAL SCALING AND SELF-SIMILARITY

A fundamental measure of a signal is the dimension of its graph. This chapter will consider only those signals that can be graphed on a plane; thus, the dimensions of these signals are no greater than 2. The concept of dimension, however, is considerably more complex than the simple Euclidian idea that the dimension of a line is one and that of a plane is 2, and so on. Furthermore, the dimension of a signal and its scaling and correlation properties are interrelated. Although our concern is with signals, some of the properties of the graph of a signal are best elucidated from a geometrical viewpoint. Thus, we initially address geometrical scaling and self-similarity.

On polar axes the function $r(\theta) = ae^{q\theta}$ describes a spiral with decreasing radius if $q < 0$ (Fig. 10.3). This function is known as a logarithmic spiral and it has an interesting self-similarity property. If the graph is rotated by an angle ϕ to create a new function $r'(\theta) = ae^{q(\theta+\phi)}$, then $r'(\theta) = e^{q\phi}r(\theta)$. That is, rotating the function is equivalent to multiplying it by the scale factor, $e^{q\phi}$. An alternative interpretation is that after scaling of $r(\theta)$ by a factor s, one can always recover the original graph simply by rotating the new graph by an angle $\phi = \ln(s)/q$. The graphs of $r(\theta)$ and $r'(\theta)$ are geo-

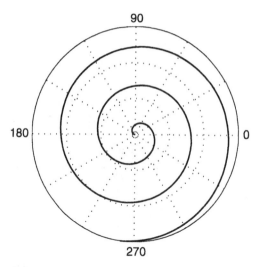

FIGURE 10.3. Graph of a logarithmic spiral, $\ln(r)$ versus θ, where r is the distance of a point from the origin and θ is the angle.

metrically similar. An object is *geometrically similar* to another object if one object can be derived by applying linear scalings, rotations, or translations to the other object. A transform that produces a similar geometrical object is called a *similarity transform*, and it comprises linear scalings, rotations, and translations (any of which might be omitted, of course). Thus, given a point (x, y) in a two-dimensional object, a similarity transform produces a point (x', y') in the new object according to the relationships

$$x' = s \cos(\theta) x - s \sin(\theta) y + d_x, \qquad y' = s \sin(\theta) x + s \cos(\theta) y + d_y, \quad (10.1)$$

where s is the scale factor, θ is the angle of rotation, and d_x and d_y are translations in the x and y directions. The object is transformed by applying these relationships to all points in the object. For example, let $s = 1/2$, $\theta = \pi/6$, $d_x = 1$, $d_y = 0$. Under this transformation the triangle of Fig. 10.4(a) becomes the smaller, rotated, and displaced triangle of Fig. 10.4(b).

One also may apply this transformation to the transformed triangle of Fig. 10.4(b), obtaining the triangle in Fig. 10.4(c). Thus, if one represents the original triangle as an object P_0, then the object of Fig. 10.4(b) may be represented as $P_1 = T[P_0]$ and that of Fig. 10.4(c) as $P_2 = T^{(2)}[P_0]$, where $T[.]$ symbolizes the transformation described by Eq. (10.1). Repeated application of $T[.]$ produces the object $P_\infty = \lim_{n \to \infty} T^{(n)}[P_0]$, which will be a very small triangle that has been translated very far along the x-axis. After every step the resultant object will be geometrically similar to the original object, P_0.

A variation on this scheme of similarity occurs when the resulting object is the union of the objects created by applying several similarity transforms to P_0. In this

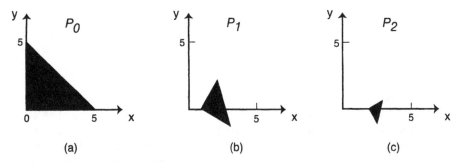

FIGURE 10.4. (a) A triangular object. (b,c) Similarity transformations of the object, P_0 (see text).

case very complex objects may result. Consider, for example, the triangle of Fig. 10.5(a), which we may call P_0. Let $T[.]$ be described by three different similarity transformations and let the output object be the union of the objects created by each transformation. Specifically, the three similarity transformations are defined by: (1) $s = 1/2$, $\theta = 0$, $d_x = d_y = 0$; (2) $s = 1/2$, $\theta = 0$, $d_x = 1/2$, $d_y = 0$; (3) $s = 1/2$, $\theta = 0$, $d_x = 0$, $d_y = 1/2$. Applying these transformations to the original triangle yields the three smaller triangles of Fig. 10.5(b), the union of which comprises $T[P_0]$, or, P_1. One

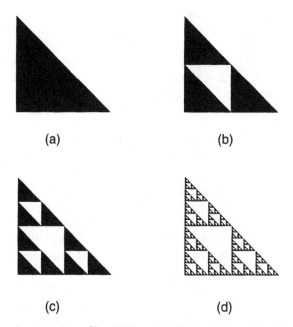

FIGURE 10.5. Steps in generating a Sierpinski gasket. (a) The starting object. (b) After one iteration. (c) After two iterations. (c) After many iterations.

may continue to iterate using this set of transformations. P_2 is drawn in Fig. 10.5(c). Finally, in the limit one obtains an object like that of Fig. 10.5(d). This limiting object is one version of the Sierpinski gasket. Sierpinski gaskets (and related objects known as Sierpinski sponges) exhibit many interesting geometrical properties that are beyond our immediate concern. The reader is referred to the book by Peitgen et al. (1992) for insightful discussions of their properties. The process by which Fig. 10.5(d) was generated is known, for obvious reasons, as an *Iterated Function System*. Planar Iterated Function Systems have important applications in image processing that lie outside of the present interest in one-dimensional signal processing, but one-dimensional Iterated Function Systems will form the basis for understanding the concept of chaos in Chapter 11.

The limit structure of the Sierpinski gasket has the property of *geometrical self-similarity*. An object is geometrically self-similar if one can recreate the object through the union of similarity transformations applied to the object. That is, given an object P, if there exists a set of similarity transformations $T_1[.], T_2[.], \ldots, T_n[.]$, such that

$$P = T_1[P] \cup T_2[P] \cup \ldots \cup T_n[P], \tag{10.2}$$

then P is geometrically self-similar. Furthermore, if one defines an operator

$$H[.] = T_1[.] \cup T_2[.] \cup \ldots \cup T_n[.], \tag{10.3}$$

then $H[P] = P$ and P is said to be *invariant* under the operator H. For example, if P is the Sierpinski gasket, then H may be defined as the union of the three similarity transforms given above for generating the Sierpinski gasket. Not only does H generate this object, but the limiting object itself is geometrically self-similar and invariant under H. It will be shown below that the Sierpinski gasket is a fractal object. Note, however, that geometrical self-similarity alone does not imply that an object is fractal. The distinction is related to the concept of the dimension of the object, as explained below.

The limit structure of the Sierpinski gasket is difficult to visualize because it contains so many "holes." Although the initial triangle is clearly a two-dimensional object in the Euclidian sense, the dimension of the Sierpinski gasket is not entirely clear, especially if we adopt the convention that any continuous line is topologically one-dimensional even if it wanders over (but does not completely fill) a plane. The question of determining the dimension of "strange" objects like the Sierpinski gasket will be considered shortly. First, consider the use of an iteration process to generate an object on a line that also contains many "holes". The iteration proceeds as follows: (1) draw a line segment over the closed interval [0, 1] (Fig. 10.6(a)); (2) remove the middle third of the line segment, creating two shorter line segments; (3) remove the middle third of each remaining line segment; (4) repeat step 3 without limit. The final object is a set of points on the original line segment known as the *middle-thirds Cantor set*, or *Cantor dust*. (There are also other forms of the Cantor set.) This Cantor set can be derived from an Iterated

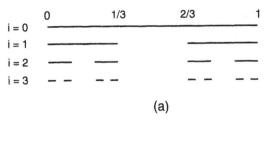

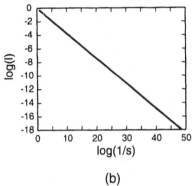

(b)

FIGURE 10.6. (a) Steps in generating the middle-thirds Cantor set. (b) *log-log* plot of the length of the middle-thirds Cantor set versus resolution.

Function System approach. Given an x in an object (such as the original line segment), let $T[x]$ comprise two transformations described by: (1) $x_{i+1} = x_i/3$; (2) $x_{i+1} = -x_i/3 + 2/3$. At each iteration the resulting object is the union of the objects generated by applying these two transformations to every point in the initial object. Again, the Cantor set has numerous important properties but our interest is in understanding how to determine the dimension of such an object. Both the Cantor set and the Sierpinski gasket are examples of objects having non-integer dimensions.

10.3 MEASURES OF DIMENSION

Consider the problem of measuring the total length of the Cantor set described above. Since all of its points lie either in the interval [0, 1/3] or in the interval [1/3, 2/3], if the minimal resolvable interval of the measuring device is one-third of a unit, then the Cantor set appears to have a length of $2(\frac{1}{3})$. But if the measuring precision is increased by reducing the minimal resolvable interval, s, to $s = (\frac{1}{3})^2 = \frac{1}{9}$, then one can resolve that the intervals (1/9, 2/9) and (7/9, 8/9) contain no points of the Cantor set. Therefore, the length, l, appears to be $l = 4(\frac{1}{3})^2$. In general, as the minimal resolvable interval of the measuring device decreases according to $s = (\frac{1}{3})^k$, the length of the Cantor set increases according to $l = 2^k(\frac{1}{3})^k = (\frac{2}{3})^k$. Defining the measurement precision as the reciprocal of the minimal resolvable interval, the length of the Cantor set is plotted as a function of the measurement precision, $1/s$, in Fig. 10.6(b). On this log-log plot, the relationship is linear and the slope of this line is

$$d = \frac{\log(l)}{\log(1/s)} = \frac{k\log(2) - k\log(3)}{k\log(3)} = \frac{\log(2)}{\log(3)} - 1 \approx -0.3691. \qquad (10.4).$$

Thus, if the equation of the line is $\log(l) = a + d\log(1/s)$, then l is related to s as $l = 10^a(1/s)^d$. This type of *power law* relationship is common for fractal objects and fractal signals. Note the seemingly paradoxical result that the apparent length of the Cantor set vanishes as the precision increases.

A similar process may be applied to measuring the area of the Sierpinski gasket. After each iteration, k, let the minimal resolvable area be $s = (1/2)^k$. Since there will be 3^k triangles, their total area will be $A = 3^k(\frac{1}{2})(\frac{1}{2})^k$. In the limit of $s \to 0$, the slope of the graph of area versus precision will be

$$d = \frac{k\log(3) - (k+1)\log(2)}{k\log(2)} = \frac{\log(3)}{\log(2)} - 1 \approx 0.5850, k \to \infty.$$

Measures of the dimension of an object typically are derived from iterative processes such as those used above to determine the length of the Cantor set and the area of the Sierpinski gasket. There are numerous definitions of the dimension of a geometrical object but we shall ignore many subtleties and focus on the two most immediately useful definitions for the present purposes, the similarity dimension and the box-counting dimension. The reader is referred to the books by Peitgen et al. (1992) and by Feder (1988), among several possibilities, for more extensive discussions of dimension.

Similarity Dimension

The similarity dimension of a self-similar object is closely related to measuring the size of the object at various precisions. Similarity dimension expresses the relationship between the number of geometrically similar pieces contained in an object and the measurement precision, as the minimal resolvable interval approaches zero. As we noted above, at the k-th iteration step we expect the total measure of the object (e.g., length or area for the previous examples) to be the product of the number of similar pieces and the resolution raised to some power, D, where (we shall argue) D is the dimension of the object. For example, if one covers a line of length L with line segments of length s, then the number of similar pieces is L/s and the measured length l will be $l = (L/s) s^1$. That is, $D = 1$. Likewise, if one covers a planar area A with squares having sides of length s, then the number of pieces will be A/s^2 and the measured area will be $a = (A/s^2) s^2$, so that $D = 2$. Generalizing from these results, we may presume that the number of similar pieces, N, will be proportional to s^{-D}, or, $N = bs^{-D}$. D may be evaluated by taking the logarithms of both sides. Thus, as s approaches 0,

$$D = -\frac{\log(N)}{\log(s)} \triangleq D_s \qquad (10.5)$$

and is called the *similarity dimension*.

Example 10.1 D_s for the Cantor set and the Sierpinski gasket   For the Cantor set the measurement length at each iteration is $s = (1/3)^k$ and the number of similar pieces is $N = 2^k$. Therefore, its similarity dimension is $D_s = -[k \log(2)/-k \log(3)] \approx 0.6309$. This result implies that the dimension of the Cantor set is greater than that of a point (i.e., zero) but less than that of a line (i.e., 1). For the Sierpinski gasket the measurement size is $s = (1/2)^k$ and the number of pieces is $N = 3^k$. Thus its similarity dimension is $D_s = -[k \log(3)/-k \log(2)] \approx 1.5850$. The Sierpinski gasket has a dimension larger than a line but less than a solid planar object.

Note that both of the above self-similar objects have non-integer similarity dimension! Self-similar objects having a non-integer similarity dimension are called *fractal objects*. (The reader should be aware that the strict definition of a fractal object is that its Hausdorff-Besicovitch dimension is greater than its topological dimension. Explication of the Hausdorff-Besicovitch dimension is beyond the present text but it is based on the concept of covering an object with small objects of variable sizes and examining the limiting relationship between the number of objects and a measure of their "size" as size approaches zero. For a self-similar object, the condition on the Hausdorff-Besicovitch dimension is satisfied if its similarity dimension is non-integer.) It is common to assume that an object is fractal if some measure of dimension is non-integer although the various definitions of dimension are not always consistent. If an object comprises pieces that are geometrically similar to the whole and is invariant in the sense of Eqs. (10.2)–(10.3), then it is fractal. In general, however, for practical purposes we shall loosen the definition and require only that self-similarity extend over a finite number of scales (especially when considering the analysis of real data). Furthermore, we shall modify the definition of self-similarity to include similarities other than geometrical self-similarity, as will be evident below in the discussion of time functions. With these looser conditions one should emphasize the indices of self-similarity and their potential implications regarding the mechanisms that created the object under study, and acknowledge that these indices create a presumption that an object is fractal rather than the proof that it is.

Comparing the similarity dimension and the length (or area) relationship for the two fractal examples above, in both cases one notes that $D_s = d + 1$. This result is general for self-similar objects. Consider a general measure of an object, u (such as length or area) measured using a minimal resolvable interval s. u is proportional to $(1/s^d)$ and we may assume that the unit of measurement is such that the proportionality constant is unity. Thus $\log(u) = d \log(1/s)$. Now the number of similar pieces at each s is $N = (1/s)^{D_s}$, so that $\log(N) = D_s \log(1/s)$. Finally, the total measure is the product of the number of pieces and the measure of each piece, the latter being s (because it was assumed that $u = 1$ when $s = 1$). Consequently, $u = N \cdot s$ and $\log(u) = \log(N) + \log(s)$, implying that $d \log(1/s) = D_s \log(1/s) - \log(1/s)$, or $D_s = d + 1$.

The above result implies that one may calculate D_s for a fractal object either through the relationship of Eq. (10.4) or via the definition of Eq. (10.5). One also may generalize the concept of dimension to objects that are not self-similar by using the approach of Eq. (10.4) and adding unity to the calculated slope. This latter result is sometimes known as the compass or divider dimension.

Box-Counting Dimension

The box-counting dimension expresses the relationsip between the number of "boxes" that contain part of an object and the size of the boxes (Fig. 10.7). The box-counting dimension is simple to evaluate. One chooses boxes of side length s, covers the object with a grid of contiguous boxes, and counts the number of them, $N(s)$, that contain at least one point of the object. This process is repeated for several (preferably many) values of s with the expectation that $N(s)$ varies as $(1/s^{D_b})$, where D_b is the box-counting dimension. Therefore, a graph of $\log(N(s))$ versus $\log(1/s)$ should be a straight line with slope D_b. If one uses box sizes $s = 2^{-k}$, then one may use base 2 logarithms and plot $\log_2(N(k))$ vs. k and evaluate the slope.

The box-counting dimension is not necessarily equal to the similarity dimension but it may be much easier to determine for an arbitrary object and may be applied to objects that are not self-similar. For instance, Peitgen et al. (1992) calculate the box-counting dimension of a map of the coast of England, arriving at a value of approximately 1.31. If one declares that a line with dimension one "fills" very little space, and if one visualizes a solid, two-dimensional, square as filling two-dimensional space, then one may say that a higher dimension corresponds to a more space-filling structure. Shortly in this chapter these same concepts will be applied to signals rather than geometrical objects.

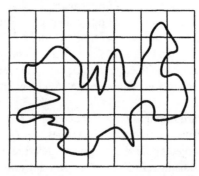

FIGURE 10.7. Example of placing a grid of contiguous boxes over an object.

Example 10.2 D_b for the Koch curve Construction of the Koch curve is indicated on Fig. 10.8. Starting with a line segment that is one unit long, one removes the middle third and replaces it by two pieces, each one-third unit long, that meet the first and third segments of the original line at 120°. In each step the middle third of every line segment is replaced in this manner. The final shape has the same large-scale structure shown in Fig. 10.8 but this structure is apparent at every magnification.

To determine the box counting dimension of the Koch curve, consider covering it with boxes having sides of length 1/3. The number of boxes required at step one will be four. If at the k-th step the box size is $s = (\frac{1}{3})^k$, then $N(s) = 4^k$. Therefore,

$$D_b = \frac{\log[N(s)]}{\log[1/s]} = \frac{k\log(4)}{k\log(3)} \approx 1.2619.$$

Random Fractals

Not all fractal objects satisfy the condition of strict geometrical self-similarity, as the following example demonstrates. Consider generating the Koch curve, as above, with one simple variation. At each step let the new triangular piece be added randomly to either side of the line segment—for example, for horizontal line segements, the triangle points either up or down with equal probability. Although the resulting object will have some visual similarity to the original Koch curve, there may be large differences between the two due to the random orientation of pieces in the former. Nonetheless, it is easy to show that both objects have the same length at each step and the same box-counting dimension, and therefore it is reasonable to conclude that both are fractal objects. One could extend this approach by making the replacement of the middle third of each line segment a random choice (with

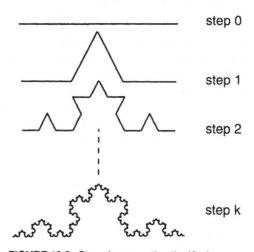

FIGURE 10.8. Steps in generating the Koch curve.

equal probability). Then at each iteration the length of the curve would increase, on average, only half as much as that of the original curve. Thus one could develop some statistical comparisons between the original Koch curve at iteration $k/2$ and the modified random Koch curve at iteration k. For example, the mean length of the line segments of the latter at step k would approach that of the former at step $k/2$ as k increases. This idea of basing similarity on statistical comparisons rather than on geometric shapes is the foundation of the following discussion regarding fractal time functions. Readers wishing more information on geometric fractals are referred to Barnsley (1993).

10.4   SELF-SIMILARITY AND FUNCTIONS OF TIME

Geometrical Self-Similarity for Time Functions

On the assumption that a graph could be viewed as a geometrical object, conceptually one may consider applying the concept of self-similarity, as described above, to the graph of a time function. This approach, however, is inappropriate because the abscissa and ordinate of such a plot have different units. On the other hand, if rotations are not allowed, one could consider a type of similarity transformation in which the two axes would be scaled by different factors, s_x and s_y. In fact, objects that satisfy conditions like those for self-similarity, but with different scalings in different directions, are said to be *self-affine*. Even so, it would be an unusual time signal that could be reconstructed exactly (in a geometrical sense) from a union of transformations of itself that are scaled in both time and amplitude. Although we will be able to calculate the box-counting (and other types of) dimension for graphs of time functions, for such functions a different concept of similarity is useful.

Statistical Similarity for Time Functions

With regard to functions of time, properties such as similarity are definable if such functions are regarded as sample functions of random processes. (At this point no assumptions are being made about stationarity.) Random processes are described by their statistical properties—in particular, their probability density functions and their joint probability density functions. A random process, $X(t)$, is *statistically self-similar* if the processes $X(t)$ and $b^{-\alpha}X(bt)$ are statistically indistinguishable, where α is a constant and b is an arbitrary time scaling factor. This definition implies that $\text{Prob}\{X(t) \leq x\} = \text{Prob}\{b^{-\alpha}X(bt) \leq x\}$. Consequently, assuming zero-mean processes, since

$$E\{(b^{-\alpha}X(bt))^2\} = E\{X^2(t)\},$$

then

$$b^{-2\alpha}E\{X^2(bt)\} = E\{X^2(t)\}$$

and

$$Var\{X(bt)\} = b^{2\alpha}Var\{X(t)\}. \tag{10.6}$$

The variance of $X(t)$ depends on the properties of the random process, of course, whereas Eq. (10.6) is a consequence of statistical self-similarity. This result will be fundamental for several methods of estimating the amplitude scaling parameter, α, of a self-similar process.

Perhaps the most familiar example of a self-similar process is one-dimensional, ordinary Brownian motion, $B(t)$. In modeling Brownian motion, one assumes that a particle is subjected to frequent, microscopic forces that cause microscopic changes in position along a line—for example, the x-axis. It is assumed that these increments of displacement—call them $\Delta(t)$—have a Gaussian distribution with mean zero and that they are statistically independent. Since it is not possible to observe every microscopic displacement, any measurement of position of the particle, $B(t)$, at the smallest possible observation interval, τ, necessarily results from a summation of displacements occurring in the time interval of length τ. Call the change in $B(t)$ over one time interval $\Delta B(\tau)$. A summation of zero-mean Gaussian random variables yields another Gaussian random variable having a mean of zero and a variance that is the sum of the individual variances. In this case, each of the original "microscopic" variables, $\Delta(t)$, is assumed to have the same density function and same variance. Thus the variance of the summed variables at the end of the interval τ will be proportional to τ and we may write that $\Delta B(\tau) = \int_0^\tau \Delta(t)dt$, and $Var\{\Delta B(\tau)\} = D\tau$, where D is a constant that we will identify as the average diffusion coefficient. Since $\Delta B(\tau)$ is Gaussian, its pdf is given by

$$p(\beta, \tau) = \frac{1}{\sqrt{4\pi D\tau}}e^{-(\beta^2/4D\tau)}, \tag{10.7}$$

where β is any particular value of $\Delta B(\tau)$. Fig. 10.9(a) shows a sample function of discrete increments of Brownian motion, $\Delta B(i \cdot \tau)$ having a mean of zero and unit variance. $B(t)$ is the summation of these increments and the corresponding sample function is graphed in Fig. 10.9(b).

To investigate the scaling properties of Brownian motion, consider the situation when the smallest observation interval is 2τ and the net total displacement in this time interval, β, is the sum of displacements in two intervals of length τ. Let β_1, β_2 be the displacement in the first and second interval, respectively. Of necessity, $\beta_2 = \beta - \beta_1$. Because we assumed that the incremental position changes are independent,

$$p(\beta, 2\tau) = \int_{-\infty}^{\beta_1} \int_{-\infty}^{\beta-\beta_1} p(\beta_1, \tau) \cdot p(\beta_2, \tau)d\beta_1 d\beta_2$$

$$= \int_{-\infty}^{\infty} p(\beta_1, \tau)p(\beta - \beta_1, \tau)d\beta_1 = \frac{1}{\sqrt{4\pi D(2\tau)}}e^{-\beta^2/4D(2\tau)}. \tag{10.8}$$

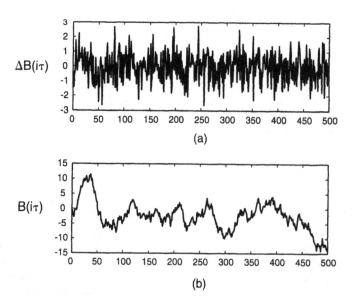

FIGURE 10.9. (a) 500 discrete increments of fractional Brownian motion (FBM). (b) The sample function of FBM whose increments are shown in (a).

The only difference from Eq. (10.7) is that the variance is twice as large. In fact, this result generalizes for an interval of any length. That is, the variance of the *increments of Brownian motion*, $\Delta B(t)$, depends on the frequency at which the increments are measured, increasing with the time between measurements. But, *for a fixed observation interval the variance is constant and independent of time.* One may simulate the example of doubling the observation interval by summing pairs of data points from Fig. 10.9(a). The resulting sequence, having half as many data points, is plotted in Fig. 10.10, along with the corresponding Brownian motion sequence. It is apparent that if one scaled Fig. 10.10(a) by $(\sqrt{2})^{-1}$, then it would "look similar" to Fig. 10.9(a). In this instance "similarity" does not mean that the two functions are identical, but that they have the same statistical properties.

The scale invariance of records of Brownian increments can be demonstrated mathematically. Making the substitutions $\tau' = b\tau$, $\beta' = b^{0.5}\beta$, the pdf may be written as

$$p(\beta', b\tau) = \frac{1}{\sqrt{4\pi Db\tau}} e^{-b(\beta)^2/4Db\tau} = b^{-0.5}p(\beta, \tau). \qquad (10.9)$$

That is, a transformation that changes the time scale by b and the length scale by $b^{0.5}$ produces a process that is equal in distribution to the original process. Equation (10.9) expresses the statistical self-similarity (or, strictly, self-affinity) of the increments of Brownian motion.

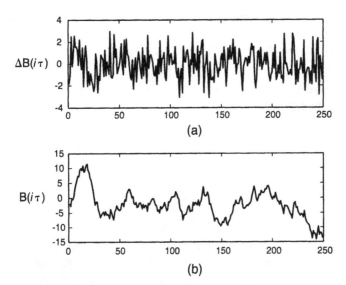

FIGURE 10.10. (a) Increments of FBM produced by summing adjacent pairs of points from Fig. 10.9(a). (b) The FBM for which the data in (a) are the increments.

The position of the particle is reflected by $B(t)$. Since $B(t)$ is essentially a summation over the interval (t_0, t) of increments having a Gaussian distribution, then the change in $B(t)$ from $B(t_0)$ is also Gaussian and, by arguments similar to those above, its probability density function is

$$p(B(t) - B(t_0)) = \frac{1}{\sqrt{4\pi D(t - t_0)}} e^{-(B(t)-B(t_0))^2/4D(t-t_0)}, \tag{10.10}$$

assuming that $t > t_0$. This function satisfies the scaling relationship

$$p(b^{0.5}[B(b(t)) - B(bt_0)]) = b^{-0.5} p(B(t) - B(t_0)). \tag{10.11}$$

Thus the Brownian motion process is also self-affine. Given these results it is easy to prove that

$$E\{B(t) - B(t_0)\} = 0 \quad \text{and}$$

$$Var\{B(t) - B(t_0)\} = 2D(t - t_0). \tag{10.12}$$

Equation (10.12) demonstrates clearly that Brownian motion is not stationary since its variance is a function of time. On the other hand the variance of the increments, $\Delta B(t)$, is constant (for a given observation interval, τ) and the process of Brownian increments is stationary.

10.5 THEORETICAL SIGNALS HAVING STATISTICAL SIMILARITY

Fractional Brownian Motion (FBM)

Fractional Brownian motion, $B_H(t)$, was introduced by Mandelbrot and Van Ness (1968) as an extension of the concept of Brownian motion. It is formally defined, for $t > 0$, by the following relations:

$$B_H(0) = b_0$$

$$B_H(t) - B_H(0) = \frac{1}{\Gamma(H + 1/2)} \left\{ \int_{-\infty}^{0} [(t - s)^{H-1/2} - (-s)^{H-1/2}]dB(s) + \int_{0}^{t} (t - s)^{H-1/2}dB(s) \right\}.$$

$$(10.13)$$

The increments of FBM, $dB(t)$, are assumed to be zero-mean, Gaussian, independent increments of ordinary Brownian motion. H is a parameter that is restricted to the range $0 < H < 1$. An equivalent mathematical form, that looks similar to a difference between two convolutions, is the following:

$$B_H(t_2) - B_H(t_1) = \frac{1}{\Gamma(H + 1/2)} \left\{ \int_{-\infty}^{t_2} (t_2 - s)^{H-1/2}dB(s) - \int_{-\infty}^{t_1} (t_1 - s)^{H-1/2}dB(s) \right\}. \quad (10.14)$$

The form of Eq. (10.14) suggests that FBM is derived from a process of weighting past values of white noise by $(t - s)^{H-1/2}$ and integrating. Note that for $H = 1/2$ FBM becomes ordinary Brownian motion. This result may be determined from Eq. (10.14), where for $H = 1/2$ the fractional exponents, $H - 1/2$, become zero and the integrals become simple integrals of the Gaussian increments.

The increments of FBM are stationary and self-similar with parameter H. That is,

$$p\{B_H(t_0 + \tau) - B_H(t_0)) = h^{-H}p\{B_H(t_0 + h\tau\} - B_H(t_0)). \quad (10.15)$$

Consequently, FBM has no intrinsic time scale. (Compare this result to geometrically self-similar objects that have no intrinsic spatial scale.) Mandelbrot and Van Ness (1969) derive the result that the variance of FBM is

$$Var\{B_H(t_0 + t\} - B_h(t_0)) \sim |t|^{2H}. \quad (10.16)$$

Therefore, like ordinary Brownian motion, fractional Brownian motion is nonstationary although it also is a zero-mean process.

The character of FBM is dependent on the value of H. For H between 0 and 1/2, a record of FBM appears to be very irregular (Fig. 10.11(a)). If $H = 1/2$, FBM becomes ordinary Brownian motion and has the appearance seen in Fig. 10.9(b). Finally, as H increases from 1/2 towards 1, then a record of FBM becomes progres-

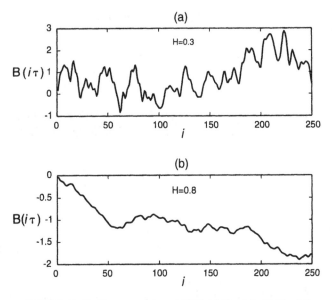

FIGURE 10.11. Two examples of FBM. (a) $H = 0.3$. (b) $H = 0.8$.

sively smoother (Fig. 10.11(b)). This changing character can be predicted on the basis of the correlation between past and future increments of FBM, which Feder (1988) derives to be

$$C(t) = 2(2^{2H-1} - 1).$$

For $H = 0.5$, this correlation vanishes, as expected, but for other values of H it is nonzero for all values of t. In particular, for $H < 0.5$ the correlation of past and future increments is negative, implying that an increasing trend in the past is more likely to be followed by a decreasing trend in the future. The opposite is true for $H > 0.5$, for which $C(t)$ is positive. These two situations are sometimes referred to as *antipersistence* and *persistence*, respectively.

These concepts may be applied to DT signals by considering the above theory for integer values of t and s. This approach thus assumes that DT FBM originates from sampling of CT FBM at a unit sampling interval. Alternatively, one may generalize to an arbitrary sampling interval, t_s. The primary effect of this alternative is to introduce a scale factor related to raising t_s to some power that is a function of H. These cases will be identified as needed below.

Of particular interest are the autocorrelation and spectral properties of FBM and of its increments. Derivation of these properties is mathematically nontrivial and here we summarize results and present heuristic arguments, referring the reader to other sources of detailed derivations. Because FBM is nonstationary, its power spectrum is not definable in the standard manner. Flandrin (1989) utilized a time-varying spectral analysis method known as the Wigner-Ville spectrum to characterize

FBM. Given a process having the autocorrelation function $r_x(t,s)$, its Wigner-Ville spectrum is

$$W_x(t, \Omega) = \int_{-\infty}^{\infty} r_x\left(t + \frac{\tau}{2}, t - \frac{\tau}{2}\right)e^{-j\Omega\tau}d\tau. \qquad (10.17)$$

After calculating the time-varying $W_x(t, \Omega)$ for FBM, Flandrin calculates the average spectrum over a time interval, T, and shows that in the limit for long intervals the average spectrum approaches

$$S_{B_H}(\Omega) = \frac{1}{|\Omega|^{2H+1}}. \qquad (10.18)$$

That is, FBM has an average $1/f$ type of spectrum.

Since ordinary Brownian motion may be properly defined as the integral of Gaussian white noise, there is the temptation to define fractional Brownian motion as the integral of a corresponding process that one might call "fractional Gaussian noise (FGN)." Since FBM is not differentiable, however, one must either formulate a smoothed, differentiable approximation to FBM or consider very small increments of FBM. Here we do the latter and define FGN as

$$B_H'(t) = \frac{B_H(t + \varepsilon) - B_H(t)}{\varepsilon} \qquad (10.19)$$

and derive the spectrum of FGN from an heuristic argument. Since FBM is approximately the integral of FGN, and for an integrator $H(\Omega) = (1/j\Omega)$, the power spectra of FBM and FGN should be related as $S_{B_H}(\Omega) = (S_{B_H'}(\Omega)/\Omega^2)$, implying that

$$S_{B_H'}(\Omega) = \frac{1}{|\Omega|^{2H-1}}. \qquad (10.20)$$

This result is consistent with the derivation of Flandrin based on a more formal mathematical argument. Note that if the variance of FGN is scaled by some parameter σ^2, then both Eq. (10.18) and (10.20) are scaled equivalently.

Because it is nonstationary, the autocorrelation function of FBM, $r_B(t,s)$, is a function of both t and s. Thus it is difficult to utilize this function for analysis of data. On the other hand, FGN is stationary and its autocorrelation function may be derived from the basic definition. Bassingthwaite and Beyer (1991) considered the case of discrete FGN, in which the FGN is viewed as small differences of FBM. For such discrete FGN, derivation of the autocorrelation between nearest neighbors— that is, $r_B[1]$—is conceptually straightforward, but the algebra is a little extensive and is omitted here. The result is

$$r_B[1] = 2^{2H-1} - 1. \qquad (10.21)$$

The derivation of the autocorrelation at lags greater than one is considerably more extensive although it again proceeds from standard concepts about independent Gaussian processes. Bassingthwaite and Beyer derive the result for discrete, unit-variance, FGN that

$$r_B[m] = 0.5[|m + 1|^{2H} - 2|m|^{2H} + |m - 1|^{2H}] \qquad (10.22)$$

This function is plotted for several values of H in Fig. 10.12, from which the long-term memory of the process is evident. Since discrete FGN is modeled here as small differences of FBM, it is possible to aggregate consecutive FGN samples—for example, add together consecutive pairs—to produce another set of differences of FBM. Note that Eq. (10.22) is independent of the degree of such aggregation so that the consistency of $\hat{r}_B[m]$ calculated from data with various degrees of aggregation provides an indication of the validity of assuming that the data are fractal.

Relating Geometrical Dimension to Statistical Self-Similarity

Although it was argued earlier that the concept of similarity dimension was inapplicable to graphs of statistically self-similar time signals, one may estimate the fractal dimension of such a record using the notion of box-counting (or simply, box) dimension. In determining box dimension one looks for a relationship between the number of boxes, $N(b)$, of size b that intersect with an object, expecting a relationship of the form $N(b) \sim b^{-D}$, where D is the dimension. For FGN, one may cover the graph with boxes of width $b\tau$ in time and length ba in amplitude. If the time span of the record is T, then one needs $T/b\tau$ segments to cover the time axis. Furthermore,

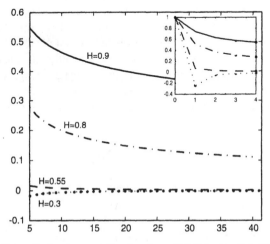

FIGURE 10.12. Theoretical autocorrelation function of FGN for $H = 0.9, 0.8, 0.55, 0.3$.

since the amplitude range is of the order $\Delta B_H(b\tau) = b^H \Delta B_H(\tau)$, one requires $b^H \Delta B_H(\tau)/ba$ boxes of height ba to cover this range. Therefore,

$$N(b) = \frac{b^H \Delta B_H(\tau)}{ba} \frac{T}{b\tau} \sim b^{H-2}.$$

But since $N(b) \sim b^{-D}$, then

$$D = 2 - H. \tag{10.23}$$

This result is intuitively sensible. We know that as H approaches 1, the signal becomes smoother and approaches a line that has a dimension of one. Conversely, as H approaches zero, the signal becomes increasingly irregular and its dimension approaches 2. The relationship of Eq. (10.23) provides a convenient means of estimating the dimension of the record of a fractal signal through estimation of H.

10.6   MEASURES OF STATISTICAL SIMILARITY FOR REAL SIGNALS

Rescaled Range Analysis

The parameter H is called the *Hurst exponent* after the Egyptian hydrologist who first noticed an empirical relationship between the range of aggregated data and the size of the aggregates for numerous natural measurements (Hurst, 1951). This method of estimating H is known as *rescaled range analysis*. It seems to be less accurate than others but the method has historical interest. Given a one-dimensional signal $x(t)$, one forms aggregates of m consecutive data points with no overlap between aggregates. Thus, if there are N total points, then one obtains N/m aggregates, $x_j[n]$, $1 \le j \le N/m$, $1 \le n \le m$. For each aggregate one calculates an accumulated sum, $v_j[n] = \sum_{k=1}^{n} x_j[k]$. Two indices are calculated for each aggregate: (1) the range, $R_j(m)$, is defined as $R_j(m) = \max(v_j[n]) - \min(v_j[n])$; (2) the standard deviation, $S(m)$, is calculated for $v_j[n]$ in the usual manner. Hurst observed for many natural variables the relationship $R(m)/S(m) \propto m^H$. Consequently, $R(m)$ and $S(m)$ are calculated for several values of m and the logarithm of their average value at each m is graphed vs. $\log(m)$. H is estimated as the slope of a regression line on this log-log plot. If a reference aggregation size, m_0, is chosen, then Hurst's equation implies that

$$\log[R(m)/S(m)] = \log[R(m_0)/S(m_0)] + H \cdot \log[m/m_0]. \tag{10.24}$$

An heuristic argument validates this empirical finding when the method is applied to FGN. Mandelbrot and Van Ness proved for FGN that the range, $R(m)$, satisfied a statistical self-similarity condition so that $R(bm)$ is equal in distribution to $b^H R(m)$. Furthermore, for a stationary process such as FGN, one expects $S(m)$ not to vary systematically with m. Therefore, Hurst's relationship is plausible for FGN.

Mandelbrot and Van Ness suggested a modification of the rescaled range analysis in which the range of each aggregate is reduced by the range of a linear regression line fit to the data of the aggregate. The accuracy of this modification, and variations on estimating the slope of the regression line, have been investigated extensively (e.g., Caccia et al., 1997).

Relative Dispersion

The relative dispersion method was suggested by Bassingthwaite (1989) as a means of estimating H from a fractional Gaussian noise signal. Although originally applied to a spatial fractal, it is equally applicable to a temporal fractal and the process of calculation is the same. Relative dispersion, RD, is defined as

$$RD(m) = SD(m)/\bar{x}_m, \tag{10.25}$$

where m indicates the size of the aggregates (or groups), $SD(m)$ is the standard deviation based on groups of size m, and the overbar indicates a mean value. The method proceeds as follows (Fig. 10.13): First, one evaluates RD for group size $m = 1$. That is, calculate the mean of the entire data set and the sample standard deviation of the data set, the latter defined as

$$SD = \frac{1}{N}\sqrt{\sum_{k=0}^{N-1}(x[k] - \bar{x}_1)^2} = \frac{1}{N}\sqrt{\sum_{k=0}^{N-1}x^2[k] - \left(\sum_{k=0}^{N-1}x[k]\right)^2}. \tag{10.26}$$

Here $\bar{x}_1 = (1/N)\sum_{k=0}^{N-1}x[n]$. For the second step aggregate adjacent data points into nonoverlapping groups of 2 (i.e., $m = 2$) and calculate their mean value. There will be $N/2$ groups. Now, *using the group mean values as data points instead of the x[n] values*, calculate SD of the group means according to Eq. (10.26). Call this $SD(2)$. Then $RD(2) = SD(2)/\bar{x}_2$, noting that $\bar{x}_2 = \bar{x}_1$. Succeeding steps repeat step 2 using $m = 4, 8, 16, \ldots$, until the number of groups is "too small." Typically one stops when

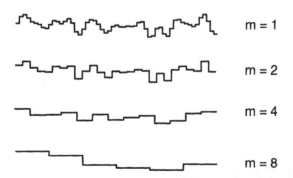

FIGURE 10.13. The first four steps in aggregating data for calculating relative dispersion.

$N/m = 4$, but note that this criterion is not absolute (Caccia et al.,1997). In addition, one also may choose values of m that are not powers of 2 (Zhang and Bruce, 2000). It will be shown below that $RD(m) \sim m^{H-1}$. Therefore, one plots $\log[(RD(m)/RD(1))]$ versus $\log(m)$ and uses linear regression to estimate the slope of this graph, from which an estimate of H is obtained as

$$H = 1 + \text{slope}. \tag{10.27}$$

In practice, since $\bar{x}_m$ is independent of m, one may alternatively evaluate the slope of a plot of

$$\log\left[\frac{SD(m)}{SD(1)}\right] \text{ vs. } \log(m).$$

To validate the proposed relationship between $RD(m)$ and m, consider the second step in which 2 original data points are averaged to produce a new data point. Let $x_1 = x[n_1]$, $x_2 = x[n_1 + 1]$ for some n_1, and let $z = \frac{1}{2}(x_1 + x_2)$. The variance of z may be found from

$$Var(z) = \tfrac{1}{4} Var(x_1) + \tfrac{1}{4} Var(x_2) + \tfrac{1}{2} Cov(x_1, x_2).$$

Now $Var(x_1) = Var(x_2) = \sigma_x^2$. The normalized covariance between adjacent samples of FGN (Eq. (10.21)) is $r_1 = 2^{2H-1} - 1$. Therefore,

$$Var(z) = \tfrac{1}{2}\sigma_x^2 + \tfrac{1}{2}\sigma_x^2 r_1 = \tfrac{1}{2}\sigma_x^2 + \tfrac{1}{2}\sigma_x^2(2^{2H-1} - 1) = \tfrac{1}{2}\sigma_x^2 2^{2H-1} = \sigma_x^2 2^{2H-2}.$$

Since $Var(z) = (RD(2))^2 \bar{x}_2^2$, and $(RD(1))^2 = \sigma_x^2 \bar{x}_1^2$, and $\bar{x}_2 = \bar{x}_1$, then

$$\frac{Var(z)}{\sigma_x^2} = \frac{(RD(2))^2}{(RD(1))^2} = 2^{2H-2}. \qquad \text{Therefore,} \qquad \frac{(RD(2))}{(RD(1))} = 2^{H-1}.$$

Consider now the aggregation of pairs of values of z to create groups of size $m = 4$. By the same argument as above,

$$\frac{(RD(4))}{(RD(2))} = 2^{H-1}. \qquad \text{Also,} \qquad \frac{RD(4)}{RD(1)} = \frac{RD(4)}{RD(2)} \cdot \frac{RD(2)}{RD(1)} = (2^{H-1})(2^{H-1}) = 4^{H-1}.$$

One may proceed in a similar fashion to show that for all levels based on pairwise averaging $(RD(m)/RD(1)) = m^{H-1}$. (Raymond and Bassingthwaite (1999) provide a more rigorous derivation of this result for arbitrary m.) From this final result one finds that $\log[RD(m)] = \log[RD(1)] + (H - 1) \log[m]$.

Example 10.3   RD analysis of a synthetic FGN signal   An example using the relative dispersion method is presented in Fig. 10.14. The signal in Fig. 10.14(a) was generated using the spectral synthesis method (SSM, see later) with $H = 0.65$.

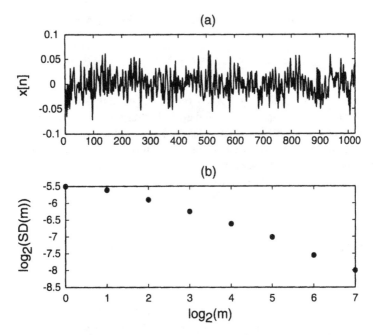

FIGURE 10.14. Relative dispersion analysis of a synthetic FGN signal having $H = 0.65$. (a) The FGN signal. (b) *log-log* plot of standard deviation versus scale. Slope of a linear regression fit to these points is -0.2692. Estimated H is 0.6308.

The slope of the log-log plot of $RD(m)$ versus m (Fig. 14(b)) was calculated using linear regression and H was estimated to be 0.6308. The difference from the specified value of H is due primarily to using a finite data record but partly to known inaccuracies in SSM.

Detrended Fluctuation Analysis (DFA)

The DFA method utilizes a scaling property of the variance of FBM, although it is usually implemented using the standard deviation. (If the signal of interest is presumed to be discrete FGN, then it is necessary to calculate a cumulative summation in order to apply the DFA method to the summed signal.) The algorithm is similar to that for relative dispersion. Given $x[n]$, $0 \le n \le N - 1$, one forms aggregates of size m as above, $x_j[k]$, numbered as $j = 1, 2, \ldots, N/m$. For each aggregate a linear regression line is fit to the data and subtracted from the data of that aggregate. If the regression line is described by $\hat{x}_j[k] = ak + b$ (where k is confined to the j-th aggregate), let the data after eliminating the linear trend be $y_j[k] = x_j[k] - \hat{x}_j[k]$. Then the standard deviation of each detrended (zero-mean) aggregate is calculated as

$$SD_j(m) = \sqrt{\frac{1}{m-1}\sum_{k=1}^{m}(y_j[k])^2}. \qquad (10.28)$$

Next determine the average SD for this m by averaging the N/m values:

$$\overline{SD}(m) = \frac{m}{N} \sum_{j=1}^{N/m} SD_j(m). \tag{10.29}$$

The final step is to plot $\log[\overline{SD}(m)]$ versus $\log[m]$ and estimate the slope of the linear portion of the graph using standard linear regression. This slope is the estimate of H. Note that it often is necessary to omit some points at both small and large m that do not fit a line well. The lack of fit occurs because there are too few points to obtain good estimates of SD for small m and too few estimates of SD to obtain a reliable average, $\overline{SD}(m)$, for large m. Raymond and Bassingthwaite (1999) formally derive the result that $\log[SD(m)] \sim H \cdot \log[m]$.

10.7 GENERATION OF SYNTHETIC FRACTAL SIGNALS

Several methods for generating synthetic fractal time series have been proposed. Some earlier methods, such as the successive random addition method, have been shown to be too inaccurate to be acceptable. One of the simpler methods, that is acceptable only if modified from its original description as discussed below, is the Spectral Synthesis Method (SSM). The basis of this method is an approximation of the $1/\Omega^\beta$ power spectrum of FBM, where $\beta = 1 + 2H$, $1 \leq \beta \leq 3$. The approximation is accomplished by generating a periodic signal having Fourier series components whose magnitudes vary with frequency as $1/\Omega^\beta$. Define a signal of length $N + 1$ as

$$x[n] = \sum_{j=1}^{N/2} \left[A_j \cos\left[\frac{2\pi j(i-1)}{N} \right] + B_j \sin\left[\frac{2\pi j(i-1)}{N} \right] \right], \quad 0 \leq i \leq N. \tag{10.30}$$

The Fourier series coefficients, A_j and B_j, are determined from the following relationships:

$$P_j = j^{-\beta/2} randn(1, 1)$$

$$Q_j = 2\pi \, rand(1, 1)$$

$$A_j = P_j \cos(Q_j), \qquad B_j = P_j \sin(Q_j),$$

where randn and rand are the Gaussian and uniform random number generators in MATLAB. The signal $x[n]$ is a (sampled) fractional Brownian motion. To obtain FGN, one calculates successive differences of $x[n]$—that is, $y[n] = x[n + 1] - x[n]$, $n = 0, 1, 2, \ldots, N - 1$.

Because the SSM approach generates a periodic signal, the autocorrelation of the generated signal does not decay appropriately for a fractal signal at long lags. This problem can be circumvented by generating a signal that is eight or more times the desired length, N, and then selecting N consecutive points beginning at a random point in the long sequence, but also avoiding data near either end of the sequence.

Two other approaches have been proposed that attempt to match the autocorrelation properties of the desired fractal signal more exactly. A method due to Lundahl et al. (1986) begins by calculating the autocorrelation values for discrete FGN according to the general result

$$r[k] = \frac{\sigma^2}{2}[|k+1|^{2H} - 2|k|^{2H} + |k-1|^{2H}], \tag{10.31}$$

where σ^2 is the desired variance of the signal to be generated, $x[n]$. One calculates as many values of $r[k]$, $0 \le k \le K$, as one wishes to match. (K may need to be quite large in order to capture the long-term correlation of FGN.) The autocorrelation matrix, R, is then decomposed into $R = L L^T$, where L is a lower triangular matrix, using Cholesky decomposition. The transformation $y = L^{-1}x$ produces a signal whose autocorrelation function is

$$E\{yy^T\} = E\{L^{-1}xx^TL^{-T}\} = L^{-1}E\{xx^T\}L^{-T} = L^{-1}RL^{-T} = (L^{-1}L)(L^TL^{-T}) = I.$$

In other words, $y[n]$ is uncorrelated with unit variance. Thus, if one generates $y[n]$, $0 \le n \le N - 1$, as unit-variance white noise, then $x[n]$ may be determined through the relationship $x = Ly$. This method produces FGN that may be summed to yield FBM. The method apparently has not been tested extensively.

Another approach to this same objective of matching the theoretical autocorrelation function of FGN has been discussed by Caccia et al. (1999). This algorithm (called fGp) has the property that the expected values of the mean and autocorrelation function of the generated signal converge to the theoretical ones. The procedure generates synthetic FGN of length N, where N is a power of 2. First one calculates the theoretical autocorrelation function values for $0 \le k \le N$ using Eq. (10.31). Then, for $M = 2N$, an autocorrelation sequence of length $M + 1$ is created in the form $\{r[0], r[1], \ldots, r[M/2 - 1], r[M/2], r[M/2 - 1], \ldots, r[1]\}$. The exact power spectrum of a process having this autocorrelation function is calculated as

$$S_p = \sum_{k=0}^{M/2} r[k]e^{-j2\pi pk/M} + \sum_{k=M/2+1}^{M-1} r[M-k]e^{-j2\pi pk/M}.$$

These spectral values should all be greater than or equal to zero. Now generate samples of Gaussian white noise having zero mean and unit variance, w_k, $0 \le k \le M - 1$. Define a sequence of randomized spectral amplitudes, V_k such that

$$V_0 = \sqrt{S_0}w_0, \qquad V_k = \sqrt{\tfrac{1}{2}S_k}(w_{2k-1} + jw_{2k}), \ 1 \le k \le M/2,$$

$$V_{M/2} = \sqrt{S_{M/2}}w_{M-1}$$

and $V_k = V^*_{M-k}$, $M/2 \le k \le M - 1$.

The final step uses the first N elements of the DFT of the sequence V_k to compute the simulated FGN time series,

$$x[n] = \frac{1}{\sqrt{M}} \sum_{k=0}^{M-1} V_k e^{-j2\pi kn/M}. \tag{10.32}$$

10.8 FRACTIONAL DIFFERENCING MODELS

Fractal Brownian motion is not the only model of long-term correlation in a signal. A fractional differencing model represents a signal as the output of a special linear system excited by a white noise process. Consider a system having the transfer function

$$H(z) = \frac{Y(z)}{X(z)} = \frac{1}{1 - z^{-1}}.$$

Its difference equation is $y[n] = y[n-1] + x[n]$. That is, the system sums the input signal, and the transfer function $(1 - z^{-1})^{-1}$ could be termed a first-order summer or a (-1)-th order differencer. This system may be generalized to be a d-th order summer by specifying its transfer function as $H(z) = (1 - z^{-1})^{-d}$, where d is not restricted to being an integer but must be in the range $-0.5 < d \le 0.5$. The impulse response of this system may be determined from a binomial expansion of $H(z)$. Thus

$$H(z) = \sum_{k=0}^{\infty} \binom{-d}{k} (-z^{-1})^k = 1 - dz^{-1} - \frac{1}{2}d(1-d)z^{-2} - \frac{1}{6}d(1-d)(2-d)z^{-3} - \cdots . \tag{10.33}$$

This expansion is valid for $|z| < 1$ and implies that

$$h[n] = (-1)^n \frac{(k+d-1)!}{k!(d-1)!} = (-1)^n \frac{d(1+d)\cdots(k-1+d)}{k!}. \tag{10.34}$$

Given a white noise input signal, $x[n] = w[n]$, the resulting output, $y[n]$, is given by

$$y[n] = h[n] * w[n] = \sum_{k=0}^{\infty} h[k]w[n-k] = \sum_{k=0}^{\infty} (-1)^k \frac{d(1+d)\cdots(k-1+d)}{k!} w[n-k]. \tag{10.35}$$

The power spectral density of $y[n]$ is $S_y(e^{j\omega}) = \sigma_w^2 |H(e^{j\omega})|^2$. To evaluate this spectrum we first express the frequency response as

$$H(e^{j\omega}) = \frac{1}{(1 - e^{-j\omega})^d} = \frac{1}{e^{-j\omega d/2}(e^{j\omega/2} - e^{-j\omega/2})^d} = \frac{(2j)^{-d}}{e^{-j\omega d/2}[\sin(\omega/2)]^d}.$$

Therefore,

$$S_y(e^{j\omega}) = \sigma_x^2 \frac{2^{-2d}}{[\sin(\omega/2)]^{2d}}. \tag{10.36}$$

Clearly, as $\omega \to 0$, $S_y(e^{j\omega}) \sim (1/\omega^{2d})$.

Consequently, this model can represent processes having long-term autocorrelation. Deriche and Tewfik (1993) have derived the exact normalized autocorrelation function of $y[n]$ to be

$$\frac{r_y[m]}{r_y[0]} = \frac{d(1+d)\cdots(m-1+d)}{(1-d)(2-d)\cdots(m-d)}. \tag{10.37}$$

For large m, Eq. (10.37) behaves asymptotically as m^{-2d-1} and exhibits the hyperbolically decaying autocorrelation characteristic of statistically self-similar processes. Examples of the normalized autocorrelation function for various values of d are presented in Fig. 10.15. One also may develop models for processes having d outside of the range $-0.5 < d \le 0.5$ by connecting a system having an integer value of d in series with one having a fractional d. For example, for an effective d of 2.2, one may cascade the system $H_1(z) = (1/(1-z^{-1})^2)$ with a system having $d = 0.2$.

Although fractionally-summed discrete white noise is a type of long-correlation process, the methods for estimating d from actual data are not well-developed and are beyond the present text—for example, see Deriche and Tewfik above. Nonetheless, these models provide a reasonable alternative as models of long-term memory processes. Furthermore, one may combine them with linear systems and analyze the

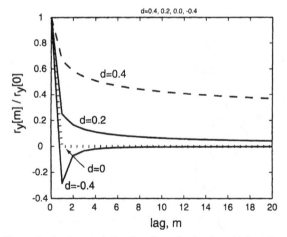

FIGURE 10.15. Theoretical autocorrelation function of a fractional integrator for d = 0.4, 0.2, 0, −0.4.

properties of the combined system in a fairly straightforward manner.

10.9 BIOMEDICAL EXAMPLES

Bassingthwaite et al. (1989) have described the spatial dispersion of measurements of local blood flow in the tissues of the heart using a fractal model. One example from a sheep heart is presented in Fig. 10.16. In this experiment a radioactive tracer was injected into the circulation and after one minute the heart was removed and separated into many small tissue pieces of approximately equal size. The amount of radioactivity in a piece of tissue was taken as an index of the blood flow to that piece, and the SD of these measurements was calculated. Then adjacent pairs of tissue pieces were pooled and the SD for the pooled data (i.e., $m = 2$) was calculated. This process of aggregating was repeated over many levels and then $RD(m)$ was plotted versus m on a log-log plot (Fig. 10.16). The fractal dimension of the local myocardial blood flow was determined from the slope of the linear regression line as $D = 2 - H = 1 - slope$. In the example of Fig. 10.16, $D = 1.297$.

Hoop et al. (1995) have utilized both rescaled range analysis and relative dispersion to characterize the fractal behavior of spontaneous phrenic nerve activity recorded from the isolated brainstem-spinal cord preparation of the neonatal rat. This activity exhibits considerable variation when averaged over 2-second intervals (Fig. 10.17(a)). Both the $\log(R/S)$ data and the $\log(RD)$ data are fit reasonably well by a straight line (Fig. 10.17(b)), the slopes of which were found to change depending on the concentration of acetylcholine (ACh) in the medium bathing the brainstem-spinal cord. From the RD plots, for which $H = 1 + slope$, H was found to be generally greater than 0.5, having a maximum near 0.85 at intermediate ACh concentrations. These findings suggest the presence of long-term correlations in

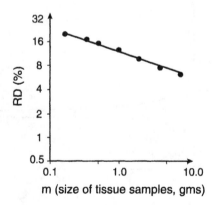

FIGURE 10.16. Relative dispersion analysis of local blood flow to a sheep heart (Bassingthwaite et al., 1989. Reprinted with permission from CIRCULATION RESEARCH).

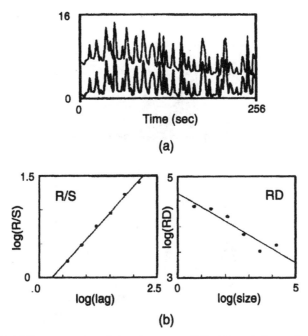

FIGURE 10.17. (a) Two examples of phrenic nerve activity (integrated over 2-second intervals) from isolated brain stem-spinal cord preparations from neonatal rats. (b) Rescaled range (*R/S*) and relative dispersion (RD) analyses of a record of phrenic nerve activity (Hoop et al., 1995. Reprinted with permission from CHAOS, 1995, American Institute of Physics).

phrenic nerve activity that are the consequence of inherent central neural mechanisms.

A direct demonstration of self-similarity in a probability density function was presented by Ivanov et al. (1996), who analyzed variations in the heart rate of human subjects. From a long ECG recording (e.g., >20,000 beats) they generated a time series of the lengths of *R–R* intervals. Using the wavelet transform they extracted a signal representing the temporal variations of this signal over time intervals of length 30–60 s. This latter signal was demodulated and a histogram constructed of the relative occurrence of various amplitudes in the demodulated signal. In Fig. 10.18(a), x represents the amplitude and $P(x)$ is the relative occurrence. Thus these curves are estimates of the probability density function of x. The curves are different for different subjects, but they become almost superimposable when they are scaled as shown in Fig.10.18(b). This scaling may be represented as $P\{ax\} = a^{-1}P\{x\}$, where $a = P_{max}^{-1}$ is the peak of an individual curve in Fig. 10.18(a). This result demonstrates that the probability density function in an individual is just a rescaled version of a fundamental *pdf*. Of course, this result itself does not imply that the derived amplitude signal for each subject is fractal because statistical self-similarity was not tested for individual signals.

A fractional Brownian motion model has been applied to the analysis of nu-

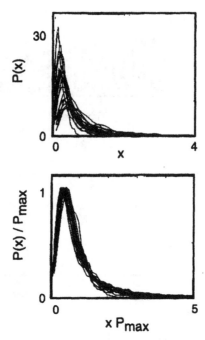

FIGURE 10.18. (a) Histograms of the frequency of occurrence of values of a demodulated heart rate signal (see text). (a) Original data. (b) After normalizing amplitude and relative occurrence scales, as shown (Ivanov et al., 1996. Reprinted with permission from NATURE (383: 323–327). MacMillan Magazines Ltd).

cleotide sequences in DNA. Peng et al. (1993) create a binary sequence $u[i]$, from a nucleotide sequence by assigning a +1 to any nucleotide position, i, that contains a pyrimidine and a –1 to any position containing a purine. Then a DNA random walk signal was defined as

$$y[l] = \sum_{i=1}^{l} u[i].$$

$y[l]$ is viewed here as a discrete-time FBM signal. Fig. 10.19(a) is an example using a rat emrbyonic skeletal myosine-heavy-chain gene (GenBank name: RATMHCG). The result of analysis of this signal using detrended fluctuation analysis is shown in Fig. 10.19(b). The regression line has a slope of 0.63, indicating the presence of long-term correlations in the nucleotide sequence that extend over a 3-decade range (i.e., about one-tenth of the sequence length).

The concept of self-similarity has been applied to prediction of the distribution and abundance of species in an ecosystem. Harte et al. (1999) postulated that the fraction of species in an area A that are also found in one-half of that area is independent of A. Specifically, if this area-independent fraction is a, after repeated

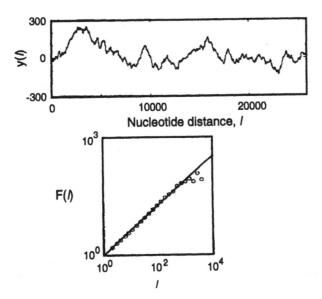

FIGURE 10.19. (a) Random walk generated by substituting +1 and −1 for pyramidine and purine bases in a DNA sequence and then cumulatively summing the resulting signal. (b) Detrended fluctuation analysis of the signal in (a) (Peng et al., 1993. Copyright 1993 by the American Physical Society).

bisection of A the number of species in the i-th bisection would be $S_i = a^i S_0$. Since the area, A_i, scales as 2^{-i}, one may write that $S_i = cA_i^z$. This latter expression is a power-law relation known in ecology as the species-area relationship (SAR). The authors then derive the probability, $P_i(n)$, that a species in area A_i contains exactly n individuals. Although unable to find a closed analytical solution, they derived the relationship $P_i(n) = 2(1 - a)P_{i+1}(n) + (1 - (2[1 - a])\Sigma_{k=1}^{n-1}P_{i+1}(n - k)P_{i+1}(k)$. Their interest focuses on $P_0(n)$, the distribution of species in the entire ecosystem. Numerical solutions to this equation were derived for various values of a and m, where the total number of individuals in the ecosystem is $S_0 = a^{-m}$ and some of these solutions for $a = 0.484$ are shown in Fig. 10.20. For large n one expects $P_0(n)$ to approach zero because the total population is limited to a^{-m}. Over several ln units, however, the function $ln[P_0(n)]$ exhibits a linear variation with $ln[n]$ that is indepedent of the initial population size, implying self-similarity in the distribution of species. This theory is supported by experimental observations. For example, the fraction of species common to two spatially-separated areas has been observed to vary as d^{-2b}, where d is the distance between the areas and b is a constant. It may be that some method such as relative dispersion could be applied successfully to estimating the possible fractal nature of species distributions from actual data.

There are numerous other examples of applying fractal and other scaling law models to biomedical data. Many of these are in the field of neuroscience.

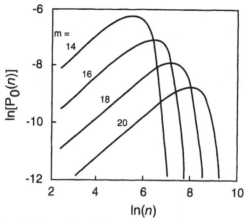

FIGURE 10.20. *log–log* plot of predicted relative fraction of total species, P_0, having n individuals in an eocsystem having a total of 0.484^{-m} individuals (Harte et al., 1999. Reprinted with permission from SCIENCE (284: 334–335). Copyright 1999, American Association for the Advancement of Science).

10.10 SUMMARY

This chapter has introduced the concepts of scaling and long-term memory. Many biomedical signals exhibit autocorrelation that decays more slowly with increasing lag than predicted by low-order ARMA models. Other types of models for such signals are often based on the concepts of scaling properties and self-similarity. These properties of geometrical objects were studied both because the concepts may be applied to the geometry of biological structures and because the graphical record of a one-dimensional signal may be characterized as a planar geometrical object. A similarity transform is a transformation of an object that may include linear scaling, translation, and rotation. An object is geometrically self-similar if it can be recreated through the union of similarity transformations applied to the object. The similarity dimension of a self-similar object is a measure of the relationship between the number of self-similar pieces comprising the object and the measurement precision, as the precision increases. Self-similar objects having non-integer similarity dimension are called fractal objects. Another measure of dimension is the box-counting dimension. This measure expresses the relationship between the number of "boxes" needed to contain the object and the size of the boxes. The strict definition of a fractal object is based on its Hausdorf-Besicovitch dimension, but it is common to assume that an object is fractal if any measure of its dimension is non-integer; however, it is more practically useful to utilize the dimension as a measure of the space-filling ability of the object and as an indication of possible self-similar scaling over some range of sizes.

One may apply the concept of geometrical dimension to a graph of a time function and interpret the dimension as a measure of roughness of the graph. Since time and amplitude have different units and may be scaled in different proportions, their graphs may be self-affine rather than self-similar.

A more general characterization for time functions is the concept of statistical self-similarity. A random process $X(t)$ is statistically self-similar if it is statistically indistinguishable from the scaled process $b^{-a}X(bt)$. Brownian motion, which is a nonstationary process derived as the summation of uncorrelated random perturbations, is statistically self-similar. Fractional Brownian motion (FBM) is an extension of Brownian motion that may be considered as a fractional integral of zero-mean, independent, Gaussian increments. FBM is characterized by a scaling exponent, H, known as the Hurst coefficient. The *increments* of FBM, known as fractional Gaussian noise (FGN), are statistically self-similar with parameter H. The character of FBM depends on H. For $H < 0.5$, FBM is very irregular. For $H > 0.5$, FBM becomes progressively smoother, and when $H = 0.5$, FBM becomes ordinary Brownian motion. FBM and FGN both possess autocorrelation functions that decay relatively slowly with increasing lag. Both have a $1/f$ type of spectrum. The geometrical dimension of a record of FGN is related to H, specifically $D = 2 - H$.

The methods for analyzing statistical self-similarity in a signal generally are based on scaling relationships of statistical measures such as the standard deviation, SD. Rescaled range analysis evaluates the ratio of the range of summed FGN over a time interval to its standard deviation. The *log* of this ratio is expected to increase in proportion to *log* of the interval length, with proportionality constant H. Relative dispersion, RD, is used to analyze FGN. RD is the ratio of SD to the mean, where both are evaluated over intervals of length m. $\log[RD(m)]$ is expected to increase linearly with $\log[m]$, with the slope of this line being $H - 1$. A method for analysis of FBM is detrended fluctuation analysis. This method evaluates the standard deviation of detrended segments of FBM as a function of the segment length, m. On a log-log plot, one anticipates a linear relationship having a slope of H.

There are two acceptable approaches to generating synthetic fractal signals having a specified value of H. The Spectral Synthesis Method (SSM) generates a Fourier series whose spectral amplitudes follow the spectral properties of FBM—that is, $1/\Omega^\beta$, $\beta = 2H + 1$. This method requires one to generate a much longer signal than desired, from which a segment is extracted that avoids both ends of the long sequence. A method due to Lundahl et al., and the fGp method both attempt to match the autocorrelation function of FGN. Currently the fGp method is probably the most precise method for generating a synthetic fractal signal.

Fractional differencing models also exhibit long-term correlations and are closely related to the fractional integration interpretation of FBM. They are based on binomial expansions of the function $(1 - z^{-1})^{-d}$. These models have infinite impulse responses and the methods for estimating d from experimental data are not well-developed.

The concepts presented in this chapter are finding increasing utility in biomedical applications. Numerous examples have appeared in the cardiovascular and neuroscience fields and others are appearing in such diverse areas as DNA sequencing, biochemical properties of membranes, and ecology.

EXERCISES

10.1   Calculate the box counting dimension for the middle-thirds Cantor set.

10.2   The even-fifths Cantor set is constructed by iterating as for the middle-thirds Cantor set, but at each iteration each line segment is divided into five sections of equal length and the second and fourth segments are removed. Determine the similarity dimension of the even-fifths Cantor set.

10.3   Calculate the box-counting dimension for the even-fifths Cantor set. Is it the same as the similarity dimension?

10.4   The Koch snowflake is constructed similarly to the Koch curve, but the starting object is an equilateral triangle. Sketch the results of the first three steps in the iteration process and determine the similarity dimension of the Koch snowflake.

10.5   The Menger sponge is constructed as follows: Given a solid cube, subdivide each face of the cube into nine equal squares. For three perpendicular faces of the original cube, cut a square hole straight through the cube by cutting along the edges of the small square at the center of the face. Next, for each of the eight remaining squares on all of the faces, divide the square into nine equal, smaller squares. Again, cut a hole straight through the original cube at the location of every central small square. The object that results from repeating this process indefinitely is known as the *Menger sponge*. Sketch the first two iterations of this procedure and determine the similarity dimension of the Menger sponge.

10.6   Consider continuous-time, ordinary Brownian motion (OBM) as the integral of a white-noise process. That is, OBM is the output of a system with frequency response $H(\Omega) = (1/j\Omega)$ when its input is a sample function of white noise. Using this model of OBM, calculate its theoretical power spectrum. Is it a $1/f$ spectrum? Also calculate its theoretical autocorrelation function, $r_{OBM}(\tau_1, \tau_2)$.

10.7   Consider discrete-time ordinary Brownian motion as the summation of discrete white noise. That is, DT OBM is the output of the system $H(z) = 1/(1 - z^{-1})$ when its input is a sample function of DT white noise. Calculate its theoretical power spectrum. Under what conditions is this spectrum of the 1/f type? Also calculate its theoretical autocorrelation function, $r_{OBM}[m_1, m_2]$.

10.8   Construction of the Peano curve begins by replacing a line segment with nine line segments as shown in Fig. 10.21. Each new line segment is then replaced in a similar fashion. Sketch the object resulting from step two of this process and determine the similarity dimension of the Peano curve.

10.9   Generate 1024 points of Gaussian white noise and estimate H. Repeat this process 10 times and calculate the mean and standard deviation of the estimates. Based on these results, if you analyzed a data set comprising 1024 points and found an estimate $\hat{H} = 0.55$, with what degree of confidence could you conclude that the data differ from Gaussian white noise?

10.10   Use the RD method to analyze the heart rate data in hrs.mat. Over what range of m do the relative dispersion values scale according to expectations? Based on your estimate of H, are these data distinguishable from Gaussian white noise?

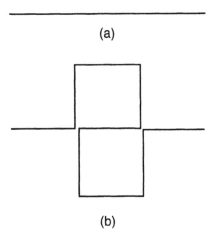

(a)

(b)

FIGURE 10.21. The first step (b) in the generation of the Peano curve from an initial straight line segment (a).

Calculate the autocorrelation function for these data. What feature do you observe in this function in addition to the hyperbolic decay that is characterisic of FGN?

10.11   Use the RD method to analyze the data file sway.mat that records the center of pressure in the antero-posterior direction of a human subject standing on a postural test system. Do the relative dispersion values scale according to expectations? Is there any range of m for which the expected scaling relationship is observed?

NONLINEAR MODELS OF SIGNALS

11.1 INTRODUCTORY EXERCISE

Figure 11.1 is a model of a biochemical reaction in which an allosteric enzyme, E, reacts with a substrate, S, to produce a product, P. Additionally the product activates the enzyme as indicated by the upper arrow. Finally, P is recycled into S by a second reaction. V is the rate of inflow, or injection rate, of S and k_s is the effective rate constant for removal of product P. This simple system has been used as a model for glycolysis in yeast and muscle (Goldbeter and Moran, 1988). This model is implemented in the Simulink file enzyme1.m which calculates the normalized concentrations of S and P, symbolized as α and γ, respectively.

Figure 11.2 presents a 4000-second record of α (i.e., 1000 points sampled at 0.25 Hz) from enzyme1. Note that this signal is cyclic but not truly periodic. Is this behavior typical or has the simulation not reached a steady state? We may answer this question by running the simulation for a longer time. (Change the maximum time in the 'Simulation/Parameters' menu. Note that the storage arrays in the 'To Workspace' blocks save only the last 1000 points of the simulation run.) If the simulation is run for a very long time (e.g., 16,000 s), the last 4000 points of the time series for α look qualitatively like the data in Fig. 11.2. This signal has the interesting property that it "looks" to be deterministic but it never reaches a steady state. On the other hand, a random process model does not seem appropriate for these data since the simulation model contains no random process. To gain more insight into the system generating this signal, the simulation may be tested with known inputs.

The Simulink model enzyme2.m is a duplicate of enzyme1.m which permits specifying various combinations of steps and sinusoids as the injection rate, V. In this model V comprises a constant level, $V0$, plus a sine wave with amplitude VA. The frequency of the sine wave can be specified via the 'Sine Generator' block labelled "Substrate rate, AC". More information and instructions can be accessed by double-clicking the "Information" and "Instructions" blocks in the Simulink block diagram.

First we test the system for homogeneity. Let $V0 = 0$ and set the frequency of the sine wave to be 0.0259—that is, its period is 242.3 seconds. Set VA to 1.0. After an

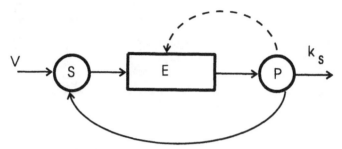

FIGURE 11.1. Model of enzyme reaction which is activated by its reaction product. S = substrate, E = enzyme, P = product, V = rate of injection of substrate. k_s = effective rate of disappearance of P. (From Goldbeter and Moran, 1988. Reprinted with permission from EUROPEAN BIOPHYSICS JOURNAL. Copyright 1988, Springer Verlag.) See the Simulink model `enzyme2.m` for additional information.

initial transient, the output α stabilizes as an ocillation with peak-to-peak amplitude approximately 80 (Fig. 11.3a). Now repeat for VA = 1.5. The output amplitude in this case is approximately 120 (Fig. 11.3b), which seems to satisfy the criterion of homogeneity. If the reader repeats this test with VA = 2.0, however, it should be apparent that this system is not linear.

Another common test of a system is to observe its responses to step inputs. Fig. 11.4 shows the responses of `enzyme2` to step inputs of three different amplitudes. When V = 0.04 $u(t)$, the output looks like the response of a first-order (or overdamped second-order) lowpass system. On the other hand, an input V = 0.06 $u(t)$ yields an output that has almost the same steady-state value as the previous step input, but the initial transient portion of the response is faster than before. Finally, in response to V = 0.08 $u(t)$ the output stabilizes as an oscillation. It is clear that this system is not linear.

When the injection rate, V, comprises a nonzero step and a sine wave of nonzero amplitude, the system exhibits the property that the very nature of the output de-

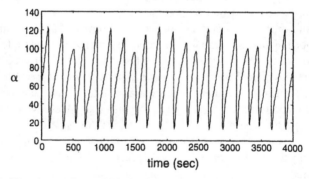

FIGURE 11.2. "Steady-state" normalized substrate concentration, α, for the model of Fig. 11.1, as implemented in the Simulink model `enzyme1.m`.

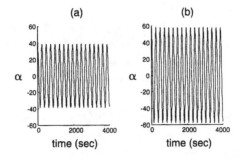

FIGURE 11.3. Steady-state responses of the enzyme reaction model (enzyme2.m) to a sinusoidal rate of delivery of substrate. Period of sinusoid is 242.3 seconds. Amplitude equals 1.0 (a) and 1.5 (b).

pends on interactions between the components of V. In Fig. 11.5 are shown the steady-state responses of α when the size of the step component is 0.607 and $VA =$ 0, 0.607, and 0.20. The frequency of the sine component is 0.0250 rad/sec (period equals 251.3 s). As seen above, the system oscillates in response to a nonzero step by itself. When the input is a large sinusoidal input added to the original step, the system oscillates at the driving frequency. Finally, for an intermediate amplitude of the sinusoidal input component, the system exhibits irregular cyclic behavior that is not periodic (Fig. 11.5, bottom).

The responses discussed above are commonly observed behaviors (in a qualitative sense) of nonlinear systems with internal feedback. The bottom response of Fig.

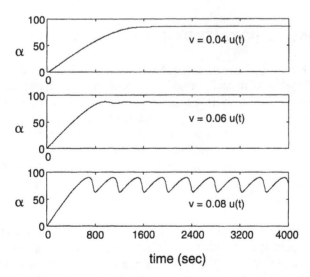

FIGURE 11.4. Responses of the enzyme reaction model to three step inputs of different amplitudes.

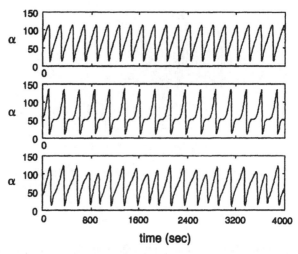

FIGURE 11.5. Steady-state responses of the enzyme reaction model to substrate input rate comprising (top) a constant level (0.607), and a constant level (0.607) plus sinusoidal inputs having periods of 242.3 sec (middle) or 251.3 sec (bottom).

11.5 is a chaotic behavior which will be discussed more fully shortly. We were able to obtain insights about this nonlinear system because we could manipulate the input. An extremely important question in biomedical applications is whether one can draw similar conclusions if one only can observe the spontaneous output signal. The remainder of this chapter will define and discuss signals generated by nonlinear systems and consider this important question.

11.2   NONLINEAR SYSTEMS AND SIGNALS: BASIC CONCEPTS

A *nonlinear system* is, by definition, any system that does not satisfy the criteria for a linear system (i.e., additive and homogeneous). For the present purposes, a *nonlinear signal* is a signal that could not have been produced by a single-input, single-output linear system except by exciting a linear system with an input which is a linearly transformed version of the original nonlinear signal. In other words, the only way to obtain the nonlinear signal as the output of a linear system is to excite the system with an input that is linearly related to the nonlinear signal. Nonlinear signals must be produced by nonlinear systems, *but under many circumstances nonlinear systems produce signals that could be generated by a linear system.*

An example is presented in Fig. 11.6, which shows a linear system in series with a unity-gain system with hard saturation limits. For small amplitude inputs having a mean level of zero, this system appears overall to be linear. Of course, for sufficiently large inputs saturation will be reached and the nonlinear nature will be obvious. Often biomedical systems exhibit saturation that is not as abrupt as that

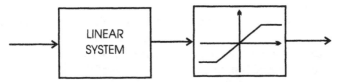

FIGURE 11.6. A simple nonlinear system consisting of a linear system in series with a saturating linear gain.

shown in the figure. Rather, the slope of the input–output response decreases progressively over a range of the input amplitude. This type of "soft saturation" often causes the apparent gain to decrease gradually as the amplitude of the input signal increases.

Because nonlinear signals must originate in nonlinear systems, it is instructive to begin by examining some basic properties of common nonlinear systems. *Unless otherwise specified, these concepts all refer to the long-term, steady-state behavior of the system after any initial transient behavior has dissipated.* A fundamental characteristic of a nonlinear system is whether it is autonomous or non-autonomous. An *autonomous system* is capable of producing a nonzero steady-state output signal in the absence of an explicit input signal (perhaps because the generator of the apparent input signal is an intrinsic part of the system). For example, the system comprising the heart and those aspects of the nervous system responsible for autonomic control of the heart could be considered an autonomous system. If a system requires an explicit input signal to produce a nonzero steady-state output, it is *non-autonomous*. The control system for the heart described above could be considered non-autonomous if one were studying respiratory sinus arrhythmia, since the signals which effect respiratory modulation of heart rate originate outside of this control system. Another fundamental measure of a nonlinear system is the number of state variables which are necessary to describe its behaviors under all possible circumstances. As with linear systems, we will define this number as the *order* of the nonlinear system and equate order with the number of (linear or nonlinear) first-order differential equations needed to describe the system. Note that this definition implies that a static nonlinearity, such as that in the righthand block of Fig. 11.6, has an order of zero. In fact the concept of order is not as useful for nonlinear systems as for linear systems because the *qualitative* behavior can be extremely sensitive to small changes in system parameters. (The only analogous situation for a linear system occurs when a feedback system switches from producing an exponentially damped oscillation that eventually decays to zero to producing a sustained oscillation at some critical value of feedback gain.) Because of the aforementioned sensitivity, the *apparent* order of a nonlinear system can vary with the values of its parameters. Consequently it is often the custom to apply measures of the complexity of the system output instead of measures of the complexity of its differential-equation structure (or difference-equation structure for discrete-time systems). Before expanding on these issues we should explore some simple examples of nonlinear systems.

Example 11.1   The model `enzyme2.m` is an example of a non-autonomous non-linear system because it produces only a zero steady-state output when it has no input, even when its initial conditions are nonzero. It has already been shown that the qualitative output depends on both the type and size of the input, V. If one considered the source of substrate, V, to be part of the system, then this system would be an autonomous system.

Example 11.2   The van der Pol system   A famous autonomous nonlinear system is the van der Pol oscillator, described by the equation

$$\ddot{x} + \mu(x^2 - 1)\dot{x} + x = 0.$$

This system is similar to a linear second-order system with nonlinear damping. That is, for positive μ the damping term is positive if $|x| > 1$ but it is negative if $|x| < 1$. In other words, if x becomes sufficiently small, then damping becomes negative and the amplitude of x subsequently increases.

The van der Pol oscillator can be written as two coupled first-order equations:

$$\dot{x} = y,$$

$$\dot{y} = -\mu(x^2 - 1)y - x.$$

This system is implemented in the `Simulink` model `vdp.m` that is distributed with MATLAB. It is instructive to run this simulation for different values of μ and observe the behavior of x after any initial transient effects have disappeared. A convenient way to test whether the output is truly periodic is to construct a phase plot of $\dot{x}$ (i.e., y) versus x for the steady-state behavior, as shown in Fig. 11.7. Each cycle exactly reproduces every other cycle on this graph, demonstrating that the behavior is periodic. A periodic oscillation is also called a *limit cycle*.

A fixed point of a CT system is any state of the CT system for which all of its derivatives are zero. Thus, when a system is at a fixed point, it remains there in the absence of any disturbances and all of its variables assume constant values. An n-th order system may be represented by n first-order differential equations in the form

$$\begin{pmatrix} \dot{x}_1 \\ \dot{x}_2 \\ \vdots \\ \dot{x}_n \end{pmatrix} = \begin{pmatrix} f_1(x_1, \ldots, x_n, t) \\ f_2(x_1, \ldots, x_n, t) \\ \vdots \\ f_n(x_1, \ldots, x_n, t) \end{pmatrix}, \tag{11.1}$$

or more compactly, as $\dot{\underline{x}} = f(\underline{x}, t)$, where the underscores indicate vector quantities. The criterion for a fixed point, $\underline{x}_p$, is that

$$\dot{\underline{x}} = f(\underline{x}_p, t) = 0. \tag{11.2}$$

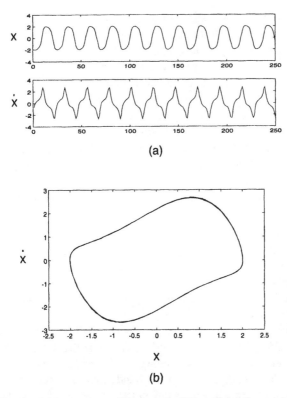

(a)

(b)

FIGURE 11.7. Steady-state responses of the van der Pol oscillator for $\mu = 1$. (a) x and dx/dt vs. time; (b) phase plot of dx/dt vs. x. Note the closed path (limit cycle) characteristic of a periodic signal.

A system may have more than one fixed point. The fixed point of the van der Pol system is easy to calculate by setting its two equations equal to zero. Thus

$$\dot{\underline{x}} = \begin{pmatrix} \dot{x} \\ \dot{y} \end{pmatrix} = \begin{pmatrix} y \\ -\mu(x^2 - 1)y - x \end{pmatrix} = 0.$$

The solution of these equations is $(x, y) = (0, 0)$. (In general, nonlinear systems have other fixed points besides the origin.)

A DT system of the form $\underline{x}_{n+1} = f(\underline{x}_n)$ is also known as an *iterated map*. An iterated map has a fixed point at $\underline{x} = \underline{x}_p$ when $\underline{x}_p = f(\underline{x}_p)$. For example, the fixed points of the logistic map,

$$x_{n+1} = \mu x_n(1 - x_n)$$

depend on the value of μ. When $0 < \mu < 1$, $x_p = 0$, but when $\mu = 1.25$, the equation $x_p = 1.25x_p(1 - x_p)$ has two solutions, $x_p = 0$ or 0.20, and the system has two fixed points.

By definition, if a system state equals a fixed point of the system, then the system will remain in that state forever. Fixed points, however, can influence system dynamics in other ways. The most important factor is whether the fixed point attracts the system variables or repels them. For an iterated map, a fixed point x_p is *attracting* if there exists a small neighborhood of diameter $\in$ such that for every point for which $\|x - x_p\| < \in$, then $\lim_{n \to \infty} f^{(n)}(x) = x_p$. Here the double vertical bars indicate some measure of distance and the superscript (n) indicates the n-th iteration of the function f. For an attracting fixed point, x_p, the set of all points for which $\lim_{n \to \infty} f^{(n)}(x) = x_p$, is known as the *basin of attraction* of the fixed point. Similarly, the fixed point is *repelling* if for some $\in$ and for every point such that

$$\|x - x_p\| < \in, \qquad \text{then } \|f(x) - x_p\| > \|x - x_p\|, \qquad \text{except for } x = x_p.$$

A simple criterion often can reveal whether a fixed point of a one-dimensional iterated map is attracting or repelling. Within the $\in$ neighborhood of x_p we may consider that $f(x) \simeq f(x_p)$ and that $x_{n+1} \simeq x_p + f'(x_p)(x_n - x_p)$. Clearly, to have $x_{n+1} \to x_p$ (i.e., x_p is an attracting fixed point), it is necessary that $|f'(x_p)| < 1$. Furthermore, for $|f'(x_p)| > 1$, the fixed point is repelling. If $|f'(x_p)| = 1$, one cannot directly conclude the nature of the fixed point.

Consider, for example, the logistic map with $\mu = 1.25$. Since $f'(x) = \mu - 2\mu x$, then $|f'(0)| = 1.25$, $|f'(0.20)| = 0.75$ and we conclude that the fixed point at zero is repelling and that at $x_p = 1.25$ is attracting. It is important to note that these properties are defined in a neighborhood of the fixed point; therefore, it is quite possible (indeed usual) that there will exist some points, x, for which their iterations do not converge to an attracting fixed point. In the present example, $x = 0$ is one such point because the iteration of zero is always zero. In n-dimensional iterated systems, with $n > 1$, there may be several attracting and several repelling fixed points. Each attracting fixed point possesses its own basin of attraction and these basins do not overlap. The basins of attraction, however, may be interdigitated so extensively that the boundaries between basins are fractal. Furthermore, in such systems fixed points may be attracting when approached from one direction and repelling when approached from another direction. This topic will be continued later when characteristic multipliers and Lyapunov exponents are discussed.

The basic definitions of attracting and repelling fixed points for CT systems parallel those for DT systems except that one must address the forward solution of the differential equation. For the system $\dot{x} = f(x)$ in the vicinity of an equilibrium point, x_p, letting $\delta x = x - x_p$, one may approximate $f(x_p + \delta x) \approx f(x_p) + D_x f(x)|_{x=x_p} \delta x$, where $D_x f(x)$ is a matrix of partial derivatives known as the *Jacobian matrix*,

$$D_x f(x) \triangleq J = \begin{pmatrix} \dfrac{\partial f_1}{\partial x_1} & \cdots & \dfrac{\partial f_1}{\partial x_n} \\ \vdots & \ddots & \vdots \\ \dfrac{\partial f_n}{\partial x_1} & \cdots & \dfrac{\partial f_n}{\partial x_n} \end{pmatrix}. \tag{11.3}$$

Since $(d/dt)(x_p + \delta x) = 0 + \dot{\delta x}$, the linearized equation of the system about the fixed point is

$$\dot{\delta x} = D_x f(x_p) \, \delta x. \tag{11.4}$$

For a first-order system, $D_x f(x_p) = a$, where a is a constant. Thus the solution to Eq. (11.4) is $\delta x(t) = ce^{at}$, where c is a constant. Clearly, if x is to converge to x_p, then $\delta x(t)$ must converge to zero, implying that $a < 0$. The extension of this topic to systems of dimension greater than one will be encountered later during the discussion of Lyapunov exponents.

Example 11.3 The Rossler system The van der Pol oscillator has a limited repertoire of behaviors. The Rossler system is a third-order system and its qualitative behavior changes remarkably as its parameter values are varied. It is described by the following equations:

$$\dot{x} = -y - z;$$
$$\dot{y} = x + ay;$$
$$\dot{z} = b + z(x - c).$$

Three-dimensional phase plots for this system (after initial transients have dissipated) are presented in Fig. 11.8 for four sets of parameters: $a = b = 0.2$, and $c = 2.5$, 3.5, 4.0, and 5.0. As the value of c is increased, the threshold for positive feedback in the third equation is raised. For $c = 2.5$ the system produces a pure oscillation. When c is increased to 3.5, the output is still a pure oscillation but the period of the oscillation has doubled. (This change is referred to as a *period-doubling bifurcation*.) The greater complexity of the oscillation is evident from the figure. But this increased complexity is merely the beginning, for when $c = 4.0$, there is another period doubling and when $c = 5.0$, the trajectory of the system in three-dimensional phase space wanders throughout a confined subspace but never crosses or duplicates itself! The signals from which these graphs were created are in the file `ross-data.m` and are named $x25$, $x35$, $x4$, and $x5$, respectively (with a similar naming convention for the y and z variables). By examining $x5$ (the x vs. t variable for $c = 5.0$) one may notice that the cyclic variations all seem to be different. One might rush to judgement and conclude that this behavior represents an oscillator with added random noise, or that it originates from a system that has not yet reached a steady state of behavior, but both conclusions would be incorrect. There is no noise in this simulation and the "steady state" is a state of continual variations, referred to as chaos. No noise-free linear system can produce this type of irregular behavior. Can biomedical systems behave thusly? There is strong evidence that some can, but since it is not possible to observe real behavior that has no noise, it is difficult to prove absolutely that a real signal originates from a nonlinear system behaving in a manner similar to this. We will examine chaotic behavior and its analysis later in

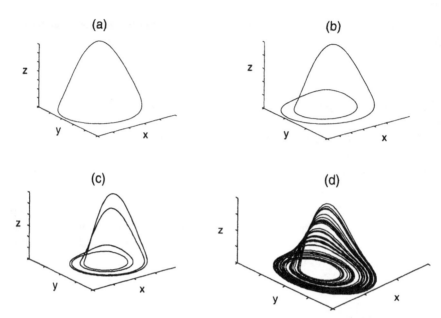

FIGURE 11.8. 3D phase plots for steady-state responses of the Rossler system. In all cases $a = b = 0.2$. The parameter c has the following values: (a) 2.5, (b) 3.5, (c) 4.0, (d) 5.0.

this chapter. To appreciate this phenomenon, the reader should use the `Simulink` model `rossler.m` and reproduce the graphs of Fig. 11.8, then examine the behaviors for intermediate values of c.

Example 11.4   A DT nonlinear system   As a final example we will consider the discrete-time system named the "logistic map," described by the difference equation

$$x_{n+1} = \mu x_n(1 - x_n), \qquad x_n \in [0, 1].$$

(Although this system will be considered to be a discrete-time system, the index "n" represents an iteration of the equation.) This system has only one parameter, μ, but the behavior of x_n as n is iterated changes dramatically with μ. Consider the case $\mu = 2.0$ shown in Fig. 11.9a. The system has an initial condition x_0 equal to 0.10. A simple way to assess the behavior of the system is to plot the logistic equation $x_{n+1} = 2.0x_n(1 - x_n)$ and the line $x_{n+1} = x_n$ on the same axes. Starting from $x_n = x_0 = 0.10$, one determines the value of x_1 from the graph of the logistic equation, as shown by the vertical line from 0.10 on the abscissa. Then moving horizontally to the identity line places one at the abscissa point $x_n = x_1$. From there move to the logistic equation curve to find x_2, then repeat this process for increasing values of "n." As the figure indicates, eventually x_n settles on the value 0.50. That is, for this value of μ the steady-state behavior is a constant that is a *fixed point* of the system.

Figure 11.9(b) shows the steady-state behavior of the logistic map for $\mu = 3.3$. x_n exhibits a *period-2 oscillation* during which it repeatedly cycles between the values

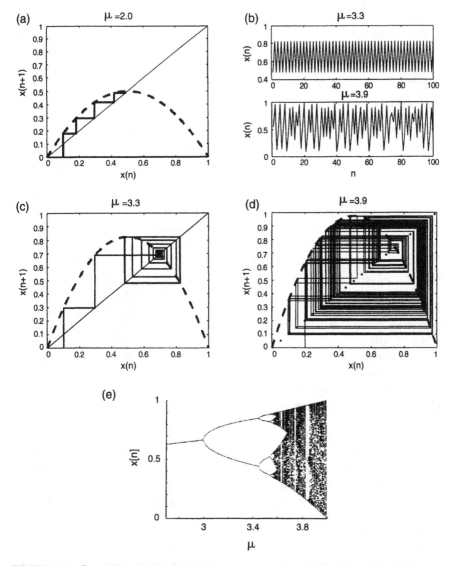

FIGURE 11.9. Examples of nonlinear discrete-time (iterated) systems. (a) Evolution of the output, $x[n]$, of the logistic map for $\mu = 2.0$. Solid diagonal line: identity line. Dashed curve: plot of logistic equation. Solid stair-step line: evolution of $x[n]$, starting at $x[0] = 0.10$. Vertical segments of stair-step end at logistic equation and indicate the next value—that is, $x[n + 1]$—of the signal. Horizontal segments move to the identity line and indicate where to place the value of $x[n + 1]$ on the $x[n]$ axis in order to initiate the next iteration. After only a few iterations the system in this example converges to a constant steady-state value of 0.500. This value represents a *fixed point* of the system and it is found where the logistic equation crosses the identity line. (b) steady-state $x[n]$ vs. n for the logistic equation for the cases $\mu = 3.3$ (top) and 3.9 (bottom). (c) Map of the evolution of $x[n]$ for $\mu = 3.3$ and $x[0] = 0.10$. After almost converging to the fixed point near 0.7, the system moves away from this point until a period-2 cycle develops, with $x[n]$ cycling approximately between 0.50 and 0.80. (d) Map of the evolution of $x[n]$ for $\mu = 3.9$ and $x[0] = 0.20$. This value of μ with this initial condition leads to chaotic behavior. (e) Bifurcation diagram for the logistic map for μ ranging from 2.5 to 4.0. The ordinate graphs all of the values assumed by $x[n]$ in the steady state for each value of μ on the abscissa.

445

0.48 and 0.83. On the graph of the logistic equation and the identity line (Fig. 11.9c), this behavior appears as the largest closed rectangle as it is continually retraced.

A small increase in μ from 3.3 to 3.9 transforms x_n into a non-fixed, non-oscillatory behavior that appears similar to random noise (Fig. 11.9(b,d)). Li and Yorke (1975) proved that this behavior never converged to a fixed point or a periodic oscillation and termed the behavior "chaos." *This behavior is completely deterministic!*

A summary of the effect of μ on the behavior of the logistic map is presented in the *bifurcation diagram* of Fig. 11.9(e). For each value of the parameter on the abscissa, all of the steady-state values assumed by x_n are plotted on the ordinate. Thus, for μ less than approximately 3.0, x_n converges to a fixed point whose value changes with μ. Near 3.0 the behavior transforms into a period-2 oscillation and the two values of x_n in this 2-cycle are graphed. When μ is around 3.45 the graph has four values on the ordinate, indicating that the behaviour has transformed from a period-2 to a period-4 oscillation, and around 3.5 it transforms to a period-8 oscillation. Each such sudden transformation of the steady-state behavior as a parameter value is changed is called a *bifurcation*. (In this case these are called *pitchfork* bifurcations because of their shape on this diagram.) Furthermore these bifurcations are a specific type known as *period-doubling bifurcations* for the obvious reason that the period of the oscillation doubles at the bifurcation. As μ continues to increase above 3.5, the bifurcations get closer together until, for certain values of the parameter, the behavior becomes aperiodic or chaotic. Note that the extreme complexity of this bifurcation diagram is not fully revealed at the scale of Fig. 11.9(e). If the regions of density of this plot were enlarged, one would observe more regions of periodic and aperiodic behavior. Also note that the sequence of period-n oscillations with $n = 1, 2, 4, 8, \ldots$, are not the only oscillations present. For example, around $\mu = 3.8282$ there is a period-3 oscillation that is the beginning of a sequence of period-doubling bifurcations which culminates in chaotic behavior at $\mu \approx 3.857$. These latter bifurcations are examples of a different type of period-doubling bifurcation, a *tangent bifurcation*. Tangent bifurcations have different dynamic properties than pitchfork bifurcations. With tangent bifurcations the behavior can almost seem to approach a fixed point for a while before it suddenly becomes very irregular again. Such behavior is termed *intermittency*.

The m-file `quadmap.m` generates bifurcation diagrams for quadratic maps after the user specifies the starting and ending values of μ, the number of equally-spaced values of μ to investigate (100 is a good number), and the initial condition (Avoid $x_0 = 0$.). The reader may use this m-file to examine the bifurcation map near $\mu \approx 3.8$ in finer detail.

11.3 POINCARÉ SECTIONS AND RETURN MAPS

The dynamics of a periodic CT system may be represented via an iterated mapping that describes the evolution of the system every T seconds, where T is the period of the system. This process may be visualized by placing a plane of section in the phase space (Fig. 11.10(a)) so that its position represents the same point within each

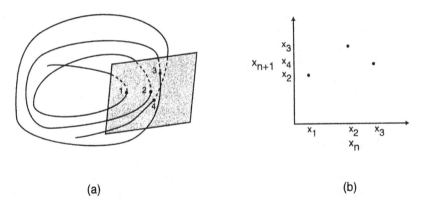

(a)   **(b)**

FIGURE 11.10. (a) A Poincaré plane showing the intersections of the plane and the trajectory of a system. (b) A first-return map of the x-axis values of the intersection points in (a).

cycle of one of the variables of the system. This plane is called a *Poincaré section.* One then constructs a map from the points at which the trajectory "pierces" the plane. Consider the point #1 in Fig. 11.10(a). After T seconds this point is mapped to point #2 by the system dynamics, and point #2 is mapped to point #3, and so on. Clearly the properties of this mapping are related intimately to the dynamics of the system. For example, if the system behavior is strictly periodic with period T, then every point should map back to itself, and the Poincaré section shows a single point. In this case, if the iterated mapping function is $x_{n+1} = F(x_n)$, then the single point is an equilibrium (i.e., fixed) point of F. If the system has period $2T$, then point #2 will differ from point #1, but point #3 should be the same as point #1. Furthermore, if there were a point #4, it would have to be the same as point #2. Thus, the Poincaré section would show two points. Note that the dimension of the Poincaré section is one less than the dimension of the phase space.

If the system is not periodic, we may augment it to produce a periodic variable by defining $\theta = 2\pi(t/T)$, where T is some desired sampling interval, and then place the Poincaré plane so that it corresponds to $\theta = 2\pi$. In effect, we are sampling the system at $f_s = 1/T$ and displaying the samples in the phase space. If T has no relationship to the system properties, however, this procedure may not be insightful.

If the points of the Poincaré section are labeled as x_n, then one may graph x_{i+1} vs. x_i to generate a *first-return map.* Often one will extract one variable x_j from the vector x_i and construct a two-dimensional graph of $x_j[i + 1]$ vs. $x_j[i]$ (Fig. 11.10(b)).

11.4 CHAOS

Some Preliminary Concepts

Before defining chaos we must recognize another important property of nonlinear systems: for many nonlinear systems their steady-state behavior depends on their initial conditions. Unlike stable linear systems, for which the only possible long-term behaviors are (1) to decay to zero, or (2) to converge to a scaled form of the steady-state behavior of the input signal, nonlinear systems can exhibit multiple possible steady-state behaviors. The behavior that is reached is determined by the transient response. If the transient behavior due to initial conditions causes the system to get "close to" a fixed point or a periodic oscillation, and if the fixed point or oscillation is stable and "attracts" the system output (as observed in Figs. 11.9(a, c)), then that behavior becomes the steady-state response. An example of a system having multiple steady-state behaviors is seen in Fig. 11.9(c), where an initial condition, x_0, of 0 or 1 produces a steady-state x_n equal to 0 instead of the period-2 oscillation discussed previously. Obviously the logistic map is only a simple example and other nonlinear systems will exhibit more complex regions of initial conditions that converge to different steady-state behaviors. In many cases the border between two such regions is not an absolute line but is fractal in structure. It is important to recognize that this effect of initial conditions represents a general property of nonlinear systems. If, for example, a nonlinear system is subject to a transient disturbance that drives it away from the region of initial conditions which converge to its current steady-state behavior, it may then converge to a *different* steady state *even though its parameter values do not change*. Carnavier et al. (1993) have demonstrated such a situation in a model of the electrochemical behavior of a neuron (Fig. 11.11).

For some nonlinear systems the initial conditions which lead to aperiodic steady-state behavior might be closely intermingled with those which lead to other steady-state behaviors. Consequently, very small changes in initial conditions can produce very large differences in eventual behaviors. But there is another critical facet to aperiodic behavior that is related to initial conditions. Two nearby initial conditions that both lead to aperiodic steady-state behaviors can (and generally will) result in

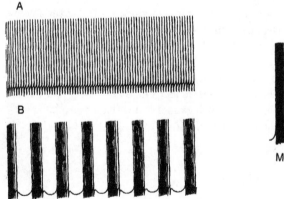

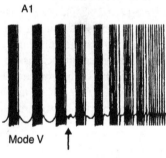

FIGURE 11.11. Depending on the initial conditions the neuron model of Carnavier et al. (1993) converges to different steady-state behaviors for the same parameter values, as shown in A and B. On the right is an example of a transition from one steady-state behavior to another induced by a transient stimulus applied to this model at the arrow.

quantitatively different aperiodic behaviors. Such is not true for an attracting periodic behavior, for example, since all initial conditions leading to this periodic behavior result in the same periodic cycle. In the latter example, even if the two behaviors eventually have different phases, the "distance" between them (in the sense of their mean square difference over a cycle, for example), remains constant. But for two different aperiodic behaviors from the same system, measures of their "distance apart" increase continuously (on average) due to the random-like nature of aperiodic signals. These latter systems are said to exhibit *sensitive dependence on initial conditions* if they exhibit such a dependence over a nontrivial range of initial conditions—i.e,, not only at a few isolated points. It is difficult to visualize how two bounded behaviors (i.e., signals) can move progressively further apart for all time, and in truth the separation distance will plateau at a value comparable to the distance between the upper and lower bounds of the signals. That is, sensitive dependence is a local property of system dynamics.

Example 11.5 Sensitive dependence on initial conditions   The Rossler system of equations was solved (using `rossler.m`) starting from two nearby initial conditions, $x_0 = 0.5000$ and $x_0 = 0.4630$, with $y_0 = z_0 = 0$. The time history of $x(t)$ is shown for several seconds for the two different values of x_0 in Fig. 11.12(a). Notice that these trajectories initially remain close together, but eventually begin to separate and then assume qualitatively similar aperiodic behaviors that are quantitatively dissimilar at most points in time.

Example 11.6 A biological example   It is difficult to "restart" biological systems from neighboring initial conditions but one can test for sensitive dependence by recognizing that the system state at any time (for an autonomous system) is indeed the initial condition for all of its future behavior. Therefore one attempts to detect two spontaneous system states that are nearly the same but separated in time, then compares the evolution of the system behavior after it leaves those two states. An example is shown in Fig. 11.12(b), in which the respiratory rhythm is modelled as a planar oscillator. Because the behavior is cyclic, one can readily find similar system states from different cycles. In this example the authors selected a group of points near the top of the graph and followed their evolutions for one average respiratory cycle duration, marking their locations at this later time by the isolated symbols. They showed that after such a short time the neighboring states were no longer neighbors, but were spread out over about 30% of the cycle (Fig. 11.12(b), right). After severing the vagus nerves (which provide afferent feedback from the lungs to the respiratory neural pattern generator in the brainstem), there was much less dispersion in a comparable time.

   An important limitation to this approach is the question of how many variables must be measured to fully describe the behavior of the system. In the present example there might be an important third variable, implying that the system behavior should be assessed in three-dimensional space. If so, the projection of two different 3D behaviors onto a plane might render them more or less similar than they are.

(a)

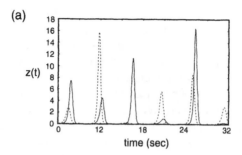

(b)

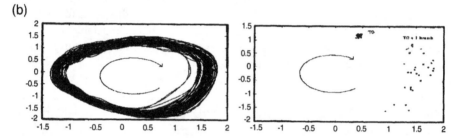

FIGURE 11.12. (a) The divergence of two trajectories of the variable $z(t)$ from the Rossler system ($a = b = 0.2$, $c = 5$) with two nearby initial conditions, $z(0)$ (0.500 and 0.463). Note the substantial differences in the responses of $z(t)$ due to the slight difference in initial conditions. (b) Divergence of nearby trajectories in the phase plot of breathing of an anesthetized rat. Airflow is plotted versus lung volume. Several points that are close to one another (but from different breaths) near the top of the graph were chosen and appear as a clump of points in the right graph. Then each was followed forward in time for the same amount of time (one average breath duration) and their final positions marked. Note that this initially close group now is spread throughout about 30% of the phase plot. From Sammon et al. (1991).

Defining Chaos

In nonlinear dynamics it is customary to consider any regularly repeating behavior (even a constant-valued signal, which is seen to be repeating the constant value) to be *periodic*. Given this definition of periodic behavior, an autonomous system is *chaotic* if, for a fixed set of parameter values: (1) it exhibits steady-state *aperiodic behavior* for many initial conditions; (2) it displays *sensitive dependence on initial conditions*; and (3) these behaviors do not depend on inputs from any random process. For our purposes we may assume that a system that exhibits sensitive dependence on initial conditions will also display aperiodic steady-state behavior for some initial conditions. (Some authors require only sensitive dependence on initial conditions as the criterion for chaos, whereas others require both conditions. In their famous paper Li and Yorke (1975) demonstrate a connection between the two criteria for iterative discrete systems.) Formally one should worry about the density of the initial conditions in the phase space of the system which lead to aperiodic behavior. Also, initial conditions may lead to periodic behavior quickly or only in the limit of $t \to \infty$, and the initial condition is said to be a *periodic point* or an *eventually periodic point*, respectively. Thus a more precise definition is that an autonomous system with no random inputs is chaotic if there exists a sufficiently dense set of initial conditions spread throughout a nontrivial subset of the phase space that are not periodic or eventually periodic, and the system exhibits sensitive dependence on initial conditions.

There is one condition for sensitive dependence on initial conditions that is necessary and sufficient: the system has a positive Lyapunov exponent (defined and discussed later in this chapter). Lyapunov exponents measure the average rate of divergence of behaviors associated with nearby initial conditions, and in practice the demonstration of a positive Lyapunov exponent for a system having a phase space of at least three dimensions is considered proof of potential chaotic behavior of the system.

It is occasionally possible to apply the preceding definitions to systems describable by exact differential or difference equations and to confirm that the system meets the criteria for the existence of chaotic behavior. (Let it be stressed again that a potentially chaotic system might not exhibit chaotic behavior during an arbitrary observation, depending on the associated initial condition.) Our concern is with signal processing, and we shall define a *chaotic signal* as a signal produced by an autonomous chaotic system in response to an initial condition that leads to aperiodic behavior. The typical application involves observing a signal from a (putative) nonlinear system and attempting to classify it as chaotic or non-chaotic and to determine some quantitative measure of the "degree" of chaos. Unfortunately these goals are especially difficult to accomplish for real data at the present time. Below we discuss some common methods and auxiliary tests which, in combination, can increase the likelihood of a correct classification and provide an approximate measure of the degree of chaos.

A simplified visualization of chaotic behavior is possible if one accepts that periodic behaviors may be *repelling* as well as attracting. That is, if the system behavior approaches closely enough to a repelling periodic behavior, then it is pushed away from this fixed point or limit cycle. An analogy is a ball balanced atop a pyramid. If one can achieve such an exact balance, the ball will remain in place indefinitely (in the absence of other external forces). But this system state is a repelling one and if the balance is not exact or is disturbed even slightly, the ball will always roll away. Some authors have suggested that one might visualize chaotic behavior as the result of a system having many repelling periodic behaviors (which is a characteristic of many chaotic systems); the ultimate system behavior evolves as a wandering that repeatedly is repelled by one or another repelling periodicity. Thus the behavior will never become periodic but for short periods of time it might assume an appearance similar to one of the periodicities which, nevertheless, will finally repel the behavior.

11.5 MEASURES OF NONLINEAR SIGNALS AND SYSTEMS

It was demonstrated above that the behavior of nonlinear systems may be altered dramatically by small changes in their parameters or in their initial conditions. This dependence of qualitative behavior on parameter values and initial conditions complicates the assessment of properties of nonlinear systems from observations of their output signals. In a practical application one might utilize the signal processing techniques to be discussed below and conclude that "the signal cannot be proven to be nonlinear." This result could mean that the signal arises from a linear system or that it originates from a nonlinear system whose parameter values or most recent initial conditions cause its behavior to be indistinguishable from that of a linear system. To proceed further, one would need either more information than that obtained from simply observing the signal or more sophisticated methods of signal analysis. Such issues are current research topics.

As a result of the potential sensitivity of nonlinear systems to changes in parameter values and to differences in initial conditions, any analysis of a signal from such a system represents the system at that moment only. Given this caveat one may then ask whether there exist measures of signals which will uniquely distinguish nonlinear signals from linear ones. While such measures do exist in theory, applying them to real-world signals has been problematic. First we will present some of these measures, then discuss their limitations relative to signals from the laboratory or the field. All of these measures are based on evaluating some index of the complexity of the signal. All assume that the signal represents a steady-state behavior and are not valid when applied to transient responses.

Dimension

There are a variety of definitions of the dimension of a signal and we shall utilize a basic definition which expresses the space-filling nature of the signal. Assume that an *autonomous* system has P, and only P, internal nodes from which P variables can be measured, and that the time history of any variable of the system can be completely specified from knowledge of these P variables. When the system is in a steady-state behavior, construct a P-dimensional space in which each axis represents the instantaneous value of one of the P variables and graph these variables. This graph is called a *phase plot* of the system and the P-dimensional space is called its *phase space*. (See, e.g., Fig. 11.8 for phase plots of the three-dimensional Rossler system.) In general, some of the P variables might be functions of the remaining variables and exhibit no independent behavior, and therefore the effective number of variables required to describe the system might be less than P but that is inconsequential to our argument. Follow the system behavior and trace out its phase plot for all of the steady-state behavior up to $t \to \infty$. The resulting shape that is drawn in P-dimensional space may be simple or very complex. The concept of dimension refers to the spatial dimension of this resulting shape, not to the dimension of the space in which it is drawn. For example, a line has one dimension whether it is drawn on a two-dimensional plane or in three-dimensional space. The objective therefore is to determine the spatial dimension of the phase plot rather than that of the phase space.

Strictly speaking, the phase plot described above exists on another mathematical object called the *attractor* of the behavior and it is the dimension of the attractor that is required. However, if the system behavior is observed for a sufficiently long time, it is almost certain that it will approach arbitrarily close to every point on the attractor and the distinction between the phase plot and the attractor becomes unimportant. The reader can anticipate one problem in applying the concept of dimension to actual data: real data has finite length and one cannot be certain that a phase plot based on a finite amount of data actually encompasses the entire attractor.

Consider a stable autonomous *linear* system that is not oscillatory. Its steady-state output and all of its internal variables must be zero. Therefore, its P-dimensional phase plot is just a point—the origin. The dimension of this phase plot is the dimension of a point—that is, zero. If the linear system produces a sustained oscillation, the graph is a closed nonintersecting curve in P-space that is retraced on every cycle (Fig. 11.13). In this case the variables are constrained so that the phase plot follows a curved line, which (like all simple lines) has a dimension of one even though it exists in P-dimensional space.

Zero and one are the only possible dimensions for the phase plot of a linear system, with one exception: If the system comprises a linearly combined set of independent oscillators having incomensurate frequencies, then the dimension of the phase space equals the number of oscillators. This result is easy to visualize (Fig. 11.14). Consider a system of two variables which produces a sinusoidal oscillation. Its phase space is two-dimensional (a plane) and its phase plot is a circle. Now add another oscillator whose two variables represent (1) the distance of the origin of this plane from an origin in three-dimensional space and (2) the angle of a line from the origin of the plane to the origin of the three-dimensional space. For the second oscillator the distance from the origin (i.e., the amplitude of the oscillation) is constant and the angle increases linearly with time but retraces itself every cycle. Therefore, the motion of the phase plot of the first oscillator (the circle) sweeps out a torus in three-space. This torus is the phase plot of a system having two independent oscillators with incommensurate frequencies. Its dimension is two because it comprises a closed, two-dimensional surface. Similar arguments apply for an increased number of oscillators.

Consider an autonomous nonlinear system for which the steady-state behavior can be a constant, an oscillation, a combination of oscillations, or an aperiodic wan-

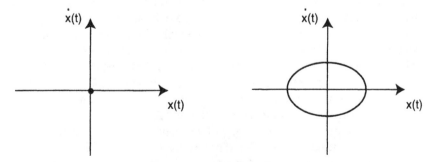

FIGURE 11.13. The two possible steady-state phase plots for a 2-variable linear system: a fixed point at the origin, or a periodic cycle.

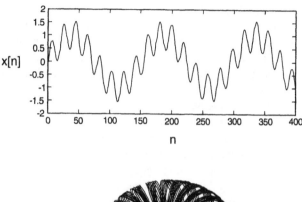

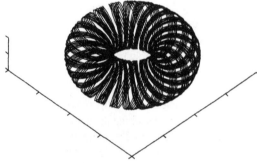

FIGURE 11.14. Top: $x[n]$ vs. n for a signal consisting of a summation of two incommensurate sine waves. Bottom: A three-dimensional phase plot of the signal above based on representing each sinusoid as a rotating vector. The horizontal plane is a described by polar coordinates describing the angle of the low-frequency sinusoid and the magnitude of the projection of the vector sum of the two sine waves. The vertical axis is the vertical component of this same vector sum. Note that the phase plot traces over the surface of a torus, whose structure is clearly evident even with only a few thousand data points.

dering. In the case of a combination of oscillations, the oscillations need not be sinusoidal, so the phase plot of an individual oscillator need not be a circle. In fact it can have any closed, non-intersecting shape. When there are two independent oscillations, one can proceed as we did for the linear oscillators but the resulting surface need not be a regular shape like a sphere or ellipsoid. It can, in theory, be any closed two-dimensional structure in three-space. The graph of the Rossler system in Fig. 11.8b is an example of the phase plot of a single nonlinear oscillator.

It is difficult to picture any sort of regular structure for the phase plot of a chaotic nonlinear system whose variables exhibit aperiodic wandering, but the phase plot should be bounded in P-space. The graph of Fig. 11.8(d) shows a phase plot for a chaotic system. At this scale it is not apparent that the structure in this graph has thickness also. In general, the phase plot of a chaotic system can appear to have a simple or complex structure but in fact if one examines the structure closely, it is always complex and frequently fractal in structure. The complete phase plot structure of a system lies on a geometric structure known as the *attractor* of the system. (Remember that a nonlinear system may converge to a different steady-state behavior, and therefore to a different attractor, for different initial conditions.) If the attractor has a fractal structure, it is called a *strange attractor*.

What is the dimension of the phase plot of a chaotic system? The answer is that it may be any positive number less than or equal to P, including non-integer values (because a fractal structure does not fill up the space in which it resides). The aperiodic wandering of a chaotic system in phase space must be bounded but the steady-state behavior cannot cross itself in phase space (otherwise two trajectories would land on the same point and merge, implying periodic behavior); therefore, when trajectories approach and pass one another, they must be separated slightly. Consequently the phase plot (strictly speaking, the attractor of the phase plot) has many thin layers, like a fractal structure. In Chapter 10 it was shown that the spatial dimension of a fractal could be non-integer, so it is intuitive that the dimension of an attractor could be non-integer although fractals in more than two dimensions were not considered there. An example of an attractor known to be fractal (from the Rossler system) is shown in Fig. 11.15.

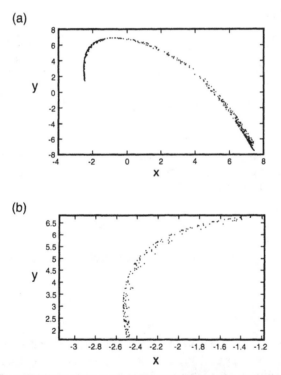

FIGURE 11.15. Cross-sections of the three-dimensional phase plot of the Rossler system for $a = b = 0.2$, $c = 5$. (a) Each point represents the data pair (x,y) when the trajectory pierces the plane $z = 0.20$. Note the apparent "thickness" of this curve, especially to the right of the graph. (b) Magnification of the upper left section of the top graph. Again note the apparent "thickness" of the curve. It has been shown that these curves are cross-sections of a three-dimensional fractal structure.

The concept of dimension, therefore, provides one means by which nonlinear signals might be distinguished from linear signals because attractors of linear signals cannot have non-integer dimension. Furthermore, the value of the dimension is also an index of the complexity of a signal in the sense that it indicates the complexity of the attractor of the phase plot of the system. On the other hand, one must keep in mind that it is not absolutely necessary that the attractor for a chaotic nonlinear system be a *strange* attractor with fractal structure; therefore, a non-integer dimension is a useful indicator but not a necessary condition for a chaotic system. There are also stochastic systems that exhibit non-integer dimension. With these caveats, we will develop two common methods of estimating dimension—the box-counting (or simply, box) dimension and the correlation dimension. First we discuss another theoretical measure of the complexity of the behavior of a system, its Lyapunov exponents.

11.6 CHARACTERISTIC MULTIPLIERS AND LYAPUNOV EXPONENTS

Characteristic Multipliers

Characteristic multipliers of nonlinear systems are closely related to eigenvalues of linear systems. Consider the n-dimensional DT system $\underline{x}_{k+1} = F(\underline{x}_k)$ having a fixed point $\underline{x}_p$. In the vicinity of the fixed point this system may be linearized as

$$\delta \underline{x}_{k+1} = D_{\underline{x}}F(\underline{x}_k)|_{\underline{x}=\underline{x}_p}\delta \underline{x}_k, \tag{11.5}$$

where $D_x F(\underline{x}_k)|_{\underline{x}=\underline{x}_p}$ is the Jacobian matrix evaluated at the fixed point, and $\delta \underline{x}_k = \underline{x}_k - \underline{x}_p$. Let $m_1, m_1, \ldots, m_n$ be the eignenvalues of the Jacobian matrix, with corresponding eigenvectors $\underline{h}_1, \underline{h}_2, \ldots, \underline{h}_n$. The trajectory of the linearized system resulting from an initial deviation from the fixed point, $\delta \underline{x}_0$, is

$$\underline{x}_k = \underline{x}_p + (D_{\underline{x}}F(\underline{x}_k)|_{\underline{x}=\underline{x}_p})^k \delta \underline{x}_0. \tag{11.6}$$

Assuming the eigenvalues are distinct, the trajectory may be expressed as

$$\underline{x}_k = \underline{x}_p + \sum_{j=1}^{n} c_j m_j^k \underline{h}_j. \tag{11.7}$$

Thus the eigenvalues of the Jacobian matrix at the fixed point determine the stability of the fixed point. If $|m_j| < 1 \; \forall j$, then the fixed point is attracting, independent of the initial orientation of the vector $\delta \underline{x}_0$. Since the eigenvectors are orthogonal, clearly one may find that the fixed point is attracting for initial points along some directions and repelling for other directions, depending on the magnitudes of the corresponding eigenvalues. If the DT system is a Poincaré section of the phase space of a CT system and the fixed point represents a periodic behavior of the CT system, then the eigenvalues, m_j, $j = 1, \ldots, n$, are known as the *characteristic multipliers* of the periodic behavior.

Turning now to a CT system, $\dot{x} = f(x, t)$, that has an equilibrium point (i.e., a fixed point) at some x_p, we may consider linearizing this system in the neighborhood of the equilibrium point. Thus, with $J(x) = D_x f(x, t)$ being the Jacobian matrix of the CT system, the linearized system is

$$\dot{\delta x} = J(x_p)\delta x. \tag{11.8}$$

The response of the linearized system to an initial displacement from the equilibrium point, δx_0, is

$$\delta x(t) = e^{J(x_p)t}\delta x_0 = \Phi(t, x_p)\delta x_0, \tag{11.9}$$

where $\Phi(t, x_p) = e^{J(x_p)t}$ is the state transition matrix of the linearized system. If l_1, l_2, $\ldots$, l_n are the eigenvalues of $J(x_p)$ and $h_1, \ldots, h_n$ are the corresponding eigenvectors, the above solution may be expressed as

$$\delta x(t) = \sum_{j=1}^{n} c_j e^{l_j t} h_j. \tag{11.10}$$

An equilibrium point may be considered a periodic solution and we may sample the system at some interval T and generate a Poincaré map having a fixed point corresponding to x_p and a set of characteristic multipliers, $\{m_i\}$. The trajectory resulting from the displacement, δx_0, will "pass through" the Poincaré section every T seconds and its value at these times will be determined by the state transition matrix evaluated at $t = T$. Therefore,

$$\delta x(kT) = \Phi(T, x_p)^k \delta x_0. \tag{11.11}$$

Comparing Eq. (11.11) with Eq. (11.6), we conclude that the eigenvalues of the state transition matrix of the linearized CT system must equal the eigenvalues of the Jacobian of the iterated mapping that describes the Poincaré map. Since the eigenvalues of $\Phi(t, x_p)$ are

$$L_i(t) = e^{l_i t}, \, i = 1, \ldots, n, \tag{11.12}$$

then

$$m_i = e^{l_i T}. \tag{11.13}$$

Note that because $|m_i| < 1$ only if $\mathrm{Re}\{l_i\} < 0$, the characteristic multipliers and the eigenvalues of the state transition matrix provide equivalent information about the stability of the equilibrium point.

Lyapunov Exponents

Lyapunov exponents generalize the concept of eigenvalues and permit assessment of the stability of behaviors other than those that are periodic. For a CT system and any initial condition, x_0, the *Lyapunov exponents* of x_0 are defined as

$$\lambda_i \triangleq \lim_{t \to \infty} \left[\frac{1}{t} \ln|L_i(t)| \right], \tag{11.14}$$

where $L_i(t)$ is defined by Eq. (11.12). From the preceding discussion, the Lyapunov exponents at an equilibrium point are

$$\lambda_i = \lim_{t \to \infty} \left[\frac{1}{t} \ln|e^{l_i t}| \right] = \lim_{t \to \infty} \left[\frac{1}{t} \ln|e^{\text{Re}\{l_i\}t} e^{j\text{Im}\{l_i\}t}| \right] = \text{Re}\{l_i\}.$$

In other words, in this case the Lyapnuov exponents are simply the real parts of the eigenvalues of the linearized system at the equilibrium point.

Although the definition of the Lyapunov exponents refers to an initial condition, the requirement that the limit be evaluated as $t \to \infty$ means that all points in the basin of attraction of the equilibrium point will have the same Lyapunov exponents.

For DT systems the Lyapunov exponents (or Lyapunov numbers) are defined as

$$m_i = \lim_{k \to \infty} |m_i(k)|^{1/k}. \tag{11.15}$$

The Lyapunov exponents provide a direct measure of the sensitivity of the system to initial conditions. For a CT system, if any Lyapunov exponent is greater than zero, then an initial difference between two points, δx_0, will grow as time progresses.

Example 11.7   Eigenvalues and stability of a two-dimensional system   The van der Pol oscillator may be expressed as a function of a real-valued, positive parameter ε in the form

$$\frac{d^2 x}{dt^2} + \varepsilon(x^2 - 1)\frac{dx}{dt} + x = 0.$$

By letting $y = (dx/dt)$, this equation may be rewritten in standard form as $\dot{x} = f(x, t)$, where

$$x = \begin{pmatrix} x \\ y \end{pmatrix}, f(x, t) = \begin{pmatrix} y \\ -\varepsilon x^2 y - x + \varepsilon y \end{pmatrix}.$$

As stated previously, the point $(x, y) = (0, 0)$ is the only equilibrium point of this system. To evaluate stability at this point, we calculate the Jacobian matrix. Thus

$$J = \begin{pmatrix} 0 & 1 \\ -2\varepsilon xy - 1 & \varepsilon - \varepsilon x^2 \end{pmatrix},$$

and at the equilibrium point

$$J = \begin{pmatrix} 0 & 1 \\ -1 & \varepsilon \end{pmatrix}.$$

Consequently, its eigenvalues are

$$\lambda_{1,2} = \frac{\varepsilon}{2} \pm \frac{\varepsilon}{2}\sqrt{2 - \frac{4}{\varepsilon^2}}.$$

For any positive real ε, if $|\varepsilon| \geq 2$, then both eigenvalues are greater than zero. If $|\varepsilon| < 2$, then the eigenvalues are complex but their real parts are positive. Therefore the origin is a repelling equilibrium point. Note that, by the Poincaré-Bendixon Theorem (see Gulick (1992)) a second-order CT system cannot exhibit chaotic behavior.

Example 11.8 Lyapunov exponents of a three-dimensional system   Consider the Lorenz system with the specific parameter values given below:

$$\dot{x} = -10x + 10y;$$
$$\dot{y} = 30x - y - xz;$$
$$\dot{z} = xy - 3z.$$

The Jacobian matrix of this system is

$$J = \begin{pmatrix} -10 & 10 & 0 \\ 30 - z & -1 & -x \\ y & x & -3 \end{pmatrix}.$$

Setting the derivatives equal to zero, one may use an iterative method to identify three equilibrium points for this sytem. These points are:

$$\begin{pmatrix} x_1 \\ y_1 \\ z_1 \end{pmatrix} = \begin{pmatrix} 0 \\ 0 \\ 0 \end{pmatrix}, \quad \begin{pmatrix} x_2 \\ y_2 \\ z_2 \end{pmatrix} = \begin{pmatrix} 9.327 \\ 9.327 \\ 29.0 \end{pmatrix}, \quad \begin{pmatrix} x_3 \\ y_3 \\ z_3 \end{pmatrix} = \begin{pmatrix} -9.327 \\ -9.327 \\ 29.0 \end{pmatrix}.$$

At each equilibrium point we may calculate the eigenvalues of the Jacobian. For example, at the origin,

$$J(0, 0, 0) = \begin{pmatrix} -10 & 10 & 0 \\ 30 & -1 & 0 \\ 0 & 0 & -3 \end{pmatrix}.$$

The eigenvalues of this matrix are

$$\begin{pmatrix} l_1 \\ l_2 \\ l_3 \end{pmatrix} = \begin{pmatrix} -23.396 \\ 12.396 \\ -3.000 \end{pmatrix}$$

and there is one positive Lyapunov exponent, 12.396. This system has sensitive dependence on initial conditions near the origin. The reader may check the Lyapunov exponents at the other equilibrium points.

11.7 ESTIMATING THE DIMENSION OF REAL DATA

Embeddings

In the preceding discussions the descriptions of nonlinear systems were based substantially on analyses of their behaviors viewed in phase space although it was assumed implicitly in the examples that one could observe chaotic behavior in every variable of a system that exhibits sensitive dependence on initial conditions and aperiodic behaviors. Thus one might ask whether it is possible to discern whether a system is chaotic given only observations of one of its signals. Intuitively one would expect that a sufficiently long observation of one signal from a chaotic system ought to contain all of the necessary information about the system. Fortunately this assumption has been demonstrated to be correct. For an autonomous system Takens (see Parker and Chua (1987)) provided a proof that certain measures of the attractor in its true phase space could be determined from an infinite observation of one signal from the system. These invariant measures include dimension and Lyapunov exponents. To permit calculation of the invariant measures it is necessary to construct an attractor in a space of sufficiently high dimension using the observed signal, a process known as *embedding*. Construction of the embedded attractor proceeds as follows: Let a signal $x(t)$ be sampled at the rate $f_s = 1/\tau_s$ such that N data points $x[n]$, $0 \leq n \leq N - 1$, are obtained. To construct an embedding in R^m, create the m-dimensional data vectors x_j, $0 \leq j \leq N - k \cdot m$, which are defined by

$$x_j = (x[j], x[j + k], x[j + 2k], \quad \ldots, \quad x[j + (m - 1)k])^T, \quad (11.16)$$

where k is a delay that is chosen as described below. Now, using each x_j to define a point in R^m, construct a pseudo-phase plot.

This method of constructing an embedding is known as the *method of delays* and the arguments of $x[n]$ used to determine each x_j are called *delay coordinates*. The objective in the choice of lag k is to ensure that the various coordinates of each x_j vector convey independent information. There are several methods for this selection but there is not general agreement regarding the most appropriate. A common approach is to calculate the autocorrelation function of $x[n]$, $r_x [l]$, and select k equal to the first value of l at which the autocorrelation function goes to zero. With this choice of k, on average the coordinates are uncorrelated if not truly independent. Another method proposed by Fraser and Swinney (1986) determines the lag k that minimizes the average mutual infomation, I, between $x[j]$ and $x[j + k]$, where I is defined as

$$I = \text{average}\left[\frac{\text{joint probability of } x[i], x[i + k]}{(\text{probability of } x[i])(\text{probability of } x[i + k])}\right].$$

The choice of the correct embedding dimension, m, also is nontrivial and again there is not general agreement regarding the best method. The most common approach utilizes the concept of invariant measures. That is, if one is attempting to estimate an invariant measure such as dimension, then the estimate should be independent of the size of the embedding space if the size of the embedding space is sufficently large. Takens showed that invariant measures for an attractor of dimension m_0 could be determined if the dimension of the embedding space was greater than $2m_0$. Therefore one can determine the correct embedding dimension by evaluating a measure such as the attractor dimension for increasing values of m until the estimate fails to increase further with m, taking the lowest value of m at which this occurs to be the proper embedding dimension.

Another useful method is the method of false nearest neighbors. This method is based on the concept that trajectories in the phase space do not cross unless the dimension chosen for the phase space is too low to allow complete separation of trajectories. On the other hand, points on the true attractor may be close together and they should remain close together at any embedding dimension. Therefore points on an attractor which are close together in R^m but move significantly further apart in R^{m+1} are likely to have been "false nearest neighbors" in R^m. Typically one uses the Euclidian distance to identify the nearest neighbor of each embedded data point in R^m and again in R^{m+1}. A data point has a false nearest neighbor in R^m if its nearest neighbor in R^{m+1} is more than some threshold distance further away than the nearest neighbor in R^m. The value of m for which the number of false nearest neighbors falls sufficiently close to zero is the proper embedding dimension for attractor reconstruction. Some computational shorcuts and guidelines for choosing the value of the threshold are given by Nayfeh and Balachandran (1995).

Now that we have established a method for constructing an "equivalent" phase space attractor from the measurement of a single signal, the question of determining some of the invariant measures of the system becomes a question of signal processing and modeling. First we discuss the issue of qualitative evaluation of the dynamics of the system from an embedding of an observed output signal.

First-Return Map

A plot in R^2 is a first-return map if it plots the current value of $x[n]$ versus its value at the previous embedding lag $n - k$. It should be noted that using a small value of k will tend to produce linear first-return maps because consecutive data samples will be correlated. On the other hand, if the system has an inherent periodicity such as a periodic forcing function or a relationship to a biological rhythm such as the diurnal rhythm, heart rate, or respiratory rate, one can use a lag equal to the average period so that the plot represents the value of $x[n]$ every time the periodic function returns near the same point in its cycle. A necessary but not sufficient condition for concluding that a signal is nonlinear is the finding of a convincingly nonlinear first-return map.

In constructing a first-return map one is representing the dynamic system by a discrete mapping from $x[n - k]$ to $x[n]$ and we can exploit some properties of discrete maps. The logistic equation of Example 11.4 is a discrete map, and it is clear from this example that a signal cannot arise from a nonlinear system if its first-return maps are linear. First-return maps for the examples of Fig. 11.9 are presented in Fig. 11.16. Note that for a discrete oscillatory signal the first-return map with $k = 1$ contains discrete points, the number of which equals the period of the discrete oscillation. First-return maps for chaotic systems are necessarily nonlinear and may contain disconnected branches.

Dimension

Knowledge of the dimension of the attractor of a phase plot provides important information about a signal and the system from which it originates. First, the higher the dimension the more spatially complex is the structure of the attractor. Consequently one infers that the degree of complexity of a signal increases with the dimension. If the dimension is non-integer, then the attractor has a fractal structure and the phase plot represents aperiodic (i.e., irregular) motion. A phase plot with *integer* dimension greater than one, however, also represents a type of aperiodic behavior. For example, a system which sums the outputs of two oscillators with incommensurate frequencies has a dimension of two and its output signal is not truly periodic (often referred to as *quasiperiodicity*). Secondly, if the attractor has fractal structure (i.e., non-integer dimension), then it is quite possible that the signal (or system) is chaotic (but this conclusion is not assured because certain stochastic systems also have been shown to have non-integer dimension). More problematic is the issue of whether the estimation of a non-integer dimension from data implies that the true dimension is non-integer. A conclusion that a true dimension is non-integer is suspect if it is based on a single data set, especially in the absence of testing of alternatives using surrogate data as described later in this chapter.

If one can obtain a long data record when dealing with real data, often a sufficiently reproducible estimate of dimension can be obtained so that one may distinguish between signals whose dimensions differ by at least one. If such discrimination is the objective, it is not required that each dimension estimate be exact; however, the question of determining statistical confidence limits for dimension estimates is still unresolved, and therefore a statistically rigorous means of comparing two estimates of dimension is lacking.

Box dimension was discussed in Chapter 10 and is reviewed briefly here. Box dimension is based on the simple geometrical question of how many "boxes" of uniform size are required to completely cover the attractor (called *tiling*) with minimal

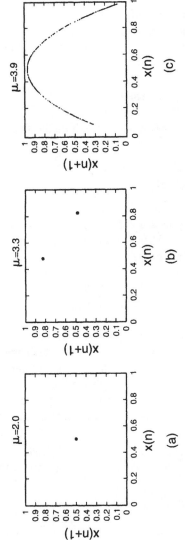

FIGURE 11.16. First-return maps (i.e., $x[n + 1]$ vs. $x[n]$) corresponding to the steady states of the system outputs shown in Fig. 11.9(a,c,d) and to $\mu = 2.0$ (a), 3.3 (b), and 3.9 (c). A constant-valued steady state corresponds to a first-return map having a single point (a), whereas a period-2 oscillation yields a map having two discrete points (b), and a chaotic behavior yields a dense nonlinear relationship (c).

463

overlapping of boxes. Consider an attractor which is a line on a two-dimensional planar surface (Fig. 11.17). A line can be covered by one-dimensional "boxes" which are line segments. For a line of length L and line segments of length ε the number of covering line segments is $N(\varepsilon) = L/\varepsilon = L(1/\varepsilon)$. Similarly a closed surface area, A, can be covered by square boxes of size ε^2 (Fig. 11.17), and the required number of boxes will be $N(\varepsilon) = A/\varepsilon^2 = A(1/\varepsilon)^2$. A comparable result applies for filling a three-dimensional solid with cubes of size ε^3, etc. In theory one can extend this approach to any shape and equate the exponent of the term $(1/\varepsilon)$ to the dimension of the shape. Thus for an unknown dimension, D, the number of covering "boxes" will be

$$N(\varepsilon) = V(1/\varepsilon)^D,$$

where V is a proportionality constant that depends on the shape being covered. Taking natural logarithms,

$$D = \frac{\ln N(\varepsilon)}{\ln V + \ln(1/\varepsilon)}.$$

Because the attractor may have small pieces or fine details, it is necessary to evaluate D in the limit as $\varepsilon \to 0$. Therefore the $\ln(V)$ term becomes insignificant and we can say that the *box dimension* is

$$D_B = \lim_{\varepsilon \to 0}\left[\frac{\ln N(\varepsilon)}{\ln (1/\varepsilon)} \right]. \tag{11.17}$$

This dimension is also called the *capacity dimension*.

The box dimension is straightforward to visualize but examples of inconsistency of box dimension with other estimates of dimension are known. Therefore, D_B is useful primarily as a measure for comparing two attractors.

Correlation dimension is based on the concept of how densely the points on an attractor aggregate around one another and its calculation can be related to the rela-

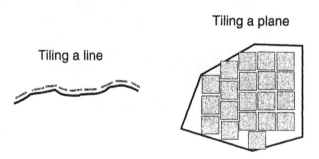

FIGURE 11.17. Examples of tiling a line with small line segments and tiling a two-dimensional object with small square "boxes."

tive frequency with which the attractor visits each covering element (which, in this case, are spheroids). Given a data time series that has been embedded in a space R^m, one first defines a distance measure in that space. Given the time series $x[n]$, $0 \leq n \leq N - 1$, which is obtained by sampling a function $x(t)$ at frequency $f_s = 1/t_s$ and from which the m-dimensional embedding vectors,

$$x_i = [x_{i1}, x_{i2}, \ldots, x_{im}] = [x[i \cdot t_s], x[(i + 1)kt_s], \quad \ldots, \quad x[(i + m - 1)kt_s]]$$

are determined, one can use the Euclidian norm as the distance measure—that is,

$$d_{ij} = \|x_i - x_j\| = \left[\sum_{k=1}^{m} (x_{ik} - x_{jk})^2 \right]^{0.5}. \tag{11.18}$$

An alternative distance measure which is much faster to calculate and seems to work equally well is the maximum distance between corresponding components of x_i and x_j—that is,

$$d_{ij} = \max_k [|x_{ik} - x_{jk}|]. \tag{11.19}$$

Using the chosen distance measure, one then evaluates a *correlation integral function*, $C^m(r)$, defined as

$$C^m(r) = \frac{1}{N^2} [\text{number of pairs of points } (x_i, x_j), \text{ such that } d_{ij} < r]. \tag{11.20}$$

Essentially, the correlation integral function is determined by making a hypersphere of radius r around every embedded data point and counting the average number of embedded data points inside these hyperspheres. This function is a type of spatial correlation because it expresses the extent to which embedded data points are close together. (One also can derive this method on the basis of the relative frequency with which the phase plot visits each hypersphere.) Using the Heaviside function, H, defined as

$$H(a) = \begin{cases} 1, & \alpha > 0 \\ 0, & \alpha \leq 0 \end{cases},$$

one can express the correlation function as

$$C^m(r) = \frac{1}{N(N/2) - N} \sum_{i=1}^{N} \sum_{j=i+1}^{N} H(r - d_{ij}). \tag{11.21}$$

This definition is not exactly equal to Eq. (11.20) but is easier to compute. It recognizes that the distance from x_i to x_j equals that from x_j to x_i, and consequently there is a factor of two difference between Eqs. (11.20) and (11.21). There is another small difference because Eq. (11.21) omits counting points for which x_j is compared with itself since it has been shown empirically that these self-comparisons bias the dimension estimate.

Now, one assumes that the number of points in the hypersphere should be proportional to the radius of the hypersphere raised to the power D_C, where D_C is the dimension estimate referred to as the *correlation dimension*. Thus $C^m(r)$ should be proportional to r^{D_C}. Setting $C^m(r) = V \cdot r^{D_C}$, where V is a proportionality constant, taking natural logs of both sides, and letting r approach zero, yields

$$D_C = \lim_{r \to 0} \left[\frac{\ln C(r)}{\ln r} \right]. \tag{11.22}$$

The estimation of D_C from data involves many limitations. First, there is a limited, and a *prior* unknown, range of r over which D_C can be evaluated because: (1) when r approaches the size of the phase plot, the calculated correlation function plateaus, and; (2) for very small r the effects of noise and finite resolution dominate. Therefore, it is necessary to calculate the correlation function over a very wide range of r and search for a more limited range over which the estimate of dimension is consistent—that is, the estimated D_C is relatively constant. This is accomplished by plotting $ln[C(r)]$ versus $ln(r)$ and finding the range for r over which this function is reasonably linear. One can best determine linearity by calculating the slope of the $ln[C(r)]$ versus $ln(r)$ function, plotting the slope versus $ln(r)$, and evaluating the range of r for which the slope is nearly constant. The average slope over this range is the estimate of D_C. The second issue is the number of data points, N, required to accurately estimate the dimension. A conservative criterion is that N should be $\geq 10^{D_C}$ but some investigators have demonstrated the ability to obtain reproducible estimates with far fewer data points. This issue is an active research area. Even when a good estimate of D_C is acquired, a third issue must be resolved: What embedding dimension, R^m, should be used? It was noted previously that invariant measures of an attractor of dimension d theoretically can be recovered completely in a space of dimension $2d + 1$. Therefore the usual practice is to evaluate D_C as a function of m as the embedding dimension is increased until the estimated dimension fails to increase as m increases further. (This convergence of estimates may occur for $d \leq m < 2d + 1$.) A typical criterion is that the estimated D_C should remain within a 10% range over three or four consecutive values of m. This criterion raises a fourth question: How does one compare values of r from different embedding dimensions? The usual approach is to normalize the distance measurements at each value of m by the maximum width of the phase plot for that embedding, which is approximated as the largest distance, d_{ij}, for that m. Therefore r ranges from zero to one at all embeddings.

A related issue is the effect of filtering to remove noise. Because uncorrelated random noise is completely unpredictable, it essentially "fills up" any space in which it is embedded, and therefore its correlation dimension is theoretically infinite. Correlated random noise is expected to have a high dimension also but dimension estimates from short data records of correlated noise can be surprisingly low. In general, passing a signal through any linear filter potentially can increase its dimension (even as it visually removes some noise). However, filtering with a linear FIR filter has been shown not to seriously corrupt the estimate of correlation dimension and an increasing number of investigators report that various nonlinear filtering schemes improve the ability to determine correlation dimension in their particular applications. The latter are often based on methods for one-step ahead nonlinear prediction and are beyond the scope of this text. Success using nonlinear radial basis functions for general functional expansions of nonlinear signals has been reported recently. This method may provide another basis for nonlinear filtering.

The concept of dimension can be generalized by defining a dimension of order q as

$$D_q = \frac{1}{q-1} \lim_{r \to 0} \frac{\ln \sum_{i=1}^{N} \{P_i(r)\}^q}{\ln(r)}, \tag{11.23}$$

where $P_i(r) = (N_i(r)/N_0)$, $N_i(r)$ is the number of embedded data points in the hypersphere of radius r centered about the i-th embedded data point, and N_0 is the total number of embedded data points. It is easy to show that D_0 is the box dimension and D_2 is the correlation dimension.

Example 11.9   Correlation dimension   Maintenance of high blood pressure in hypertensive subjects may be related in part to malfunction in mechanisms which regulate salt and water excretion. Part of this regulatory mechanism is the feedback control of pressures and flows in nephrons of the kidney. Jensen et al. (1987) have measured the pressure in the proximal tubule of nephrons of spontaneously hypertensive (SH) rats (Fig. 11.18) and have estimated a correlation dimension from these data. To further substantiate the potential for chaotic behavior in this biological system, they also developed a mathematical model of regulation of proximal tubule pressure and estimated the correlation dimension for the noise-free pressure signal from the model. Figure 11.18 presents some of their analyses of simulation data (top) and real data (bottom). Several typical features of these calculations are evident. First, it is easier to identify a useable range of r (referred to as the *scaling region*) from the derivatives rather than directly from the correlation function (which also is called the *correlation integral*). Second, as $ln(r)$ decreases from zero there is a range for which $C(r)$ changes rapidly as the calculation becomes more reflective of average local properties of the phase plot. At small values of r, $C(r)$ becomes quite variable even for simulated data (reflecting both noise and roundoff errors). For both simulated and actual data there are intervals of width 2.5–3 log units (i.e., spatial scales ranging over a factor of approximately 8:1) for which the slopes (and therefore the dimension estimates) are reasonably constant for each embedding and also consistent over three or more embedding dimensions. The average slopes in these intervals, somewhat > 2.0, are the estimated D_C values. Note that they are similar for the simulated and the actual data. Note also that the degree of convergence of slopes to a scaling range evidenced in this example is typical. Often it is difficult to determine a scaling range.

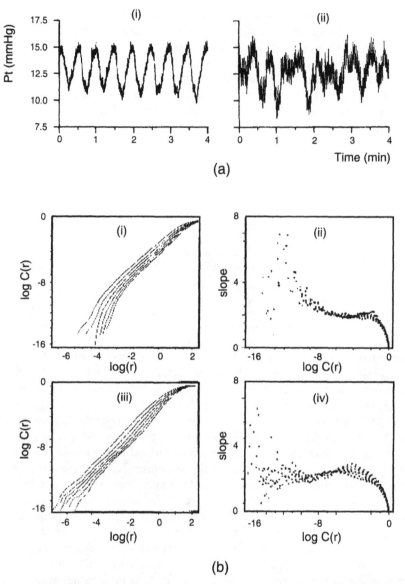

FIGURE 11.18. Graphs of renal tubule pressure versus time from a normal rat (i) and a spontaneously hypertensive rat (ii). From Jensen et al. (1987). (c) Correlation integral curves (left) and their corresponding derivatives versus log(r) (right) are shown for six different embedding dimensions (i, ii) for simulations of renal tubule pressure variations from a model and (iii, iv) for experimental data . From Jensen et al. (1987). Reprinted with permission from CHAOS IN BIOLOGICAL SYSTEMS. 1987, Plenum Publishing Corp.

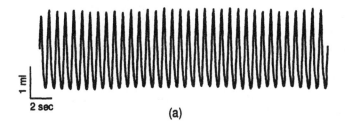

(a)

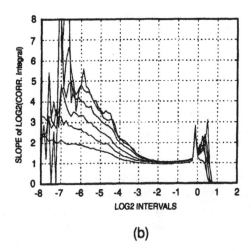

(b)

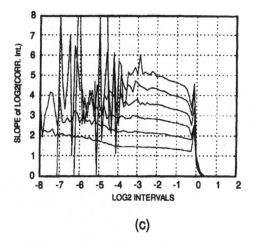

(c)

FIGURE 11.19. (a) Lung volume recording from an anesthetized rat during quiet breathing (4000 points sampled at 150 Hz). (b) Slope of $C^m(r)$ versus $\ln(r)$ for the original lung volume data. (c) Slope of $C^m(r)$ versus $\ln(r)$ for the lung volume data after random shuffling.

Obviously all biomedical data are not chaotic and it is instructive to examine another example. Figure 11.19(a) displays a tracing of the lung volume changes of an anesthetized rat during quiet breathing. This data (comprising 4000 data points sampled at 150 Hz) was analyzed for its correlation dimension using a lag value of $k = 20$. Over embedding dimensions 2–7 the slope of the correlation integral curves converged very near 1.0 for the interval range $2^{-0.5}$–2^{-3} (Fig. 11.19(b)), implying that the signal could be modeled as a periodic signal. As above, at smaller scales the noise dominates the correlation integral. To demonstrate that this convergence was due to the temporal ordering of the data, the original 4000 data points were randomly shuffled and the analysis repeated. The apparent correlation dimension of the shuffled data increased with embedding dimension, as expected for random noise.

Lyapunov Exponents

Since the hallmark of chaos is sensitive dependence on initial conditions, it is desirable to devise a direct measure of this property. The Lyapunov exponents provide one such measure. Consider the path of a system in phase space after a slight disturbance. This path differs from the one the system would have followed, and one can measure the distance between the two paths as a function of time. For a short time the separation distance can be modeled as following an exponential time course, such that if the initial disturbance is δ_0, then after a sufficient time the separation distance will be dominated by the largest Lyapunov exponent and may be approximated as $\|\delta(t)\| \approx \|\delta_0\| e^{\lambda_1 t}$. In order to exhibit sensitive dependence on initial conditions, at least one Lyapunov exponent must be positive. Therefore, if the estimated λ_1 is not positive, there can be no chaotic behavior. The Lyapunov exponents of a system directly measure the average rate of divergence of system trajectories in various directions in the (small) neighborhood of a point on an attractor. Often, however, one assumes that these measures are invariant on the attractor and obtains values averaged over all data points. Nonetheless, it is important to remember that Lyapunov exponents may vary significantly at different sites on an attractor, especially if the motion of the system is irregular and there are numerous stable and unstable periodic cycles near the attractor.

Given a data set embedded in R^m, the usual algorithmic approach for calculating the largest Lyapunov exponent involves selecting two neighboring points from the phase plot, moving ahead in time by a small Δt, and calculating the change in separation of the two points in this time. This process may be repeated for larger Δt's, and the entire process is repeated for many (if not virtually all) of the embedded data points. Finally some sort of averaging of the ln of the separation distances divided by Δt is used to calculate an estimate of λ_1.

Typically it has been very difficult to calculate even one Lyapunov exponent from noisy data without obtaining very many data points, and the usefulness of this calculation for biomedical data has often been questioned. Nevertheless some investigators have reported "reasonable" values based on the algorithms of Wolfe et al. (1985) or Rosenstein et al. (1993). Lyapunov exponents can be estimated more successfully from noise-free data generated by models of biological behavior and many such examples are present in the literature.

Other Measures Related to Nonlinear Properties of Signals

Entropy can be considered a measure of the disorder in a system or as a measure of the amount of information required in order to specify the state of a system. Therefore entropy should be dependent on the precision with which the state, or degree of disorder, can be specified. The Kolmogorov entropy of the attractor is defined as

$$K = -\lim_{\tau \to 0} \lim_{\varepsilon \to 0} \lim_{d \to \infty} \sum_{i=i_1}^{i_d} P(i_1, \ldots, i_d) \ln P(i_1, \ldots, i_d), \qquad (11.24)$$

where $P(i_1, \ldots, i_d)$ is the joint probability that the trajectory is in neighborhood i_1 at time τ and in neighborhood i_2 at time 2τ, and so forth, up to neighborhood i_d and time $d\tau$, and $\in$ is the radius of the neighborhood. K is a positive constant different from zero for a chaotic system and is infinite for a stochastic system. Grassberger and Procacia (1983(b)) defined an approximation to the Kolmogorov entropy, K_2, that can be estimated from the correlation integral used in the calculation of the correlation dimension. Essentially for a fixed value of r (in the scaling range), $K_2(r)$ is proportional to the difference between $C^m(r)$ and $C^{m+1}(r)$, such that

$$K_2(r) = \frac{1}{\tau} \ln\left[\frac{C^m(r)}{C^m(r+1)} \right]. \qquad (11.25)$$

An estimate of $K_2(r)$ is obtained by averaging over the scaling range for r. The authors suggest evaluating this estimate for progressively increasing values of m and using its asymptotic value as a measure of $K_2(r)$. Furthermore, they demonstrated that $K \geq K_2$, implying that a positive K_2 means the signal is chaotic. Sammon et al. (1991) have applied this analysis to show a difference in entropy of respiratory flow patterns of anesthetized rats before and after bilateral sectioning of the vagus nerves. A variation of this measure, named *approximate entropy*, has been used to differentiate heart rate signals from healthy and apneic premature human infants.

Power spectrum. Neglecting noise, the power spectrum of the steady-state output of an autonomous linear system can only contain discrete peaks representative of periodic behavior. For a chaotic behavior, on the other hand, the power spectrum may exhibit broadened peaks, representing nearly periodic behavior associated with repelling periodicities, but also contains power at frequencies between peaks associated with the aperiodic wandering of the signal. In the presence of noise it is difficult to distinguish between linear and chaotic signals on this basis alone however.

Bispectrum is an extension of the concept of the power spectrum. Whereas the power spectrum of a random process is defined as the Fourier transform of the (second-order) autocorrelation function of the process, the bispectrum, $B(\omega_1, \omega_2)$, is defined as

$$B(\omega_1, \omega_2) = \Im^2\{r_{x_1x_2x_3}[n, m]\}, \tag{11.28}$$

where $\Im^2\{.\}$ refers to the two-dimensional Fourier transform, and

$$r_{x_1x_2x_3}[n, m] = E\{x[k] \cdot x[k + n] \cdot x[k + m]\} \tag{11.29}$$

is the third-order autocorrelation function. The bispectrum is a measure of the correlation between frequency components at the two frequencies, ω_1 and ω_2. For a linear system, a frequency component at one frequency cannot be correlated to a frequency component at another frequency unless that correlation exists in the input signal. Therefore, if $B(\omega_1, \omega_2)$ differs from zero anywhere, then the signal *could not* have arisen from a linear system excited by Gaussian white noise. Although this test often is interpreted to mean that the signal under study arose from a nonlinear system, one should keep in mind the alternative that the signal could have originated from a linear system excited by a non-Gaussian signal. Nevertheless the bispectrum can provide a useful test for possible nonlinearity. The estimated bispectrum can be calculated either by estimating the third-order autocorrelation function first or by calculating the Fourier transform of the signal and evaluating it at various frequencies ω_1, ω_2, $\omega_1 + \omega_2$, as

$$\hat{B}(\omega_1, \omega_2) = \frac{1}{N^2}X^*(\omega_1 + \omega_2)X(\omega_1)X(\omega_2),$$

where $X(\omega)$ is the DTFT of $x[n]$. Note that several computational difficulties and caveats about interpreting the bispectrum have been ignored here.

11.8 TESTS OF NULL HYPOTHESES BASED ON SURROGATE DATA

Overview

We saw in the examples that it is difficult to estimate properties such as dimension and Lyapunov exponents even using noise-free data. The presence of noise, especially correlated noise, further complicates these analyses because of the statistical uncertainty that noise introduces. Furthermore it has been demonstrated numerically that short data records of both correlated noise and noise-free linear signals can produce estimates of invariant measures that mimic those of nonlinear signals. Consequently it is essential to test whether the result of a particular data analysis could have been produced by noise or linear signals having properties similar to those of the data signal. There are several methods for generating surrogate signals which exhibit some of the properties of the data signal. The analysis is then repeated on these surrogates and the estimated invariant measure is compared with that calculated from the data signal. Typically one attempts to generate several surrogates of each type so that it is possible to determine the mean and standard deviation of the invariant measure estimated from each type of surrogate signal. From this information one can test statistically whether the estimate from the original data record is within the 95% confidence interval of the estimates from the surrogates, in which case one would conclude that it is not possible to distinguish the data signal from the surrogates.

Surrogate signals. There are several common types of surrogate signals that can be created from a data signal. Each has a particular value for testing null hypotheses. Assume that we are interested in estimating some invariant measure, M, and have observed the data signal $x[n]$, $0 \leq n \leq N - 1$, with mean of zero and variance s_x^2, and an estimate of M—call it M^*— has been calculated from the original data signal. The following surrogates can be generated for testing various null hypotheses regarding the source of the data signal:

a. *Uncorrelated noise.* The very first null hypothesis to test is that the data could have arisen from a white-noise process. Using a Gaussian random number generator such as `randn` in MATLAB, generate a sequence of N normal variates having a variance equal to s_x^2. By generating a large number of such sequences (e.g., 20) one can estimate the mean and variance of the estimate of M from the surrogate data. One then tests whether M^* lies within the 95% confidence interval of the estimates from the surrogate signals.

b. *Uncorrelated noise having the same amplitude histogram as the data.* It is possible that the distribution of amplitudes of the actual data signal skews the density of points on the phase plot and biases the estimate of M. To reproduce this situation in the surrogate signals one tests the null hypothesis that the data could have originated from an uncorrelated random process having the same amplitude distribution as the data. The procedure is simply to randomly shuffle the original data to generate the surrogate signal. Again, this procedure may be repeated 20 or more times to ascertain a mean and variance for the estimate of M based on these surrogates. The m-file `xscrambl.m` generates such surrogates.

c. *Surrogates having the same second-order autocorrelation function as the data signal.* One test of whether M^* could have resulted from analysis of a linear signal can be achieved by generating surrogates having the same second-order autocorrelation function as the data signal. Surrogates are created by calculating the DTFT of the data signal, then multiplying all the phases of the DTFT by $e^{j\theta}$, where θ is chosen independently for each frequency and is uniform on the interval $[0, 2\pi]$. If one makes sure to satisfy the symmetry properties of phase of the DTFT for real-valued signals (i.e., $\theta(\omega) = -\theta(-\omega)$), then the inverse DTFT produces a real-valued signal having the same frequency-dependent magnitude of its DTFT as the original data. Consequently both $x[n]$ and the surrogate have the same power spectrum and, since the power spectrum is the Fourier transform of the autocorrelation function, they have the same autocorrelation function. (This result exemplifies the property that there is not a unique signal that can be determined for each autocorrelation function.) As above, it is possible to generate 20 or more such surrogates. The m-file `phscram.m` implements this method. This method does not produce good surrogates of purely periodic signals. It also can introduce high-frequencies if one is not careful to ensure that $x[0]$ approximately equals $x[N-1]$.

d. *Surrogates from a monotonic nonlinear transformation of a linear Gaussian process.* These surrogate data test the hypothesis that the original data arose from a linear Gaussian process and subsequently were modified via a static nonlinear filter. A limitation of this method is that it assumes that the nonlinear filter is invertible. The process proceeds as follows: First a Gaussian time series, $y[n]$, is generated. Then $y[n]$ is reordered so that, for every $n = m$, if $x[m]$ is the k-th smallest value of the original time series, then $y[m]$ is similarly the k-th smallest value of the time series $y[n]$. Then a surrogate for the reordered $y[n]$—call it $y'[n]$—is created using the DTFT method described immediately above. Finally, $x[n]$ is reordered so that it follows the amplitudes of $y'[n]$. The reordered original time series—call it $x'[n]$—has the same amplitude histogram as the original time series whereas the "underlying" time series (the reordered $y[n]$ and $y'[n]$) are Gaussian and have the same Fourier power spectrum.

Hypothesis testing. As described above, the most common approach involves evaluating a sufficient number of estimates of the desired measure, M, based on surrogate data so that one can obtain a reasonable evaluation of their variance. One then determines whether the estimate M^* from the original data lies within the 95% confidence interval of those from the surrogate data. The statistical characterization of most of these analyses of nonlinear signal properties remains to be developed, however, and obtaining multiple data records for analysis is desirable; however, obtaining multiple data records often is an unattainable goal for biomedical applications.

11.9   OTHER BIOMEDICAL APPLICATIONS

There are many examples of biomedical applications of the methods developed in this chapter and a discussion of them will evolve into a cautionary tale as well as providing some enlightenment. The most successful use of these methods involves their application to mathematical models, and we start with two such examples.

Bifurcations in a Model of a Cardiac Pacemaker Cell

For many years experimentalists have observed that the application of a periodic stimulus (e.g., by electrical stimulation) to a biological neural oscillator (including cardiac sinus node pacemaker cells) can produce complex behavior patterns in the neural oscillator that are strongly dependent on the frequency of stimulation. These complex behaviors include $M:N$ phase-locking, in which there are N cycles of the neural oscillator for every M stimuli, as well as very irregular cycling of the neural oscillator. Guevera and Glass (1982) explored the behaviors of a mathematical model of an oscillator in order to understand the genesis of these complex behaviors.

The equations of the model were rather simple:

$$\frac{d\Phi}{dt} = 2\pi;$$

$$\frac{dr}{dt} = ar(1-r).$$

Note that r can be interpreted as the length of a rotating vector that is attached at the origin of a planar coordinate system, and Φ is then the angle of the vector with the positive x-axis. This model is sometimes referred to as a *radial isochron clock*. The authors mathematically applied to the rotating vector a perturbation of length b parallel to the x-axis every τ sec. The system comprising the generator of the disturbances and the oscillator can be considered an autonomous system, and one can determine bifurcation diagrams as parameters of this system are varied. For example, Fig. 11.20 shows the stable fixed points of Φ as b is varied and τ is held fixed at 0.35. Observe the series of period-doubling bifurcations over the range of b. In the blank space on the right of the graph are periodic orbits of all periods except four and three, implying the potential presence of chaotic dynamics.

One can define a first-return map by sampling the system at the time interval τ since this interval is an inherent periodic interval of the system. The first-return maps of Φ for three values of b (τ is held at 0.35) are shown in Fig. 11.21. The curvilinearity of this map is a necessity for chaotic behavior to occur but does not guarantee that it does. In fact, knowing the equations of the system the authors could determine the exact functional relationship for this map and show that the specific conditions for a period-doubling bifurcation to occur are met. (This calculation is beyond the scope of this text.) Thus the authors demonstrated the potential

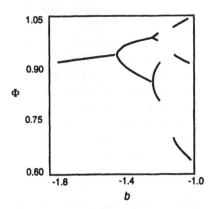

FIGURE 11.20. Bifurcation diagram for the model of Guevara and Glass (1982) for b ranging from -1.8 to -1.0 and $\tau = 0.35$. Between the period-4 behavior and the period-3 behavior is a range in which behaviors of all other periods occur. (Reprinted with permission from JOURNAL OF MATHEMATICAL BIOLOGY. 1982, Springer Verlag)

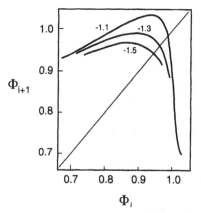

FIGURE 11.21. First-return maps for the model of Guevara and Glass (1982) for $b = -1.1$, -1.3, -1.5. (Reprinted with permission from JOURNAL OF MATHEMATICAL BIOLOGY. 1982, Springer Verlag)

for complex behaviors to arise from periodically stimulated neural oscillators although their model was not presented as a model of a specific biological oscillator. This type of analysis has been applied subsequently to models of sinus node pacemakers, models of electrically stimulated cardiac myocytes, and models of other neural pattern generators for rhythms such as locomotion.

Chaos in a Model of a Neuron

Guckenheimer et al. (1993) developed a model of the anterior burster (AB) neuron of the stomatogastric ganglion of the lobster. This neuron displays various types of rhythmic behavior, including aperiodic behaviors, when various neuromodulators or channel-blocking toxins are applied to it. Their model is an extension of the Hodgkin-Huxley formalism for describing the relationships between ionic channel currents and electrical activity of a neuron. It includes equations for voltage-activated sodium current, voltage-dependent calcium current, voltage-dependent potassium current, calcium-modulated potassium current, a potassium current called the A current, and a leak current. Each current depends on an ion-specific conductance which is modulated by activation and inactivation functions. The entire model comprises six differential and ten algebraic equations. Finally, the authors also recorded the activities of real AB neurons for comparison with the modeling results.

The behaviors of the model for various sets of parameter values are shown in Fig. 11.22. The two upper graphs represent two types of periodic behavior, whereas the two lower graphs are chaotic behaviors. The response of calcium concentration, instead of membrane voltage, is given in Fig. 11.22(c) because it shows the aperiodicity more clearly. The activity pattern in Fig. 11.22(d) is chaotic because of the variable number of action potentials per burst. Comparisons of model behaviors with similar

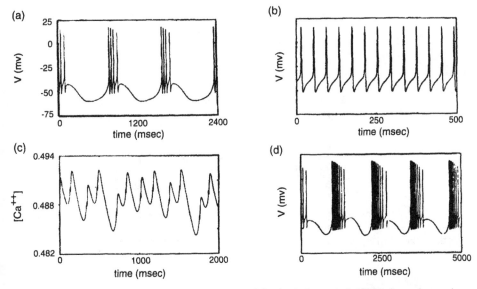

FIGURE 11.22. Behaviors of the neuron model of Guckenheimer et al. (1993), for various values of conductances of the *A* and K_{Ca} currents. (a) and (b) are periodic whereas (c) and (d) are chaotic. Reprinted with permission from PHILOSOPHICAL TRANSACTIONS OF THE ROYAL SOCIETY, LONDON. 1993, The Royal Society.

behaviors observed in an actual AB neuron are presented in Fig. 11.23. These graphs show a sequence of responses when increasing concentrations of the blocker 4-aminopyridine are applied to the AB neuron (right) and when the maximum A current conductance is reduced in the model (left). The visual similarities are obvious. Note that these results were intended to demonstrate the feasibility of the model, not to define any equivalence between model parameters and the experimental preparation.

Correlation Dimension of the EEG

The electroencephalogram (EEG), which was modeled as a random process in Chapter 9, is highly irregular in some states of consciousness and more regular (but still not periodic) in other states. Various investigators have attempted to discern whether the EEG can be modeled as a chaotic signal whose dimension changes with state of consciousness. Early reports in this field concluded that the correlation dimension of the EEG signal was around five in wakefulness and decreased to about four in quiet sleep. However, the current criteria for adequacy of data were not yet established and it is clear that these criteria were not always met in these studies. Nor had the concepts of surrogate data been fully established. Consequently, more recent studies have been more exhaustive in their data analyses and more circumspect in their conclusions. These studies provide an example of the need for a cautious approach in applying the techniques of this chapter to real biomedical data.

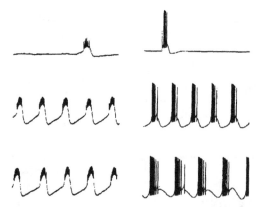

FIGURE 11.23. Comparisons of the behaviors of an actual AB neuron (left) subjected to increasing levels of the blocker 4-AP and simulation results (right), providing evidence that the model captures many features of the actual behaviors. From Guckenheimer et al. (1993).

Chaos in Respiratory Patterns

Although the pattern of breathing is usually almost periodic, in the REM state of sleep and in some diseases it can be irregular. Sammon et al. (1991) presented fairly strong evidence of a decrease in the correlation dimension of respiration of anesthetized rats after severing the vagus nerves in the neck (which interrupts the flow of afferent information from the lungs to the respiratory areas of the brainstem). Before severing the vagus neves the correlation dimension ranged between 2.5 and 3.5, and after nerve sectioning it was approximately 1.0. At the time of that study, however, the concepts of surrogate data analysis were not well established. A subsequent report of fractional correlation dimensions and positive Lyapunov exponents for the respiration of awake humans has motivated others to investigate more intensely the possibility of modeling breathing as a chaotic process. Hughson et al. (1996) conducted a more thorough analysis of respiratory data from humans and found no convincing differences between these data and surrogates. On the other hand, a recent study (Hoyer et al. 1997) reports positive Lyapunov exponents for both respiration and heart rate from conscious rabbits. This latter study applied surrogate data methods extensively and also showed a decrease in these exponents after anesthesia. These studies provide another example of the need for a cautious approach in applying the techniques of this chapter to real biomedical data.

Correlation Dimension of Optokinetic Nystagmus

Optokinetic nystagmus (OKN) is a type of eye movement in which the eye moves slowly in one direction then rapidly returns toward, but not exactly to, its initial position. This pattern is repeated as long as the eyes are either tracking a moving target or staring at a fixed target. An example of OKN is shown in Fig. 11.24(a). Shelhamer (1992) noted the aperiodic nature of OKN and analyzed records of these eye

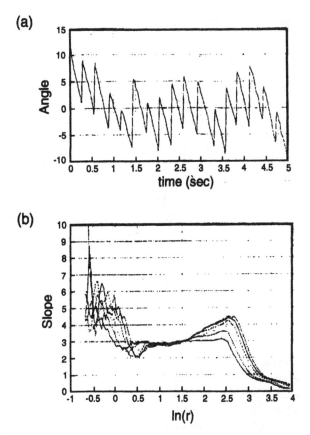

FIGURE 11.24. (a) Sample of horizontal optokinetic nystagmus from a normal human. (b) Slope of the log of the correlation integral versus log[$C(r)$] for the data of (a). Embedding dimension, m, equals 5, 6, 7, 8, 9, 10. Samples size was 7300 data points. From Shelhamer (1992). © 1992 IEEE

movements for correlation dimension. He acquired 7300 data points by sampling at 100 Hz and embedded in dimensions $5 \leq m \leq 10$. The delay, k, was determined as three times the lag at which the autocorrelation function fell to $1/e$ of its peak value. Typically $k \sim 0.6$ s. The author also generated surrogate data which had the same mean and variances for amplitude and timing of the fast and slow phases as the original OKN data. An example of the correlation integral slope plotted versus $\log(r)$ is given in Fig. 11.24(b), where a scaling region is observed near $\log(r) = 1.0$. In this example the calculated correlation dimension is 2.87, which suggests that the attractor may have a fractal structure. Such a finding is only the first step in a process of assessing whether OKN may be modeled as a nonlinear process that can exhibit chaotic behavior and one anticipates further studies of Lyapunov exponents or other potential measures of chaotic behaviors. One should note, however, that these results may be useful as indices of pathophysiology if the estimated dimension changes with disease whether or not it is established firmly that the behavior is chaotic.

Other Examples

To provide an understanding of the variety of biomedical applications in which the present methods have been utilized, some additional examples will be mentioned briefly. There have been studies which report that the yearly variations in epidemics of childhood diseases are chaotic. There are many studies of the dynamics of neural systems that address the possibility of chaotic behavior. Ion channel kinetics have been modeled as chaotic processes. The nonlinear nature of immunological responses is well recognized and some reports have investigated possible chaos in these systems. The tremor of Parkinsons disease has been reported to be chaotic as have the dynamics of arterial blood pressure and renal blood flow. Postural sway has been analayzed as chaotic behavior but there is evidence that a random process model may be appropriate. All of these application areas are fields in which current research is continually refining our understanding.

11.10   SUMMARY

This chapter introduced the concept of a nonlinear dynamical system (and its output, a nonlinear signal). Nonlinear signals are signals which could not have been produced by a linear system except by exciting the linear system with a linearly transformed version of the nonlinear signal. Whereas in the steady state an autonomous linear system can exhibit either a constant output or a periodic one and it always converges to the same behavior for any initial condition, nonlinear systems may converge to different steady-state outputs for different initial conditions. Some nonlinear systems also can produce continuously aperiodic behavior known as chaos. A hallmark of chaos is sensitive dependence on initial conditions. This sensitivity implies that even when the system is started from two nearly identical initial conditions (both of which will converge to aperiodic behavior), the resulting outputs will diverge.

Nonlinear signals (or the systems which generate them) can be described by several measures (which are equally applicable to linear systems). Dimension is a measure of the minimum number of coordinates needed to describe the set of points forming the attractor of the steady-state behavior of the signal. Lyapunov exponents describe the average rate of divergence of nearby trajectories. Entropy is a measure of the rate of loss of information about past states of the system. All of these measures can be estimated using embeddings constructed from observations of a signal but it is imperative to compare the calculated results with those determined using surrogate signals. Several types of surrogates should be evaluated including white-noise signals, randomized signals with the same amplitude histogram as the data, signals having the same autocorrelation function, and nonlinear signals having the same amplitude variation in time.

There are many examples in which investigators have applied these concepts to biomedical signals. In the case of mathematical models of biomedical systems, exact analyses and well-founded conclusions are often possible. Where real data are involved, the conclusions are often debated in the literature and the utilization of tools from nonlinear dynamics to characterize biomedical signals is an active research topic.

EXERCISES

11.1 Plot $x[n]$ versus n and $x[n+1]$ versus $x[n]$ for the logistic equation, using the last 100 of 400 iterates, for the following values of μ and interpret these plots. Let $x[0] = 0.5$.
  a. $\mu = 2.8$;
  b. $\mu = 3.2$;
  c. $\mu = 3.83$;
  d. $\mu = 3.664$;
  e. Determine the autocorrelation function $r_x[m]$ for each of the signals above and compare them.

11.2 Find the theoretical period-2 points of the logistic equation for $\mu = 3.2$.

11.3 Determine the Lyapunov exponents for the Lorenz system of Example 11.8 at the two equilibrium points that were not evaluated in the example.

11.4 Generate a bifurcation diagram for the logistic equation (known also as the quadratic family $Q_\mu(x)$). Generate 400 iterates for each value of μ and graph the last 100 iterates. Identify all attracting n-cycles that you observe. Do this for $x[0] = 0.25$ and for 100 values of μ in each of the following ranges:
  a. $2 \le \mu \le 3.9$;
  b. $3.5 \le \mu \le 3.7$;
  c. $3.662 \le \mu \le 3.664$.

11.5 Find the fixed points and eigenvalues of the following systems:
  a. $\dot{x} = -x + x^3$, $\qquad \dot{y} = -2y$;
  b. $x[k+1] = 0.1x[k](1 - x[k] - y[k])$, $\qquad y[k+1] = 0.1x[k]y[k]$.

11.6 Given the nonlinear, continuous-time system described by $\ddot{x} + 0.5\dot{x} + x^3 - x = 0$, determine its fixed points. Evaluate the eigenvalues of the linearized system at each fixed point.

11.7 Given the real-valued, two-dimensional mapping

$$F\binom{x}{y} = \binom{0.5xy}{1 - 0.5x^2 - 0.5y},$$

find any fixed point of F. Evaluate the eigenvalues of the linearized system at the fixed point.

11.8 The Li-Yorke theorem says that "period-3 implies chaos"—that is, a nonlinear mapping exhibiting a period-3 behavior will exhibit all other periodic behaviors and aperiodic chaotic behavior. But when behaviors of the logistic equation family, $Q_\mu(x)$, are calculated numerically, one finds only period-n cycles for many values of μ. Explain why this result is not incompatible with the Li-Yorke theorem.

11.9 Using the model `enzyme2.m` reproduce the three steady-state behaviors shown in Figure 11.5. The parameters for the substrate rate, V, are given in the legend of the figure.

a. Show that the first two behaviors are indeed periodic by plotting γ versus α and demonstrating that these graphs are limit cycles.

b. Construct a similar plot for the third behavior and interpret it.

c. Find the power spectrum of α from part b. What can you conclude from this power spectrum regarding the presence or absence of periodic behaviors?

11.10 In this problem you are going to examine the properties of a simplified model of a predator population interacting with a prey population. One model of this type is

$$x_{k+1} = ax_k(1 - x_k - y_k), \qquad 2 < a \le 4$$
$$y_{k+1} = bx_ky_k, \qquad 2 < b \le 4.$$

This model says that at each generation (k) the prey population, x, decreases when the predator population, y, increases, and that the prey population otherwise obeys in inverse quadratic rule (i.e., it increases when it is small and decreases when it is large). The predator population increases when the prey population (its food source) is large and also increases when its own population is large (because there are more to reproduce). A special case is the situation in which $b = a$. In that case $0 < a \le 4$. This is the case you will examine. In all cases let the initial values be: $x_0 = 0.2$, $y_0 = 0.6$.

a. Implement these equations and determine how many iterations are necessary in order to achieve convergence to a steady-state behavior. For this part use a=3.03. You should find a period-n cycle as the steady-state behavior. What is the value of n?

b. Now generate a bifurcation diagram to examine the steady-state behavior over a range of parameter values. Let a range from 0.1 to 4.0, and use increments of 0.1 just to get a broad picture.

c. You should notice a bifurcation near $a = 3.0$. Generate another bifurcation diagram to examine the range $2.9 < a < 3.5$. What do you think happens at 3.0?

d. Another bifurcation appears to occur near 3.4. Generate a bifurcation diagram for $3.35 < a < 3.45$. What appears to happen at this bifurcation?

e. Check your answer to part d by making another bifurcation diagram for $3.375 < a < 3.80$.

f. Make steady-state "time series" plots *and* first-return maps of x for the following values of a: 2.997, 3.03, 3.3755, 3.3775, 3.430. Compare these results to your bifurcation diagrams and explain any apparent discrepancies.

g. Determine the autocorrelation functions for $x[n]$ for the solutions in part f.

11.11 In this problem you are to calculate the correlation dimension of some known signals. In all cases you should follow these steps: (1) predict the correlation dimension and explain your prediction; (2) choose a sampling interval (t_s), decide how long to sample the signal, and create the sampled signal using MATLAB; (3) analyze the signal using a program that you will write to calculate correlation dimension. Recall that one method to determine a lag to use for constructing an embedding is to find the first zero-crossing of the autocovariance function of the data. If your calculated dimension deviates significantly from your prediction above, then repeat the exercise using different analysis parameters (e.g., more data points, or different sampling frequency) until the calculated dimension is closer to what you expect. Also, after you have achieved a satisfactory calculation of D_C, then scramble the data (using xscrambl.m) and repeat the calculation of D_C to show that convergence of the dimension estimate does not occur after you whiten the data. Do all of this for the following signals:

a. $x_1(t) = \sin(2\pi t) + \sin(4\pi t + \pi/6)$;

b. $x_2(t) = \sin(2t) + \sin(4\pi t)$;

c. $x_3(t) = \sin(2t)\sin(4\pi t)$;

d. $x_4(t) = $ Gaussian white noise.

11.12 The Lorenz equations have been used to describe temperature variations and convective motion in a fluid medium. The equations of this model are:

$$\dot{x} = \sigma(y-x), \qquad \dot{y} = rx - y - xz, \qquad \dot{z} = xy - bz,$$

where x, y, z are the system variables and σ, r, b are parameters.
For this problem let $\sigma = 10$, $b = 8/3$.

a. Use the MATLAB file lorenz.m, determine the trajectory for $r = 10$. (You may let the initial conditions be zero.) You should find a stable fixed point.

b. Simulate the Lorenz system for $r = 21$ and show that for some (actually many) initial conditions this system exhibits transient, chaotic-like behavior before converging to a stable fixed point. This behavior arises because the system contains many non-attracting periodic and aperiodic behaviors for these parameter values.

c. Simulate the Lorenz system for $r = 32$. You should find chaotic behavior. Determine the power spectrum of $x(t)$.

d. For $r = 32$ calculate the theoretical values of the Lyapunov exponents of the Lorenz system.

11.13 In Chapter 9 you analyzed the data file montage.mat from a random signal perspective. Reconsider these data using the methods of the present chapter. Recall that channels 2–8 of this file contain EEG recordings.

a. Use the first 4000 data points of channel 2 as a data signal. To determine a lag for embedding, evaluate the autocovariance function of this signal and find the lag corresponding to its first zero-crossing.

b. Write a program to calculate its correlation dimension. You will find that it is much more difficult to define a scaling region over which to determine the correlation dimension than it was for the synthetic signals of the previous problem.

c. Once you feel that you have "the best" estimate of correlation dimension you can obtain from these data, generate a white-noise signal having the same variance as the data signal and compare its correlation dimension to that of the data signal.

d. To obtain a loose sense of the variability of the estimated correlation dimension of the white-noise process, generate four more sample functions of white noise and compare the correlation dimension of the data with the range of values obtained for the white noise.

e. To generate an uncorrelated signal that has the same amplitude distribution as the original data, pass the original data signal (i.e., the 4000 points that you analyzed) through the filter `xscramble.m`. Now pass the output of the filter through the filter a second time and then evaluate the correlation dimension for this doubly-scrambled signal. Is it different from that of the original data? How could you determine whether it is significantly different?

f. As a final test of surrogate data, write an m-file that determines the DTFT of the data signal, randomizes its phases (but retains antisymmetry of the phase, which is necessary for real signals), and then takes the inverse DTFT to generate a surrogate signal having the same power spectrum as the data. Compare the correlation dimension of this signal to that of the original data.

12

ASSESSING STATIONARITY
AND REPRODUCIBILITY

12.1   INTRODUCTION

Given a novel data signal, the first task in signal processing is to decide to which category (i.e., deterministic, stochastic, fractal, or chaotic) it should be assigned. In cases in which the objective is merely to filter the signal (e.g., to minimize aliasing), then the question of categorization may assume less importance. On the other hand, if the properties of the signal are changing with time, then some of the analyses that have been discussed can give erroneous results (the exception being the methods for studying deterministic signals). If one cannot assume that a signal is deterministic, therefore, it is necessary to assess the stationarity of signal properties. Furthermore, if one is seeking to draw inferences about the mechanisms that generated a signal, then assessing stationarity of signal properties may be a key step.

Another fundamental question about a novel signal is whether it carries more information than (and therefore can be distinguished from) uncorrelated white noise. Once a signal has been determined to be white noise, then the only available information is its variance and its probability density function. The determination that a signal may be white noise does not preclude applying other analysis methods, but it does mean that one must be circumspect in interpreting those analyses. In such cases it is important to evaluate the range of outputs from a particular analysis that one might observe if the analysis were applied to sample functions of true white noise. By doing so, one may determine whether the results of analyzing a novel data signal fall within this range. A related question is the expected range of outputs from an analysis if other data signals having properties similar to the novel data signal were analyzed. For example, if one models a signal as an autoregressive process and calculates its power spectrum, how much must the spectrum differ from that of another signal in order to conclude that the underlying processes are different? Equivalently one might ask how precisely one can specify the parameters of a signal based on a particular analysis. This chapter will address such questions.

Consider, for example, the question of whether the EMG signal recorded from the tongue of a human subject differs from white noise. The file tongue_emg.mat provides an EMG signal from the tongue, recorded while the subject pressed his tongue against his incisor teeth with the mouth closed. This signal is presented in Fig. 12.1(a). Visually the left and right halves of the data record appear to have different amplitudes, implying that the properties of the random process generating this EMG signal might be changing with time (i.e., nonstationary). Therefore, the power spectrum of each half was determined using the MATLAB function psd and an FFT length of 256 points. The resulting spectra are shown in Fig. 12.1(b). These spectra appear to confirm the impression that the amplitudes differ between the two halves of the data and also suggest that a spectral peak around $\omega = 0.5\pi$ is absent from the second half of the data. Since the second spectrum is almost uniformly lower than the first, one might feel comfortable concluding that the two spectra truly differ. The question of a "missing" spectral peak, however, is more perplexing since it is possible that any particular feature of a random process might be absent from any one finite-length sample function from that random process. One might conclude that the spectra "probably" are different.

Assuming that the signal is at least wide-sense stationary over each half of the data, one may consider the question of whether the EMG signal from the first half of the data (call it *emg1*) can be differentiated from white noise. The theoretical power spectrum of white noise is constant and, for white noise having the same mean power as the *emg1* signal, its spectrum would be that given by the solid line in Fig. 12.1(c). On the other hand, because of the finite data length one expects the calculated power spectrum of a sample function of white noise to vary about its theoretical level. Therefore, the fact that the spectrum of *emg1* differs from the theoretical spectrum of noise is not convincing evidence that *emg1* differs from white noise. One might be tempted to utilize the variance of the *emg1* power spectrum as an estimate of the variance of the white noise spectrum, but one expects that approximately 5% of the power spectrum of *emg1* would lie outside of +/– 2 standard deviations (SD) from the mean. Given (in this example) that the spectrum comprises 128 frequency samples, one might expect six to seven points to lie more than two SD from the mean. Thus the observation that five points lie above the dotted line representing the mean + 2 SD (Fig. 12.1(c)) is not very informative.

The preceding discussions have led to "soft" conclusions because we lack the theoretical foundations for assessing stationarity of signals and reproducibility of signal processing results. This chapter will develop these foundations. *Unless noted otherwise, it is assumed that signals are being analyzed in discrete-time form.*

12.2 ASSESSING STATIONARITY OF A RANDOM PROCESS FROM A SAMPLE FUNCTION

In general, it is not possible to test rigorously for strict stationarity; however, if one can show that data are Gaussian and wide-sense stationary, then one may conclude strict-sense stationarity because of the properties of Gaussian probability density

(a)

1 sec

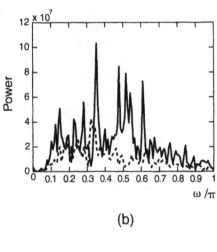

(b)

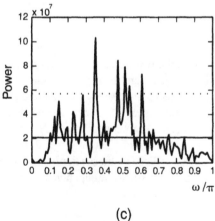

(c)

FIGURE 12.1. (a) EMG signal (sampled at 500 Hz) from the file `tongue_emg.mat`. (b) Periodograms of the left (solid) and right (dashed) halves of the tongue EMG signal. (c) Periodogram of the left half of the tongue EMG signal, the mean level of the periodogram (solid line) and two standard deviations above the mean level (dotted line).

functions. Therefore, we focus on testing for wide-sense stationarity. There are two general approaches, the first being a direct statistical test of the data and the second being an assessment of the reproducibility of analyses of the data.

The *runs test* is a measure of time invariance of statistical measures of the data. To perform this test, one segments the total data points of a record, N, into k non-overlapping groups of N/k consecutive data points. The statistic of interest—for example, the mean value—is calculated for each of the k groups. Represent these values as p_i, $1 \leq i \leq k$. Next p_{med}, the median value of these p_i statistics, is calculated. Then another sequence, $d_i = p_i - p_{med}$, is generated and the number of sign changes in this latter sequence, n_s, is counted. The number of runs is one plus n_s. Marmorelis and Marmorelis (1978) have tabulated the bounds on a stationary Gaussian random process for the number of runs for a given value of k, on the theoretical basis that the mean is $k/2 + 1$ and its variance is $k(k - 2)/(4(k - 1))$. If at a given significance level the number of runs lies outside of these bounds, one concludes that the data are not stationary. This test may be applied to both the means of the segments and their variances in order to test both first-order and second-order moments of the data. Note, however, that demonstrating time-invariance of the mean and the variance does not, by itself, prove wide-sense stationarity. In applying the runs tests one needs to select k so that there are enough samples of the test statistic to provide a meaningful result (e.g., $k \geq 10$) but also small enough that the number of points per segment, N/k, provides a meaningful calculation of the variance.

The second approach to testing for stationarity involves dividing the data record into two halves and calculating various measures for each half. If one can determine statistical confidence limits for these measures, then it is possible to test the hypothesis that the measures from the two halves of the data are equal. Means, variances, autocovariance functions, and power spectra all may be tested for equality. If no differences are found between the two halves of the data record, one has greater confidence that the data are wide-sense stationary during the interval of observation.

Tests for equality of means and variances are standard statistical procedures that may be found in most textbooks of statistics. Let the data signal $x[n]$, $0 \leq n \leq N - 1$, be divided into two equal segments of length N_1 and N_2 (which either will be equal or will differ by one if N is odd). Let $\overline{m}_j$, $j = 1, 2$, be the sample means of the two data segments and $\hat{\sigma}_j^2$, $j = 1, 2$, be their sample variances. The test statistic for testing the equality of the means of two independent samples of data is a t statistic given by

$$T = \frac{\overline{m}_1 - \overline{m}_2}{\sqrt{s_{12}^2}}, \tag{12.1}$$

where s_{12}^2 is a pooled variance estimate given by

$$s_{12}^2 = \left(\frac{1}{N_1} + \frac{1}{N_2} \right) \left(\frac{N_1 \hat{\sigma}_1^2 + N_2 \hat{\sigma}_2^2}{N_1 + N_2 - 2} \right). \tag{12.2}$$

The statistic T has a Student's t distribution with $v = N_1 + N_2 - 2$ degrees of freedom. For a given confidence level, α, (typically 0.05), if T falls within the range

$$t_{v,1-\alpha/2} \leq T \leq t_{v,\alpha/2},$$

then one concludes that the two means are not different. Note that this test assumes that the two sample variances are equal. Therefore one should compare these variances, as described in the next paragraph, both to determine stationarity of the variances and to validate the test for equality of the means.

Equality of variances is testable using a Fisher statistic. This statistic is defined as

$$F = \frac{\hat{\sigma}_1^2\left(\dfrac{N_1}{v_1}\right)}{\hat{\sigma}_2^2\left(\dfrac{N_2}{v_2}\right)}, \tag{12.3}$$

where $v_1 = N_1 - 1$, $v_2 = N_2 - 1$, and the sample variances are unbiased (i.e., when calculating variances, divide by $N_1 - 1$ or $N_2 - 1$ instead of N_1 or N_2). The F statistic has two indices of freedom, and the confidence interval for F is

$$F_{v_1,v_2;1-\alpha/2} \leq F \leq F_{v_1,v_2;\alpha/2},$$

where fractional points of the F distribution for various degrees of freedom are available from tables in standard statistics texts. If F lies within this confidence interval, one concludes that the variances do not differ.

Example 12.1   Testing for stationarity   A runs test was applied to the tongue EMG signal of Fig. 12.1(a) by sectioning the data into 10 segments and calculating the standard deviation of each segment. Figure 12.2 shows the deviations from the median standard deviation for the 10 segments. There is only one sign change and thus the number of runs equals two. From the table in Marmorelis and Marmorelis (1978) the 95% confidence interval for the number of runs, R, when $k = 10$ is $3 \leq R \leq 8$. Therefore we conclude that the data are not stationary.

Comparing the means and variances of the first and second halves of this data set also indicates a lack of stationarity. The two mean values are -14.09 and -14.14, and the T value from Eq. (12.1) is not significant, implying that the means do not differ statistically. On the other hand, the two variances are $s_1^2 = 4772$, $s_2^2 = 3114$. The F statistic of Eq. (12.3) is 2.3486, whereas the 95% confidence limit for F is $F_{1023,1023;.025} = 0.8846 \leq F \leq F_{1023,1023;.975} = 1.1305$. Since F does not lie within this range, one concludes that the variances from the two halves of the data are unequal and the data are not stationary.

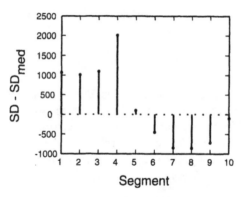

FIGURE 12.2. Runs test applied to the tongue EMG data. Differences of the standard deviation (SD) of each of 10 data segments from the median SD (SD_{med} = 3692) are plotted vs. segment number.

12.3   STATISTICAL PROPERTIES OF AUTOCOVARIANCE ESTIMATORS

In order to compare autocovariance functions that have been estimated from two halves of the data record, it is necessary to evaluate the bias and variance of the estimates. The bias is the difference between the expected value of an estimator and the true value of the parameter being estimated. For the estimated autocovariance function calculated from a finite observation of one sample function, $y[n]$, $0 \leq n \leq N-1$ of a zero-mean random process,

$$E\{\hat{c}_y[m]\} = E\left\{\frac{1}{N} \sum_{n=0}^{N-1-|m|} y[n]y[n+m]\right\}.$$

It is easy to show that

$$E\{\hat{c}_y[m]\} = \left[1 - \frac{|m|}{N}\right]c_y[m]. \tag{12.4}$$

That is, unless $N \to \infty$, the expected value of the autocovariance function is not equal to the true autocovariance. If we focus on a small range of m such as $|m| \leq 0.1N$, then this bias error is small.

The other type of uncertainty in the estimated autocovariance function, its variance, is more relevant to the comparison of two estimates when each is based on a similar amount of data. By definition (for $m > 0$),

$$\text{Var}\{\hat{c}_y[m]\} = E\{\hat{c}_y^2[m]\} - E^2\{\hat{c}_y[m]\}$$

$$= E\left\{\frac{1}{N^2} \sum_{n=0}^{N-1-m} \sum_{k=0}^{N-1-k} y[n]y[n+m]y[k]y[k+m]\right\} - \left(1 - \frac{m}{N}\right)^2 c_y^2[m]. \tag{12.5}$$

By linearity, the expectation operator on the right-hand side of Eq. (12.5) may be moved inside the summations. For processes that are fourth-order stationary and Gaussian, this expectation may be expressed as (Shiavi, 1991; p. 196)

$$E\{y[n]y[n + m]y[k]y[k + m]\} = c_y^2[m] + c_y^2[n - k] + c_y[n - k + m]c_y[n - k - m]. \quad (12.6)$$

Note that in Eq. (12.6) the indices of the two summations always occur as $n - k$; therefore, the double summation may be written as a single summation, with the result that

$$\text{Var}\{\hat{c}_y[m]\} = \frac{1}{N} \sum_{r=-N+m+1}^{N-1-m} (1 - \frac{|r|}{N})(c_y^2[r] + c_y[r + m]c_y[r - m]). \quad (12.7)$$

For large values of N this equation simplifies to

$$\text{Var}\{\hat{c}_y[m]\} \approx \frac{1}{N} \sum_{r=-\infty}^{\infty} (c_y^2[r] + c_y[r + m]c_y[r - m]). \quad (12.8)$$

Thus $\lim_{N\to\infty} \text{Var}\{\hat{c}_y[m]\} = 0$ if $|c_y[0]| < \infty$ and $\lim_{m\to\infty} c_y[m] = 0$. Recall that these two conditions are the criteria for a random process to be autocorrelation ergodic.

Note that an alternative definition of the autocovariance function that is sometimes encountered utilizes $1/(N - |m|)$ in front of the summation instead of $(1/N)$. This change renders the expected value of the estimated autocovariance function equal to the true autocovariance function. This desirable consequence, however, has too high a price because the $(1/N)$ term in Eq. (12.8) now becomes $1/(N - |m|)$, and the variance of the estimate becomes very large for all but small values of m. Therefore, it is usual to employ the biased estimator.

According to Eq. (12.8), in order to calculate the variance of the estimated autocovariance function, it is necessary to know the true autocovariance function. In most situations, however, one must utilize the estimated values of the autocovariance function to calculate an estimate of the variance. For practical purposes one may calculate the 95% confidence interval of the autocovariance estimate at each lag, m, as the estimate plus or minus 1.96 times the square root of the variance at lag m. When comparing two estimated autocovariance functions over a range of lags, it is difficult to compare them quantitatively because of the problem of statistically comparing multiple correlated estimates having different variances. A practical guideline is that the autocovariance estimates are likely to be different at some lag if their 95% confidence intervals do not overlap.

Example 12.2 Comparing autocovariance functions To demonstrate the statistical testing of autocovariance functions, the tongue EMG signal of Fig. 12.1(a) was divided into two non-overlapping 1024-point signals, $x_1[n]$ and $x_2[n]$. The biased autocovariance functions for these two signals were evaluated using the MATLAB command xcov and are plotted in Fig. 12.3 along with the 95% confidence inter-

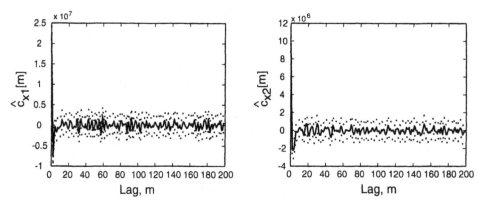

FIGURE 12.3. Estimated autocovariance functions from the two halves ($x_1[n]$ and $x_2[n]$) of the tongue EMG data, +/− 2 standard deviations (dotted).

vals calculated as $\hat{c}_x[m] \pm 2\sqrt{\text{Var}\{\hat{c}_x[m]\}}$. The autocovariance functions are somewhat different over the first few lags but, in fact, the confidence intervals overlap. For greater lags the confidence intervals overlap extensively. Thus there is a suggestion that these functions differ but the clearest difference is at zero lag—that is, corresponding to the variances, which we have shown to be unequal by a more stringent statistical test.

12.4 STATISTICAL PROPERTIES OF THE PERIODOGRAM

Consider a zero-mean, w.s.s., random process for which one has observed one sample function $y[n]$, $0 \leq n \leq N - 1$. Previously we defined the periodogram spectral estimator in two equivalent ways:

$$\hat{P}_y(e^{j\omega}) = \frac{|X(e^{j\omega})|^2}{N} = DTFT\{\hat{r}_y[m]\}, \tag{12.9}$$

where $\hat{r}_y[m]$ is the estimated autocorrelation function of $y[n]$ (which equals $\hat{c}_y[m]$ since it was assumed that the process has a mean of zero).

Using the second definition,

$$E\{\hat{P}_y(e^{j\omega})\} = \sum_{m=-N+1}^{N-1} E\{\hat{r}_y[m]\}e^{-j\omega m} = \sum_{m=-N+1}^{N-1} \left(1 - \frac{|m|}{N}\right)r_y[m]e^{-j\omega m}.$$

The term inside the parentheses of the preceding equation is the equation for a Bartlett window. Using the frequency-domain convolution theorem for the DTFT and recalling that the true power spectrum of a random process is the DTFT of its autocorrelation function, we may write

$$E\{\hat{P}_y(e^{j\omega})\} = \sum_{m=-\infty}^{\infty} r_y[m]w_B[m]e^{-j\omega m} = \frac{1}{2\pi}P_y(e^{j\omega}) * W_B(e^{j\omega}), \quad (12.10)$$

where $w_B[m]$, $W_B(e^{j\omega})$ refer to the Bartlett window in the time and frequency domains, respectively, and $P_y(e^{j\omega})$ is the true power spectrum of the random process. Thus the periodogram is a biased estimator because its expected value does not equal the true power spectrum. Note that the Bartlett window in the frequency domain is

$$W_B(e^{j\omega}) = \frac{1}{N}\left[\frac{\sin(N\omega/2)}{\sin(\omega/2)}\right]^2.$$

For large N this function approaches an impulse and the expected value of the periodogram approaches the true power spectrum.

For comparing two periodograms based on approximately the same amount of data, the important statistical measure is the variance of the periodogram. To derive this measure we first determine the variance of the periodogram when $y[n]$ arises from a Gaussian white-noise process having variance σ_y^2. This periodogram is

$$\hat{P}_y(e^{j\omega}) = \frac{1}{N}\left|\sum_{k=0}^{N-1} y[k]e^{-j\omega k}\right|^2 = \frac{1}{N}\sum_{k=0}^{N-1}\sum_{l=0}^{N-1} y[k]y^*[l]e^{-j(k-l)\omega}. \quad (12.11)$$

The next step is to evaluate $E\{\hat{P}_y(e^{j\omega_1})\hat{P}_y(e^{j\omega_2})\}$. From Eq. (12.11) it is apparent that this expectation will involve the product of four terms involving $y[n]$. Following Hayes (1996), this term may be written as

$$E\{y[k]y^*[l]y[m]y^*[n]\} = E\{y[k]y^*[l]\}E\{y[m]y^*[n]\} + E\{y[k]y^*[n]\}E\{y[m]y^*[l]\}.$$
$$(12.12)$$

The first term on the r.h.s. is zero except when $k = l$ and $m = n$, when it equals σ_y^4. Likewise, the second term is nonzero only when $k = n$ and $m = l$. Applying these conditions on the indices, one may show that

$$E\{\hat{P}_y(e^{j\omega_1})\hat{P}_y(e^{j\omega_2})\} = \sigma_y^4\left(1 + \left[\frac{\sin[N(\omega_1 - \omega_2)/2]}{N\sin[(\omega_1 - \omega_2)/2]}\right]^2\right). \quad (12.13)$$

We may now derive the covariance between periodogram estimates at two different frequencies as

$$\text{Cov}\{\hat{P}_y(e^{j\omega_1})\hat{P}_y(e^{j\omega_2})\} = E\{\hat{P}_y(e^{j\omega_1})\hat{P}_y(e^{j\omega_2})\} - E\{\hat{P}_y(e^{j\omega_1})\}E\{\hat{P}_y(e^{j\omega_2})\}. \quad (12.14)$$

For the case of white noise, the expected value of the periodogram is approximately equal to the value of the true power spectrum, which is σ_y^2. Therefore, setting $\omega_1 = \omega_2 = \omega$, the variance of the periodogram of white noise is

$$\text{Var}\{\hat{P}_y(e^{j\omega})\} = \sigma_y^4. \quad (12.15)$$

We observe that the variance equals the square of the true power spectrum of the white noise.

To generalize the above result for arbitrary (but w.s.s., zero-mean) random processes is difficult, but an approximation can be derived by assuming that $y[n]$ is the output of an LSI system excited by unit-variance white noise, $w[n]$. Assume the frequency response of the system is $H(e^{j\omega})$. Then the power spectrum of $y[n]$ is

$$P_y(e^{j\omega}) = \sigma_w^2 |H(e^{j\omega})|^2 = |H(e^{j\omega})|^2.$$

Now for simultaneously observed sample functions, $y[n]$, $w[n]$, $0 \leq n \leq N-1$, by neglecting end effects we may say that $y[n] \approx h[n] * w[n]$. Consequently, $|Y(e^{j\omega})|^2 \approx |H(e^{j\omega})|^2|W(e^{j\omega})|^2$. Dividing both sides of this latest equation by N yields

$$\hat{P}_y(e^{j\omega}) \approx |H(e^{j\omega})|^2\hat{P}_w(e^{j\omega}). \qquad (12.16)$$

This result implies that

$$\mathrm{Var}\{\hat{P}_y(e^{j\omega})\} \approx (|H(e^{j\omega})|^2)^2 \, \mathrm{Var}\{\hat{P}_w(e^{j\omega})\} \approx P_y^2(e^{j\omega}), \qquad (12.17)$$

since by Eq. (12.15) the variance of the periodogram of white noise equals the square of the white-noise variance, which was specified to be unity. Therefore, in the general case the variance of the periodogram at any frequency is approximately equal to the square of the periodogram value at that frequency. Note that this result is true even as $N \rightarrow \infty$, implying that accumulating more data in a single data record will not reduce the variance of the periodogram! (In practice, one acquires as many different data records as possible, finds the periodogram for each one, and averages the periodogram values at each frequency. There are many variations on this approach and other subtleties that may be applied to reduce the variance. The reader is referred to an advanced textbook such as Hayes (1996).)

Example 12.3 *Comparison of two periodograms*   The tongue EMG signal was tested for stationarity by comparing the periodograms of the first and second halves of the data set. Fig. 12.4 presents the two periodograms and the upper 95% confidence limit for the smaller periodogram (which is from the second half of the data). The confidence limit was determined by evaluating 1.96 times the square root of the variance (Eq. (12.17)). (Note that for clarity the periodograms were smoothed in the frequency domain by a five-point moving average.) The other periodogram lies within this confidence limit except in the range 0.46π–0.64π, suggesting that spectral features in the first half of the data are absent from the second half. This result suggests nonstationarity of the data but is not conclusive. A more rigorous test was implemented by comparing averaged periodograms. Each data set of 1024 points was segmented into 15 data segments of 128 points with adjacent segments overlapping by 64 points. This approach achieves considerable reduction of the variance of the periodogram because of averaging, albeit at the expense of frequency resolution. The variance of the averaged spectrum is not one-fifteenth of that of one spectrum

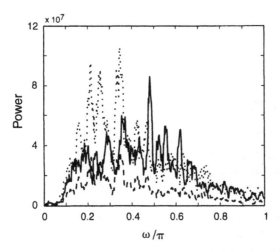

FIGURE 12.4. Periodograms of the tongue EMG segments ($x_1[n]$, solid; $x_2[n]$, long dashes). Dotted line: upper 95% confidence limit for the periodogram of $x_2[n]$.

because the data segments are correlated (due to overlapping). For 50% overlap (as here) the variance of the averaged spectrum is approximately 0.0078 of that of one spectrum (Hayes, 1996). One also could approximate the variance of the averaged spectrum by calculating the variance of the 15 spectral estimates at each frequency.

The trade-off of poorer frequency resolution for reduced variance is beneficial for testing stationarity, as Fig. 12.5 indicates. It is likely that the two averaged peri-

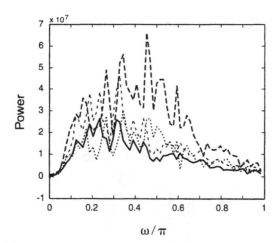

FIGURE 12.5. Periodograms of the tongue EMG segments ($x_1[n]$, long dashes; $x_2[n]$, solid) based on averaging fifteen, half-overlapping, 128-point subsegments each. Dotted: lower 95% confidence limit for periodogram of $x_1[n]$. Short dashes: upper 95% confidence limit for periodogram of $x_2[n]$.

odograms in this figure differ near $\omega/\pi = 0.50$ since the two confidence intervals do not overlap.

12.5 ANALYSIS OF NONSTATIONARY SIGNALS

It is important to recognize that one may utilize methods for deterministic signals if one is confident that any random process component in the signal is trivially small. For example, under some circumstances the control signal to a prosthesis might be modeled appropriately as a deterministic signal and one may apply any of the methods for modeling and analysis of deterministic signals without concern for stationarity. Our emphasis here is on the situation in which one must distinguish between stochastic, fractal, and chaotic signals.

Often in experimental data a signal is nonstationary because its mean is changing with time due to incomplete control of some additional variable during the measurement. Consider, for example, the measurement of diurnal variations in oxygen consumption of pigeons discussed in Chapter 5. If the environmental temperature had been allowed to increase slowly throughout the day (possibly due to heating of the cage by the metabolic heat production of the pigeon), then one would expect a corresponding increase in oxygen consumption associated with increasing demands of thermoregulation. In such situations, so long as one expects that the underlying process controlling diurnal variations in oxygen consumption is not altered by the thermoregulatory effects, it is permissible to remove the best estimate of the nonstationary event from the data in order to produce a stationary data signal. Often the best estimate is achieved by fitting a linear regression line to the original data. One generates the modified data signal by subtracting the regression line from the original signal. That is, given an original signal $x[n]$, the modified signal is

$$x_r[n] = x[n] - (an + b), \tag{12.18}$$

where

$$a = \frac{\displaystyle\sum_{n=0}^{N-1} nx[n] - \frac{1}{N}\left(\sum_{n=0}^{N-1} n\right)\left(\sum_{n=0}^{N-1} x[n]\right)}{\displaystyle\sum_{n=0}^{N-1} n^2 - \frac{\left(\displaystyle\sum_{n=0}^{N-1} n\right)^2}{N}} \tag{12.19}$$

and

$$b = \frac{\displaystyle\sum_{n=0}^{N-1} x[n]}{N} - a\frac{\displaystyle\sum_{n=0}^{N-1} n}{N} = \bar{x} - a\bar{n}. \tag{12.20}$$

One may test the statistical significance of this regression by calculating 95% confidence limits on the slope, a. Given the variance of the original data about the regression line,

$$s_e^2 = \frac{1}{N} \sum_{n=0}^{N-1} (x[n] - (an + b))^2,$$

using the Student's t-test one may evaluate the confidence limits on a as $a \pm 1.96 s_e$. The reader is referred to any standard statistical text (e.g., Zar (1998)) for further details. If the 95% confidence limits do not include zero, then one concludes that the regression line is statistically significant and should be removed from the data. Of course, the nonstationarity may differ from a linear change. While one may use polynomial regression to remove more complex nonstationary signal components, often such an approach will be expected to remove some frequency components of the desired signal as well if the order of the polynomial is significantly greater than unity. On the other hand, a polynomial of too low an order might fail to fit certain features of the nonstationary part of the signal and emphasize other temporal fluctuations in the modified signal because the terms in a polynomial function are not orthogonal.

If a signal is found to be nonstationary, one might ask whether an FBM model is appropriate. An initial test is to determine whether its first differences are stationary and whether the power spectrum of this latter signal is of the $1/f$ type. If the data are limited, these two tests may be satisfied but the results may not be compelling. The next step would be to apply the method of relative dispersion to the signal of first differences and examine the goodness of fit to linear scaling on the *log-log* plot. If this fit is acceptable, one may conclude that the data exhibit scaling behavior consistent with the original signal having the characteristics of FBM.

Example 12.4   Effect of a linear trend   Even a small linear trend may corrupt a spectrum noticably, as this example demonstrates. Two signals were generated. $d_1[n]$ is the numerical derivative of the pressure transducer under test in Fig. 7.18 while $d_2[n]$ is the same signal with an added linear trend that has an amplitude of 0.0025 at $n = 0$ and declines linearly to zero by the end of the graph in Fig. 12.6(a). This small trend produces a significant change in the spectrum at low frequencies compared to the spectrum of $d_1[n]$, even altering the height of the major peak near $\omega = 0.03\pi$ (Fig. 12.6(b)).

12.6   NONSTATIONARY SECOND-ORDER STATISTICS

If $r_x[m_1, m_2]$, the autocorrelation function of a random process, $x[n]$, is not strictly a function of $m_1 - m_2$, then the process is nonstationary. For example, if one compares the autocovariance function from two halves of a data record and finds that they dif-

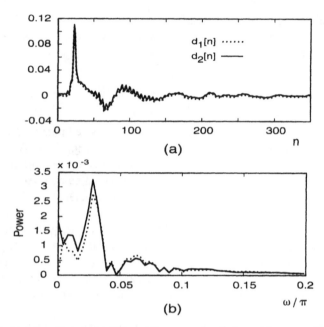

FIGURE 12.6. (a) A transient signal $d_1[n]$ and the same signal after adding a small linear trend $d_2[n]$. (b) Periodograms of $d_1[n]$ (dotted) and $d_2[n]$ (solid).

fer, then this function (and almost certainly the autocorrelation function) depends on the time at which the data were acquired. Consequently the process cannot be stationary. In such cases the power spectrum as we have defined it cannot be evaluated. Of course, one may calculate the periodogram as a measure of the deterministic power spectrum by ignoring the stochastic nature of the signal, but the lack of reproducibility of the data under such circumstances, which will be reflected in the periodogram, may prevent meaningful use of the data. It should be recognized that time-dependence of the autocorrelation function is independent of whether or not the mean is time-dependent. Next we consider some methods for analysis of nonstationary signals.

Complex Demodulation

If a signal contains one or a few oscillatory components whose amplitudes and phases (but not frequencies) are changing slowly with time, these temporal changes may be detected using a process known as complex demodulation. Let the random process be described as a summation of cosine terms with time-varying amplitudes plus white noise. Thus

$$x[n] = \sum_{k=1}^{K} A_k[n] \cos[\omega_k n + \phi_k] + w[n], \qquad (12.21)$$

where it is assumed that the ω_k are ordered in terms of increasing frequency. If the frequencies, ω_k, are known and are sufficiently separated, then the time-varying amplitudes, $A_k[n]$, may be determined as follows: First, multiply $x[n]$ by $e^{-j\omega_1 n}$. Consider an arbitrary term in Eq. (12.21) after this multiplication. Thus

$$A_k[n]e^{-j\omega_1 n}\cos[\omega_k n + \phi_k] = 2A_k[n]e^{j(\omega_k-\omega_1)n}e^{j\phi k} + e^{-j(\omega_k+\omega_1)n}e^{-j\phi k}], \qquad 1 \le k \le K.$$

Observe that (for $k = 1$) the modified signal will have a component $2A_1[n]e^{j\phi_1}$, plus other components at frequencies $\omega_k \pm \omega_1$. If $A_k[n]$ varies slowly with n so that its frequency content is well below the smaller of $\omega_2 - \omega_1$ and $2\omega_1$, then the signal $y[n] = e^{-j\omega_1 n}x[n]$ may be passed through a narrow lowpass filter to extract $A_1[n]e^{j\phi_1}$. The magnitude and phase of the filter output are estimates of the time-varying amplitude and phase of the cosine component at ω_1. This method may be applied to determine the time-varying amplitude and phase of the k-th cosine component by multiplying $x[n]$ by $e^{-j\omega_k n}$.

The method of complex demodulation may be sucessfully applied even if the frequencies are known only approximately. Consider a signal that has only one time-varying cosine component,

$$x[n] = A_1[n]\cos[\omega_1 n + \phi_1].$$

Now multiply this signal by $e^{-j\omega_0 n}$, where $\omega_0 \sim \omega_1$, and filter the modified $x[n]$ signal by a noncausal FIR filter having impulse response $g[n]$ such that $g[-n] = g[n]$. The frequency response of this filter,

$$G(e^{j\omega}) = \sum_{n=-G_1}^{G_1} g[n]e^{-j\omega n}$$

will be a real, even function of frequency. The output of the filter will be

$$y[n] = e^{-j\omega_0 n}x[n] * g[n] = \sum_{u=-G_1}^{G_1} g[u]x[n-u]e^{-j\omega_0[n-u]}. \qquad (12.22)$$

Assuming that $A_1[n]$ changes slowly over any time interval of length $2G_1$, we may assume that $A_1[n]$ is constant relative to Eq. (12.22). Therefore,

$$y[n] = \frac{1}{2}A_1[n]\sum_{u=-G_1}^{G_1} g[u][e^{-j(\omega_0-\omega_1)(n-u)-\phi_1} - e^{-j(\omega_0+\omega_1)(n-u)+\phi_1}]$$

$$= \frac{1}{2}A_1[n][G(e^{-j(\omega_0-\omega_1)})e^{-j(\omega_0-\omega_1)n-\phi_1} - G(e^{j(\omega_0+\omega_1)})e^{-j(\omega_0+\omega_1)n+\phi_1}]. \qquad (12.23)$$

If the filter meets the condition that $G(e^{j\omega})|_{|\omega|>2\omega_0} \approx 0$, then the second term in Eq. (12.23) vanishes. Thus, recognizing that $y[n]$ will be complex-valued, we have the result that

$$A_1[n] = \frac{2|y[n]|}{G(e^{j(\omega_0 - \omega_1)})}, \qquad \theta[n] = \measuredangle y[n] = (\omega_0 - \omega_1)n - \phi_1. \qquad (12.24)$$

Therefore, multiplying $x[n]$ by the complex exponential $e^{-j\omega_0 n}$ whose frequency is close to that of $x[n]$, then filtering through an appropriately designed lowpass filter, can permit recovery of the time variations of the amplitude, $A_1[n]$. Furthermore, one may plot $\theta[n]$ vs. n and look for a linear variation with n. The slope of this linear variation should be $\omega_0 - \omega_1$, thereby permitting identification of the unknown frequency, ω_1, and its offset provides an estimate of ϕ_1.

Example 12.5 Complex demodulation   To demonstrate the capabilities of this method, a heart rate signal was simulated under two different conditions. In both simulations the heart rate was represented as the R-R interval in milliseconds, and the R-R interval was modulated by sine waves at 0.03 Hz and 0.10 Hz to reflect known physiological modulators. R-R interval also is modulated by breathing, and in the first instance the simulated R-R interval signal also was modulated by a sinusoid at frequency 0.30 Hz having an amplitude of 9 msec. The resulting simulated R-R interval signal is shown in Fig. 12.7(a). It was desired to extract the amplitude and phase of the modulation at 0.30 Hz using the complex demodulation technique. When a demodulating signal at exactly 0.30 Hz was used, the estimate of the amplitude, $A[n]$, was variable over a small range (Fig. 12.7(b)) and close to the actual amplitude used in the simulation (i.e., 9 msec). Likewise, the phase estimate, $\theta[n]$, (Fig. 12.7(c)) was close to the actual value of -0.678π.

In the second simulation the modulating signal at 0.30 Hz was itself modulated

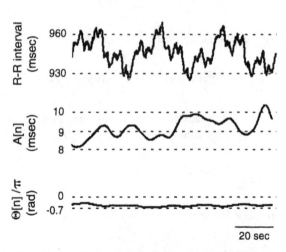

FIGURE 12.7. Estimating the amplitude and phase of a modulating, 0.30 Hz, sinusoid using complex demodulation. R-R interval: the simulated heart rate signal (see text). $A[n]$: the demodulated amplitude. $\theta[n]$: the demodulated phase. (This example courtesy of Dr. Abhijit Patwardhan.)

so that its amplitude was time-varying. This amplitude variation was sinusoidal with a period of 100 seconds and an amplitude of 9 msec. The simulated R-R interval signal, the signal modulating the amplitude of the 0.30 Hz sine wave, and the amplitude and phase of the demodulated signal (again demodulating at the exact frequency of 0.30 Hz), are shown in Fig.12.8. $A[n]$ follows the amplitude of the signal $A_1[n]$ fairly closely whereas $\theta[n]$ clearly refects the changes in sign of this latter signal. Note that $\theta[n]$ is offset by the mean phase of the 0.30 Hz signal (i.e., -0.678π).

The Short-Time Fourier Transform (STFT)

A simple approach to estimation of a time-varying spectrum is to choose time segments of the desired resolution and apply the periodogram to each time segment. If each segment is L points in length, one may overlap adjacent segments by $L - 1$ points but such a fine increment of time probably is not warranted. An overlap of M points, where $L/4 \le M \le L/2$, is a reasonable compromise. Therefore, given a signal $x[n]$, $0 \le n \le N - 1$, and the assumption that the data are w. s. s. over

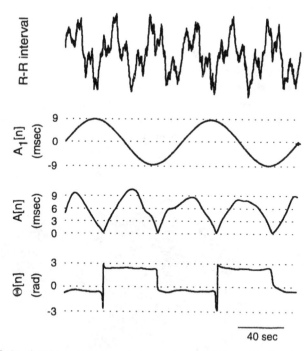

FIGURE 12.8. Estimating the amplitude and phase of a modulating, 0.30 Hz, sinusoid whose amplitude is also varying with time, using complex demodulation. R-R interval: the simulated heart rate signal (see text). $A_1[n]$: signal modulating the amplitude of the sinusoid that is modulating heart rate. $A[n]$: the demodulated amplitude. $\theta[n]$: the demodulated phase. (Example courtesy of Dr. Abhijit Patwardhan.)

some time interval L, one chooses an overlap of M points and segments the data into sub-records

$$x_j[n], \quad j(L-M) \le n \le j(L-M)+L-1, \quad 0 \le j \le \frac{N-L}{L-M} = J. \quad (12.25)$$

The total number of time segments is $J + 1$. For each $x_j[n]$ one calculates the periodogram in the usual manner. It is typical to plot these spectra as if they were two-dimensional slices of a function $\hat{P}_y(e^{j\omega}, n)$ using a three-dimensional graph having frequency and time as the two independent axes.

Serious limitations to this approach are the large variance of the periodogram and the poor frequency resolution that one achieves using short data records. The variance problem may be reduced by convolving $\hat{P}_y(e^{j\omega}, n)$ with a time-frequency window function. But because this convolution further reduces the frequency resolution, an alternative approach is suggested. After segmenting the data as descibed above, one may apply an AR spectral estimation method to each data segment instead of the periodogram. This approach was utilized in the m-file tfar.m to produce the time-frequency spectral plot of Fig. 9.2(b). The order of the AR model may be adjusted to minimize sensitivity to the length of the data segments; the actual time resolution, however, is still determined by L.

Example 12.6 Short-time AR spectral estimation The file eegsim.m was used to generate a 6000-point, simulated EEG data record sampled at 50 Hz. Its frequency content changed abruptly at $t = 40$ sec. This signal was analyzed using a fifth order autoregressive model in tfar.m by segmenting the signal into non-overlapping records of 250 data points each. A three-dimensional plot of the estimated time-varying spectrum is presented in Fig. 12.9. Each spectrum is plotted at

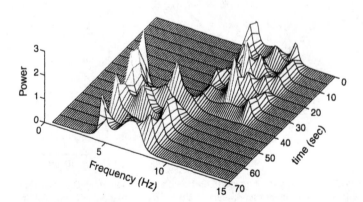

FIGURE 12.9. Short-time AR spectra of a simulated EEG signal whose spectral characteristics change abruptly at $t = 40$ sec. An AR(5) model was fit to consecutive, nonoverlapping, segments of 250 data points.

the beginning of the 5 second time interval that it represents. Note that the abrupt change in the spectral characteristics was not detected until 5 seconds after it oc-curred.

AR Spectral Estimation via Recursive Least Squares

The essence of this method stems from the recognition that one may determine the coefficients of an AR(p) model using a least squares approach instead of the Yule-Walker equations. (In fact, the two methods turn out to be equivalent computation-ally.) It is convenient to approach the method from the viewpoint of estimating the coefficients of a Linear Prediction Filter (LPF) instead of an AR model. The two models are equivalent, as seen by comparing their system equations. For an AR(p) model of a process $x[n]$, the predicted value, $\hat{x}[n]$, is

$$\hat{x}[n] = -\sum_{i=1}^{p} a[i]x[n-i], \tag{12.26}$$

whereas, by definition, an LPF of order p predicts the next data point on the basis of p prior points as

$$\hat{x}[n] = -\sum_{i=1}^{p} c[i]x[n-i]. \tag{12.27}$$

Clearly the only difference between the two models is a sign difference, and thus $a[i] = -c[i]$.

Given $x[n]$, $0 \le n \le N-1$, one may write Eq. (12.27) for each time point n, thereby generating $N + p$ equations in p unknowns. Using the fact that we desire $\hat{x}[n] = x[n]$, these equations may be written in matrix form as

$$\begin{bmatrix} x[0] & 0 & \cdots & & 0 \\ x[1] & x[0] & 0 & \cdots & 0 \\ \vdots & \vdots & \vdots & \vdots & \vdots \\ x[p-1] & \cdots & & & x[0] \\ x[p] & x[p-1] & \cdots & & x[1] \\ \vdots & \vdots & \vdots & \vdots & \vdots \\ x[N-1] & \cdots & & & x[N-p] \\ 0 & x[N-1] & \cdots & & x[N-p+1] \\ \vdots & \vdots & \vdots & \vdots & \vdots \\ 0 & 0 & \cdots & 0 & x[N-1] \end{bmatrix} \begin{bmatrix} c[0] \\ c[1] \\ \vdots \\ c[p] \end{bmatrix} = \begin{bmatrix} x[1] \\ x[2] \\ \vdots \\ x[p] \\ x[p+1] \\ \vdots \\ x[N] \\ 0 \\ \vdots \\ 0 \end{bmatrix}. \tag{12.28}$$

Of course, one is unlikely to find coefficients $c[i]$, $1 \le i \le p$, to achieve exact equality for all of these equations and therefore the optimal coefficients may be es-timated using an ordinary least squares (OLS) approach. The optimal coefficients may be found by minimizing the sum of squared errors, $e[n]$, where $e[n] = \hat{x}[n] - x[n]$, and the sum of squared errors is

$$J = \sum_{n=1}^{N+p} e^2[n] = \sum_{n=1}^{N+p} (\hat{x}[n] - x[n])^2.$$ (12.29)

Equation (12.29) is in standard form for an ordinary least squares approach. Rewriting Equation (12.28) as

$$X_p \underline{c}_p = \underline{x}_1,$$ (12.30)

where X_p is the matrix of Eq. (12.28) and

$$\underline{c}_p = [c[1], c[2], \ldots, c[p]]^T, \quad \underline{x}_1 = [x[1], \quad x[2], \quad \ldots, \quad x[N-1], \quad 0, \ldots, 0]^T.$$

The least squares solution to Eq. (12.30) is

$$\underline{c}_p = (X_p^T X_p)^{-1} X_p^T \underline{x}_1.$$ (12.31)

If one defines an autocorrelation matrix R such that $R(i,j) = r_x[j-i]$, $1 \le i,j \le p$, where, similar to the definition in Chapter 2 (but omitting dividing by N),

$$r_x[i,j] = \sum_{k=0}^{N-1} x[k-i]x[k-j] = \sum_{k=0}^{N-1} x[k]x[k-|j-i|],$$ (12.32)

then $R = X_p^T X_p$. Furthermore, $X_p^T \underline{x}_1$ is a vector of autocorrelation values which we may define as $\underline{g} = [r_x[1], r_x[2], \ldots, r_x[p]]^T$ and rewrite Eq. 12.31 as

$$\underline{c}_p = R^{-1} \underline{g}.$$ (12.33)

The solution provided by Eq. (12.33) applies to a stationary signal, $x[n]$. The coefficients of an AR(p) model for $x[n]$ may be found from the relationship $a[i] = -c[i]$, from which the power spectrum may be calculated in the usual manner (as discussed in Chapter 9). This AR(p) model is equivalent to that found by the Yule-Walker method. To apply this method to nonstationary data, two modifications are necessary. First we develop a recursive method for solving Eq. (12.33) when $x[n]$ is augmented by one new data point, then we explain how to "forget" past data in order to allow for time-varying model coefficients.

Assume that the solution given by Eq. (12.33) based on acquiring n data points, $x[k]$, $0 \le k \le n-1$, is available. If we now acquire one more data point, $x[n]$, it is possible to solve for the new estimates of the coefficients by applying a "correction" to the original estimates. At time step $n-1$ the matrix R_{n-1} and the vector $\underline{g}_{n-1}$ based on the original n data points are known and the solution at this time is $\underline{c}_p^{(n-1)} = R_{n-1}^{-1} \underline{g}_{n-1}$. Now note that R_n may be written as

$$R_n = \sum_{k=0}^{n} \underline{x}_k \underline{x}_k^T = R_{n-1} + \underline{x}_n \underline{x}_n^T,$$ (12.34)

where $\underline{x}_k = [x[k], x[k-1], x[k-p+1]]^T$. Similarly,

$$\underline{g}_n = \sum_{k=0}^{n} x[k]\underline{x}_{k-1} = \underline{g}_{n-1} + x[n]\underline{x}_{n-1}. \qquad (12.35)$$

To calculate the solution at step n, however, requires R_n^{-1}. By expressing R_n in the form given in Eq. (12.34), we may use the Matrix Inversion Lemma to find R_n^{-1}. The Matrix Inversion Lemma states that if A, B are square matrices and $B = A + \underline{u}\underline{u}^T$, then the inverse of B may be calculated as

$$B^{-1} = A^{-1} - \frac{A^{-1}\underline{u}\underline{u}^T A^{-1}}{1 + \underline{u}^T A^{-1}\underline{u}}.$$

Letting $B = R_n$, $A = R_{n-1}$, $\underline{u} = \underline{x}_n$, the inverse of R_n is found to be

$$R_n^{-1} = R_{n-1}^{-1} - \frac{R_{n-1}^{-1}\underline{x}_n \underline{x}_n^T R_{n-1}^{-1}}{1 + \underline{x}_n^T R_{n-1}^{-1}\underline{x}_n}. \qquad (12.36)$$

Now

$$\underline{c}_p^{(n)} = R_n^{-1}\underline{g}_n = \left[R_{n-1}^{-1} - \frac{R_{n-1}^{-1}\underline{x}_n \underline{x}_n^T R_{n-1}^{-1}}{1 + \underline{x}_n^T R_{n-1}^{-1}\underline{x}_n}\right][\underline{g}_{n-1} + x[n]\underline{x}_n]. \qquad (12.37)$$

When the r.h.s. of Eq. (12.37) is expanded, the first term will be $R_{n-1}^{-1}\underline{g}_{n-1}$, which equals $\underline{c}_p^{(n-1)}$. Defining the error in predicting $x[n]$ based on the coefficients at step $n-1$ as

$$e[n|n-1] = x[n] - (\underline{c}_p^{(n-1)})^T \underline{x}_{n-1}, \qquad (12.38)$$

after some algebra we may express the new estimate of the vector of coefficients as

$$\underline{c}_p^{(n)} = \underline{c}_p^{(n-1)} + \alpha[n]e[n|n-1]R_{n-1}^{-1}\underline{x}_n, \qquad (12.39)$$

where

$$\alpha[n] = \frac{1}{1 + \underline{x}_n^T R_{n-1}\underline{x}_n}. \qquad (12.40)$$

The above procedure for determining $\underline{c}_p^{(n)}$ from $\underline{c}_p^{(n-1)}$ and R_{n-1}^{-1} is known as the Recursive Least Squares (RLS) method. To summarize, this method proceeds as follows: First one chooses a small starting value for the inverse of the autocorrelation matrix, $R_{-1}^{-1} = s^2 I$, where I is the p-th order identity matrix and $s^2 \ll 1$. Then for $n = 0, 1, 2, \ldots$, one acquires $x[n]$ and calculates the vector of coefficient estimates by calculating $e[n|n-1]$ using Eq. (12.38), $\alpha[n]$ using Eq. (12.40), and $\underline{c}_p^{(n)}$ using Eq. (12.39). Finally, one calculates R_n^{-1} from Eq. (12.36) since it will be needed in the succeeding step. For this calculation Eq. (12.36) may be rewritten in the form

$$R_n^{-1} = R_{n-1}^{-1} - \alpha[n] R_{n-1}^{-1} \underline{x}_n \underline{x}_n^T R_{n-1}^{-1}. \tag{12.41}$$

This iteration is repeated until all of the data points, $x[n]$, have been acquired, at which point the final solution will be exactly identical to the result of a batch analysis of the entire data set using the ordinary least squares method.

Up to this point the RLS method has no advantage with respect to identifying spectra that vary slowly with time because each new data point is aggregated with all of the previous data. To adapt to changes in signal properties, we utilize a sum of squared errors that weights errors on "older" data less than errors on recently acquired data, in effect "forgetting" $x[k]$ when k is sufficiently far from n. To accomplish this objective the squared error function is revised to be

$$J = \sum_{k=0}^{n} \lambda^{n-k} e^2[k], \tag{12.42}$$

where λ is a positive number less than unity known as a *forgetting factor*. Typically λ is chosen in the range 0.90–0.99, with a smaller value implying that older data are "forgotten" more quickly. This modification of the RLS method is known as exponentially weighted RLS (EWRLS). In the formulation of the recursive structure of the EWRLS method, we now define R_n and g_n as follows:

$$R_n = \sum_{k=0}^{n} \lambda^{n-k} \underline{x}_k \underline{x}_k^T; \tag{12.43}$$

$$\underline{g}_n = \sum_{k=0}^{n} \lambda^{n-k} x[k] \, \underline{x}_k. \tag{12.44}$$

The recursion equations (Eq. (12.41) and Eq. (12.40)) become:

$$R_{n-1} = \frac{1}{\lambda} [R_{n-1}^{-1} - \alpha[n] R_{n-1}^{-1} \underline{x}_n \underline{x}_n^T R_{n-1}^{-1}]; \tag{12.45}$$

$$\alpha[n] = \frac{1}{\lambda + \underline{x}_n^T R_{n-1} \underline{x}_n}. \tag{12.46}$$

With these two substitutions the EWRLS method proceeds as described for the RLS method. At each time point n one may obtain an AR(p) model from the relationships

$$a_n[0] = 1, \qquad a_n[i] = -c_n[i], \qquad 1 \le i \le p.$$

To estimate the power spectrum at each step it is necessary to estimate the noise variance for the AR model. The mean square error from the least squares estimation is taken as the estimated noise variance. This may be derived from the LPF model as

$$\hat{\sigma}_w^2 = r_n[0] - \sum_{i=1}^{p} c_n[i] r_n[i], \tag{12.47}$$

in which $r_n[i]$ is defined by Eq. (12.32) and $c_n[i]$ is the i-th element of $\underline{c}_p^{(n)}$. One may evaluate the $r_n[i]$ recursively also via the relationship

$$R_n = R_{n-1} + \underline{x}_n \underline{x}_n^T.$$

The power spectrum is calculated at step n in the usual manner as

$$\hat{S}_x^{(n)}(e^{j\omega}) = \frac{\hat{\sigma}_w^2}{\left| 1 + \sum_{k=1}^{p} a_n[k] e^{-jk\omega} \right|^2}. \tag{12.48}$$

MATLAB includes a recursive implementation for AR model estimation in the function rarx and one of its options implements the EWRLS method. The EWRLS method is perhaps the most direct extension of the standard Yule-Walker approach to spectral estimation. It is more complex to calculate than approaches based on the LMS algorithm (see Hayes (1996)) but is less prone to instability. Least squares methods have the advantage of possessing one global minimum and one may tune the value of λ to achieve adequate resolution for slowly varying spectra. Obviously this method will not immediately detect rapid changes in the spectrum. Other methods based on generalized concepts of time-dependent autocorrelation functions, such as the Wigner-Ville Transform, or on wavelets are more appropriate for such applications, but these methods are beyond the present text. The reader should refer to the numerous available books (e.g., Akay (1997)) and review papers (e.g., Cohen (1996)) on these latter topics.

Example 12.7 Spectral estimation using EWRLS   An actual EEG signal (channel 7 from the file montage.mat)was subsampled to an effective sampling rate of 50 Hz and the MATLAB function rarx was used to estimate a time-varying AR(4) model by the EWRLS method with $\lambda = 0.975$. This function provides a new model for every data point. At every 20th data point the model parameters from five consecutive data points were averaged and a spectrum was calculated using the averaged parameters. These spectra are plotted in Fig. 12.10. Note that the low-frequency range of the spectrum exhibits a clear variation with time, with increases and decreases in amplitude over the temporal range of the data. The reader may confirm that these results are sensitive to the choice of forgetting factor in that the temporal changes become obscured if λ is less than 0.95.

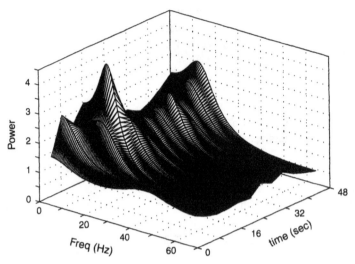

FIGURE 12.10. Spectral analysis of an EEG signal using the EWRLS method with $p = 4$ and $\lambda = 0.975$.

12.7 SUMMARY

This chapter has discussed topics related to assessing stationarity of signals and reproducibility of analyses. Stationarity of the mean and variance of a sample function may be tested using the runs test if a sufficient amount of data is available. Another approach is to separate a data record into two halves and test for the equality of the two means and variances using standard statistical methods. To compare the estimated autocovariance functions or the estimated power spectra from two halves of the data record, it is necessary to evaluate confidence limits on these estimates. Approximate values were derived for the variances of both sample autocovariance functions and periodogram spectral estimates. It was noted, in particular, that the variance of the periodogram is quite large, being approximately equal to the square of the periodogram value at any frequency.

If a signal is found to be nonstationary, one may attempt to generate a stationary signal by removing any linear trend from the original signal using the method of linear regression. If the signal is still nonstationary, then removal of a polynomial regression curve may be attempted so long as one compares the effects on stationarity and signal spectra of several different polynomial orders. When the autocorrelation function or power spectrum is found to be time-dependent, there are some basic approaches to evaluating the time-varying frequency components of the signal. Complex demodulation is useful when the signal contains only a few time-varying oscillations, each having a fixed frequency. Another alternative is to break the data into short segments and apply the periodogram or an AR spectral method to each segment. When used with an AR method, this approach can successfully follow slowly-varying spectra. This latter method may be extended as a formalized recursive algorithm for estimating AR spectra known as the Recursive Least Squares method. To adapt to time-varying spectra, the RLS method must be implemented with a forgetting factor, λ, that gradually reduces the contribution of past data to the squared error function.

EXERCISES

12.1. In Fig. 12.1(c) the mean level of the periodogram was plotted as an estimate of the true spectrum of white noise that has the same average power as the EMG signal. Determine the upper 95% confidence limit for estimates of the noise spectrum assuming that the line plotted in Fig. 12.1(c) is the true white noise spectrum.

12.2. The autocovariance function of a discrete-time Poisson process having an average level of L events per second is $r[m] = L\delta[m]$. Calculate approximate confidence limits on a periodogram based on 512 samples of this process. Repeat for 5120 samples.

12.3. Find the confidence limits for the autocovariance function in Exercise 12.2.

12.4. The file sigtype.m generates samples of four different types of signals each time it is invoked. Generate one such set of signals and use the runs test to determine whether the mean level of each signal is stationary.

12.5. A signal comprising 256 samples is divided into two, non-overlapping segments of 128 points each. The mean levels for the two segments are 0.00745 and 0.00386, and their standard deviations are 0.01120 and 0.01340, respectively. Are their means and variances statistically equal?

12.6. Statistically compare averaged periodograms for the two halves of the EEG signal in channel 2 of the file montage.mat. Choose an appropriate number of segments to average.

12.7. Use the signal in the file ma3sig.mat to create a signal having a linear trend by adding to the signal in the file a line that starts at zero amplitude and has a final value equal to the standard deviation of this signal. Calculate the periodogram of the signal with the trend in three ways: (1) with no trend removal; (2) after removing the trend by linear regression; (3) after removing the trend by filtering with a highpass filter of any design you choose. Compare the resulting periodograms and explain why highpass filtering does not work well to remove a linear trend.

12.8. The two signals in the file cdmsigs.mat each contain a sinusoidal component having a frequency near 56 Hz. Use complex demodulation to extract the (constant) amplitudes of these components. On what basis can one discern that the frequencies are not exactly at 56 Hz?

12.9. Determine whether the heart rate signal in the file hrv1.mat is stationary.

12.10. A heart rate signal representing 1025 heart beats from a human subject is contained in the file hrs.mat. Use short-term AR estimation to calculate the time-varying spectrum of this signal. Test several AR model orders and several degrees of segmenting the data until you are confident that the results follow the changing spectrum adequately without introducing spurious spectral peaks.

12.11.   Run the m-file `sigtype.m` and plot the signal s_4 that will be in the workspace. It is likely that the spectrum of this signal is not stationary. Use the EWRLS method (see the MATLAB function `rarx`) to track the spectrum of this signal. Test various values of the forgetting factor in order to determine what spectral features are consistent and what ones depend on the value of λ.

BIBLIOGRAPHY

BOOKS

Akay, M. *Biomedical Signal Processing*. Academic Press, San Diego, 1994.

Akay, M. (ed.) *Time Frequency and Wavelets in Biomedical Signal Processing*, IEEE Press, 1997.

Barnsley, M. F. *Fractals Everywhere*, Academic Press, Boston, 1988.

Bassingthwaite, J., L. Liebovich, and B. West. *Fractal Physiology*. Oxford University Press, New York, 1994.

Bendat, J. S. *Nonlinear System Analysis and Identification from Random Data*. Wiley, New York, 1990.

Degn, H., A. Holden, and L. Olsen. *Chaos in Biological Systems*. Plenum Press, New York, 1987.

Feder, J. *Fractals*, Plenum Press, New York, 1988.

Glass, L., and M. Mackey. *From Clocks to Chaos: The Rhythms of Life*. Princeton University Press, Princeton, NJ, 1988.

Gulick, D. *Encounters with Chaos*, McGraw Hill, New York, 1992.

Hayes, M. *Statistical Digital Signal Processing and Modeling*. Wiley, New York, 1996.

Jackson, L. *Signals, Systems, and Transforms*, Addison Wesley, Reading, MA, 1997.

Kamen, E. W. *Introduction to Signals and Systems*, Macmillan, New York, 1990, pp. 180–181.

Kamen, E. and B. Heck. *Fundamentals of Signals and Systems Using MATLAB*, Prentice-Hall, Englewood Cliffs, NJ, 1997.

Ljung, L. *System Identification: Theory for the User*, Prentice-Hall, Englewood Cliffs, NJ, 1987.

Mandelbrot, B. B. *The Fractal Geometry of Nature*, W. H. Freeman and Co., New York, 1977.

Marmorelis, P., and V. Z. Marmorelis. *Analysis of Physiological Systems: The White-Noise Approach*. Plenum Press, New York, 1978.

McGillam, C., and G. Cooper. *Continuous and Discrete Signal and System Analysis*, Saunders, Philadelphia, 1991.

Nayfeh, A., and B. Balachandran. *Applied Nonlinear Dynamics*, Wiley, New York, 1995.

Ott, E., T. Sauer, and J. A. Yorke. *Coping with Chaos: Analysis of Chaotic Data and the Exploitation of Chaotic Systems*, Wiley, New York, 1994.

Parker, T. S., and L. O. Chua. *Practical Numerical Algorithms for Chaotic Systems*, Springer Verlag, New York, 1989.

Peitgen, H-O., H. Jurgens, and D. Saupe. *Chaos and Fractals: New Frontiers of Science*, Springer, New York, 1992.

Priestley, M. B. *Spectral Analysis and Time Series*, Academic Press, Orlando, FL, 1982.

Proakis, J. G., and D. G. Manolakis. *Digital Signal Processing, Principles, Algorithms, and Applications* (3rd ed.), Prentice-Hall, Englewood Cliffs, NJ, 1996.

Ruch, T. C., and H. D. Patton. *Physiology and Biophysics*, W. B. Saunders, Philadelphia, 1965, p. 384.

Shiavi, R. *Introduction to Applied Statistical Signal Analysis*, Aksen Publishers, Boston, 1991.

Strogatz, S. *Nonlinear Dynamics and Chaos*, Addison Wesley, Reading, MA, 1994.

Zar, J. H. B. *Biostatistical Analysis* (4th ed.), Prentice-Hall, Englewood Cliffs, NJ, 1998.

JOURNAL ARTICLES AND REPORTS

Allum, J., V. Dietz, and H-J. Freund. Neuronal mechanisms underlying physiological tremor. *J. Neurophysiol.* 41: 557–571, 1978.

Babloyantz, A, J. Salazar, and C. Nicolis. Evidence of chaotic dynamics of brain activity during the sleep cycle. *Phys. Lett.* 111A: 152–156, 1985.

Bassingthwaighte, J. B., R. B. King, and S. A. Roger. Fractal nature of regional myocardial blood flow heterogeneity. *Circ. Res.* 65: 578–590, 1989.

Bassingthwaighte, J. B., and G. Raymond. Evaluation of the dispersional analysis method for fractal time series. *Ann. Biomed. Engr.* 23: 491–505, 1995.

Bassingthwaighte, J. B., and R. P. Beyer. Fractal correlation in heterogeneous systems. *Physica D* 53: 71–84, 1991.

Berger, R. J., and N. H. Phillips. Regulation of energy metabolism and body temperature during sleep and circadian torpor. In: R. Lydic and J. F. Biebuyck, eds.; *Clinical Physiology of Sleep*, Amer. Physiol. Soc., Bethesda, MD, 1988, pp. 171–189.

Bernotas, L. A., P. E. Crago, and H. J. Chizeck. A discrete-time model of electrically stimulated muscle. *IEEE Trans. Biomed. Eng.*, 33: 829–838, 1986.

Bower, J. S., T. G. Sandercock, E. Rothman, P. H. Abbrecht, and D. R. Dantzler. Time domain analysis of diaphragmatic electromyogram during fatigue in men. *J. Appl. Physiol.: Respirat. Environ. Exercise Physiol.* 57(3): 913–916, 1984.

Caccia, D. C., D. Percival, M. J. Cannon, G. Raymond, and J. B. Bassingthwaighte. Analyzing exact fractal time series: evaluating dispersional analysis and rescaled range methods. *Physica A* 246: 609–632, 1997.

Canavier, C., D. Baxter, J. Clark, and J. Byrne. Nonlinear dynamics in a model neuron provide a novel mechanism for transient synaptic inputs to produce long-term alterations of postsynaptic activity. *J. Neurophysiol.*, 69: 2252–2257, 1993.

Cannon, M. J., D. Percival, D. C. Caccia, G. Raymond, and J. B. Bassingthwaighte. Evaluating scaled windowed variance methods for estimating the Hurst coefficient of time series. *Physica A* 241: 606–626, 1997.

Chen, T., Y-F. Chen, C-H. Lin, and T-T. Tsai. Quantification analysis for saccadic eye movements. *Annals Biomed. Eng.* 26: 1065–1071, 1998.

Chen, J., J. Vandewalle, W. Sanson, et al. Adaptive spectral analysis of cutaneous electrogastric signals using autoregressive moving average modeling. *Med. Biol. Engr. Comput.*, 28: 531–536, 1990.

Chen, et al., Frequency components of the electrogastrogram and their correlations with gastrointestinal contractions in humans. *Med. Biol. Eng. Comput.* 31:60–67, 1993.

Cohen, L. Preface to the special issue on time-frequency analysis. *Proc. IEEE*, 84: 1197–1198, 1996.

Collins, J., and I. Stewart. Symmetry-breaking bifurcation: A possible mechanism for 2:1 frequency-locking in animal locomotion. *J. Math. Biology*, 30: 827–838, 1992.

Deriche, M., and A. Tewfik. Signal modeling with discrete fractional noise processes. *IEEE Trans. Sig. Proc.* 41: 2839–2849, 1993.

Flandrin, P. On the spectrum of fractional Brownian motions. *IEEE Trans. Inform. Theory*, 35: 197–199, 1989.

Fraser, A. M., and H. L. Swinney. Independent coordinates for strange attractors from mutual information. *Phys. Rev. A*, 33:1134–1140, 1986.

Goldbeter, A., and F. Moran. Dynamics of a biochemical system with multiple oscillatory domains as a clue for multiple modes of neuronal oscillations. *Eur. Biophys. J.*, 15: 277–287, 1988.

Grassberger, P., and I. Procaccia. Measuring the strangeness of strange attractors. *Physica D*, 9D: 189–208, 1983(a).

Grassberger, P., and I. Procaccia. Estimation of the Kalomogorov entropy from a chaotic signal. *Phys. Rev. A*, 28: 2591–2593, 1983(b).

Guckenheimer, J., S. Gueron, and R. Harris-Warwick. Mapping the dynamics of a bursting neuron. *Phil. Trans. Roy. Soc., London*, B221: 345–359, 1993.

Guevara, M., and L. Glass. Phaselocking, period doubling bifurcations and chaos in a mathematical model of a periodically driven oscillator: A theory for the entrainment of biological oscillators and the generation of cardiac dysrhythmias. *J. Math. Biol.*, 14: 1–23, 1982.

Gdowski, G., and N. F. Voight. Intrinsic oscillations and discharge regularity of units in the dorsal cochlear nucleus (DCN) of the barbiturate anesthetized gerbil. *Ann. Biomed. Eng.* 26: 473–487, 1998

Furler, S. M., E. W. Kraeger, R. H. Smallwood, and D. J. Chisholm. Blood glucose control by intermittent loop closure in the basal mode: Computer simulation studies with a diabetic model. *Diabetes Care* 8(6): 553–561, 1985.

Harte, J., A. King, J. Green. Self-similarity in the abundance and distribution of species. *Science* 284: 334–336, 1999.

Hosking, J. R. M. Fractional differencing. *Biometrika* 68 (1): 165–176, 1981.

Hartford, C. G., J. van Schalkwyk, G. G. Rogers, and M. J. Turner. Predicting air-balloon and water-filled infant catheter frequency responses. *IEEE Eng. Med. Biol. Mag.* 16(3): 27–33, 1997.

Hathorn, M. K. S. Analysis of periodic changes in ventilation in new-born infants. *J. Physiol. (L)* 285: 85–99, 1978.

Hoop, B., M. D. Burton, and H. Kazemi. Correlation in stimulated respiratory neural noise. *CHAOS* 5: 609–612, 1995.

Hoyer, D., K. Schmidt, R. Bauer, et al. Nonlinear analysis of heart rate and respiratory dynamics. *IEEE Eng. Med. Biol. Mag.* 16: 31–39, 1997.

Hughson, R., Y. Yamamoto, J.-O. Fortrat, R. Leask, and M. Fofana. Possible fractal and/or chaotic breathing patterns in resting humans. In: *Bioengineering Approaches to Pulmonary Physiology and Medicine*, M. Khoo (ed.); Plenum Press, New York, 1996, pp. 187–196.

Hurst, H. E. Long-term storage capacity of reservoirs. *Trans. Amer. Soc. Civ. Engrs.* 116: 770–808, 1951.

Ivanov, P. Ch., M. G. Rosenblum, C.-K. Peng, J. Mietus, S. Havlin, H. E. Stanley, and A. L. Goldberger. Scaling behavior of heartbeat intervals obtained by wavelet-based time-series analysis. *Nature* 383: 323–327, 1996.

Jensen, K., N.-H. Holstein-Rathlou, P. Leyssec, E. Mosekilde, and D. Rasmussen. Chaos in a system of interacting nephrons. In: Degn, H., A. Holden, and L. Olsen. *Chaos in Biological Systems*. Plenum Press, New York, 1987, pp. 23–32.

Lewis, T., and M. Gueverra. Chaotic dynamics in an ionic model of the propogated cardiac action potential. *J. Theor. Biol.,* 146: 407–432, 1990.

Li, T.-Y., and J. Yorke. Period three implies chaos. *Amer. Math. Monthly,* 82: 985–992, 1975.

Liebovitch, L., and T. Toth. A model of ion channel kinetics using deterministic chaotic rather than stochastic processes. *J. Theor. Biology,* 148: 243–267, 1991.

Lin, Z. Y., and J. D. Chen. Time-frequency representation of the electrogastrogram-application of the exponential distribution. *IEEE Trans. Biomed. Engr.,* 41: 267–275, 1994.

Linkens, D., and S. Datardina. Estimation of frequencies of gastrointestinal electrical rhythms using autoregressive modeling. *Med. Biol. Engr. Comput.,* 16: 262–268, 1972.

Lundahl, T., W. J. Ohley, S. M. Kay, and R. Siffert. Fractional Brownian motion: A maximum likelihood estimator and its application to image texture. *IEEE Trans. on Medical Imaging* MI-5 (3): 152–161, 1986.

Mandelbrot, B. and J. W. Van Ness. Fractional Brownian motions, fractional noises and applications. *SIAM Rev.,* 10: 422–437, 1968.

Mao, H. and L. Wong. Fluorescence and laser photon counting: Measurements of epithelial $[Ca^{2+}]_i$ or $[Na^+]_i$ with ciliary beat frequency. *Ann. Biomed. Engr.* 26: 666–678, 1998.

Okazaki, M. M., et al., Damped oscillation analysis of natural and artificial periodontal membranes, *Ann. Biomed. Eng.* 24: 234–240, 1996.

Parker, T., and L. Chua. Chaos: A tutorial for engineers. *Proc. IEEE,* 75: 982–1008, 1987.

Peng, C.-K., S. Havlin, H. E. Stanley, and A. L. Goldberger. Quantification of scaling exponents and crossover phenomena in nonstationary heartbeat time series. *CHAOS* 5(1): 82–87, 1995.

Peng, C.-K., S. V. Buldyrev, A. L. Goldberger, S. Havlin, M. Simons, and H. E. Stanley. Finite-size effects on long-range correlations: implications for analyzing DNA sequences. *Phys. Rev. E* 47(5): 3730–3733, 1993.

Peslin, R., C. Duvivieer, J. Didelon, and C. Gallina. Respiratory impedance measured with head generator to minimize upper airway shunt. *J. Appl. Physiol.* 59(6): 1790–1795, 1985.

Raymond, G. and J. B. Bassingthwaighte. Deriving dispersional and scaled windowed variance analyses using the correlation function of discrete fractional Gaussian noise. *Physica A* 262: 85–96, 1998.

Rosenstein, M. T., J. J. Collins, and C. De Luca. A practical method for calculating the largest Lyapunov exponents from small data sets. *Physica D,* 65: 117–134, 1993.

Sammon, M., and E. Bruce. Vagal afferent activity increases dynamical dimension of respiration in rats. *J. Appl. Physiol.,* 70: 1748–1762, 1991.

Sammon, M., and F. Curley. Nonlinear systems identification: autocorrelation vs. autoskewness. *J. Appl. Physiol.* 83: 975–993, 1997.

Schaffer, W. Chaos in biology and epidemiology. In: Degn, H., A. Holden, and L. Olsen. *Chaos in Biological Systems.* Plenum Press, New York, 1987, pp. 233–248.

Schmid-Schoenbein, G. W., and Y. C. Fung. Forced perturbation of the respiratory system (A). The traditional model. *Ann. Biomed. Eng.* 6: 194–211, 1978.

Shelhamer, M. Correlation dimension of optokinetic nystagmus as evidence of chaos in the oculomotor system. *IEEE Trans. Biomed. Eng.* 39: 1319–1321, 1992.

Theiler, J., S. Eubank, A. Longtin, B. Galdrikin, and J. Doyne Farmer. Testing for nonlinearity in time series: The method of surrogate data. *Physica D,* 58: 77–94, 1992.

Wolf, A., J. B. Swift, H. L. Swinney, and J. A. Vastano. Determining Lyapunov exponents from a time series. *Physica D,* 16: 285, 1985.

Zhang, X., and E. N. Bruce. Fractal characteristics of end-expiratory lung volume in anesthetized rats. *Ann. Biomed. Eng. 28:* 94–101, 2000.

INDEX